Chaos in Complex Dynamical System: Theory and Application

Shutang Liu · Ping Liu · Cuimei Jiang ·
Changan Liu

Chaos in Complex Dynamical System: Theory and Application

Springer

Shutang Liu
School of Control Science and Engineering
Shandong University
Jinan, Shandong, China

Cuimei Jiang
School of Mathematics and Statistics
Qilu University of Technology (Shandong Academy of Sciences)
Jinan, Shandong, China

Ping Liu
Shandong Key Laboratory of Gardening Machinery and Equipment
College of Mechanical and Electronic Engineering
Shandong Agricultural University
Tai'an, Shandong, China

Changan Liu
Department of Systems Biomedicine
School of Basic Medical Sciences
Shandong University
Jinan, Shandong, China

ISBN 978-981-95-7761-3 ISBN 978-981-95-7762-0 (eBook)
https://doi.org/10.1007/978-981-95-7762-0

This Springer imprint is published by the registered company Springer Nature Singapore Pte Ltd.
The registered company address is: 152 Beach Road, #21-01/04 Gateway East, Singapore 189721, Singapore

Preface

Chaos is an external, seemingly irregular and complex motion generated by the internal causes of deterministic systems. Its research has become one of the key areas in nonlinear science. The chaos theory, together with the theory of relativity and quantum mechanics, is hailed as one of the three major discoveries in natural sciences of the twentieth century. It reveals the complexity that is ubiquitous in both nature and human society. It embodies the unity of order and disorder, as well as the unity of determinacy and randomness. It broadens people's horizons and deepens their understanding of the objective world. At present, chaos science and other sciences are mutually infiltrating, and chaos has been widely applied in fields such as biology, physics, electronics, and information science.

Complex chaotic systems are typical complex dynamical systems where state variables are in a complex space. Compared with real chaotic systems, due to the existence of the imaginary part of the state variables, the dynamical behavior of complex chaotic systems is more complex. Since Fowler et al. proposed the complex Lorenz equation in 1982, complex chaotic systems have been used to describe numerous problems in fields such as physics, biology, and chemistry. They play an important role in many branches of physics, especially in secure communication. The complex variables not only increase the content of the transmitted information but also improve the confidentiality effect. The research on complex chaotic systems expands previous research results, better promotes the development of chaos theory, and lays a theoretical foundation for its wide application. It has important research significance and application value and is one of the hot topics in current research. Therefore, this book introduces in detail the characteristics, synchronization control of complex chaotic systems, and their application in communication.

The latest research achievements of complex chaotic systems have been concentrated in this book, and the achievements form the overall content of the book. Furthermore, the existence of such research achievements is the first time in the international. In this book, the dynamical behavior of different complex chaotic systems, chaos synchronization and applications in communication are discussed by using different control methods engineering technology in the light of the practical engineering and technical.

It is noteworthy that this book can be used as a textbook for college students, postgraduates and doctoral students in the polytechnic university, also be used as the reference book for engineers and professional researchers. This book will promote the development of theoretical research and applied research in this academic field.

The structure of this book is as follows: Chap. 1 introduces the basic knowledge of chaotic systems, the research status quo, and commonly used mathematical theorems. In Chaps. 2–12, we mainly discuss the dynamical characteristics and various synchronization types of complex chaotic systems, including anti-synchronization, projective synchronization, hybrid projective synchronization, modified function projective synchronization, complex modified projective synchronization, generalized combination complex synchronization, and so on. In Chaps. 13–18, we mainly investigate the dynamical behaviors and various synchronization types of fractional-order complex chaotic systems. Chapters 19 and 20 introduce the application of chaotic complex systems in secure encryption.

We would like to express our appreciation for the great efforts of Drs. Fangfang Zhang and Jian Liu. Finally, we are grateful for the sequential support of the National Natural Science Foundation of China-Shandong joint fund (Number U1806203) and the Key Program of the National Natural Science Foundation of China (Number 61533011), the Foundation for the Author of National Excellent Doctoral Dissertation of China (Number 200444), the National Natural Science Foundation of China (Numbers 60472112, 60372028, 600874009, 10971120, 61273088), the Shandong Province Natural Science Foundation (Numbers ZR2010FM010, ZR2010MF073, ZR2023MF089, ZR2022MA057, Y98-A02005), Shandong Province Small and Medium-sized Enterprise Innovation Capacity Improvement Project (2023TSGC0621, 2024TSGC0645), the China Postdoctoral Science Foundation (Number 2004035036), Project supported by the Scientific Research Staring Foundation for the Returned Overseas Chinese Scholars, Ministry of Education of China, and the K. C. Wong Education Foundation, Hong Kong, China, which ensure the smooth developments of this research project.

Jinan, China — Shutang Liu
Tai'an, China — Ping Liu
Jinan, China — Cuimei Jiang
Jinan, China — Changan Liu

Contents

Symbols

$\mathbb{R}$, $\mathbb{R}^n$, $\mathbb{R}^{n \times m}$	Set of real numbers, set of n-dimensional real vectors, and set of $n \times m$ real matrices
$\mathbb{C}$, $\mathbb{C}^n$, $\mathbb{C}^{n \times m}$	Set of complex numbers, set of n-dimensional complex vectors, and set of $n \times m$ complex matrices
$\mathbb{R}^+$	Set of non-negative real numbers
$\mathbb{Z}^+$	Set of non-negative integers
$Re(s)$ and s^r	Real part of $s \in \mathbb{C}$
$Im(s)$ and s^i	Imaginary part of $s \in \mathbb{C}$
A^T	Transpose of matrix A
A^{-1}	Inverse of matrix A
I_n	$n \times n$ identify matrix (the subscript is omitted if no confusion will occur)
$j = \sqrt{-1}$	The imaginary unit
$\dot{z}$	Denotes variable z derivatives with respect to time
$\overline{z}$	Represents the complex conjugate of z
$diag\{A_1, \ldots, A_n\}$	Diagonal matrix with A_i as its ith diagonal element
$A > 0$ (<0)	Symmetric positive (negative) definite matrix
$A \geq 0$ (≤ 0)	Symmetric positive (negative) semi-definite matrix
$det(A)$	Determinant of matrix A
$Tr\{A\}$	Trace of matrix A
$\lambda(A)$	Eigenvalue of matrix A
$arg(\lambda(A))$	The argument of the eigenvalue λ of matrix A
$\lambda_{max}(A)$	Largest eigenvalue of matrix A
$\lambda_{min}(A)$	Smallest eigenvalue of matrix A
$\sigma_{max}(A)$	Largest singular value of matrix A
3D (4D)	3-dimensional (4-dimensional)
$\lvert \cdot \rvert$	Absolute value (or modulus)
$\lVert \cdot \rVert$	Euclidean norm of a vector or spectral norm of a matrix
$\lVert \cdot \rVert_\infty$	Included l_∞-norm
$\forall$	For all
$\in$	Belongs to

$\exists$	There exists
$\subseteq$	Is a subset of
$\cup$	Union
$\cap$	Intersection
$\rightarrow$	Tends toward or is mapped into (case sensitive)
$\Rightarrow$	Implies
$\frac{\partial V}{\partial t}$	Partial derivative of function V
grad V	Gradient of function V
$\prod_{i=1}^{n} \lambda_i$	Product of $\lambda_1, \lambda_2, \ldots, \lambda_n$
ΔA	Amount of change in matrix A
CS	Complete synchronization
AS	Anti-synchronization synchronization
PS	Projective synchronization
MPS	Modified projective synchronization
MGPS	Modified generalized projective synchronization
CMPS	Complex modified projective synchronization
CMHPS	Complex modified hybrid projective synchronization
FSHPS	Full state hybrid projective synchronization
LS	Lag synchronization
STDS	Self-time-delay synchronization
$\square$	End of proof

Chapter 1
Introduction and Preliminary Knowledge

1.1 Introduction

As a branch of dynamic system, the chaotic system is closely connected with nature and people's life [1–4]. At present, the chaotic system has been widely used in biology, finance and engineering and other disciplines [5–7]. It is closely related to fractal geometry, mixed and pure theory, bifurcation theory and other fields, and is the focus of current research [8, 9]. Synchronization of the chaotic system and its application to secure communication is an interesting research area [10–12], and it has attracted broad attention [13]. Some preliminary knowledge such as chaos definition and characters, chaos synchronization, chaos control and some important theorems are introduced in this chapter.

1.2 Preliminary Knowledge

1.2.1 Chaos Definition and Its Characters

Chaos refers to the unpredictable and stochastic motion of deterministic dynamic systems due to their sensitivity to initial values [14]. The chaos word comes from Greek [15]. Its original meaning is the scene before the beginning of the universe. Its basic meaning mainly refers to the state of chaos and disorder.

The certainty of dynamic system is a mathematical concept, which means that the state of the system at any time is determined by the initial state. Although the motion state at any future time can be calculated according to the initial state data and motion law of the motion, because the measurement of the initial data can not be completely accurate, the prediction results must have errors or even be unpredictable. The predictability of motion is a physical concept. Even if a motion is deterministic, it can still be unpredictable, and the two are not contradictory. The success of Newtonian

S. Liu et al., *Chaos in Complex Dynamical System: Theory and Application*,
https://doi.org/10.1007/978-981-95-7762-0_1

mechanics, especially its success in predicting Neptune, leads to misunderstanding to a certain extent. It equates certainty with predictability and thinks that deterministic motion must be predictable. Studies after the 1970s have shown that in a large number of nonlinear systems, although the system is deterministic, there is a seemingly random and unpredictable motion state, chaotic motion, which is very sensitive to the initial value of motion state [16, 17].

Chaos is a kind of seemingly irregular complex motion in the real world. The common feature is that the original orderly movement form following the simple physical law suddenly deviates from the expected regularity and becomes a disordered form under certain conditions. Chaos can occur in a wide range of deterministic dynamic systems [18]. Chaos is similar to stochastic process in statistical characteristics and is considered as an intrinsic randomness in deterministic systems [19–23].

1.2.2 Chaotic Synchronization and Control

Synchronization is a phenomenon that widely exists in nature. Since Pecora and Carroll published the concept and phenomenon of synchronization in two coupled chaotic oscillators in 1990 [24], chaotic synchronization has attracted extensive and in-depth attention. Many types of synchronization phenomena have been found in a variety of chaotic systems, such as complete synchronization [24–28], anti-synchronization [29–31], projective synchronization [32–34], lag synchronization [35–39], phase synchronization [40, 41], etc. In the following, several typical types of synchronization are briefly introduced.

The real drive system is described as

$$\dot{x} = F(x), \tag{1.2.1}$$

the real response system is

$$\dot{y} = G(y) + u(x, y), \tag{1.2.2}$$

where $x = (x_1, x_2, \ldots, x_n)^T$ and $y = (y_1, y_2, \ldots, y_n)^T \in \mathbb{R}^n$ are state variables of the drive system and the response system, respectively; $F(x)$ and $G(y)$ denote nonlinear vector functions, $u(x, y)$ is a controller.

(1) Complete synchronization (CS)

Complete synchronization is the simplest form of synchronization, where two interacting chaotic systems with different initial values evolve over time until their behaviors become entirely consistent. If there exist a controller u such that $\lim_{t\to\infty} \|y(t) - x(t)\| = 0$, then we call them complete synchronization.

(2) Anti-synchronization (AS)

If there exist a controller u such that $\lim_{t\to\infty} \|y(t) + x(t)\| = 0$, then system (1.2.1) and system (1.2.2) can realize anti-synchronization.

(3) Projective synchronization (PS)

If there exists a constant matrix $\mathbf{H} = diag\{h, h, \ldots, h\} \in \mathbb{R}^{n\times n}$ such that $\lim_{t\to\infty} \|y(t) - Hx(t)\| = 0$, then systems (1.2.1) and (1.2.2) achieve projective synchronization, where matrix $\mathbf{H}$ is called the scaling matrix.

(4) Modified projective synchronization (MPS)

If there exists a constant matrix $\mathbf{H} = diag\{h_1, h_2, \ldots, h_n\} \in \mathbb{R}^{n\times n}$ such that $\lim_{t\to\infty} \|y(t) - Hx(t)\| = 0$, then systems (1.2.1) and (1.2.2) realize modified projective synchronization. It is worth noting that full state hybrid projective synchronization essentially refers to modified projective synchronization [42, 43].

(5) Modified function projective synchronization (MFPS)

If there exists a constant matrix $\mathbf{H(t)} = diag\{h_1(t), h_2(t), \ldots, h_n(t)\} \in \mathbb{R}^{n\times n}$ such that $\lim_{t\to\infty} \|y(t) - Hx(t)\| = 0$, then systems (1.2.1) and (1.2.2) achieve modified function projective synchronization. Notably, MFPS is also called generalized function projective synchronization (GFPS) [44–47].

(6) Lag synchronization (LS)

Lag synchronization, also known as delayed synchronization, is characterized by the coincidence between two chaotic systems with a constant time lag. If there exist a controller u such that $\lim_{t\to\infty} \|y(t) - x(t-\tau)\| = 0$, then system (1.2.1) and system (1.2.2) can realize lag synchronization.

Table 1.1 shows the relationships among the above synchronization types. CS, AS, PS, and MPS are special cases of MFPS.

Table 1.1 Types of synchronization

Settings the matrix $\mathbf{H}$	Type of synchronization
$\mathbf{H(t)} = diag\{h_1(t), h_2(t), \ldots, h_n(t)\} \in \mathbb{R}^{n\times n}$	MFPS
$\mathbf{H} = diag\{h_1, h_2, \ldots, h_n\} \in \mathbb{R}^{n\times n}$	MPS
$\mathbf{H} = diag\{h, h, \ldots, h\} \in \mathbb{R}^{n\times n}$	PS
$\mathbf{H} = diag\{-1, -1, \ldots, -1\} \in \mathbb{R}^{n\times n}$	AS
$\mathbf{H} = diag\{1, 1, \ldots, 1\} \in \mathbb{R}^{n\times n}$	CS
$\mathbf{H} = diag\{0, 0, \ldots, 0\}$	Control problem

1.2.3 Local Lipschitz Condition

Theorem 1.1 ([48, 49]) ***Lipschitz Condition***. *We say f satisfies a Lipschitz condition on a set H if there is a constant L such that*

$$|f(x, y_1) - f(x, y_2)| \leq L|y_1 - y_2|$$

for all

$$(x, y_1), (x, y_2) \in H.$$

Theorem 1.2 ([48, 49]) ***Local existence and uniqueness*** *Suppose that f is continuous in $H \subseteq \mathbb{R}$, $(x_0, y_0) \in H$. Assume f is locally Lipschitzian with respect to y at (x_0, y_0). Then there exists a closed interval I containing x_0 in its interior that the initial value problem has a unique solution, defined in I.*

Lemma 1.1 *Let H be the rectangle, $\mathbf{R} = \{(x, y) : x \in [a, b], |y - A| \leq B\}$ or the infinite strip*

$$S = \{(x, y) : x \in [a, b], |p| < \infty\}$$

if $f : D \to \mathbb{R}$ and $\frac{\partial f}{\partial y}$ are continuous and there is a constant $\mathbf{k} \geq 0$ such that

$$| \frac{\partial f}{\partial y}(x, y) | \leq \mathbf{k}$$

for all $(x, y) \in \Omega$, then Lipschitz condition holds with $L = k$.

1.2.4 Stability Theorem

Chaos stability theorem refers to that a class of chaotic systems can achieve stable periodic orbits after a series of specific control operations [50, 51]. This theorem provides a way to stabilize chaotic systems by means of control. The core idea of chaos stability theorem is to change the behavior of chaotic system from original chaotic state to periodic orbit by applying appropriate control signals or control methods. Specifically, the chaos stability theorem states that for some chaotic systems, when appropriate external control signals are applied or the system parameters are adjusted, the system can be transformed from a chaotic state to a stable periodic orbit.

For the equation

$$\dot{x} = f(x, t), x(t_0) = x_0. \tag{1.2.3}$$

We say $x = 0$ is the stable equilibrium point of the equation if there exists $\delta(t_0, \varepsilon)$ such that

$$|x_0| < \delta(t_0, \varepsilon) \to |x(t)| < \varepsilon, \forall t \geq t_0$$

for all

$$t_0 \geq 0, \varepsilon < 0,$$

where $x(t)$ is the solution of this equation from x_0 at time t_0.

1.2.5 Fractional Calculus

There are many definitions of fractional derivatives [54], such as Riemann-Liouville, Grünwald-Letnikov and Caputo definitions. The definition of the Riemann-Liouville derivative is given as follows

$$D^{\alpha} f(t) = \frac{d^m}{dt^m} J^{m-\alpha} f(t), \quad \alpha > 0, \tag{1.2.4}$$

where $m = [\alpha]$, i.e. m is the first integer which is not less than α, J^{β} is the β-order Riemann-Liouville integral operator as described by

$$J^{\beta} f(t) = \frac{1}{\Gamma(\beta)} \int_{t_0}^{t} \frac{f(\tau)}{(t-\tau)^{1-\beta}} d\tau, \quad 0 < \beta \leq 1, \tag{1.2.5}$$

where $\Gamma(\cdot)$ denotes the gamma function $\Gamma(x) = \int_0^{\infty} t^{x-1} e^{-x} dt$.

Here, the following Caputo definition is introduced:

$$D_*^{\alpha} f(t) = \frac{d^m}{dt^m} J^{m-\alpha} f(t) = \begin{cases} \frac{1}{\Gamma(m-\alpha)} \int_{t_0}^{t} \frac{f^{(m)}(\tau)}{(t-\tau)^{\alpha-m+1}} d\tau, & m-1 < \alpha < m, \\ \frac{d^m}{dt^m} f(t), & \alpha = m, \end{cases} \tag{1.2.6}$$

where D_*^{α} is generally called α-order Caputo differential operator.

An advantage of using the Caputo derivative is that the initial conditions for the fractional-order systems are in the same form as the integer-order systems. So, throughout this paper, the fractional-order chaotic systems will be described by using Caputo definition with lower limit of integral $t_0 = 0$ and the order $0 < \alpha < 1$.

In the following, we introduce the properties of Caputo fractional derivative.

Lemma 1.2 ([55]) *Suppose that $x(t) \in \mathbb{R}$ is a continuous and derivable function. Then, for any time instant $t \geq t_0$,*

$$\frac{1}{2} D_*^{\alpha} x^2(t) \leq x(t) D_*^{\alpha} x(t), \quad \forall \alpha \in (0, 1). \tag{1.2.7}$$

Lemma 1.3 ([55]) *Suppose that $x(t) \in \mathbb{R}^n$ is a continuous and derivable vector function. Then, for $\forall \alpha \in (0, 1)$ and $\forall t \geq t_0$,*

$$\frac{1}{2} D_*^{\alpha} x^T(t) x(t) \leq x^T(t) D_*^{\alpha} x(t). \tag{1.2.8}$$

Lemma 1.4 ([56]) *Let $\alpha \in (0, 1]$, $x(t) = (x_1(t), x_2(t), \ldots, x_n(t))^T \in \mathbb{R}^n$ and $x_k(t)$ $(k = 1, 2, \ldots, n)$ be continuous and derivable functions. Then for any time instant $t > 0$,*

$$D_*^\alpha \left(x^T(t)Px(t)\right) \le (D_*^\alpha x(t))^T Px(t) + x^T(t)PD_*^\alpha x(t), \tag{1.2.9}$$

where $P \in \mathbb{R}^{n\times n}$ is a positive definite matrix.

Lemma 1.5 ([57]) *Let $V(t)$ be a continuous function on $[0, \infty)$ and satisfy*

$$D_*^\alpha V(t) \le \theta V(t),$$

where θ is a constant. Then

$$V(t) \le V(0)E_\alpha(\theta t^\alpha), \quad t \ge 0. \tag{1.2.10}$$

Lemma 1.6 ([58]) *Let $W(t) = x^T(t)x(t)/2 + y^T(t)y(t)/2$, where $x(t), y(t) \in \mathbb{R}^n$ have continuous derivatives. If there exists a constant $h_0 > 0$ such that*

$${}_{t_0}D_t^\alpha W(t) \le -h_0 x^T(t)x(t),$$

then $||x(t)||$ and $||y(t)||$ are bounded for all $t > 0$ and $x(t)$ converges to zero asymptotically, where $||\cdot||$ is the Euclidean norm.

1.2.6 The Stability of the Fractional-Order Systems

Stability is the most basic and important research content of the control systems. In 1996, Matignon studied the asymptotic stability of linear fractional-order systems.

Theorem 1.3 ([59]) *For the fractional-order autonomous system*

$$D_*^\alpha x(t) = Ax(t), \quad x(0) = x_0, \tag{1.2.11}$$

where $x \in \mathbb{R}^n$, the matrix $A \in \mathbb{R}^{n\times n}$, α denotes the fractional orders. System (1.2.11) is
(i) asymptotically stable if

$$|arg(\lambda_l(A))| > \alpha\pi/2, \quad for\ l = 1, 2, 3\ldots, \tag{1.2.12}$$

where $arg(\lambda_l(A))$ stands for the argument of the eigenvalue λ_l of A. For this case, the component of the state decay converges to 0 as $t^{-\alpha}$.
(ii) stable if

$$|arg(\lambda_l(A))| \ge \alpha\pi/2, \quad for\ l = 1, 2, 3\ldots, \tag{1.2.13}$$

and those critical eigenvalues with $|arg(\lambda_l(A))| = \alpha\pi/2$ *have geometric multiplicity one.*

Theorem 1.4 ([60]) *For a given linear fractional-order system*

$$D_*^{\alpha} x(t) = Ax(t), \quad x(0) = x_0, \tag{1.2.14}$$

where $x \in \mathbb{R}^n$, *the matrix* $A \in \mathbb{R}^{n\times n}$, $\alpha = [\alpha_1, \alpha_2, \ldots, \alpha_n]^T$ *denotes the fractional orders with* $\alpha_k \neq \alpha_l$ $(k \neq l)$. *Thus, the above system can be described as*

$$\begin{cases} D_*^{\alpha_1} x_1(t) = a_{11}x_1(t) + a_{12}x_2(t) + \ldots + a_{1n}x_n(t), \\ D_*^{\alpha_2} x_2(t) = a_{21}x_1(t) + a_{22}x_2(t) + \ldots + a_{2n}x_n(t), \\ \vdots \\ D_*^{\alpha_n} x_n(t) = a_{n1}x_1(t) + a_{n2}x_2(t) + \ldots + a_{nn}x_n(t), \end{cases} \tag{1.2.15}$$

where $\alpha_l \in (0, 1)$ *are rational numbers and* $\alpha_l = k_l/h_l$, $(k_l, h_l) = 1$, m *is the least common multiple of* h_l $(l = 1, 2, \ldots, n)$. *Then system* (1.2.15) *is asymptotically stable if all roots* λ *of the equation* $\det(diag(\lambda^{m\alpha_1}, \lambda^{m\alpha_2}, \ldots, \lambda^{m\alpha_n}) - A) = 0$ *satisfy* $|\arg(\lambda)| > \gamma\pi/2$.

Theorem 1.5 ([61, 62]) *Let* $x(t) = 0$ *be the equilibrium point of the following fractional-order system*

$$D^{\alpha} x(t) = f(x, t), \quad x \in \mathbb{R}^n, \tag{1.2.16}$$

where $0 < \alpha < 1$. *Assume that there exist a Lyapunov function* $V(t, x(t))$ *and class-K* γ_k $(k = 1, 2, 3)$ *satisfying*

$$\gamma_1(||x||) \le V(t, x(t)) \le \gamma_2(||x||), \tag{1.2.17}$$

$$D^{\alpha} V(t, x(t)) \le -\gamma_3(||x||). \tag{1.2.18}$$

Then the equilibrium point $x = 0$ *of system* (1.2.16) *is asymptotically stable.*

1.2.7 Numerical Algorithms for the Fractional-Order Systems

Unlike the numerical algorithm for solving an ordinary differential equation, the numerical simulation of a fractional differential equation is not so easy. In the literatures of fractional chaos field, there are two approximation methods, which can be used to solve numerically fractional-order differential equations. The first one is based on the approximation of the fractional-order system behavior in the frequency domain [63]. The other algorithm to find an approximation for fractional-order systems is based on the predictor-correctors scheme [64, 65]. This method is an improved version of Adams-Bashforth-Moulton algorithm. Here, we choose the Adams-Bashforth-Moulton predictor correctors for numerical computation. Next, we interpret the approximate solution of nonlinear fractional-order differential equations using this algorithm in the following way.

The differential equation can be described by

$$\begin{cases} D_*^\alpha x(t) = f(t, x(t)), & 0 \le t \le T, \\ x^{(k)}(0) = x_0^{(k)}, & k = 0, 1, \ldots, [\alpha] - 1, \end{cases} \tag{1.2.19}$$

which is equivalent to the Volterra integral equation

$$x(t) = \sum_{k=0}^{[\alpha]-1} x_0^{(k)} \frac{t^k}{k!} + \frac{1}{\Gamma(\alpha)} \int_0^t (t-\tau)^{(\alpha-1)} f(\tau, x) d\tau. \tag{1.2.20}$$

Now, set $h = T/N, t_j = jh$ $(j = 0, 1, \ldots, N)$. Then, the corrector formula for the latter equation can be discretized as follows:

$$x_h(t_{n+1}) = \sum_{k=0}^{[\alpha]-1} x_0^{(k)} \frac{t_{n+1}^k}{k!} + \frac{h^\alpha}{\Gamma(\alpha+2)} f(t_{n+1}, x_h^p(t_{n+1})) + \frac{h^\alpha}{\Gamma(\alpha+2)} \sum_{j=0}^{n} \beta_{j,n+1} f(t_j, x_h(t_j)),$$

where predicted value $x_h^p(t_{n+1})$ is determined by the following formula:

$$x_h^p(t_{n+1}) = \sum_{k=0}^{[\alpha]-1} x_0^{(k)} \frac{t_{n+1}^k}{k!} + \frac{1}{\Gamma(\alpha)} \sum_{j=0}^{n} \gamma_{j,n+1} f(t_j, x_h(t_j)), \tag{1.2.21}$$

$$\beta_{j,n+1} = \begin{cases} n^{\alpha+1} - (n-\alpha)(n+1)^\alpha, & j = 0, \\ (n-j+2)^{\alpha+1} + (n-j)^{\alpha+1} \\ -2(n-j+1)^{\alpha+1}, & 1 \le j \le n, \\ 1, & j = n+1, \end{cases} \tag{1.2.22}$$

and

$$\gamma_{j,n+1} = \frac{h^\alpha}{\alpha} \left((n-j+1)^\alpha - (n-j)^\alpha \right), \quad 1 \le j \le n. \tag{1.2.23}$$

The error estimate is $\max_{j=0,1,\ldots,N} |x(t_j) - x_h(t_j)| = O(h^p)$, where $p = \min(2, 1+\alpha)$.

Chapter 2
Anti-synchronization of Complex Chaotic Systems

Numerous studies have been conducted on the anti-synchronization of nonlinear real dynamic systems [66–70], Notably, complex chaotic systems possess more abundant dynamic characteristics, rendering the research on their anti-synchronization of great significance. Firstly, this chapter proposes a novel complex chaotic system and investigates its anti-synchronization (AS) performance. Secondly, it explores the adaptive anti-synchronization problem for a class of chaotic complex nonlinear systems with unknown parameters. Based on the Lyapunov stability theory [71], an adaptive control scheme and adaptive laws of parameters are developed to anti-synchronize two chaotic complex systems.

2.1 A New Complex Chaotic System and Its Anti-synchronization

2.1.1 A New Complex Chaotic System

Wang et al. [72] proposed a new real chaotic nonlinear system in 2006 through the extension of the Lorenz system, which is briefly referred to as the Wang system and is described by the following equation

$$\begin{cases} \dot{x}_1 = a_1(x_1 - x_2), \\ \dot{x}_2 = -c_1 x_2 + x_1 x_3, \\ \dot{x}_3 = -b_1 x_3 + d_1 x_1 x_2, \end{cases} \tag{2.1.1}$$

where x_1, x_2 and x_3 are real state variables, a_1, b_1, c_1 and d_1 are real parameters. System (2.1.1) exhibits chaotic attractors when $a_1 = 20,\ b_1 = 2,\ c_1 = 28,\ d_1 = 1$ and also displays periodicity for $a_1 = 20,\ b_1 = 8,\ c_1 = 28,\ d_1 = 1$ or $a_1 = 7,\ b_1 =$

S. Liu et al., *Chaos in Complex Dynamical System: Theory and Application*,
https://doi.org/10.1007/978-981-95-7762-0_2

2, $c_1 = 28$, $d_1 = 1$. Extensive investigations on the control and synchronization of this system have been reported in Refs. [73, 74].

The complex extension of the Wang system (2.1.1) is expressed by

$$\begin{cases} \dot{z_1} = a_1(z_1 - z_2), \\ \dot{z_2} = -c_1 z_2 + z_1 z_3, \\ \dot{z_3} = -b_1 z_3 + \frac{1}{2} d_1(\bar{z}_1 z_2 + z_1 \bar{z}_2), \end{cases} \tag{2.1.2}$$

where $z_1 = z_1^r + jz_1^i = u_{11} + ju_{21}$ and $z_2 = z_2^r + jz_2^i = u_{31} + ju_{41}$ are complex state variables, $z_3 = u_{51}$ is a real state variable.

In the following section, we present the complex behaviors of the complex Wang system (2.1.2). Separating real and imaginary parts in system (2.1.2), we obtain the real version as

$$\begin{cases} \dot{u}_{11} = a_1(u_{11} - u_{31}), \\ \dot{u}_{21} = a_1(u_{21} - u_{41}), \\ \dot{u}_{31} = -c_1 u_{31} + u_{11} u_{51}, \\ \dot{u}_{41} = -c_1 u_{41} + u_{21} u_{51}, \\ \dot{u}_{51} = -b_1 u_{51} + d_1(u_{11} u_{31} + u_{21} u_{41}). \end{cases} \tag{2.1.3}$$

2.1.2 Dynamical Properties of a New Chaotic Complex System

System (2.1.3) exhibits the following fundamental dynamical properties:
(1) Dissipation
In this regard, the formula

$$\nabla F = \sum_{p=1}^{5} \frac{\partial \dot{u}_{p1}}{\partial u_{p1}} = 2a_1 - b_1 - 2c_1 \tag{2.1.4}$$

denotes the rate of change of phase space volume of system (2.1.3). When $2a_1 - b_1 - 2c_1 < 0$ and the initial volume element is $V(0)$, $V(t) = V(0)e^{-(2c_1+b_1-2a_1)t}$. That is, when $t \to \infty$, every volume element containing trajectories of system (2.1.3) shrinks to zero as exponential convergent rate.
(2) Chaotic behavior and attractors
System (2.1.3) in vector notation can be written as

$$U(t) = H(U(t); \eta), \tag{2.1.5}$$

where $U(t) = [u_{11}, u_{21}, u_{31}, u_{41}, u_{51}]^T$ is the state space vector, $H = [h_1, h_2, h_3, h_4, h_5]^T$, η is a set of real parameters. The linearized (variational) equations describing small deviations δU from the trajectory $U(t)$ are

$$\delta \dot{U}(t) = J_{pq}(U(t); \eta)\delta U, \ (p, q = 1, 2, 3, 4, 5), \tag{2.1.6}$$

where $J_{pq} = \frac{\partial h_p}{\partial u_q}$ is the Jacobian matrix of system (2.1.3) and is denoted as

$$J_{pq} = \begin{pmatrix} a_1 & 0 & -a_1 & 0 & 0 \\ 0 & a_1 & 0 & -a_1 & 0 \\ u_{51} & 0 & -c_1 & 0 & u_{11} \\ 0 & u_{51} & 0 & -c_1 & u_{21} \\ d_1 u_{31} & d_1 u_{41} & d_1 u_{11} & d_1 u_{21} & -b_1 \end{pmatrix}. \tag{2.1.7}$$

According to the definition of Lyapunov exponents [75], we get the Lyapunov exponents of system (2.1.3) as

$$\lambda_p = \lim_{t \to \infty} \frac{1}{t} \log \frac{\|\delta u'_{p1}(t)\|}{\|\delta u'_{p1}(0)\|}, \ (p = 1, 2, 3, 4, 5). \tag{2.1.8}$$

To get λ_p, equations (2.1.5) and (2.1.6) must be numerically solved with the fourth-order Runge-Kutta method.

For the case of $a_1 = 20$, $b_1 = 2$, $c_1 = 28$, $d_1 = 1$ with initial values $(u_{11}(0), u_{21}(0), u_{31}(0), u_{41}(0), u_{51}(0)) = (2, 1, -5, 3, 4)$, we calculate the Lyapunov exponents which are $\lambda_1 = 2.22$, $\lambda_2 = \lambda_3 = 0$, $\lambda_4 = -8.05$ and $\lambda_5 = -12.67$. This means that system (2.1.3) for such choice of a_1, b_1, c_1, d_1 is chaotic since one of the Lyapunov exponents is positive. The chaotic attractor of system (2.1.3) using the same choice of the parameters and initial conditions is plotted in Fig. 2.1 in different phase planes and projections.

The maximum Lyapunov exponent is defined as Ref. [75]:

$$\lambda = \lim_{t \to \infty} \frac{1}{t} \log \frac{\|\delta U(t)\|}{\|\delta U(0)\|}. \tag{2.1.9}$$

For the same parameters and initial values as in Fig. 2.1, the numerical calculation of the maximal Lyapunov exponent versus t of the attractor shown in Fig. 2.1 using the formula (2.1.9) is plotted as shown in Fig. 2.2. Obviously, the maximum Lyapunov exponent of this attractor is clearly positive, indicating that the motion on the attractor is chaotic.

(3) Fixed points and their stability

The fixed points of system (2.1.3) can be calculated by solving equations $\dot{u}_{p1} = 0 \ (p = 1, 2, \ldots, 5)$. If $b_1 c_1 d_1 > 0$, we obtain five fixed points of the system (2.1.3) as

$$\begin{aligned} E_1 &= (0, 0, 0, 0, 0), \\ E_{2,3} &= \left(\pm\sqrt{\frac{b_1 c_1}{2d_1}}, \pm\sqrt{\frac{b_1 c_1}{2d_1}}, \pm\sqrt{\frac{b_1 c_1}{2d_1}}, \pm\sqrt{\frac{b_1 c_1}{2d_1}}, c_1 \right), \end{aligned}$$

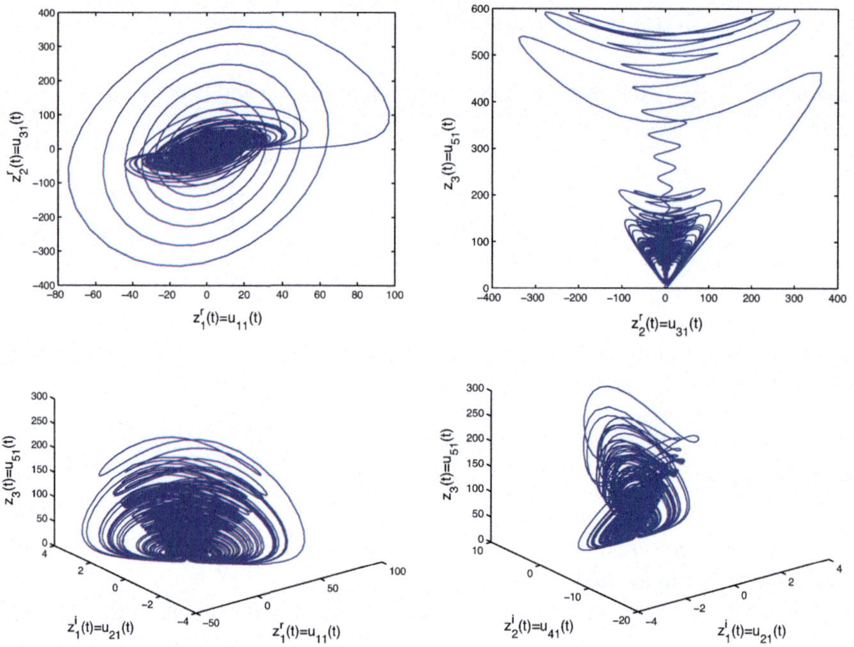

Fig. 2.1 Chaotic behavior and attractors of system (2.1.3) in different phase planes and projections

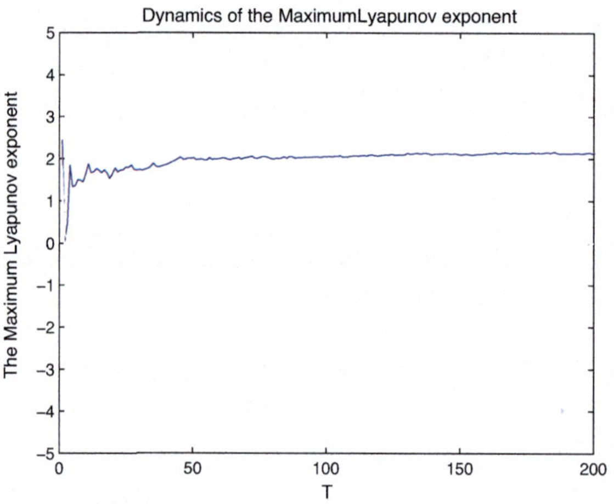

Fig. 2.2 The Maximum Lyapunov exponent of system (2.1.3) for the same parameters and initial conditions as in Fig. 2.1

$$E_{4,5} = \left(\pm\sqrt{\frac{b_1c_1}{2d_1}}, \mp\sqrt{\frac{b_1c_1}{2d_1}}, \pm\sqrt{\frac{b_1c_1}{2d_1}}, \mp\sqrt{\frac{b_1c_1}{2d_1}}, c_1\right).$$

To study the stability of E_1, the characteristic polynomial of the Jacobian matrix J_{pq} of system (2.1.3) at point E_1 is

$$(\delta - a_1)^2(\delta + c_1)^2(\delta + b_1) = 0.$$

Then the eigenvalues of the corresponding linearized system (2.1.5) at point E_1 are $\delta_1 = \delta_2 = a_1$, $\delta_3 = \delta_4 = -c_1$ and $\delta_5 = -b_1$. According to the linear system theory, if and only if real parts of all eigenvalues of the linearized system (2.1.5) at the fixed points are negative, the fixed points are stable. Therefore, when $a_1 < 0,\ b_1 > 0$ and $c_1 > 0$, the fixed point E_1 of system (2.1.3) is stable, otherwise it becomes unstable.

The stability of fixed points $E_{2,3}$ and $E_{4,5}$ can be also analyzed in a similar manner and thus are omitted herein. The necessary conditions for system (2.1.3) to exhibit chaos are that formula (2.1.4) is satisfied and all fixed points E_m $(m = 1,\ 2,\ 3,\ 4,\ 5)$ are unstable. To satisfy these conditions, there are many choices of parameters $a_1,\ b_1$ and c_1.

(4) Symmetry

System (2.1.3) is symmetrical about the u_{51}-axis, due to invariance of system (2.1.3) under the transformation $(u_{11}, u_{21}, u_{31}, u_{41}, u_{51}) \to (-u_{11}, -u_{21}, -u_{31}, -u_{41}, u_{51})$. Specially, the symmetry about the u_{51}-axis for any chosen parameters a_1, b_1, c_1 and d_1 is accurate. Figure 2.1 also shows the symmetry of the motion trajectory of system (2.1.3) about the u_{51}-axis.

2.1.3 The Anti-synchronization of the New Complex Chaotic System

(1) **The method of active control**

The driving system is chosen as the complex Lorenz system [76, 77], whose mathematical expression is given as follows:

$$\begin{cases} \dot{w}_1 = a_2(w_2 - w_1), \\ \dot{w}_2 = c_2 w_1 - w_1 w_3 - w_2, \\ \dot{w}_3 = -b_2 w_3 + \frac{1}{2}(\bar{w}_1 w_2 + w_1 \bar{w}_2), \end{cases} \tag{2.1.10}$$

where a_2, b_2 and c_2 are real constants, $w_1 = w_1^r + jw_1^i = u_{12} + ju_{22}$ and $w_2 = w_2^r + jw_2^i = u_{32} + ju_{42}$ are complex variables, $w_3 = u_{52}$ is a real variable.

By separating the real and imaginary parts of the driving system (2.1.10), we obtain the following real system

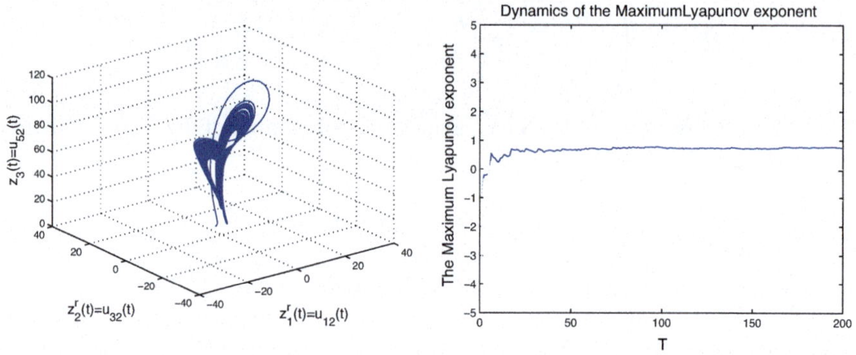

Fig. 2.3 The attractor of complex Lorenz system (2.1.10) in different projections and its maximum Lyapunov exponent

$$
\begin{cases}
\dot{u}_{12} = a_2(u_{32} - u_{12}),\\
\dot{u}_{22} = a_2(u_{42} - u_{22}),\\
\dot{u}_{32} = c_2 u_{12} - u_{12}u_{52} - u_{32},\\
\dot{u}_{42} = c_2 u_{22} - u_{22}u_{52} - u_{42},\\
\dot{u}_{52} = -b_2 u_{52} + u_{12}u_{32} + u_{22}u_{42}.
\end{cases} \tag{2.1.11}
$$

System (2.1.10) or (2.1.11) exhibits chaotic behavior under the parameter configurations $a_2 = 35, b_2 = 8/3, c_2 = 55$ with initial conditions $u_{12}(0) = 1,\ u_{22}(0) = 2,\ u_{32}(0) = 5,\ u_{42}(0) = 4,\ u_{52}(0) = 3$. The corresponding chaotic attractor and its maximum Lyapunov exponent are illustrated in Fig. 2.3.

The complex Wang system is adopted as the response system and defined as follows:

$$
\begin{cases}
\dot{z}_1 = a_1(z_1 - z_2) + I_1^r + jI_2^i,\\
\dot{z}_2 = -c_1 z_2 + z_1 z_3 + I_3^r + jI_4^i,\\
\dot{z}_3 = -b_1 z_3 + \frac{1}{2} d_1(\bar{z}_1 z_2 + z_1 \bar{z}_2) + I_5,
\end{cases} \tag{2.1.12}
$$

where $I_1^r + jI_2^i$ and $I_3^r + jI_4^i$ are complex variables input; I_5 is a real variable input. The real version of the response system (2.1.12) reads

$$
\begin{cases}
\dot{u}_{11} = a_1(u_{11} - u_{31}) + I_1^r,\\
\dot{u}_{21} = a_1(u_{21} - u_{41}) + I_2^i,\\
\dot{u}_{31} = -c_1 u_{31} + u_{11}u_{51} + I_3^r,\\
\dot{u}_{41} = -c_1 u_{41} + u_{21}u_{51} + I_4^i,\\
\dot{u}_{51} = -b_1 u_{51} + d_1(u_{11}u_{31} + u_{21}u_{41}) + I_5.
\end{cases} \tag{2.1.13}
$$

In line with the definition of anti-synchronization [78], the state error variables are defined as follows:

$$\begin{cases} s_1^r = u_{11} + u_{12}, \\ s_2^i = u_{21} + u_{22}, \\ s_3^r = u_{31} + u_{32}, \\ s_4^i = u_{41} + u_{42}, \\ s_5 = u_{51} + u_{52}. \end{cases} \tag{2.1.14}$$

In order to obtain the anti-synchronization error system, we add system (2.1.11) to system (2.1.13) using the formula (2.1.14) and get

$$\begin{cases} \dot{s}_1^r = -a_2 s_1^r - a_1 s_3^r + (a_1 + a_2)(u_{11} + u_{32}) + I_1^r, \\ \dot{s}_2^i = -a_2 s_2^i - a_1 s_4^i + (a_1 + a_2)(u_{21} + u_{42}) + I_2^i, \\ \dot{s}_3^r = -c_1 s_3^r - u_{52} s_1^r + u_{11} s_5 + c_2 u_{12} + (c_1 - 1)u_{32} + I_3^r, \\ \dot{s}_4^i = -c_1 s_4^i - u_{52} s_2^i + u_{21} s_5 + c_2 u_{22} + (c_1 - 1)u_{42} + I_4^i, \\ \dot{s}_5 = -b_1 s_5 + d_1(s_1^r u_{31} + s_4^i u_{21}) + s_2^i u_{42} + s^r(3)u_{12} + (b_1 - b_2)u_{52} \\ \qquad - (d_1 + 1)(u_{12}u_{31} + u_{21}u_{42}) + I_5. \end{cases} \tag{2.1.15}$$

Choose a Lyapunov function as

$$V(t) = \frac{1}{2} S^T(t)S(t) = \frac{1}{2}[(s_1^r(t))^2 + (s_2^i(t))^2 + (s_3^r(t))^2 + (s_4^i(t))^2 + (s_5(t))^2],$$

where $S = [s_1^r,\ s_2^i,\ s_3^r,\ s_4^i,\ s_5]^T$. The derivative of $V(t)$ along the solution of the error system (2.1.15) is

$$\begin{aligned} \dot{V}(t) &= \dot{s}_1^r s_1^r + \dot{s}_2^i s_2^i + \dot{s}_3^r s_3^r + \dot{s}_4^i s_4^i + \dot{s}_5 s_5 \\ &= [-a_2 s_1^r - a_1 s_3^r + (a_1 + a_2)(u_{11} + u_{32}) + I_1^r] s_1^r + [-a_2 s_2^i - a_1 s_4^i \\ &\quad + (a_1 + a_2)(u_{21} + u_{42}) + I_2^i] s_2^i + [-c_1 s_3^r - u_{52} s_1^r + u_{11} s_5 + c_2 u_{12} \\ &\quad + (c_1 - 1)u_{32} + I_3^r] s_3^r + [-c_1 s_4^i - u_{52} s_2^i + u_{21} s_5 + c_2 u_{22} \\ &\quad + (c_1 - 1)u_{42} + I_4^i] s_4^i + [-b_1 s_5 + d_1(s_1^r u_{31} + s_4^i u_{21}) + s_2^i u_{42} \\ &\quad + s_3^r u_{12} + (b_1 - b_2)u_{52} - (d_1 + 1)(u_{12}u_{31} + u_{21}u_{42}) + I_5] s_5. \end{aligned} \tag{2.1.16}$$

The control input functions I_1^r, I_2^i, I_3^r, I_4^i and I_5 are selected as

$$\begin{cases} I_1^r = a_1 s_3^r - (a_1 + a_2)(u_{11} + u_{32}), \\ I_2^i = a_1 s_4^i - (a_1 + a_2)(u_{21} + u_{42}), \\ I_3^r = u_{52} s_1^r - u_{11} s_5 - c_2 u_{12} - (c_1 - 1)u_{32}, \\ I_4^i = u_{52} s_2^i - u_{21} s_5 - c_2 u_{22} - (c_1 - 1)u_{42}, \\ I_5 = -d_1(s_1^r u_{31} + s_4^i u_{21}) - s_2^i u_{42} - s_3^r u_{12} - (b_1 - b_2)u_{52} \\ \qquad + (d_1 + 1)(u_{12}u_{31} + u_{21}u_{42}), \end{cases} \tag{2.1.17}$$

then equation (2.1.16) becomes

$$\dot{V}(t) = -[a_2(s_1^r)^2 + a_2(s_2^i)^2 + c_1(s_3^r)^2 + c_1(s_4^i)^2 + b_1(s_5)^2] < 0. \tag{2.1.18}$$

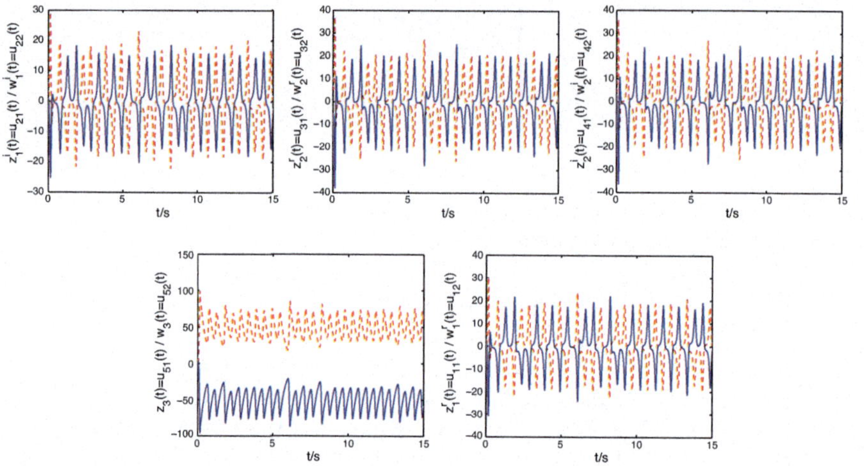

Fig. 2.4 The anti-synchronization of complex system (2.1.13) and complex Lorenz system (2.1.11) with the control inputs (2.1.17)

Since $V(t)$ is a positive definite function and its derivative is negative in the neighborhood of zero solution of the error system (2.1.15), then Lyapunov's methods imply that the equilibrium point $(s_1^r, s_2^i, s_3^r, s_4^i, s_5) = (0, 0, 0, 0, 0)$ of the error system (2.1.15) is asymptotically stable, namely,

$$\lim_{t\to\infty} \|S(t)\| = \lim_{t\to\infty} \sum_{l=1}^{5} \|s_l\| = 0,$$

then the anti-synchronization between systems (2.1.11) and (2.1.13) is globally asymptotically achieved, that is, complex Lorenz system (2.1.10) and the complex Wang system (2.1.12) are globally asymptotically anti-synchronized under the control input functions (2.1.17).

To validate the realization of global asymptotic anti-synchronization between the complex Lorenz System (2.1.10) and the complex Wang System (2.1.12), the fourth-order Runge-Kutta integration method is employed to solve systems (2.1.11) and (2.1.13) with the control inputs (2.1.17). The parameters of system (2.1.12) and system (2.1.10) are selected as $a_1 = 20, b_1 = 2, c_1 = 28, d_1 = 1$ and $a_2 = 1, b_2 = 2, c_2 = 10$ and assume that initial values of two systems are $(u_{11}(0), u_{21}(0), u_{31}(0), u_{41}(0), u_{51}(0)) = (2, 1, -4, 3, 4)$ and $(u_{12}(0), u_{22}(0), u_{32}(0), u_{42}(0), u_{52}(0)) = (1, 2, 3, 4, 3)$ respectively. The simulation results are depicted in Figs. 2.4 and 2.5.

Figure 2.4 shows that globally chaos anti-synchronization of systems (2.1.11) and (2.1.13) is indeed achieved after a very small interval in time t. In Fig. 2.5, it can be seen that the anti-synchronization errors $s_1^r(t), s_2^i(t), s_3^r(t), s_4^i(t)$ and $s_5(t)$

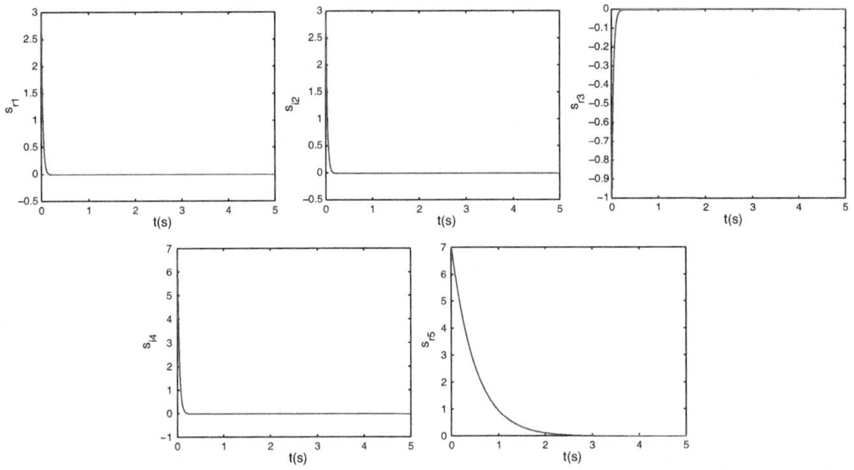

Fig. 2.5 Dynamics of anti-synchronization errors between systems (2.1.11) and (2.1.13) under the controller (2.1.17)

converge asymptotically to zero as $t \to \infty$, as expected from the above analytical considerations.

(2) **The method of nonlinear feedback control**

In this section, we discuss the anti-synchronization between complex Wang system (2.1.2) and complex Lü system via nonlinear control. The method is different from the active control discussed in above section, since the Lyapunov function is not required for calculation and the speed of achieving anti-synchronization of each state can also be controlled.

In 2007, Mahmoud et al. introduced the complex Lü system and investigated its chaotic properties and synchronization [79]. We take the complex Lü system

$$\begin{cases} \dot{v}_1 = a_3(v_2 - v_1), \\ \dot{v}_2 = c_3 v_2 - v_1 v_3, \\ \dot{v}_3 = -b_3 v_3 + \frac{1}{2}(\bar{v}_1 v_2 + v_1 \bar{v}_2), \end{cases} \tag{2.1.19}$$

as a drive system, where $v_1 = v_1^r + j v_1^i = u_{13} + j u_{23}$ and $v_2 = v_2^r + j v_2^i = u_{33} + j u_{43}$ are complex state variables, $v_3 = u_{53}$ is a real state variable.

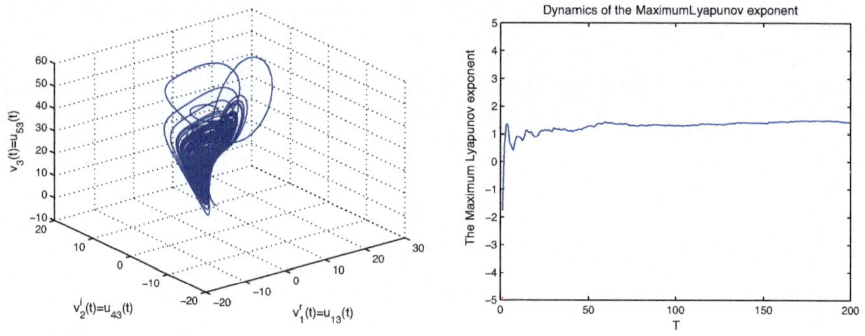

Fig. 2.6 The attractor and its maximum Lyapunov exponent of complex Lü system (2.1.19)

Separating real and imaginary parts of the complex Lü system (2.1.19), we have

$$\begin{cases} \dot{u}_{13} = a_3(u_{33} - u_{13}), \\ \dot{u}_{23} = a_3(u_{43} - u_{23}), \\ \dot{u}_{33} = c_3 u_{33} - u_{13}u_{53}, \\ \dot{u}_{43} = c_3 u_{43} - u_{23}u_{53}, \\ \dot{u}_{53} = -b_3 u_{53} + u_{13}u_{33} + u_{23}u_{43}. \end{cases} \tag{2.1.20}$$

When the parameters are set as $a_3 = 29$, $b_3 = 2$, $c_3 = 21$ and initial values $u_{13}(0) = 4$, $u_{23}(0) = -2$, $u_{33}(0) = 3$, $u_{43}(0) = 2$, $u_{53}(0) = -0.5$, the attractor and its maximum Lyapunov exponent of complex Lü system (2.1.19) are illustrated in Fig. 2.6.

The complex Wang system (2.1.2) is adopted as the response system and its real version is obviously the same as system (2.1.13). The response system (2.1.13) tracks anti-synchronously system (2.1.20) via the controllers I_i $(i = 1, 2, \ldots, 5)$ to be designed.

Similarly, anti-synchronous error systems are obtained as follows

$$\begin{cases} \dot{s}_1^r = a_3(s_3^r - s_1^r) + (a_1 + a_3)(u_{11} - u_{31}) + I_1^r, \\ \dot{s}_2^i = -a_3(s_4^i - s_2^i) + (a_1 + a_3)(u_{21} - u_{41}) + I_2^i, \\ \dot{s}_3^r = -c_1 s_3^r + (c_1 + c_3)u_{33} + u_{11}u_{51} - u_{13}u_{53} + I_3^r, \\ \dot{s}_4^i = -c_1 s_4^i + (c_1 + c_3)u_{43} + u_{21}u_{51} - u_{23}u_{53} + I_4^i, \\ \dot{s}_5 = -b_1 s_5 + (b_1 - b_3)u_{53} + u_{13}u_{33} + u_{23}u_{43} + d_1(u_{11}u_{31} + u_{21}u_{41}) + I_5. \end{cases} \tag{2.1.21}$$

Accordingly, the controllers for system (2.1.20) are designed as follows:

$$\begin{cases} I_1^r = a_3(s_1^r - s_3^r) - (a_1 + a_3)(u_{11} - u_{31}) - k_1 s_1^r, \\ I_2^i = a_3(s_2^i - s_4^i) - (a_1 + a_3)(u_{21} - u_{41}) - k_2 s_2^i, \\ I_3^r = -(c_1 + c_3)u_{33} + (c_1 - k_3)s_3^r + u_{13}u_{53} - u_{11}u_{51}, \\ I_4^i = -(c_1 + c_3)u_{43} + (c_1 - k_4)s_4^i + u_{23}u_{53} - u_{21}u_{51}, \\ I_5 = (b_1 - k_5)s_5 + (b_3 - b_1)u_{53} - u_{13}u_{33} - u_{23}u_{43} - d_1(u_{11}u_{31} - u_{21}u_{41}), \end{cases} \tag{2.1.22}$$

where $k_n > 0$ $(n = 1, 2, \ldots, 5)$ are real numbers and used to control the speed of achieving anti-synchronization.

By substituting the controllers (2.1.22) into the error system (2.1.21), the resulting error system is obtained as follows:

$$\begin{cases} \dot{s}_1^r = -k_1 s_1^r, \\ \dot{s}_2^i = -k_2 s_2^i, \\ \dot{s}_3^r = -k_3 s_3^r, \\ \dot{s}_4^i = -k_4 s_4^i, \\ \dot{s}_5 = -k_5 s_5. \end{cases} \tag{2.1.23}$$

The error system (2.1.23) can be written $\dot{S} = KS$, where $S = [s_1^r,\ s_2^i,\ s_3^r,\ s_4^i,\ s_5]^T$, $K = diag(-k_1, , -k_2,\ -k_3,\ -k_4,\ -k_5)$. It is clear that all eigenvalues of the closed loop system (2.1.23) are $-k_1,\ -k_2,\ -k_3,\ -k_4,\ -k_5$ and are negative real parts, hence anti-synchronization between complex Wang system (2.1.2) and complex Lü system (2.1.19) is realized.

The control coefficients $k_1 = 1.5, k_2 = 1, k_3 = 0.5, k_4 = 2$ and $k_5 = 1.5$ are taken and systems (2.1.20) and (2.1.13) with the controllers (2.1.22) are solved numerically for $a_3 = 29$,$b_3 = 2$,$c_3 = 21$ of system (2.1.20), $a_1 = 20$,$b_1 = 2$,$c_1 = 28$,$d_1 = 1$ of system (2.1.3) and their initial values $(u_{13}(0), u_{23}(0), u_{33}(0), u_{43}(0), u_{53}(0)) = (4, -2, 3, 2, -0.5)$, $(u_{11}(0), u_{21}(0), u_{31}(0), u_{41}(0), u_{51}(0)) = (2, 1, -5, 3, 4)$. The simulation results are qualitatively very similar to those demonstrated in Figs. 2.4 and 2.5. The solutions of systems (2.1.20) and (2.1.13) display that chaos anti-synchronization is again achieved after very small time intervals. As expected, the anti-synchronization errors $s_1^r(t), s_2^i(t), s_3^r(t), s_4^i(t)$ and $s_5(t)$ are also found to asymptotically converge to zero in this case.

On the other hand, the control coefficients $k_n (n = 1, 2, \ldots, 5)$ can be selected appropriately in advance to regulate the speed of achieving anti-synchronization between systems (2.1.3) and (2.1.20). Specifically, the larger the values of these coefficients, the faster the anti-synchronization is achieved.

2.2 Adaptive Anti-synchronization of Chaotic Complex Systems with Unknown Parameters

2.2.1 System Mathematical Models and Problem Descriptions

We consider a drive chaotic complex nonlinear system given by:

$$\dot{\mathbf{z}} = \mathbf{F}(\mathbf{z})\mathbf{A} + \mathbf{f}(\mathbf{z}), \tag{2.2.1}$$

where $\mathbf{z} = [z_1, z_2, \ldots, z_n]^T$ is a state complex vector ,$\mathbf{z} = \mathbf{z}^r + j\mathbf{z}^i$, and $\mathbf{z}^r = (u_1, u_3, \ldots, u_{2n-1})$, $\mathbf{z}^i = (u_2,\ u_4,\ \ldots,\ u_{2n})^T$. $\mathbf{f} = (f_1,\ f_2,\ \ldots,\ f_n)^T$ is a vector of complex nonlinear functions, $\mathbf{F}(\mathbf{z})$ is an $n \times n$ complex matrix and the elements of this matrix are functions of state complex variables, $\mathbf{A}$ is an $n \times 1$ real (or complex) vector of system parameters. On the other hand, the response chaotic complex system is assumed as

$$\dot{\mathbf{w}} = \mathbf{G}(\mathbf{w})\mathbf{B} + \mathbf{g}(\mathbf{w}) + \mathbf{v}, \tag{2.2.2}$$

where $\mathbf{w} = [w_1, w_2, \ldots, w_n]^T$ is a state complex vector, $\mathbf{w} = \mathbf{w}^r + j\mathbf{w}^i$, $\mathbf{w}^r = (u'_1, u'_3, \ldots, u'_{2n-1})$, $\mathbf{w}^i = (u'_2, u'_4, \ldots, u'_{2n})^T$. $\mathbf{g} = (g_1, g_2, \ldots, g_n)^T$ is a vector of complex nonlinear functions, $\mathbf{G}(\mathbf{w})$ is an $n \times n$ complex matrix and the elements of this matrix are functions of state complex variables, $\mathbf{B}$ is an $n \times 1$ real (or complex) vector of system parameters and the designed controller $\mathbf{v} = \mathbf{v}^r + j\mathbf{v}^i$, where $\mathbf{v}^r = (v_1, v_3, \ldots, v_{2n-1})$ and $\mathbf{v}^i = (v_2, v_4, \ldots, v_{2n})$.

Let $\mathbf{e}(t) = \mathbf{z}(t) + \mathbf{w}(t)$ denote the anti-synchronization error vector. The primary objective is to design a controller $\mathbf{v}$ such that the anti-synchronization error vector $\mathbf{e}(t) \to 0$ as $t \to \infty$, namely, the two coupled systems achieve asymptotic anti-synchronization.

$$\begin{aligned}
\lim_{t\to\infty} \|\mathbf{e}^r(t)\|^2 &= \lim_{t\to\infty} \|\mathbf{z}^r(t, \mathbf{z_0}^r) + \mathbf{w}^r(t, \mathbf{w_0}^r)\|^2 \\
&= \lim_{t\to\infty} \Big(\sum_{k=1}^{n} \|u_{2k-1}(t, u_{2k-1}(0)) + u'_{2k-1}(t, u'_{2k-1}(0))\|\Big)^2 \\
&= 0, \\
\lim_{t\to\infty} \|\mathbf{e}^i(t)\|^2 &= \lim_{t\to\infty} \|\mathbf{z}^i(t, \mathbf{z_0}^i) + \mathbf{w}^i(t, \mathbf{w_0}^i)\|^2 \\
&= \lim_{t\to\infty} \Big(\sum_{k=1}^{n} \|u_{2k}(t, u_{2k}(0)) + u'_{2k}(t, u'_{2k}(0))\|\Big)^2 \\
&= 0,
\end{aligned}$$

and

$$\lim_{t\to\infty} \|\mathbf{e}(t)\| = \lim_{t\to\infty} \sqrt{\|\mathbf{e}^r(t)\|^2 + \lim_{t\to\infty} \|\mathbf{e}^i(t)\|^2} = 0,$$

where $\mathbf{z_0}^r = [u_1(0), u_3(0), \ldots, u_{2n-1}(0)]^T$ and $\mathbf{z_0}^i = [u_2(0), u_4(0), \ldots, u_{2n}(0)]^T$ are initial values of real and imaginary parts of the complex system (2.2.1), while $\mathbf{w_0}^r = [u'_1(0), u'_3(0), \ldots, u'_{2n-1}(0)]^T$ and $\mathbf{w_0}^i = [u'_2(0), u'_4(0), \ldots, u'_{2n}(0)]^T$ are ones of the complex system (2.2.2).

2.2.2 Adaptive Anti-synchronization Controller Design

Theorem 2.1 *If nonlinear controller is taken as*

$$\begin{aligned}\mathbf{v} = \mathbf{v}^r + j\mathbf{v}^i &= -\mathbf{F}(\mathbf{z})\hat{\mathbf{A}} - \mathbf{f}(\mathbf{z}) - \mathbf{G}(\mathbf{w})\hat{\mathbf{B}} - \mathbf{g}(\mathbf{w}) - k\mathbf{e}\\ &= -\mathbf{F}^r(u_1, u_3, \ldots, u_{2n-1})\hat{\mathbf{A}} - \mathbf{f}^r(u_1, u_3, \ldots, u_{2n-1})\\ &\quad -\mathbf{G}^r(u_1', u_2', \ldots, u_{2n-1}')\hat{\mathbf{B}} - \mathbf{g}^r(u_1', u_3', \ldots, u_{2n-1}') - k\mathbf{e}^r\\ &\quad +j[-\mathbf{F}^i(u_2, u_4, \ldots, u_{2n})\hat{\mathbf{A}} - \mathbf{f}^i(u_2, u_4, \ldots, u_{2n})\\ &\quad -\mathbf{G}^i(u_2', u_4', \ldots, u_{2n}')\hat{\mathbf{B}} - \mathbf{g}^i(u_2', u_4', \ldots, u_{2n}') - k\mathbf{e}^i]\end{aligned} \tag{2.2.3}$$

and adaptive laws of parameters are selected as

$$\begin{cases}\dot{\hat{\mathbf{A}}} = [(\mathbf{F}^r(u_1, u_3, \ldots, u_{2n-1}))^T, (\mathbf{F}^i(u_2, u_4, \ldots, u_{2n}))^T]\begin{bmatrix}\mathbf{e}^r\\ \mathbf{e}^i\end{bmatrix},\\ \dot{\hat{\mathbf{B}}} = [(\mathbf{G}^r(u_1', u_3', \ldots, u_{2n-1}'))^T, (\mathbf{G}^i(u_2', u_4', \ldots, u_{2n}'))^T]\begin{bmatrix}\mathbf{e}^r\\ \mathbf{e}^i\end{bmatrix},\end{cases} \tag{2.2.4}$$

then the response system (2.2.2) can anti-synchronize the drive system (2.2.1) asymptotically, where $k > 0$ is a constant, the parameters of vectors $\hat{\mathbf{A}}$ and $\hat{\mathbf{B}}$ are, respectively, estimations of the unknown parameters of vectors $\mathbf{A}$ and $\mathbf{B}$.

Proof From chaotic complex systems (2.2.1) and (2.2.2), we get the error dynamical system as follows

$$\begin{aligned}\dot{\mathbf{e}} &= \dot{\mathbf{e}}^r + j\dot{\mathbf{e}}^i\\ &= \mathbf{F}^r(u_1, u_3, \ldots, u_{2n-1})\mathbf{A} + \mathbf{f}^r(u_1, u_3, \ldots, u_{2n-1}) + \mathbf{G}^r(u_1', u_2', \ldots, u_{2n-1}')\mathbf{B}\\ &\quad +\mathbf{g}^r(u_1', u_3', \ldots, u_{2n-1}') + \mathbf{v}^r + j[\mathbf{F}^i(u_2, u_4, \ldots, u_{2n})\mathbf{A} + \mathbf{f}^i(u_2, u_4, \ldots, u_{2n})\\ &\quad +\mathbf{G}^i(u_2', u_4', \ldots, u_{2n}')\mathbf{B} + \mathbf{g}^i(u_2', u_4', \ldots, u_{2n}') + \mathbf{v}^i].\end{aligned} \tag{2.2.5}$$

Thus, substituting equations (2.2.1), (2.2.2) and (2.2.3) into (2.2.5) and separating real and imaginary parts in these equations, the error system (2.2.5) can be written as

$$\begin{cases}\dot{\mathbf{e}}^r = \mathbf{F}^r(u_1, u_3, \ldots, u_{2n-1})(\mathbf{A} - \hat{\mathbf{A}}) + \mathbf{G}^r(u_1', u_2', \ldots, u_{2n-1}')(\mathbf{B} - \hat{\mathbf{B}}) - k\mathbf{e}^r,\\ \dot{\mathbf{e}}^i = \mathbf{F}^i(u_1, u_3, \ldots, u_{2n-1})(\mathbf{A} - \hat{\mathbf{A}}) + \mathbf{G}^i(u_1', u_2', \ldots, u_{2n-1}')(\mathbf{B} - \hat{\mathbf{B}}) - k\mathbf{e}^i.\end{cases} \tag{2.2.6}$$

We define the parameter errors as $\tilde{\mathbf{A}} = \hat{\mathbf{A}} - \mathbf{A}$, $\tilde{\mathbf{B}} = \hat{\mathbf{B}} - \mathbf{B}$ and consider a Lyapunov function candidate as

$$V(\mathbf{e}^r, \mathbf{e}^i, \tilde{\mathbf{A}}, \tilde{\mathbf{B}}) = \frac{1}{2}[(\mathbf{e}^r)^T\mathbf{e}^r + (\mathbf{e}^i)^T\mathbf{e}^i + \tilde{\mathbf{A}}^T\tilde{\mathbf{A}} + \tilde{\mathbf{B}}^T\tilde{\mathbf{B}}],$$

then time derivative of V along the trajectory of the error system (2.2.6) is as follows

$$\dot{V} = (\dot{\mathbf{e}}^r)^T\mathbf{e}^r + (\dot{\mathbf{e}}^i)^T\mathbf{e}^i + \tilde{\mathbf{A}}^T\dot{\hat{\mathbf{A}}} + \tilde{\mathbf{B}}^T\dot{\hat{\mathbf{B}}}$$

$$
\begin{aligned}
&= [\mathbf{F}^r(u_1, u_3, \ldots, u_{2n-1})(\mathbf{A}-\hat{\mathbf{A}}) + \mathbf{G}^r(u_1', u_2', \ldots, u_{2n-1}')(\mathbf{B}-\hat{\mathbf{B}}) - k\mathbf{e}^r]^T\mathbf{e}^r \\
&\quad + [\mathbf{F}^i(u_1, u_3, \ldots, u_{2n-1})(\mathbf{A}-\hat{\mathbf{A}}) + \mathbf{G}^i(u_1', u_2', \ldots, u_{2n-1}')(\mathbf{B}-\hat{\mathbf{B}}) - k\mathbf{e}^i]^T\mathbf{e}^i \\
&\quad + \tilde{\mathbf{A}}^T[(\mathbf{F}^r(u_1, u_3, \ldots, u_{2n-1}))^T\mathbf{e}^r + (\mathbf{F}^i(u_2, u_4, \ldots, u_{2n}))^T\mathbf{e}^i] \\
&\quad + \tilde{\mathbf{B}}^T[(\mathbf{G}^r(u_1', u_3', \ldots, u_{2n-1}'))^T\mathbf{e}^r + (\mathbf{G}^i(u_2', u_4', \ldots, u_{2n}'))^T]\mathbf{e}^i \\
&= -k[(\mathbf{e}^r)^T\mathbf{e}^r + (\mathbf{e}^i)^T\mathbf{e}^i] \le 0.
\end{aligned}
$$

Since V is positive definite, and $\dot{V}$ is negative semi-definite, we cannot immediately obtain that the origin of error system (2.2.5) is asymptotically stable. In fact, since $\dot{V} \le 0$, then $e_1, e_3, \ldots, e_{2n-1} \in L_\infty$, $e_2, e_4, \ldots, e_{2n} \in L_\infty$, $\tilde{a}_1, \tilde{a}_2, \ldots, \tilde{a}_n \in L_\infty$ and $\tilde{b}_1, \tilde{b}_2, \ldots, \tilde{b}_n \in L_\infty$. From the fact that

$$
\int_0^t ((\mathbf{e}^r)^T\mathbf{e}^r + (\mathbf{e}^i)^T\mathbf{e}^i)dt = \int_0^t \|\mathbf{e}^r\|^2 + \|\mathbf{e}^i\|^2 dt = -(1/k)(V(t) - V(0)) \le V(0)/k,
$$

we can easily know that $e_1, e_3, \ldots, e_{2n-1} \in L_2$ and $e_2, e_4, \ldots, e_{2n} \in L_2$. From Eq.(2.2.6), we have $\dot{e}_1, \dot{e}_3, \ldots, \dot{e}_{2n-1} \in L_\infty$ and $\dot{e}_2, \dot{e}_4, \ldots, \dot{e}_{2n} \in L_\infty$. According to the Barbalats lemma, we get, as $t \to \infty$, $e_1, e_3, \ldots, e_{2n-1} \to 0$ and $e_2, e_4, \ldots, e_{2n} \to 0$, namely

$$
\begin{aligned}
&\lim_{t\to\infty} \|\mathbf{e}^r(t)\|^2 + \|\mathbf{e}^i(t)\|^2 \\
&= \lim_{t\to\infty} [(\sum_{k=1}^{n} \|e_{2k-1}(t, e_{2k-1}(0))\|)^2 + (\sum_{k=1}^{n} \|e_{2k}(t, e_{2k}(0))\|)^2] \\
&= \lim_{t\to\infty} [(\sum_{k=1}^{n} \|u_{2k-1}(t, u_{2k-1}(0)) + u_{2k-1}'(t, u_{2k-1}'(0))\|)^2 \\
&\quad + (\sum_{k=1}^{n} \|u_{2k}(t, u_{2k}(0)) + u_{2k}'(t, u_{2k}'(0))\|)^2] \\
&= 0,
\end{aligned}
$$

therefore

$$
\lim_{t\to\infty} \|\mathbf{e}(t)\| = \lim_{t\to\infty} \sqrt{\|\mathbf{e}^r(t)\|^2 + \|\mathbf{e}^i(t)\|^2} = 0.
$$

From the fact that

$$
\int_0^t \|\mathbf{e}^r\| + \|\mathbf{e}^i\| = -\frac{1}{k}(V(t) - V(0)) \le \frac{V(0)}{k},
$$

we can easily know that $\mathbf{e}^r$, $\mathbf{e}^i \in L_2$. From equation (2.2.6), we have $\dot{\mathbf{e}}^r$, $\dot{\mathbf{e}}^i \in L_\infty$. According to the Barbalats lemma, we get

$$
\lim_{t\to\infty} \mathbf{e}^r + \mathbf{e}^i = 0,
$$

namely,

$$
\begin{aligned}
&\lim_{t\to\infty} \|\mathbf{e}^r(t)\| + \|\mathbf{e}^i(t)\| \\
&= \lim_{t\to\infty} \|\mathbf{z}^r(t,\ \mathbf{z_0}^r) + \mathbf{w}^r(t,\ \mathbf{w_0}^r) + \|\mathbf{z}^i(t,\ \mathbf{z_0}^i) + \mathbf{w}^i(t,\ \mathbf{w_0}^i)\| \\
&= \lim_{t\to\infty} \|u_{2n-1}(t,\ u_{2n-1}(0)) + u'_{2n-1}(t,\ u'_{2n-1}(0))\| + \|u_{2n}(t,\ u_{2n}(0)) + u'_{2n}(t,\ u'_{2n}(0))\| \\
&= 0.
\end{aligned}
$$

Therefore, the response complex system (2.2.2) can asymptotically anti-synchronize the drive complex system (2.2.1) under the proposed controllers (2.2.3) and the parameter estimation update laws (2.2.4). This completes the proof.

Remark 2.1 Most of typical chaotic complex systems can be written as system (2.2.1), such as complex Lorenz system [77], complex Chen system [79], complex Lü system [79], complex Van der Pol oscillator [80, 81] and complex Duffing system [82, 83].

Remark 2.2 If systems (2.2.1) and (2.2.2) satisfy $\mathbf{F}(\mathbf{z}) = \mathbf{G}(\mathbf{z})$ and $f(\mathbf{z}) = g(\mathbf{z})$, then the structure of systems (2.2.1) and (2.2.2) is identical. Hence, Theorem 2.1 is also applied to anti-synchronize two identical chaotic complex systems with unknown parameters.

2.2.3 Examples of Numerical Simulation

(1) Adaptive anti-synchronization of two identical chaotic complex systems

In order to observe the anti-synchronization behavior between two identical complex Lorenz systems, assume that we have two complex Lorenz systems where the drive system is denoted with the complex variable z and the response system having identical equations is denoted by the complex variable w. Therefore, we define the drive and response systems as follows

$$
\begin{cases}
\dot{z}_1 = a_1(z_2 - z_1), \\
\dot{z}_2 = a_2 z_1 - z_2 - z_1 z_3, \\
\dot{z}_3 = -a_3 z_3 + \frac{1}{2}(\bar{z}_1 z_2 + z_1 \bar{z}_2)
\end{cases}
\tag{2.2.7}
$$

and

$$
\begin{cases}
\dot{w}_1 = a_1(w_2 - w_1) + v_1 + j v_2, \\
\dot{w}_2 = a_2 w_1 - w_2 - w_1 w_3 + v_3 + j v_4, \\
\dot{w}_3 = -a_3 w_3 + \frac{1}{2}(\bar{w}_1 w_2 + w_1 \bar{w}_2) + v_5.
\end{cases}
\tag{2.2.8}
$$

First, we rewrite the complex systems (2.2.7) and (2.2.8) into the respective forms of systems (2.2.1) and (2.2.2) as follows

$$
\dot{\mathbf{z}} = \mathbf{F}(\mathbf{z})\mathbf{A} + \mathbf{f}(\mathbf{z}),
\tag{2.2.9}
$$

where

$$\mathbf{F}(\mathbf{z}) = \begin{pmatrix} z_2 - z_1 & 0 & 0 \\ 0 & z_1 & 0 \\ 0 & 0 & -z_3 \end{pmatrix}, \quad \mathbf{f}(\mathbf{z}) = \begin{pmatrix} 0 \\ -z_2 - z_1 z_3 \\ \frac{1}{2}(\bar{z}_1 z_2 + z_1 \bar{z}_2) \end{pmatrix}.$$

Similarly,

$$\dot{\mathbf{w}} = \mathbf{G}(\mathbf{w})\mathbf{B} + \mathbf{g}(\mathbf{w}), \tag{2.2.10}$$

$$\mathbf{G}(\mathbf{w}) = \begin{pmatrix} w_2 - w_1 & 0 & 0 \\ 0 & w_1 & 0 \\ 0 & 0 & -w_3 \end{pmatrix}, \quad \mathbf{g}(\mathbf{w}) = \begin{pmatrix} 0 \\ -w_2 - w_1 w_3 \\ \frac{1}{2}(\bar{w}_1 w_2 + w_1 \bar{w}_2) \end{pmatrix},$$

where $\mathbf{z} = (z_1, z_2, z_3)^T$ and $\mathbf{w} = (w_1, w_2, w_3)^T$ are the state vectors of the drive system (2.2.7) and response system (2.2.8) respectively, $\mathbf{A} = \mathbf{B} = (a_1, a_2, a_3)^T$ is the unknown parameter vector $v_i (i = 1, 2, \dots, 5)$ are five control functions to be designed.

According to Theorem 2.1, the controller is chosen as

$$\begin{aligned}
\mathbf{v} &= -\mathbf{F}(\mathbf{z})\hat{\mathbf{A}} - \mathbf{f}(\mathbf{z}) - \mathbf{G}(\mathbf{w})\hat{\mathbf{B}} - \mathbf{g}(\mathbf{w}) - k\mathbf{e} \\
&= \begin{pmatrix} -(z_2 - z_1)\hat{a}_1 - (w_2 - w_1)\hat{b}_1 - k(e_1 + je_2) \\ -(z_1\hat{a}_2 - z_2 - z_1 z_3) - (w_1\hat{b}_2 - w_2 - w_1 w_3) - k(e_3 + je_4) \\ z_3\hat{a}_3 - \frac{1}{2}(\bar{z}_1 z_2 + z_1\bar{z}_2) - (-w_3\hat{a}_3 + \frac{1}{2}(\bar{w}_1 w_2 + w_1\bar{w}_2)) - ke_5 \end{pmatrix} \\
&= \begin{pmatrix} -\hat{a}_1(u_3 - u_1 + u_3' - u_1') - ke_1 \\ -\hat{a}_2(u_1 + u_1') + u_3 + u_1 u_5 + u_3' + u_1' u_5' - ke_3 \\ \hat{a}_3(u_5 + u_5') - \frac{1}{2}(u_1 u_3 + u_2 u_4 + u_1' u_3' + u_2' u_4') - ke_5 \end{pmatrix} \\
&\quad + j \begin{pmatrix} -\hat{a}_1(u_4 - u_2 + u_4' - u_2') - ke_2 \\ -\hat{a}_2(u_2 + u_2') + u_4 + u_2 u_5 + u_4' + u_2' u_5' - ke_4 \\ 0 \end{pmatrix}
\end{aligned} \tag{2.2.11}$$

and adaptive laws of parameters are chosen as

$$\begin{cases} \dot{\hat{\mathbf{A}}} = [(\mathbf{F}^r)^T, (\mathbf{F}^i)^T] \begin{bmatrix} \mathbf{e}^r \\ \mathbf{e}^i \end{bmatrix}, \\ \dot{\hat{\mathbf{B}}} = [(\mathbf{G}^r)^T, (\mathbf{G}^i)^T] \begin{bmatrix} \mathbf{e}^r \\ \mathbf{e}^i \end{bmatrix}, \end{cases}$$

since $\mathbf{A} = \mathbf{B} = (a_1,\ a_2,\ a_3)^T$, then

$$\dot{\hat{\mathbf{A}}} = \dot{\hat{\mathbf{B}}} = \frac{1}{2} \left[(\mathbf{F}^r)^T + (\mathbf{G}^r)^T \ (\mathbf{F}^i)^T + (\mathbf{G}^i)^T \right] \begin{bmatrix} \mathbf{e}^r \\ \mathbf{e}^i \end{bmatrix}$$

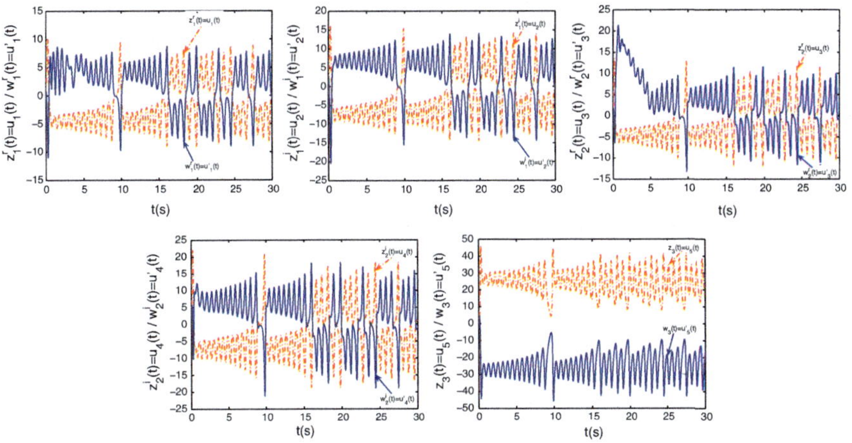

Fig. 2.7 The anti-synchronization of two identical complex Lorenz systems (2.2.7) and (2.2.8) with the controller (2.2.11) and the adaptive law (2.2.12)

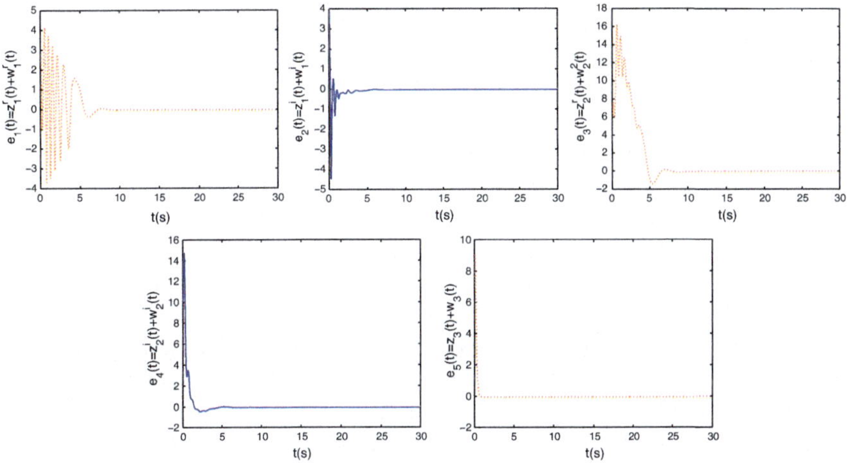

Fig. 2.8 The error dynamic of anti-synchronization between two identical complex Lorenz systems (2.2.7) and (2.2.8) with the controller (2.2.11) and the adaptive law (2.2.12)

$$
\begin{aligned}
&= \begin{pmatrix} (u_3 - u_1 + u_3' - u_1')e_1 + (u_4 - u_2 + u_4' - u_2')e_2 \\ (u_1 + u_1')e_3 + (u_2 + u_2')e_4 \\ -(u_5 + u_5')e_5 \end{pmatrix} \\
&= \begin{pmatrix} (e_3 - e_1)e_1 + (e_4 - e_2)e_2 \\ e_1 e_3 + e_2 e_4 \\ -e_5^2 \end{pmatrix},
\end{aligned}
\tag{2.2.12}
$$

where $e_m = u_m + u_m'$, $(m = 1, 2, \ldots, 5)$.

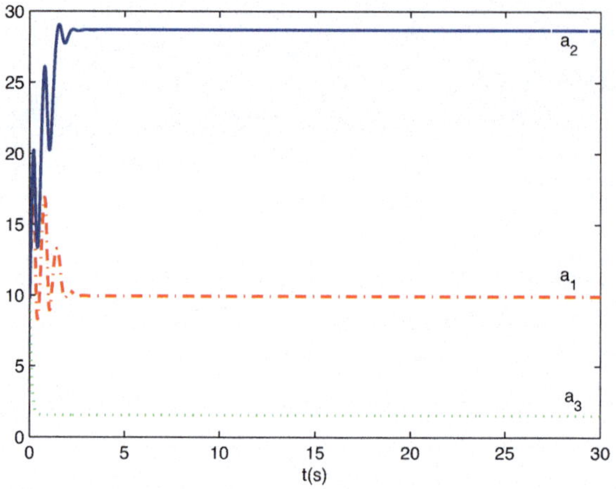

Fig. 2.9 Changing parameters $\hat{a}_1$, $\hat{a}_2$ and $\hat{a}_3$ of the complex Lorenz systems with time t

To verify the effectiveness of the proposed anti-synchronization scheme, we present the numerical simulation results for the anti-synchronization between two identical complex-valued Lorenz systems (2.2.7) and (2.2.8). In the numerical simulations, the unknown parameters are chosen as $a_1 = 10$, $a_2 = 28$, $a_3 = 8/3$, system (2.2.7) (or (2.2.8)) exhibits chaotic behavior. The initial condition of the drive system state vector, the initial value of the response system state vector, the initial value of estimate for unknown parameter vector $\hat{\mathbf{A}}$ and the constant k are taken as $(z_1(0),\ z_2(0),\ z_3(0))^T = (2+1j,\ 1.5+5j,\ 3)^T$, $(w_1(0),\ w_2(0),\ w_3(0))^T = (0.8 + 3j,\ 2+5j,\ 4)^T$, $(a_1(0),\ a_2(0),\ a_3(0))^T = (10,\ 10,\ 10)^T$ and $k = 0.5$. Hence, the initial value of the error vector is $(e_1(0),\ e_2(0),\ e_3(0))^T = (2.8+4j,\ 3.5+ 10j,\ 7)^T$. The anti-synchronization between the systems (2.2.7) and (2.2.8) via the controller (2.2.11) and parameters adaptive laws (2.2.12) are shown in Figs. 2.7, 2.8 and 2.9. Figures 2.7 and 2.8 display the state response and anti-synchronization errors of systems (2.2.7) and (2.2.8). Figure 2.9 shows that the estimates $\hat{a}_1$, $\hat{a}_2$, $\hat{a}_3$ of the unknown parameters converge to $a_1 = 10$, $a_2 = 28$, $a_3 = 8/3$ as $t \to \infty$, respectively.

(2) Adaptive anti-synchronization of two different chaotic complex systems

In order to observe the anti-synchronization behavior between two different chaotic complex systems, we assume that complex Lü system with three unknown parameters drives the complex Chen system [84, 85]. Therefore, the drive and response systems are defined as follows

$$\begin{cases} \dot{z}_1 = a_1(z_2 - z_1), \\ \dot{z}_2 = a_2 z_2 - z_1 z_3, \\ \dot{z}_3 = -a_3 z_3 + \frac{1}{2}(\bar{z}_1 z_2 + z_1 \bar{z}_2), \end{cases} \tag{2.2.13}$$

and

$$\begin{cases} \dot{w}_1 = b_1(w_2 - w_1) + v_1 + jv_2, \\ \dot{z}_2 = (b_2 - b_1)w_1 + b_2 w_2 - w_1 w_3 + v_3 + jv_4, \\ \dot{z}_3 = -b_3 w_3 + \frac{1}{2}(\bar{w}_1 w_2 + w_1 \bar{w}_2) + v_5. \end{cases} \tag{2.2.14}$$

First, we rewrite the complex systems (2.2.13) and (2.2.14) in the form of systems (2.2.1) and (2.2.2) as follows

$$\dot{\mathbf{z}} = \mathbf{F}(\mathbf{z})\mathbf{A} + \mathbf{f}(\mathbf{z}), \tag{2.2.15}$$

where

$$\mathbf{F}(\mathbf{z}) = \begin{pmatrix} z_2 - z_1 & 0 & 0 \\ 0 & z_2 & 0 \\ 0 & 0 & -z_3 \end{pmatrix}, \quad \mathbf{f}(\mathbf{z}) = \begin{pmatrix} 0 \\ -z_1 z_3 \\ \frac{1}{2}(\bar{z}_1 z_2 + z_1 \bar{z}_2) \end{pmatrix},$$

and $\mathbf{z} = (z_1, z_2, z_3)^T$ denotes the state vector of the drive system (2.2.13), $\mathbf{A} = (a_1, a_2, a_3)^T$ denotes the unknown parameter vector of the drive system (2.2.13).

$$\dot{\mathbf{w}} = \mathbf{G}(\mathbf{w})\mathbf{B} + \mathbf{g}(\mathbf{w}), \tag{2.2.16}$$

where

$$\mathbf{G}(\mathbf{w}) = \begin{pmatrix} w_2 - w_1 & 0 & 0 \\ -w_1 & w_1 + w_2 & 0 \\ 0 & 0 & -w_3 \end{pmatrix}, \quad \mathbf{g}(\mathbf{z}) = \begin{pmatrix} 0 \\ -w_1 w_3 \\ \frac{1}{2}(\bar{w}_1 w_2 + w_1 \bar{w}_2) \end{pmatrix},$$

and $\mathbf{w} = (w_1, w_2, w_3)^T$ is the state vector of the response system (2.2.14), $\mathbf{B} = (b_1, b_2, b_3)^T$ is the unknown parameter vector of the response system (2.2.14), and $v_i (i = 1, 2, \ldots, 5)$ are control input to be chosen.

According to Theorem 2.1, the controller is taken as

$$\begin{aligned} \mathbf{v} &= -\mathbf{F}(\mathbf{z})\hat{\mathbf{A}} - \mathbf{f}(\mathbf{z}) - \mathbf{G}(\mathbf{w})\hat{\mathbf{B}} - \mathbf{g}(\mathbf{w}) - k\mathbf{e} \\ &= \begin{pmatrix} (z_1 - z_2)\hat{a}_1 - (w_2 - w_1)\hat{b}_1 - k(e_1 + je_2) \\ z_1 z_3 - z_2 \hat{a}_2 + w_1 \hat{b}_1 - (w_1 + w_2)\hat{b}_2 + w_1 w_3 - k(e_3 + je_4) \\ z_3 \hat{a}_3 - \frac{1}{2}(\bar{z}_1 z_2 + z_1 \bar{z}_2) + w_3 \hat{a}_3 - \frac{1}{2}(\bar{w}_1 w_2 + w_1 \bar{w}_2) - ke_5 \end{pmatrix} \\ &= \begin{pmatrix} \hat{a}_1(u_1 - u_3) + \hat{b}_1(u_1' - u_3') - ke_1 \\ u_1 u_5 - \hat{a}_2 u_3 + \hat{b}_1 u_1' - (u_1' + u_3')\hat{b}_2 + u_1' u_5' - ke_3 \\ \hat{a}_3 u_5 - (u_1 u_3 + u_2 u_4) + \hat{b}_3 u_5' - (u_1' u_3' + u_2' u_4') - ke_5 \end{pmatrix} \\ &\quad + j \begin{pmatrix} \hat{a}_1(u_2 - u_4) + \hat{b}_1(u_2' - u_4') - ke_2 \\ u_2 u_5 - \hat{a}_2 u_4 + u_2' \hat{b}_1 - (u_2' + u_4')\hat{b}_2 + u_2' u_5' - ke_4 \\ 0 \end{pmatrix} \end{aligned} \tag{2.2.17}$$

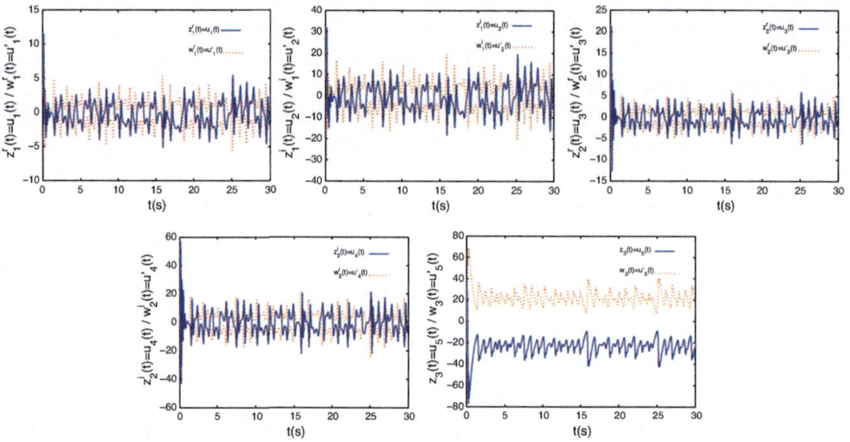

Fig. 2.10 The anti-synchronization between complex systems (2.2.13) and (2.2.14) with the controller (2.2.17) and the adaptive laws (2.2.18) and (2.2.19)

and the adaptive laws of parameters are chosen as

$$\dot{\hat{\mathbf{A}}} = \begin{pmatrix} \dot{\hat{a}}_1 \\ \dot{\hat{a}}_2 \\ \dot{\hat{a}}_3 \end{pmatrix} = \begin{pmatrix} e_1(z_2^r - z_1^r) + e_2(z_2^i - z_1^i) \\ z_2^r e_3 + z_2^i e_4 \\ -z_3 e_5 \end{pmatrix} = \begin{pmatrix} (u_3 - u_1)e_1 + (u_4 - u_2)e_2 \\ u_3 e_3 + u_4 e_4 \\ -u_5 e_5 \end{pmatrix} \tag{2.2.18}$$

and

$$\begin{aligned} \dot{\hat{\mathbf{B}}} = \begin{pmatrix} \dot{\hat{b}}_1 \\ \dot{\hat{b}}_2 \\ \dot{\hat{b}}_3 \end{pmatrix} &= \begin{pmatrix} (w_2^r - w_1^r)e_1 - w_1^r e_3 + (w_2^i - w_1^i)e_2 - w_1^i e_4 \\ (w_2^r + w_1^r)e_3 + (w_2^i + w_1^i)e_4 \\ -w_3 e_5 \end{pmatrix} \\ &= \begin{pmatrix} (u_3' - u_1')e_1 - u_1' e_3 + (u_4' - u_2')e_2 - u_2' e_4 \\ (u_3' + u_1')e_3 + (u_4' + u_2')e_4 \\ -u_5' e_5 \end{pmatrix}. \end{aligned} \tag{2.2.19}$$

In this simulation, the true values of unknown parameter vectors for the drive and response systems are set to $\mathbf{A} = [a_1,\ a_2,\ a_3]^T = [29,\ 21,\ 2]^T$ and $\mathbf{B} = [b_1,\ b_2,\ b_3]^T = [27,\ 23,\ 1]^T$, respectively. The initial value of the drive system (2.2.13), the initial value of the response system (2.2.14), the initial value of estimate for unknown parameter vector $\hat{\mathbf{A}}$, the initial value of estimate for unknown parameter vector $\hat{\mathbf{B}}$ and constant k are chosen as $\mathbf{z}(0) = [-3-2j,\ -2-7j,\ -5]^T$, $\mathbf{w}(0) = [4+15j,\ -1-5j,\ 30]^T$, $\hat{\mathbf{A}}(0) = [10,\ 10,\ 10]^T$, $\hat{\mathbf{B}}(0) = [15,\ 15,\ 15]^T$ and $k = 2.5$. While the initial value of the error vector is $[e_1(0),\ e_2(0),\ e_3(0)]^T = [1+13j,\ -3-12j,\ 25]^T$. As illustrated in Fig. 2.10, the response system anti-synchronizes the drive

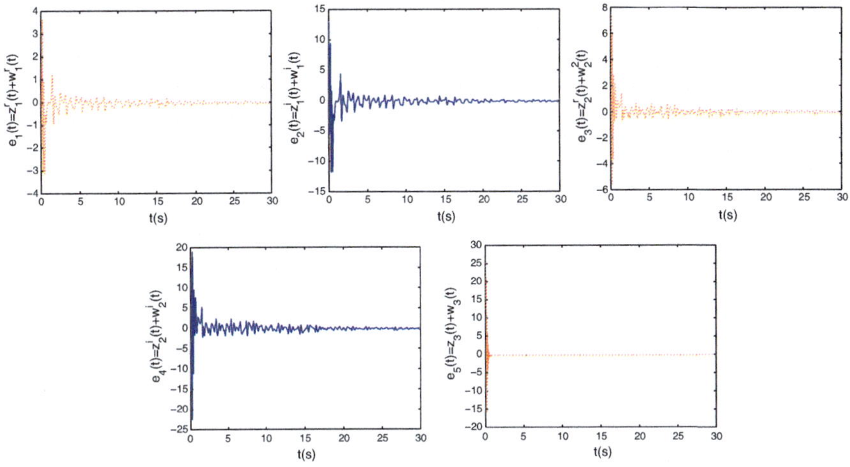

Fig. 2.11 The error dynamic of anti-synchronization between complex systems (2.2.13) and (2.2.14) with the controller (2.2.17) and the adaptive laws (2.2.18) and (2.2.19)

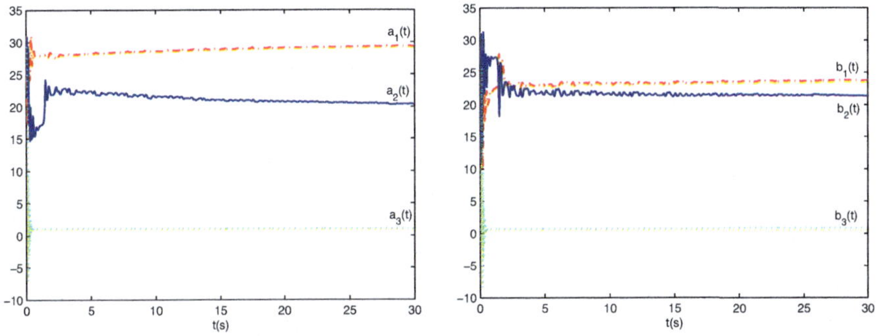

Fig. 2.12 Changing parameters $\hat{a}_1$, $\hat{a}_2$, $\hat{a}_3$ and $\hat{b}_1$, $\hat{b}_2$, $\hat{b}_3$ with time t

system. Figure 2.11 demonstrates that the anti-synchronization errors converge to zero as t increases. The changing parameters of $\hat{\mathbf{A}}$ and $\hat{\mathbf{B}}$ are shown in Fig. 2.12. Obviously, the estimations of parameters converge to some constants.

Chapter 3
Observe-Based Complete Synchronization of Chaotic Complex Systems

In recent years, engineering approaches have been extensively employed in the synchronization control of chaotic systems [86–91]. Among these methods, the state observer-based control strategy stands out, as it eliminates the need to construct a synchronous conditional Lyapunov function and does not require the initial states of the two chaotic systems to be synchronized within the same attractor domain, thereby garnering significant attention from researchers [92–95]. This chapter adopts the state observer design-based control method to investigate complete synchronization (CS) of complex chaotic systems and provides a specific case of synchronization achieved via this approach. Section 3.1 focuses on the realization of observer-based complete synchronization. Section 3.2 presents several simulation examples to verify the proposed method.

3.1 Implementation of Observe-Based CS

Consider the chaotic complex nonlinear system described by the following form

$$\dot{\mathbf{z}} = \mathbf{A}\mathbf{z} + \mathbf{B}\mathbf{f}(\mathbf{z}) + \mathbf{C}, \tag{3.1.1}$$

where $\mathbf{z} = \mathbf{z}^r + j\mathbf{z}^i$=$[z_1, z_2, \ldots, z_n]^T$ is the state complex vector, $\mathbf{z}^r = (u_1, u_3, \ldots, u_{2n-1})$, $\mathbf{z}^i = (u_2, u_4, \ldots, u_{2n})^T$.$\mathbf{f} = (f_1, f_2, \ldots, f_n)^T$ is a vector of complex nonlinear functions, $\mathbf{A}$ is an $n \times n$ real matrix, $\mathbf{B}$ is an $n \times m$ real matrix, $\mathbf{C}$ is an $n \times 1$ real (or complex) vector.

The observer of the complex system (3.1.1) is defined as

$$\dot{\hat{\mathbf{z}}} = \mathbf{A}\hat{\mathbf{z}} + \mathbf{B}\mathbf{f}(\hat{\mathbf{z}}) + \mathbf{C} + \mathbf{B}(\mathbf{s}(\mathbf{z}) - \mathbf{s}(\hat{\mathbf{z}})), \tag{3.1.2}$$

where $\mathbf{s}(\mathbf{z})$ is constructed output. In the sequel, we define the CS error as follows:

S. Liu et al., *Chaos in Complex Dynamical System: Theory and Application*,
https://doi.org/10.1007/978-981-95-7762-0_3

$$e(t) = \hat{\mathbf{z}}(\mathbf{t}) - \mathbf{z}(\mathbf{t}).$$

For achieving synchronization, we suppose system (3.1.1) with the output

$$\mathbf{s}(\mathbf{z}) = \mathbf{f}(\mathbf{z}) + \mathbf{kz}, \tag{3.1.3}$$

where $\mathbf{k}$ is an $n \times m$ feedback gain matrix.

Therefore, by decomposing the complex variables into their real and imaginary components, the complex system given in (3.1.1) can be reformulated as

$$\begin{pmatrix} \dot{\mathbf{z}}^{\mathbf{r}} \\ \dot{\mathbf{z}}^{\mathbf{i}} \end{pmatrix} = \begin{pmatrix} \mathbf{A} & \mathbf{0} \\ \mathbf{0} & \mathbf{A} \end{pmatrix} \begin{pmatrix} \mathbf{z}^{\mathbf{r}} \\ \mathbf{z}^{\mathbf{i}} \end{pmatrix} + \begin{pmatrix} \mathbf{B} & \mathbf{0} \\ \mathbf{0} & \mathbf{B} \end{pmatrix} \begin{pmatrix} \mathbf{f}^{\mathbf{r}}(\mathbf{z}^{\mathbf{r}}) \\ \mathbf{f}^{\mathbf{i}}(\mathbf{z}^{\mathbf{i}}) \end{pmatrix} + \begin{pmatrix} \mathbf{C}^{\mathbf{r}} \\ \mathbf{C}^{\mathbf{i}} \end{pmatrix}, \tag{3.1.4}$$

while the output (3.1.3) can also be written as

$$\begin{pmatrix} \dot{\mathbf{s}}^{\mathbf{r}(\mathbf{z}^{\mathbf{r}})} \\ \dot{\mathbf{s}}^{\mathbf{i}(\mathbf{z}^{\mathbf{i}})} \end{pmatrix} = \begin{pmatrix} \mathbf{f}^{\mathbf{r}}(\mathbf{z}^{\mathbf{r}}) \\ \mathbf{f}^{\mathbf{i}}(\mathbf{z}^{\mathbf{i}}) \end{pmatrix} + \begin{pmatrix} \mathbf{k} & \mathbf{0} \\ \mathbf{0} & \mathbf{k} \end{pmatrix} \begin{pmatrix} \mathbf{z}^{\mathbf{r}} \\ \mathbf{z}^{\mathbf{i}} \end{pmatrix}. \tag{3.1.5}$$

Similarly, the observer of system (3.1.4) is

$$\begin{aligned} \begin{pmatrix} \dot{\hat{\mathbf{z}}}^{\mathbf{r}} \\ \dot{\hat{\mathbf{z}}}^{\mathbf{i}} \end{pmatrix} &= \begin{pmatrix} \mathbf{A} & \mathbf{0} \\ \mathbf{0} & \mathbf{A} \end{pmatrix} \begin{pmatrix} \hat{\mathbf{z}}^{\mathbf{r}} \\ \hat{\mathbf{z}}^{\mathbf{i}} \end{pmatrix} + \begin{pmatrix} \mathbf{B} & \mathbf{0} \\ \mathbf{0} & \mathbf{B} \end{pmatrix} \begin{pmatrix} \mathbf{f}^{\mathbf{r}}(\hat{\mathbf{z}}^{\mathbf{r}}) \\ \mathbf{f}^{\mathbf{i}}(\hat{\mathbf{z}}^{\mathbf{i}}) \end{pmatrix} \\ &+ \begin{pmatrix} \mathbf{C}^{\mathbf{r}} \\ \mathbf{C}^{\mathbf{i}} \end{pmatrix} + \begin{pmatrix} \mathbf{B} & \mathbf{0} \\ \mathbf{0} & \mathbf{B} \end{pmatrix} \begin{pmatrix} \mathbf{s}^{\mathbf{r}}(\mathbf{z}^{\mathbf{r}}) - \mathbf{s}^{\mathbf{r}}(\hat{\mathbf{z}}^{\mathbf{r}}) \\ \mathbf{s}^{\mathbf{i}}(\mathbf{z}^{\mathbf{i}}) - \mathbf{s}^{\mathbf{i}}(\hat{\mathbf{z}}^{\mathbf{i}}) \end{pmatrix}. \end{aligned} \tag{3.1.6}$$

Based on system (3.1.1) and system (3.1.2), we obtain the error dynamical system

$$\dot{\mathbf{e}} = \dot{\hat{\mathbf{z}}} - \dot{\mathbf{z}} = \mathbf{Ae} + \mathbf{Bf}(\hat{\mathbf{z}}) - \mathbf{Bf}(\mathbf{z}) + \mathbf{B}(\mathbf{s}(\mathbf{z}) - \mathbf{s}(\hat{\mathbf{z}})) = \mathbf{Ae} - \mathbf{Bke}. \tag{3.1.7}$$

From systems (3.1.4), (3.1.5), (3.1.6), and (3.1.7), we obtain

$$\dot{\mathbf{e}} = \dot{\mathbf{e}}^{\mathbf{r}} + j\dot{\mathbf{e}}^{\mathbf{i}} = \mathbf{Ae}^{\mathbf{r}} - \mathbf{Bke}^{\mathbf{r}} + j(\mathbf{Ae}^{\mathbf{i}} - \mathbf{Bke}^{\mathbf{i}}), \tag{3.1.8}$$

namely,

$$\begin{cases} \dot{\mathbf{e}}^{\mathbf{r}} = \mathbf{Ae}^{\mathbf{r}} - \mathbf{Bke}^{\mathbf{r}} = (\mathbf{A} - \mathbf{Bk})\mathbf{e}^{\mathbf{r}}, \\ \dot{\mathbf{e}}^{\mathbf{i}} = \mathbf{Ae}^{\mathbf{i}} - \mathbf{Bke}^{\mathbf{i}} = (\mathbf{A} - \mathbf{Bk})\mathbf{e}^{\mathbf{i}}. \end{cases} \tag{3.1.9}$$

If system (3.1.9) is controllable, then systems (3.1.7) or (3.1.8) are also controllable. In order to make system (3.1.9) controllable, the feedback gain matrix $\mathbf{k}$ will be appropriately chosen. According to Lyapunov stable theory, $\mathbf{e}^{\mathbf{r}} \to 0$ and $\mathbf{e}^{\mathbf{i}} \to 0$ ($\mathbf{e} = \mathbf{e}^{\mathbf{r}} + j\mathbf{e}^{\mathbf{i}} \to 0$) as $t \to \infty$ when the real parts of all eigenvalues of $\mathbf{A} - \mathbf{Bk}$ in system (3.1.9) are negative, that is, complex systems (3.1.1) and (3.1.2) will achieve CS. The error system is governed by linear differential equations, briefly denoted as

$$\dot{\mathbf{e}} = \mathbf{A}\mathbf{e} + \mathbf{B}\mathbf{v}, \tag{3.1.10}$$

here $\mathbf{v} = -\mathbf{k}\mathbf{e}$ is the state feedback. Moreover, if the controllable matrix $[\mathbf{A}, \mathbf{A}\mathbf{B}, \ldots, \mathbf{A}^{\mathbf{n}-1}\mathbf{B}]$ is in full rank, the complex system (3.1.2) is the global observer of system (3.1.1). Hence, the pole placement technique [96] is usually used to determine the feedback gain matrix $\mathbf{k}$ for any eigenvalues chosen of matrix $\mathbf{A} - \mathbf{B}\mathbf{k}$.

3.2 Simulation Examples

3.2.1 Synchronization of the Complex Lorenz System

As is well known, the well-recognized complex Lorenz system is described as

$$\begin{cases} \dot{z_1} = a(z_2 - z_1), \\ \dot{z_2} = cz_1 - z_1 z_3 - z_2, \\ \dot{z_3} = -bz_3 + \frac{1}{2}(\bar{z}_1 z_2 + z_1 \bar{z}_2), \end{cases} \tag{3.2.1}$$

where a, b, c are real constants. System (3.2.1) is chaotic in a wide parameter range.

We rearrange the complex system (3.2.1) into the structure consistent with system (3.1.1), as presented below.

$$\begin{pmatrix} \dot{z}_1 \\ \dot{z}_2 \\ \dot{z}_3 \end{pmatrix} = \begin{pmatrix} -a & a & 0 \\ c & -1 & 0 \\ 0 & 0 & -b \end{pmatrix} \begin{pmatrix} z_1 \\ z_2 \\ z_3 \end{pmatrix} + \begin{pmatrix} 0 & 0 \\ 1 & 0 \\ 0 & 1 \end{pmatrix} \begin{pmatrix} -z_1 z_3 \\ \frac{1}{2}(\bar{z}_1 z_2 + z_1 \bar{z}_2) \end{pmatrix} + \begin{pmatrix} 0 \\ 0 \\ 0 \end{pmatrix}. \tag{3.2.2}$$

Comparing system (3.2.2) with system (3.1.1), we can get

$$\mathbf{A} = \begin{pmatrix} -a & a & 0 \\ c & -1 & 0 \\ 0 & 0 & -b \end{pmatrix}, \ \mathbf{B} = \begin{pmatrix} 0 & 0 \\ 1 & 0 \\ 0 & 1 \end{pmatrix}, \ \mathbf{C} = \mathbf{0}, \ \mathbf{f}(\mathbf{x}) = \begin{pmatrix} -z_1 z_3 \\ \frac{1}{2}(\bar{z}_1 z_2 + z_1 \bar{z}_2) \end{pmatrix}.$$

If the output of system (3.2.2) is $\mathbf{s}(\mathbf{z}) = \mathbf{f}(\mathbf{z}) + \mathbf{k}\mathbf{z}$ with a gain matrix $\mathbf{k} \in \mathbb{R}^{2\times 3}$, then the observer of system (3.1.7) can be written as

$$\begin{aligned} \begin{pmatrix} \dot{\hat{z}}_1 \\ \dot{\hat{z}}_2 \\ \dot{\hat{z}}_3 \end{pmatrix} = \begin{pmatrix} -a & a & 0 \\ c & -1 & 0 \\ 0 & 0 & -b \end{pmatrix} \begin{pmatrix} \hat{z}_1 \\ \hat{z}_2 \\ \hat{z}_3 \end{pmatrix} + \begin{pmatrix} 0 & 0 \\ 1 & 0 \\ 0 & 1 \end{pmatrix} \begin{pmatrix} -\hat{z}_1 \hat{z}_3 \\ \frac{1}{2}(\bar{\hat{z}}_1 \hat{z}_2 + \hat{z}_1 \bar{\hat{z}}_2) \end{pmatrix} \\ + \begin{pmatrix} 0 & 0 \\ 1 & 0 \\ 0 & 1 \end{pmatrix} (\mathbf{s}(\mathbf{z}) - \mathbf{s}(\hat{\mathbf{z}})). \end{aligned} \tag{3.2.3}$$

The error is defined as $\mathbf{e} = \hat{\mathbf{z}} - \mathbf{z} = \mathbf{e}^r + j\mathbf{e}^i$, and the corresponding error system can be described as

$$\dot{\mathbf{e}} = (\mathbf{A} - \mathbf{B}\mathbf{k})(\mathbf{e}^r + j\mathbf{e}^i) = \left(\begin{pmatrix} -a & a & 0 \\ c & -1 & 0 \\ 0 & 0 & -b \end{pmatrix} - \begin{pmatrix} 0 & 0 \\ 1 & 0 \\ 0 & 1 \end{pmatrix} \mathbf{k} \right) (\mathbf{e}^r + j\mathbf{e}^i), \quad (3.2.4)$$

namely,

$$\dot{\mathbf{e}^r} = \left(\begin{pmatrix} -a & a & 0 \\ c & -1 & 0 \\ 0 & 0 & -b \end{pmatrix} - \begin{pmatrix} 0 & 0 \\ 1 & 0 \\ 0 & 1 \end{pmatrix} \mathbf{k} \right) \mathbf{e}^r, \quad (3.2.5)$$

$$\dot{\mathbf{e}^i} = \left(\begin{pmatrix} -a & a & 0 \\ c & -1 & 0 \\ 0 & 0 & -b \end{pmatrix} - \begin{pmatrix} 0 & 0 \\ 1 & 0 \\ 0 & 1 \end{pmatrix} \mathbf{k} \right) \mathbf{e}^i. \quad (3.2.6)$$

Since the matrix $[\mathbf{B}, \mathbf{AB}, \mathbf{A}^2\mathbf{B}]$ is full rank, system (3.2.3) serves as a global observer for system (3.2.1), ensuring that the complex state vector $\hat{\mathbf{z}}$ tends to $\mathbf{z}$ as $t \to \infty$ regardless of the initial values of systems (3.2.2) and (3.2.3). Moreover, the feedback gain matrix $\mathbf{k}$ can be derived via the pole placement technique.

The initial conditions of systems (3.2.6) and (3.2.7) are set as $(z_1(0), z_2(0), z_3(0)) = (1 + 2j, 0.5 + 5j, 3)$ and $(\hat{z}_1(0), \hat{z}_2(0), \hat{z}_3(0)) = (0.8 + 3j, 2 + 5j, 4)$. So the initial values of error system (3.2.4) are $(0.2 - 1j, -1.5, -1)$. The parameters are selected as $a = 10$, $b = 8/3$, $c = 28$. The eigenvalues of $\mathbf{A} - \mathbf{Bk}$ in system (3.2.4) or systems (3.2.5) and (3.2.7) are assigned as $(-1, -1, -1)$, then the gain matrix

$$\mathbf{k} = \begin{pmatrix} 46 & -5 & 0 \\ 0 & 0 & 2 \end{pmatrix}$$

is obtained. Hence, z and $\hat{z}$ achieve complete synchronization between systems (3.2.6) and (3.2.7) as shown in Fig. 3.1 and the error $e(t)$ has been asymptotically stabilized to zero as demonstrated by Fig. 3.2.

3.2.2 *Synchronization of the Complex Two Coupled Dynamos System*

The complex two coupled dynamos system [97] is described as follows

$$\begin{cases} \dot{z_1} = -\alpha z_1 + z_2(z_3 + \beta), \\ \dot{z_2} = -\alpha z_2 + z_1(z_3 - \beta), \\ \dot{z_3} = 1 - \frac{1}{2}(\bar{z}_1 z_2 + z_1 \bar{z}_2), \end{cases} \quad (3.2.7)$$

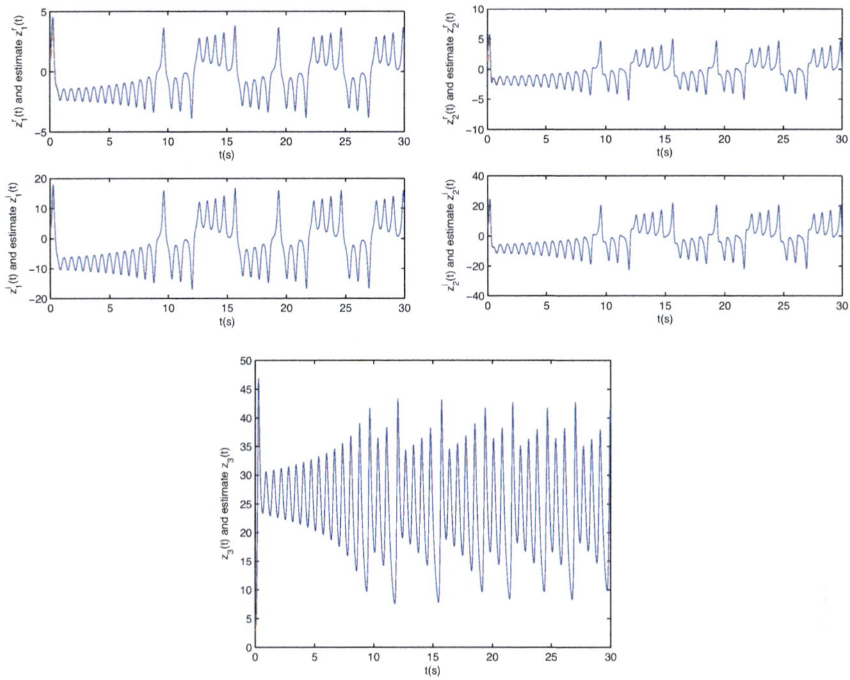

Fig. 3.1 Synchronization process between chaotic complex systems (3.2.2) and (3.2.3) with $a = 10,\ b = 8/3$ and $c = 28$

where $\alpha,\ \beta$ are real constants. System (3.2.7) can be rewritten as

$$\begin{pmatrix} \dot{z}_1 \\ \dot{z}_2 \\ \dot{z}_3 \end{pmatrix} = \begin{pmatrix} -\alpha & \beta & 0 \\ -\beta & -\alpha & 0 \\ 0 & 0 & 0 \end{pmatrix} \begin{pmatrix} z_1 \\ z_2 \\ z_3 \end{pmatrix} + \begin{pmatrix} 1 & 0 & 0 \\ 0 & 1 & 0 \\ 0 & 0 & 1 \end{pmatrix} \begin{pmatrix} z_2 z_3 \\ z_1 z_3 \\ -\frac{1}{2}(\bar{z}_1 z_2 + z_1 \bar{z}_2) \end{pmatrix} + \begin{pmatrix} 0 \\ 0 \\ 1 \end{pmatrix}. \tag{3.2.8}$$

Comparing system (3.2.8) with system (3.1.1), we have

$$\mathbf{A} = \begin{pmatrix} -\alpha & \beta & 0 \\ -\beta & -\alpha & 0 \\ 0 & 0 & 0 \end{pmatrix}, \quad \mathbf{B} = \begin{pmatrix} 1 & 0 & 0 \\ 0 & 1 & 0 \\ 0 & 0 & 1 \end{pmatrix}, \mathbf{C} = \begin{pmatrix} 0 \\ 0 \\ 1 \end{pmatrix}, \quad \mathbf{f}(\mathbf{z}) = \begin{pmatrix} z_2 z_3 \\ z_1 z_3 \\ -\frac{1}{2}(\bar{z}_1 z_2 + z_1 \bar{z}_2) \end{pmatrix}.$$

Given the output as $\mathbf{s}(\mathbf{z}) = \mathbf{f}(\mathbf{z}) + \mathbf{k}\mathbf{z}$ with a gain matrix $\mathbf{k} \in \mathbb{R}^{3\times 3}$, the observer (3.1.7) becomes

$$\begin{aligned} \begin{pmatrix} \dot{\hat{z}}_1 \\ \dot{\hat{z}}_2 \\ \dot{\hat{z}}_3 \end{pmatrix} &= \begin{pmatrix} -\alpha & \beta & 0 \\ -\beta & -\alpha & 0 \\ 0 & 0 & 0 \end{pmatrix} \begin{pmatrix} \hat{z}_1 \\ \hat{z}_2 \\ \hat{z}_3 \end{pmatrix} + \begin{pmatrix} 1 & 0 & 0 \\ 0 & 1 & 0 \\ 0 & 0 & 1 \end{pmatrix} \begin{pmatrix} \hat{z}_2 \hat{z}_3 \\ \hat{z}_1 \hat{z}_3 \\ -\frac{1}{2}(\bar{\hat{z}}_1 \hat{z}_2 + \hat{z}_1 \bar{\hat{z}}_2) \end{pmatrix} + \begin{pmatrix} 0 \\ 0 \\ 1 \end{pmatrix} \\ &\quad + \begin{pmatrix} 1 & 0 & 0 \\ 0 & 1 & 0 \\ 0 & 0 & 1 \end{pmatrix} (\mathbf{s}(\mathbf{z}) - \mathbf{s}(\hat{\mathbf{z}})). \end{aligned} \tag{3.2.9}$$

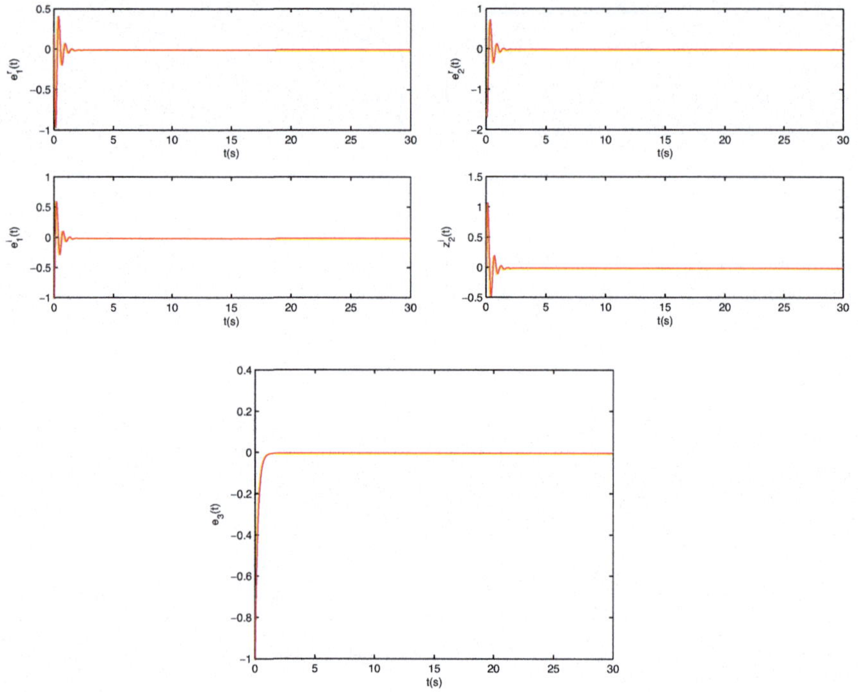

Fig. 3.2 Synchronization errors of chaotic complex systems (3.2.2) and (3.2.3) with $a = 10,\ b = 8/3$ and $c = 28$

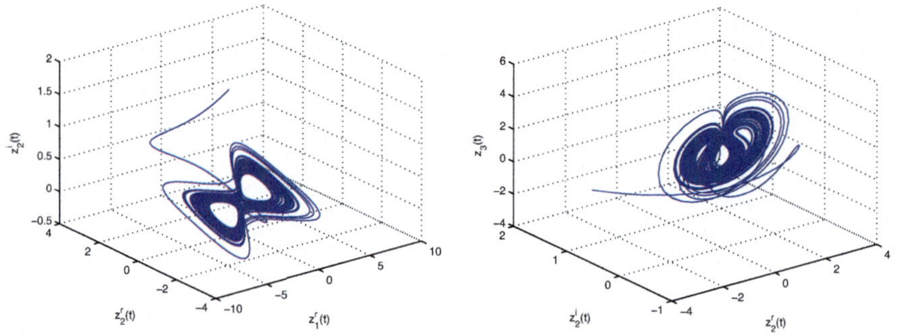

Fig. 3.3 Chaotic attractor of systems (3.2.8) for $\alpha = 0.8$ and $\beta = 1.8$

Similarly, we get the error system as follows

$$\dot{\mathbf{e}} = (\mathbf{A} - \mathbf{B}\mathbf{k})(\mathbf{e}^r + j\mathbf{e}^i) = \left(\begin{pmatrix} -\alpha & \beta & 0 \\ -\beta & -\alpha & 0 \\ 0 & 0 & 0 \end{pmatrix} - \begin{pmatrix} 1 & 0 & 0 \\ 0 & 1 & 0 \\ 0 & 0 & 1 \end{pmatrix} \mathbf{k} \right) (\mathbf{e}^r + j\mathbf{e}^i). \tag{3.2.10}$$

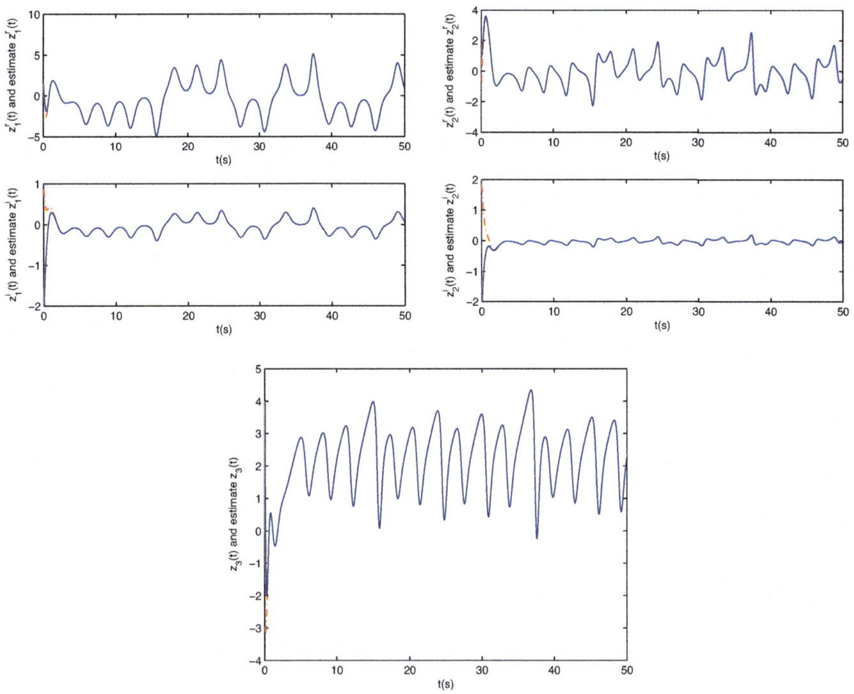

Fig. 3.4 Synchronization process between chaotic complex systems (3.2.8) and (3.2.9) with $\alpha = 0.8$ and $\beta = 1.8$

The parameters are configured as $\alpha = 0.8$ and $\beta = 1.8$, and the chaotic attractor of system (3.2.8) is verified by Fig. 3.3. Owing to the full rank (rank=3) of the matrix $[\mathbf{B},\ \mathbf{AB},\ \mathbf{A}^2\mathbf{B}]$, system (3.2.9) constitutes the global observer corresponding to system (3.2.8). For the error system (3.2.10), the eigenvalues of $(\mathbf{A} - \mathbf{Bk})$ are assigned as $(-2,\ -2,\ -2)$, and such an assignment is attainable via the pole placement technique. The eigenvalues of $(\mathbf{A} - \mathbf{Bk})$ for error system (3.2.10) are assigned as $(-2,\ -2,\ -2)$. By the pole placement technique, we get the gain matrix

$$\mathbf{k} = \begin{pmatrix} 5 & 0 & 0 \\ 0 & 2 & 0 \\ 0 & 0 & 10 \end{pmatrix},$$

which leads to synchronization between systems (3.2.8) and (3.2.9) as shown in Fig. 3.4. Figure 3.5 shows that error $e(t)$ tends to zero $t \to \infty$, which means system (3.2.9) is synchronized with system (3.2.8).

Moreover, comparing Figs. 3.2 and 3.5 respectively, we can see that the bigger the absolute values of the negative eigenvalues in $\mathbf{A} - \mathbf{Bk}$, the faster the convergence rate of errors is. Hence, the convergence rate can be flexibly adjusted by selecting the eigenvalues.

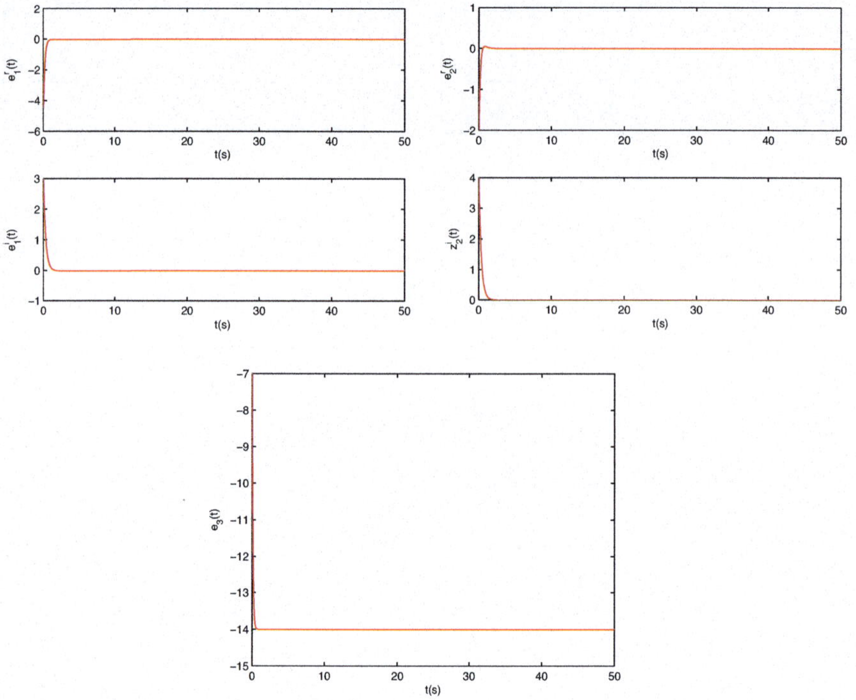

Fig. 3.5 Synchronization errors of chaotic complex systems (3.2.8) and (3.2.9) with $\alpha = 0.8$ and $\beta = 1.8$

Chapter 4
Robust Adaptive Modified Projective Synchronization of Chaotic Complex Systems with Disturbances

The modified projective synchronization (MPS), is also called the full state hybrid projective synchronization (FSHPS), which was proposed in [98]. The Lyapunov's direct method has been used to realize MPS of two identical chaotic [98] and hyperchaotic [99] real systems. Notably, in many practical situations, the system parameters or even system models may be perturbed and cannot be known exactly in priority. Therefore, the adaptive control scheme [100–107], which has been widely used for chaos synchronization, is an effective method to synchronize two uncertain chaotic systems in sense of MPS. For example, the adaptive control method has been used to achieve MPS of chaotic (hyperchaotic) real systems with unknown parameters [102] and synchronized two different uncertain chaotic (hyperchaotic) real systems with the same and different order in sense of MPS [103], separately. This section is divided into two subsections, covering the definition of MPS for chaotic complex systems and the realization of robust adaptive MPS.

4.1 Definition of MFPS of Chaotic Complex Systems

First, with reference to the definition of MFPS for real-valued chaotic systems [102], we extend and present the definition of MFPS between two chaotic complex systems, namely, for two complex dynamical systems:

$$\begin{cases} \dot{\mathbf{z}} = \mathbf{H}(\mathbf{z}) \leftarrow \textit{drive system}, \\ \dot{\mathbf{w}} = \mathbf{L}(\mathbf{w}) + \mathbf{u}(\mathbf{z},\ \mathbf{w}) \leftarrow \textit{response system}, \end{cases}$$

where $\mathbf{z} = [z_1,\ z_2,\ \ldots,\ z_n]^T$ and $\mathbf{w} = [w_1,\ w_2,\ \ldots,\ w_n]^T$ are the state complex vectors.

S. Liu et al., *Chaos in Complex Dynamical System: Theory and Application*,
https://doi.org/10.1007/978-981-95-7762-0_4

Definition 4.1 If there exists a real constant matrix $\mathbf{H} = diag\{h_1, h_2, \ldots, h_n\}$, such that $\lim_{t\to\infty} \|\mathbf{w}(t) - \mathbf{Hz}(t)\| = 0$, then $\mathbf{z}$ synchronizes $\mathbf{w}$ up to a scaling matrix $\mathbf{H}$, such synchronization is called as modified projective synchronization (MFPS). Similarly, the matrix $\mathbf{H}$ is called as scaling matrix, $h_1, h_2, \ldots, h_n$ are scaling factors.

Remark 4.1 CS, AS and PS are the special cases of MPS, where $h_1 = h_2 = \cdots = h_n = 1$, $h_1 = h_2 = \cdots = h_n = -1$ and $h_1 = h_2 = \cdots = h_n = \alpha(\alpha$ is a constant), respectively. Specially, when $h_1 = h_2 = \cdots = h_n = 0$, the synchronization problem is turned into a control problem of chaotic complex systems.

A category of drive chaotic complex systems affected by external disturbances is described by the following form

$$\dot{\mathbf{z}} = \mathbf{f}(\mathbf{z}) + \mathbf{F}(\mathbf{z})\mathbf{A} + \mathbf{d}_1, \tag{4.1.1}$$

where $\mathbf{z} = (z_1, z_2, \ldots, z_n)^T \in \mathbb{C}^n$ is the complex state vector , and $\mathbf{z} = \mathbf{z}^r + j\mathbf{z}^i$. Define $z_1 = z_1^r + jz_1^i$, $z_2 = z_2^r + jz_2^i$, $\ldots$, $z_n = z_n^r + jz_n^i$, then $\mathbf{z}^r = (z_1^r, z_2^r, \ldots, z_n^r)$, $\mathbf{z}^i = (z_1^i, z_2^i, \ldots, z_n^i)^T$; $\mathbf{f} : \mathbb{C}^n \to \mathbb{C}^n$ is a complex vector function, $\mathbf{F} : \mathbb{C}^n \to \mathbb{C}^{n\times p}$ is a function matrix, $\mathbf{A} = (a_1, a_2, \ldots, a_n)^T$ is a real (or complex) uncertain parameter vector. $\mathbf{d}_1$ is an $n \times 1$ matrix, which represents the external disturbances, with the elements $d_{1l}(l = 1, 2, ..., n)$ satisfying $|h_l d_{1l}| \leq \rho_{1l}$. Here assume ρ_{1l} are given and note $\rho_1 = [\rho_{11}, \rho_{12}, \ldots, \rho_{1n}]^T$. The response chaotic complex system subject to external disturbances is assumed as

$$\dot{\mathbf{w}} = \mathbf{g}(\mathbf{w}) + \mathbf{G}(\mathbf{w})\mathbf{B} + \mathbf{u} + \mathbf{d}_2, \tag{4.1.2}$$

where $\mathbf{w} = [w_1, w_2, \ldots, w_n]^T$ is a complex vector, and $\mathbf{w} = \mathbf{w}^r + j\mathbf{w}^i$, $\mathbf{w}^r = (w_1^r, w_2^r, \ldots, w_n^r)$, $\mathbf{w}^i = (w_1^i, w_2^i, \ldots, w_n^i)^T$; $\mathbf{g} : \mathbb{C}^n \to \mathbb{C}^n$ is a complex vector function, $\mathbf{G} : \mathbb{C}^n \to \mathbb{C}^{n\times p}$ is a function matrix, $\mathbf{B} = (b_1, b_2, \ldots, b_n)^T$ is a real (or complex) uncertain parameter vector, $\mathbf{u} \in C^n$ is a control input complex vector. $\mathbf{d}_2$ is an $n \times 1$ matrix, which represents the external disturbances, with the elements $d_{2l}(l = 1, 2, ..., n)$ satisfying $|d_{2l}| \leq \rho_{2l}$, and $\rho_{2l} \in \rho_2 = [\rho_{21}, \rho_{22}, \ldots, \rho_{2n}]^T$.

For chaotic adaptive MFPS, the core aim is to construct the controller $\mathbf{u}$ to ensure that the response complex (4.1.2) synchronizes asymptotically with the drive complex system (4.1.1) under the MFPS framework with nonidentical initial vectors, namely, the complex error vector $\mathbf{e} = \mathbf{e}^r + j\mathbf{e}^i = \mathbf{w} - \mathbf{Hz}$ will satisfy

$$\begin{aligned}
\lim_{t\to\infty} \|\mathbf{e}^r(t)\|^2 &= \lim_{t\to\infty} \|\mathbf{w}^r(t, \mathbf{w_0}^r) - \mathbf{Hz}^r(t, \mathbf{z_0}^r)\|^2 \\
&= \lim_{t\to\infty} (\sum_{l=1}^{n} \|w_l^r(t, w_l^r(0)) - h_l z_l^r(t, z_l^r(0))\|)^2 = 0, \\
\lim_{t\to\infty} \|\mathbf{e}^i(t)\|^2 &= \lim_{t\to\infty} \|\mathbf{w}^i(t, \mathbf{w_0}^i) - \mathbf{Hz}^i(t, \mathbf{z_0}^i)\|^2 \\
&= \lim_{t\to\infty} (\sum_{l=1}^{n} \|w_l^i(t, w_l^i(0)) - h_l z_l^i(t, z_l^i(0))\|)^2 = 0,
\end{aligned}$$

then

$$\lim_{t\to\infty}\|\mathbf{e}(t)\| = \lim_{t\to\infty}(\|\mathbf{e}^r(t)\|^2 + \|\mathbf{e}^i(t)\|^2)^{\frac{1}{2}} = 0,$$

where $\mathbf{z_0}^r = [z_1^r(0),\ z_2^r(0),\ \ldots,\ z_n^r(0)]^T$ and $\mathbf{z_0}^i = [z_1^i(0),\ z_2^i(0),\ \ldots,\ z_n^i(0)]^T$ are initial values of real and imaginary parts of the drive chaotic complex system (4.1.1), while $\mathbf{w_0}^r = [w_1^r(0),\ w_2^r(0),\ \ldots,\ w_n^r(0)]^T$ and $\mathbf{w_0}^i = [w_1^i(0),\ w_2^i(0),\ \ldots,\ w_n^i(0)]^T$ are ones of the response chaotic complex system (4.1.2), so the initial values of real and imaginary parts of error systems are $\mathbf{e_0}^r = [w_1^r(0) - h_1z_1^r(0),\ w_2^r(0) - h_2z_2^r(0),\ \ldots,\ w_n^r(0) - h_nz_n^r(0)]^T$ and $\mathbf{e_0}^i = [w_1^i(0) - h_1z_1^i(0),\ w_2^i(0) - h_2z_2^i(0),\ \ldots,\ w_n^i(0) - h_nz_n^i(0)]^T$.

4.2 Realization of Robust Adaptive MFPS

Considering that chaotic complex systems (4.1.1) and (4.1.2) are linearly parameterized, the function $\mathbf{F}(\cdot)$ allows for complete separation into real and imaginary components, denoted as $\mathbf{F}^r(\cdot)$ and $\mathbf{F}^i(\cdot)$. Here, $\mathbf{F}^r(\cdot)$ is a function of the real state variables $z_1^r,\ z_2^r,\ \ldots,\ z_n^r$, while $\mathbf{f}^i(\cdot)$ is a function of the imaginary state variables $z_1^i,\ z_2^i,\ \ldots,\ z_n^i$. In contrast, the functions $\mathbf{f}^r(\cdot)$ and $\mathbf{f}^i(\cdot)$ are defined over the joint variable set $z_1^r,\ z_1^i,\ \ldots,\ z_n^r,\ z_n^i$, whose structures are governed by the characteristics of the specific complex systems. As a result, the real and imaginary parts of $\mathbf{f}(\cdot)$ cannot be decoupled entirely. This identical trait is exhibited by the functions $\mathbf{G}^r(\cdot),\ \mathbf{G}^i(\cdot)$ and $\mathbf{g}^r(\cdot),\ \mathbf{g}^i(\cdot)$.

Theorem 4.1 *Suppose that the complex controller vector* $\mathbf{u}$ *is designed as*

$$\begin{aligned}\mathbf{u} &= \mathbf{u}^r + j\mathbf{u}^i = \mathbf{H}\mathbf{f}(\mathbf{z}) + \mathbf{H}\mathbf{F}(\mathbf{z})\hat{\mathbf{A}} - \mathbf{g}(\mathbf{w}) - \mathbf{G}(\mathbf{w})\hat{\mathbf{B}} - k\mathbf{e} - \delta\\ &= \mathbf{H}\mathbf{f}^r(z_1^r,\ z_1^i,\ \ldots,\ z_n^r,\ z_n^i) + \mathbf{H}\mathbf{F}^r(z_1^r,\ z_2^r,\ \ldots,\ z_n^r)\hat{\mathbf{A}} - \mathbf{g}^r(w_1^r,\ w_1^i,\ \ldots,\ w_n^r,\ w_n^i)\\ &\quad -\mathbf{G}^r(w_1^r,\ w_2^r,\ \ldots,\ w_n^r)\hat{\mathbf{B}} - k\mathbf{e}^r - \delta^r\\ &\quad +j(\mathbf{H}\mathbf{f}^i(z_1^r,\ z_1^i,\ \ldots,\ z_n^r,\ z_n^i) + \mathbf{H}\mathbf{F}^i(z_1^i,\ z_2^i,\ \ldots,\ z_n^i)\hat{\mathbf{A}} - \mathbf{g}^i(w_1^r,\ w_1^i,\ \ldots,\ w_n^r,\ w_n^i)\\ &\quad -\mathbf{G}^i(w_1^i,\ w_2^i,\ \ldots,\ w_n^i)\hat{\mathbf{B}} - k\mathbf{e}^i - \delta^i),\end{aligned} \tag{4.2.1}$$

the update laws of the estimations of the unknown parameters are chosen as

$$\begin{cases}\dot{\hat{\mathbf{A}}} = [-(\mathbf{F}^r(z_1^r,\ z_2^r,\ \ldots,\ z_n^r))^T\mathbf{H},\ -(\mathbf{F}^i(z_1^i,\ z_2^i,\ \ldots,\ z_n^i))^T\mathbf{H}]\begin{bmatrix}\mathbf{e}^r\\ \mathbf{e}^i\end{bmatrix},\\ \dot{\hat{\mathbf{B}}} = [(\mathbf{G}^r(w_1^r,\ w_2^r,\ \ldots,\ w_n^r))^T,\ (\mathbf{G}^i(w_1^i,\ w_2^i,\ \ldots,\ w_n^i))^T]\begin{bmatrix}\mathbf{e}^r\\ \mathbf{e}^i\end{bmatrix},\end{cases} \tag{4.2.2}$$

and the compensator δ *is taken as*

$$\delta = \frac{1}{2\xi}[(\rho_{11} + \rho_{21})^2e_1,\ (\rho_{12} + \rho_{22})^2e_2,\ \ldots,\ (\rho_{1n} + \rho_{2n})^2e_n]^T, \tag{4.2.3}$$

then the error dynamical system will be stabilized at the zero equilibrium asymptotically with a very small predictable bound, i.e., the response chaotic complex system (4.1.2) *can synchronize the drive chaotic complex system* (4.1.1) *in the sense of MFPS asymptotically with an arbitrarily small error bound, where* $k > 0$ *is a constant, the parameter vectors* $\hat{\mathbf{A}}$ *and* $\hat{\mathbf{B}}$ *are, respectively, estimates of the unknown parameters vectors* $\mathbf{A}$ *and* $\mathbf{B}$*, and the compensator* δ *is introduced to eliminate the influence of external disturbances.*

Proof The error dynamical system between the drive chaotic complex system (4.1.1) and the response chaotic complex system (4.1.2) is

$$\begin{aligned}
\dot{\mathbf{e}} &= \dot{\mathbf{e}}^r + j\dot{\mathbf{e}}^i = \dot{\mathbf{w}} - \mathbf{H}\dot{\mathbf{z}} \\
&= \mathbf{g}(\mathbf{w}) + \mathbf{G}(\mathbf{w})\mathbf{B} + \mathbf{u} + \mathbf{d}_2 - \mathbf{H}(\mathbf{f}(\mathbf{z}) + \mathbf{F}(\mathbf{z})\mathbf{A} + \mathbf{d}_1 \\
&= \mathbf{g}^r(w_1^r, w_1^i, \ldots, w_n^r, w_n^i) + \mathbf{G}^r(w_1^r, w_2^r, \ldots, w_n^r)\mathbf{B} + \mathbf{d}_2^r - \mathbf{H}(\mathbf{f}^r(z_1^r, z_1^i, \ldots, z_n^r, z_n^i) \\
&\quad + \mathbf{F}^r(z_1^r, z_2^r, \ldots, z_n^r)\mathbf{A} + \mathbf{d}_1^r + \mathbf{u}^r + j[\mathbf{g}^i(w_1^r, w_1^i, \ldots, w_n^r, w_n^i) + \mathbf{G}^i(w_1^i, w_2^i, \ldots, w_n^i)\mathbf{B} \\
&\quad + \mathbf{d}_2^i - \mathbf{H}(\mathbf{f}^i(z_1^r, z_1^i, \ldots, z_n^r, z_n^i) + \mathbf{F}^i(z_1^i, z_2^i, \ldots, z_n^i)\mathbf{A} + \mathbf{d}_1^i + \mathbf{u}^i].
\end{aligned} \tag{4.2.4}$$

Thus, by substituting Eq. (4.2.1) into Eq. (4.2.4) and decomposing the resulting equations into real and imaginary parts, we obtain the error system as follows

$$\begin{cases}
\dot{\mathbf{e}}^r = -\mathbf{H}\mathbf{F}^r(z_1^r, z_2^r, \ldots, z_n^r)(\mathbf{A} - \hat{\mathbf{A}}) + \mathbf{G}^r(w_1^r, w_2^r, \ldots, w_n^r)(\mathbf{B} - \hat{\mathbf{B}}) \\
\qquad - k\mathbf{e}^r - \delta^r + \mathbf{d}_2^r - \mathbf{H}\mathbf{d}_1^r, \\
\dot{\mathbf{e}}^i = -\mathbf{H}\mathbf{F}^i(z_1^i, z_2^i, \ldots, z_n^i)(\mathbf{A} - \hat{\mathbf{A}}) + \mathbf{G}^i(w_1^i, w_2^i, \ldots, w_n^i)(\mathbf{B} - \hat{\mathbf{B}}) \\
\qquad - \mathbf{e}^i - \delta^i + \mathbf{d}_2^i - \mathbf{H}\mathbf{d}_1^i.
\end{cases} \tag{4.2.5}$$

Define the parameter errors as $\tilde{\mathbf{A}} = \mathbf{A} - \hat{\mathbf{A}}$, $\tilde{\mathbf{B}} = \mathbf{B} - \hat{\mathbf{B}}$ and consider a Lyapunov function candidate as

$$V = \frac{1}{2}[(\mathbf{e}^r)^T\mathbf{e}^r + (\mathbf{e}^i)^T\mathbf{e}^i + \tilde{\mathbf{A}}^T\tilde{\mathbf{A}} + \tilde{\mathbf{B}}^T\tilde{\mathbf{B}}]. \tag{4.2.6}$$

Then the time derivative of V along the trajectory of the error system (4.2.5) is

$$\dot{V} = (\dot{\mathbf{e}}^r)^T\mathbf{e}^r + (\dot{\mathbf{e}}^i)^T\mathbf{e}^i - \tilde{\mathbf{A}}^T\dot{\hat{\mathbf{A}}} - \tilde{\mathbf{B}}^T\dot{\hat{\mathbf{B}}}. \tag{4.2.7}$$

By inserting Eqs.(4.2.2), (4.2.3) and (4.2.5) into Eq.(4.2.7) yields

$$\begin{aligned}
\dot{V} &= (-\mathbf{H}\mathbf{F}^r(z_1^r, z_2^r, \ldots, z_n^r)\tilde{\mathbf{A}} + \mathbf{G}^r(w_1^r, w_1^i, \ldots, w_n^r)\tilde{\mathbf{B}} - k\mathbf{e}^r - \delta^r + \mathbf{d}_2^r - \mathbf{H}\mathbf{d}_1^r)^T\mathbf{e}^r \\
&\quad + (-\mathbf{H}\mathbf{F}^i(z_1^i, z_2^i, \ldots, z_n^i)\tilde{\mathbf{A}} + \mathbf{G}^i(w_1^i, w_2^i, \ldots, w_n^i)\tilde{\mathbf{B}} - k\mathbf{e}^i - \delta^i + \mathbf{d}_2^i - \mathbf{H}\mathbf{d}_1^i)^T\mathbf{e}^i \\
&\quad - \tilde{\mathbf{A}}^T\dot{\hat{\mathbf{A}}} - \tilde{\mathbf{B}}^T\dot{\hat{\mathbf{B}}} \\
&= -k({\mathbf{e}^r}^T\mathbf{e}^r + {\mathbf{e}^i}^T\mathbf{e}^i) + (\mathbf{d}_2^r - \mathbf{H}\mathbf{d}_1^r - \delta^r)^T\mathbf{e}^r + (\mathbf{d}_2^i - \mathbf{H}\mathbf{d}_1^i - \delta^i)^T\mathbf{e}^i \\
&= -k({\mathbf{e}^r}^T\mathbf{e}^r + {\mathbf{e}^i}^T\mathbf{e}^i) + (\mathbf{d}_2^r - \mathbf{H}\mathbf{d}_1^r)^T\mathbf{e}^r \\
&\quad - \frac{1}{2\xi}[(\rho_{11} + \rho_{21})^2 e_1^r, (\rho_{12} + \rho_{22})^2 e_2^r, \ldots, (\rho_{1n} + \rho_{2n})^2 e_n^r]^T\mathbf{e}^r + (\mathbf{d}_2^i - \mathbf{H}\mathbf{d}_1^i)^T\mathbf{e}^i \\
&\quad - \frac{1}{2\xi}[(\rho_{11} + \rho_{21})^2 e_1^i, (\rho_{12} + \rho_{22})^2 e_2^i, \ldots, (\rho_{1n} + \rho_{2n})^2 e_n^i]^T\mathbf{e}^i - \frac{n\xi}{2} - \frac{n\xi}{2} + n\xi.
\end{aligned} \tag{4.2.8}$$

From $|h_l d_{1l}| = |h_l d^r_{1l} + j h_l d^i_{1l}| \le \rho_{1l}$, we get $|h_l d^r_{1l}| < |\sqrt{(h_l d^r_{1l})^2 + (h_l d^i_{1l})^2}| \le \rho_{1l}$ and $|h_l d^r_{1l}| < \rho_{1l}$. Similarly, $|d^r_{2l}| < \rho_{2l}$ and $|d^i_{2l}| < \rho_{2l}$ are obtained. Using Young's inequality, we obtain

$$\frac{(\rho_{1l} + \rho_{2l})^2 (e^r_l)^2}{2\xi} + \frac{\xi}{2} \ge (\rho_{1l} + \rho_{2l})|e^r_l| \ge (\rho_{1l} + \rho_{2l}) e^r_l \tag{4.2.9}$$

and

$$\frac{(\rho_{1l} + \rho_{2l})^2 (e^i_l)^2}{2\xi} + \frac{\xi}{2} \ge (\rho_{1l} + \rho_{2l}) e^i_l. \tag{4.2.10}$$

Hence,

$$\begin{aligned}\dot{V} &< -k(\mathbf{e}^{rT}\mathbf{e}^r + \mathbf{e}^{iT}\mathbf{e}^i) + (\rho_1 + \rho_2)^T(\mathbf{e}^r + \mathbf{e}^i) - \sum_{l=1}^{n} (\rho_{1l} + \rho_{2l})(e^r_l + e^i_l) + n\xi \\ &< -k(\mathbf{e}^{rT}\mathbf{e}^r + \mathbf{e}^{iT}\mathbf{e}^i) + n\xi. \end{aligned} \tag{4.2.11}$$

The $(\mathbf{e}^r)^T\mathbf{e}^r + (\mathbf{e}^i)^T\mathbf{e}^i = \phi$ is taken, and k can also be taken as any positive constant, so we choose $k \ge \frac{n\xi}{\phi}$, then $\dot{V} < 0$. Also $(\mathbf{e}^r)^T\mathbf{e}^r + (\mathbf{e}^i)^T\mathbf{e}^i = \phi$ is an invariant set, and according to the LaSalle invariant principle, then the error resides in a ball with the radius $n\xi/k$. We can select appropriate ξ and k to make the error bound as small as possible. Hence, the error dynamical system (4.2.5) will be asymptotically stabilized to nearly the zero equilibrium with a very small error bound. Namely, $\lim_{t\to\infty}(\|e^r\|^2 + \|e^i\|^2)^{\frac{1}{2}} = 0$ and the response chaotic complex system (4.1.2) will asymptotically synchronize the drive chaotic complex system (4.1.1) in the sense of MFPS. This completes the proof. □

Corollary 4.1 *If the structures of general chaotic complex systems* (4.1.1) *and* (4.1.2) *without external disturbances are identical, i.e.,* $\mathbf{F}(\mathbf{z}) = \mathbf{G}(\mathbf{w})$ *and* $\mathbf{f}(\mathbf{z}) = \mathbf{g}(\mathbf{w})$, *then the controller vector is designed as*

$$\begin{aligned}\mathbf{v} &= [\mathbf{HF}(\mathbf{z}) - \mathbf{F}(\mathbf{w})]\hat{\mathbf{A}} + \mathbf{Hf}(\mathbf{z}) - \mathbf{f}(\mathbf{w}) - k\mathbf{e} - \delta \\ &= [\mathbf{HF}^r(z^r_1, z^r_2, \ldots, z^r_n) - \mathbf{F}^r(w^r_1, w^r_2, \ldots, w^r_n)]\hat{\mathbf{A}} + \mathbf{Hf}^r(z^r_1, z^i_1, \ldots, z^r_n, z^i_n) \\ &\quad -\mathbf{f}^r(w^r_1, w^i_1, \ldots, w^r_n, w^i_n) - k\mathbf{e}^r - \delta^r \\ &\quad +j[(\mathbf{HF}^i(z^i_1, z^i_2, \ldots, z^i_n) - \mathbf{F}^i(w^i_1, w^i_2, \ldots, w^i_n)\hat{\mathbf{A}} + \mathbf{Hf}^i(z^r_1, z^i_1, \ldots, z^r_n, z^i_n) \\ &\quad -\mathbf{f}^i(w^r_1, w^i_1, \ldots, w^r_n, w^i_n) - k\mathbf{e}^i - \delta^i], \end{aligned} \tag{4.2.12}$$

where $k > 0$ *is a constant, the parameters adaptive laws are chosen as*

$$\begin{aligned}\dot{\hat{\mathbf{A}}} &= [\mathbf{F}^r(w^r_1, w^r_2, \ldots, w^r_n) - \mathbf{HF}^r(z^r_1, z^r_2, \ldots, z^r_n)]^T\mathbf{e}^r \\ &\quad +[\mathbf{F}^i(w^i_1, w^i_2, \ldots, w^i_n) - \mathbf{HF}^i(w^i_1, w^i_2, \ldots, w^i_n)]^T\mathbf{e}^i, \end{aligned} \tag{4.2.13}$$

and the compensator δ *is taken as*

$$\delta = \frac{1}{2\xi}[(\rho_{11} + \rho_{21})^2 e_1, (\rho_{12} + \rho_{22})^2 e_2, \ldots, (\rho_{1n} + \rho_{2n})^2 e_n]^T, \quad (4.2.14)$$

thus, the MFPS of two identical general uncertain chaotic complex systems is also achieved with an arbitrarily small error bound.

Remark 4.2 If some parameters from parameter matrix $\mathbf{A}$ or $\mathbf{B}$ are unknown, Theorem 4.1 and Corollary 4.1 can also be applied to realize MFPS of chaotic complex systems with partly unknown parameters and external disturbances. Specially, when parameter matrices $\mathbf{A}$ and $\mathbf{B}$ are known, then Theorem 4.1 and Corollary 4.1 can also be applied to achieve MFPS of chaotic complex systems with external disturbances.

Remark 4.3 When $\mathbf{z}^i = \mathbf{w}^i = \mathbf{e}^i = 0$, our scheme is suitable to achieve MFPS of different uncertain chaotic systems with real variables.

Remark 4.4 If $\mathbf{A}$ is a complex parameter vector, the $\mathbf{A}$ can be rewritten as $\mathbf{A} = \mathbf{A}^r + j\mathbf{A}^i$. Then the chaotic complex system (4.1.1) becomes

$$\dot{\mathbf{z}} = \mathbf{F}(\mathbf{z})(\mathbf{A}^r + j\mathbf{A}^i) + \mathbf{f}(\mathbf{z}) + \mathbf{d}_1', \quad (4.2.15)$$

which can also be rewritten as

$$\dot{\mathbf{z}} = \mathbf{F}(\mathbf{z})\mathbf{A}^r + \mathbf{f}_1(\mathbf{z}) + \mathbf{d}_1'' \quad (4.2.16)$$

and

$$\dot{\mathbf{z}} = \mathbf{F}(\mathbf{z})\mathbf{A}^i - j\mathbf{f}_2(\mathbf{z}) - j\mathbf{d}_1''', \quad (4.2.17)$$

where $\mathbf{f}_1(\mathbf{z}) + \mathbf{f}_2(\mathbf{z}) = \mathbf{f}(\mathbf{z})$, and $\mathbf{d}_1'' + \mathbf{d}_1''' = \mathbf{d}_1'$. Similarly, when $\mathbf{B}$ is a complex parameter vector, then chaotic complex system (4.1.2) also become

$$\dot{\mathbf{w}} = \mathbf{G}(\mathbf{w})(\mathbf{B}^r + j\mathbf{B}^i) + \mathbf{g}(\mathbf{w}) + \mathbf{d}_2 + \mathbf{u}', \quad (4.2.18)$$

then

$$\dot{\mathbf{w}} = \mathbf{G}(\mathbf{w})\mathbf{B}^r + \mathbf{g}_2(\mathbf{w}) + \mathbf{d}_2'' + \mathbf{u}'' \quad (4.2.19)$$

and

$$\dot{\mathbf{w}} = \mathbf{G}(\mathbf{w})\mathbf{B}^i - j\mathbf{g}_2(\mathbf{w}) - j\mathbf{d}_2'' + \mathbf{u}''', \quad (4.2.20)$$

where $\mathbf{g}_1(\mathbf{w}) + \mathbf{g}_2(\mathbf{w}) = \mathbf{g}(\mathbf{w})$, and $\mathbf{d}_1'' + \mathbf{d}_2''' = \mathbf{d}_2'$. The structures of complex subsystems (4.2.16), (4.2.17), (4.2.19) and (4.2.20) are the same as ones of the complex system (4.1.1) and (4.1.2), respectively. So the design of the controller to achieve MFPS between subsystems (4.2.16) and (4.2.19)(or (4.2.17) and (4.2.20)) is the same as one of the complex systems (4.1.1) and (4.1.2). If the subsystems (4.2.16) and

(4.2.19), and the subsystems (4.2.17) and (4.2.20) realize MFPS, respectively, the chaotic complex system (4.2.15) and the system (4.2.18) also accomplish MFPS. The controllers $\mathbf{u}''$, $\mathbf{u}'''$ and the estimates $\hat{\mathbf{A}}^r$, $\hat{\mathbf{B}}^r$, $\hat{\mathbf{A}}^i$, $\hat{\mathbf{B}}^i$ are designed to realize MFPS of the subsystems (4.2.16) and (4.2.19), and the subsystems (4.2.17) and (4.2.20), respectively. So the system (4.2.15) and system (4.2.18) achieve MFPS, and the controller $\mathbf{u}' = \mathbf{u}'' + j\mathbf{u}'''$ and the estimates $\hat{\mathbf{A}} = \hat{\mathbf{A}}^r + j\hat{\mathbf{A}}^i$, $\hat{\mathbf{B}} = \hat{\mathbf{B}}^r + j\hat{\mathbf{B}}^i$ are obtained.

The approach to designing controllers for complex parameter vectors $\mathbf{A}$ and $\mathbf{B}$ is consistent with that for real parameter vectors. For simplicity, we thus adopt complex systems with real parameter vectors as illustrative examples to verify the proposed method, detailed as follows.

4.3 Simulation Examples

To illustrate the feasibility and effectiveness of the proposed scheme, two representative examples are selected and numerical simulations are performed. All simulations presented hereinafter are carried out in MATLAB, with the fourth-order Runge-Kutta algorithm (step size set to 0.001) adopted as the standard numerical solution method throughout this work.

4.3.1 MFPS of Two Different Chaotic Complex Systems with Fully Uncertain Parameters and External Disturbances

In this subsection, we designate the MFPS between the complex Chen system and the complex Lü system (featuring fully uncertain parameters and external disturbances) as a representative example. The expression of the complex Chen system is as follows:

$$\begin{cases} \dot{z}_1 = a_1(z_2 - z_1), \\ \dot{z}_2 = (a_3 - a_1)z_1 - z_1 z_3 + a_3 z_2, \\ \dot{z}_3 = \frac{1}{2}(\bar{z}_1 z_2 + z_1 \bar{z}_2) - a_2 z_3, \end{cases} \tag{4.3.1}$$

where z_1, z_2 are complex state variables, z_3 is a real state variable, and a_1, a_2, a_3 are system parameters.

The complex Lü system is written in the form:

$$\begin{cases} \dot{w}_1 = b_1(w_2 - w_1), \\ \dot{w}_2 = -w_1 w_3 + b_3 w_2, \\ \dot{w}_3 = \frac{1}{2}(\bar{w}_1 w_2 + w_1 \bar{w}_2) - b_2 w_3, \end{cases} \tag{4.3.2}$$

where $w_1,\ w_2$ are complex state variables and the state variable z_3 is real; $b_1,\ b_2,\ b_3$ are system parameters.

The complex Chen system (4.3.1) with external disturbances is considered as the drive system and it can be rewritten as in the form of system (4.1.1) as follows

$$\begin{pmatrix} \dot{z}_1 \\ \dot{z}_2 \\ \dot{z}_3 \end{pmatrix} = \begin{pmatrix} 0 \\ -z_1 z_3 \\ \frac{1}{2}(\bar{z}_1 z_2 + z_1 \bar{z}_2) \end{pmatrix} + \begin{pmatrix} z_2 - z_1 & 0 & 0 \\ -z_1 & 0 & z_1 + z_2 \\ 0 & -z_3 & 0 \end{pmatrix} \begin{pmatrix} a_1 \\ a_2 \\ a_3 \end{pmatrix} + \begin{pmatrix} d_{11} \\ d_{12} \\ d_{13} \end{pmatrix}, \tag{4.3.3}$$

where $\mathbf{d}_1 = [d_{11},\ d_{12},\ d_{13}]^T$ represents external disturbances. And assume the complex Lü system (4.3.2) with external disturbances as the response system and it can be described in the form of the system (4.1.2)

$$\begin{pmatrix} \dot{w}_1 \\ \dot{w}_2 \\ \dot{w}_3 \end{pmatrix} = \begin{pmatrix} 0 \\ -w_1 w_3 \\ \frac{1}{2}(\bar{w}_1 w_2 + w_1 \bar{w}_2) \end{pmatrix} + \begin{pmatrix} w_2 - w_1 & 0 & 0 \\ 0 & 0 & w_2 \\ 0 & -w_3 & 0 \end{pmatrix} \begin{pmatrix} b_1 \\ b_2 \\ b_3 \end{pmatrix} + \begin{pmatrix} d_{21} \\ d_{22} \\ d_{23} \end{pmatrix} + \begin{pmatrix} u_1 \\ u_2 \\ u_3 \end{pmatrix}, \tag{4.3.4}$$

where $\mathbf{d}_2 = [d_{21},\ d_{22},\ d_{23}]^T$ represents external disturbances and $\mathbf{u} = [u_1,\ u_2,\ u_3]^T$ is the control input vector to be designed.

Then, according to the error vector $\mathbf{e} = \mathbf{w} - \mathbf{Hz}$, we obtain the following error dynamical system:

$$\begin{cases} \dot{e}_1 = b_1(w_2 - w_1) + d_{21} - h_1(a_1(z_2 - z_1) + d_{11}) + u_1, \\ \dot{e}_2 = -w_1 w_3 + b_3 w_2 + d_{22} - h_2((a_3 - a_1)z_1 - z_1 z_3 + a_3 z_2 + d_{12}) + u_2, \\ \dot{e}_3 = \frac{1}{2}(\bar{w}_1 w_2 + w_1 \bar{w}_2) - b_2 w_3 + d_{23} - h_3(\frac{1}{2}(\bar{z}_1 z_2 + z_1 \bar{z}_2) - a_2 z_3 + d_{13}) + u_3. \end{cases} \tag{4.3.5}$$

We give the controllers $u_l (l = 1,\ 2,\ 3)$ as follows:

$$\begin{cases} u_1 = h_1(z_2 - z_1)\hat{a}_1 - (w_2 - w_1)\hat{b}_1 - ke_1 - \frac{1}{2\xi}(\rho_{11} + \rho_{21})^2 e_1, \\ u_2 = -h_2 z_1 z_3 - h_2 z_1 \hat{a}_1 + h_2(z_1 + z_2)\hat{a}_3 + w_1 w_3 - w_2 \hat{b}_3 - ke_2 - \frac{1}{2\xi}(\rho_{12} + \rho_{22})^2 e_2, \\ u_3 = \frac{h_3}{2}(\bar{z}_1 z_2 + z_1 \bar{z}_2) - h_3 z_3 \hat{a}_2 - \frac{1}{2}(\bar{w}_1 w_2 + w_1 \bar{w}_2) + w_3 \hat{b}_2 - ke_3 \\ \quad - \frac{1}{2\xi}(\rho_{13} + \rho_{23})^2 e_3, \end{cases} \tag{4.3.6}$$

where $\hat{a}_1,\ \hat{a}_2,\ \hat{a}_3,\ \hat{b}_1,\ \hat{b}_2$ and $\hat{b}_3$ are the estimations of parameters $a_1,\ a_2,\ a_3,\ b_1,\ b_2$ and b_3, separately. And the update laws of the parameter estimations are given by

$$\begin{cases} \dot{\hat{a}}_1 = -(z_2^r - z_1^r)h_1 e_1^r + z_1^r h_2 e_2^r - (z_2^i - z_1^i)h_1 e_1^i + z_1^i h_2 e_2^i, \\ \dot{\hat{a}}_2 = z_3 h_3 e_3, \\ \dot{\hat{a}}_3 = -(z_1^r + z_2^r)h_2 e_2^r - (z_1^i + z_2^i)h_2 e_2^i, \end{cases} \tag{4.3.7}$$

and

$$\begin{cases} \dot{\hat{b}}_1 = (w_2^r - w_1^r)e_1^r + (w_2^i - w_1^i)e_1^i, \\ \dot{\hat{b}}_2 = w_3 e_3, \\ \dot{\hat{b}}_3 = w_2^r e_2^r + w_2^i e_2^i. \end{cases} \tag{4.3.8}$$

Hence, the two uncertain chaotic complex systems (4.3.3) and (4.3.4) will be synchronized in the sense of MFPS asymptotically with a very small error bound.

In number simulation, we take $h_1 = 0.5,\ h_2 = 1,\ h_3 = -1$ and the external disturbances as $d_{11} = \sin(0.25\pi t) + j\cos(\pi t),\ d_{12} = 0.3\sin(\pi t) + j0.5\cos(\pi t),$ $d_{13} = 0.2\sin(0.5\pi t)$ and $d_{21} = 0.2cos(1.5\pi t) + j0.5\sin(2\pi t),$ $d_{22} = 0.5\sin(0.5\pi t) + j0.3\sin(\pi t), d_{23} = 0.8\sin(1.5\pi t)$, then we get the bound as $\rho_{11} = 0.5, \rho_{12} = 0.8, \rho_{13} = 0.2$ and $\rho_{21} = 0.7, \rho_{22} = 0.8, \rho_{23} = 0.8$. The unknown parameters are taken as $a_1 = 42,\ a_2 = 4,\ a_3 = 26$ and $b_1 = 40,\ b_2 = 5,\ b_3 = 22$, so that the complex Chen and Lü systems have chaotic attractors respectively as shown in Figs. 4.1 and 4.2. The initial states of the drive system (4.3.3) and response system (4.3.4) are selected as $z_1(0) = -1 + 2j,\ z_2(0) = 3 + 5j,\ z_3(0) = -4$ and $w_1(0) = 5 + 4j,\ w_2(0) = 2 + 3j,\ w_3(0) = 0.5$, separately, then the initial conditions of error

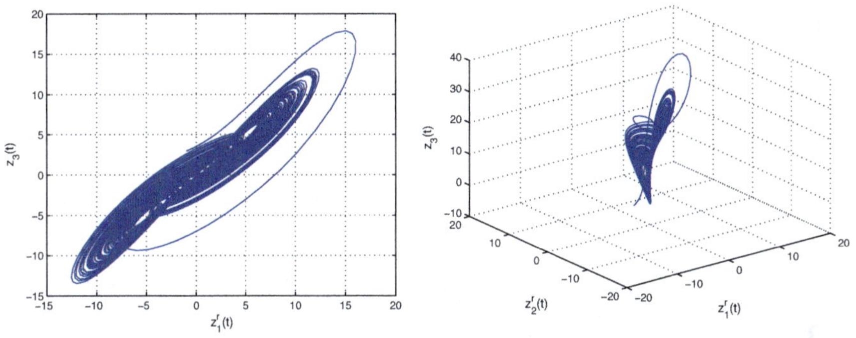

Fig. 4.1 Chaotic behavior and attractors of the complex Chen system (4.3.1)

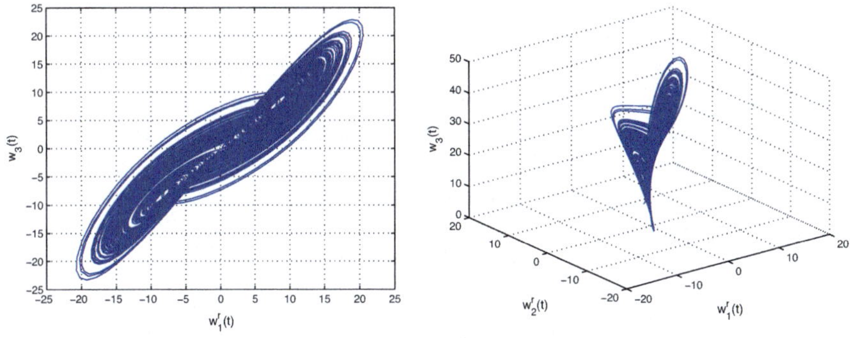

Fig. 4.2 Chaotic behavior and attractors of the complex Lü system (4.3.2) in different phase planes and projections

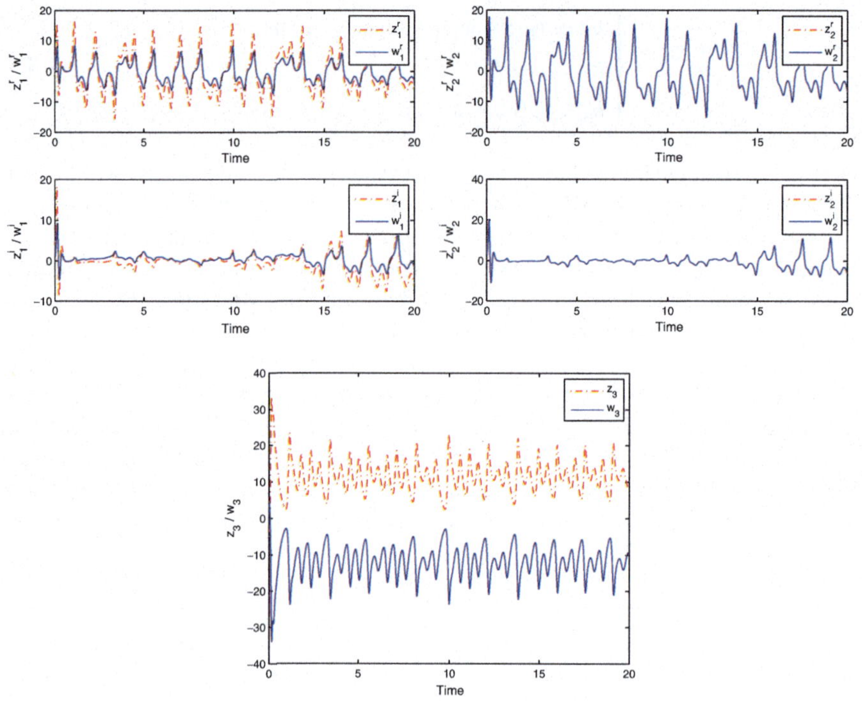

Fig. 4.3 Chaotic MFPS between the drive system (4.3.3) and response system (4.3.4) with the controllers (4.3.6) and the update laws of the estimations (4.3.7) and (4.3.8)

dynamical system (4.3.5) are $e_1(0) = 5.5 + 3j$, $e_2(0) = -1 + 0.5j$, $e_3(0) = 2.5$. And the estimations of the parameters have the initial values $\hat{a}_1(0) = 20$, $\hat{a}_2(0) = 15$, $\hat{a}_3(0) = 20$ and $\hat{b}_1(0) = 20$, $\hat{b}_2(0) = 10$, $\hat{b}_3(0) = 15$. Also the $k = 2$ and $\xi = 0.1$ are chosen.

We assume that the controllers (4.3.6) are switched on at $t = 0s$, and the simulation results are shown in Figs. 4.3, 4.4 and 4.5. Figure 4.3 displays the time synchronization responses of the real and imaginary parts of the complex states for drive system (4.3.3) and response system (4.3.4), from which it is evident that the response system (4.3.4) synchronizes asymptotically with the drive system (4.3.3) in the MFPS framework. The real and imaginary components of the MFPS errors are respectively plotted in Figs. 4.4(a)–(c), indicating that each error signal quickly converges to approximately zero via the controllers (4.3.6). Moreover, Fig. 4.5 presents the estimates of the unknown parameters for the two systems, which all tend to constant values as time approaches infinity.

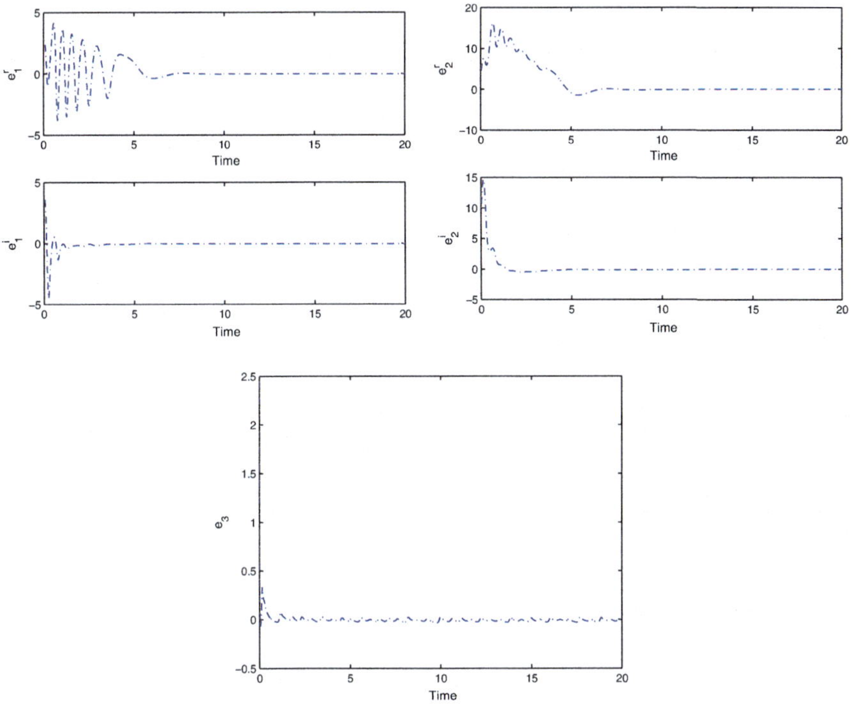

Fig. 4.4 Error dynamics of between the drive chaotic complex system (4.3.3) and response chaotic complex system (4.3.4)

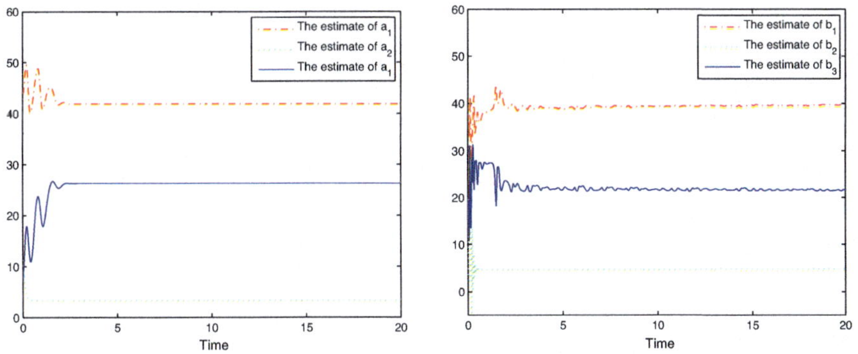

(a) a_1, a_2 and a_3.

(b) b_1, b_2 and b_3.

Fig. 4.5 The estimates of uncertain parameters

4.3.2 MFPS of Two Identical Hyperchaotic Complex Lorenz Systems with Fully Uncertain Parameters and External Disturbances

In order to validate the robust adaptive MFPS scheme for two identical uncertain autonomous hyperchaotic complex Lorenz systems [108] affected by external disturbances, we consider a set of two such systems. The drive system is denoted with the complex variable $\mathbf{z}$, and the response system-possessing an identical structural configuration-is represented by the complex variable $\mathbf{w}$. Hence, the drive system is expressed as

$$\begin{cases} \dot{z}_1 = a_1(z_2 - z_1) + jz_4 + d_{11}, \\ \dot{z}_2 = a_2 z_1 - z_2 - z_1 z_3 + jz_4 + d_{12}, \\ \dot{z}_3 = -a_3 z_3 + (1/2)(\bar{z}_1 z_2 + z_1 \bar{z}_2) + d_{13}, \\ \dot{z}_4 = -a_4 z_4 + (1/2)(\bar{z}_1 z_2 + z_1 \bar{z}_2) + d_{14}, \end{cases} \tag{4.3.9}$$

where $z_1,\ z_2$ are complex state variables and $z_3,\ z_4$ are real state variables; $a_1,\ a_2,\ a_3,\ a_4$ are system parameters; $\mathbf{d}_1 = [d_{11},\ d_{12},\ d_{13}]^T$ represents external disturbances. The response system is

$$\begin{cases} \dot{w}_1 = a_1(w_2 - w_1) + jw_4 + d_{21} + u_1, \\ \dot{w}_2 = a_2 w_1 - w_2 - w_1 w_3 + jw_4 + d_{22} + u_2, \\ \dot{w}_3 = -a_3 w_3 + (1/2)(\bar{z}_1 w_2 + w_1 \bar{z}_2) + d_{23} + u_3, \\ \dot{w}_4 = -a_4 w_4 + (1/2)(\bar{z}_1 w_2 + w_1 \bar{z}_2) + d_{24} + u_4, \end{cases} \tag{4.3.10}$$

where $\mathbf{d}_2 = [d_{21},\ d_{22},\ d_{23}]^T$ represents external disturbances and $\mathbf{v} = [u_1,\ u_2,\ u_3]^T$ is the control input vector needed to be designed.

Based on equations (4.2.12) and (4.2.13) in Corollary 4.1, the controller vector is designed as

$$\begin{aligned} \mathbf{v} &= [\mathbf{H}F(\mathbf{z}) - \mathbf{F}(\mathbf{w})]\hat{\mathbf{A}} + \mathbf{Hf}(\mathbf{z}) - \mathbf{f}(\mathbf{w}) - k\mathbf{e} - \delta \\ &= \begin{pmatrix} [h_1(z_2 - z_1) - (w_2 - w_1)]\hat{a}_1 + j(h_1 z_4 - w_4) - ke_1 - \frac{1}{2\xi}(\rho_{11} + \rho_{21})^2 e_1 \\ (h_2 z_1 - w_1)\hat{a}_2 - (z_2 + z_1 z_3 - jz_4)h_2 + w_2 + w_1 w_3 - jw_4 - ke_2 - \frac{1}{2\xi}(\rho_{12} + \rho_{22})^2 e_2 \\ (w_3 - h_3 z_3)\hat{a}_3 + \frac{1}{2}h_3(\bar{z}_1 z_2 + z_1 \bar{z}_2) + \frac{1}{2}(\bar{w}_1 w_2 + w_1 \bar{w}_2) - ke_3 - \frac{1}{2\xi}(\rho_{13} + \rho_{23})^2 e_3 \\ (w_4 - h_4 z_4)\hat{a}_4 + \frac{1}{2}h_4(\bar{z}_1 z_2 + z_1 \bar{z}_2) + \frac{1}{2}(\bar{w}_1 w_2 + w_1 \bar{w}_2) - ke_4 - \frac{1}{2\xi}(\rho_{14} + \rho_{24})^2 e_3 \end{pmatrix} \end{aligned} \tag{4.3.11}$$

and the adaptive laws of parameters are given as

$$\begin{aligned} \dot{\hat{\mathbf{A}}} = {} & [\mathbf{F}^r(w_1^r,\ w_2^r,\ \ldots,\ w_n^r) - \mathbf{HF}^r(z_1^r,\ z_2^r,\ \ldots,\ z_n^r)]^T \mathbf{e}^r \\ & + [\mathbf{F}^i(w_1^i,\ w_2^i,\ \ldots,\ w_n^i) - \mathbf{HF}^i(z_1^i,\ z_2^i,\ \ldots,\ z_n^i)]^T \mathbf{e}^i \end{aligned}$$

$$
= \begin{pmatrix} [(w_2^r - w_1^r) - h_1(z_2^r - z_1^r)]e_1^r + [(w_2^i - w_1^i) - h_1(z_2^i - z_1^i)]e_2^i \\ (w_1^r - h_2 z_1^r)e_2^r + (w_1^i - h_2 z_1^i)e_2^i \\ (h_3 z_3 - w_3)e_3 \\ (h_4 z_4 - w_4)e_4 \end{pmatrix}. \tag{4.3.12}
$$

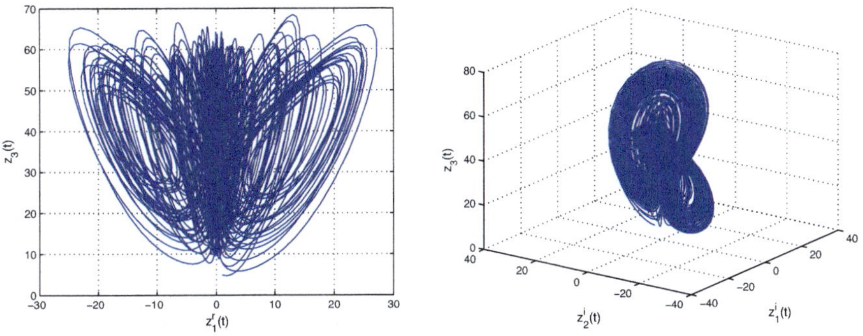

Fig. 4.6 Attractors of hyperchaotic complex Lorenz system

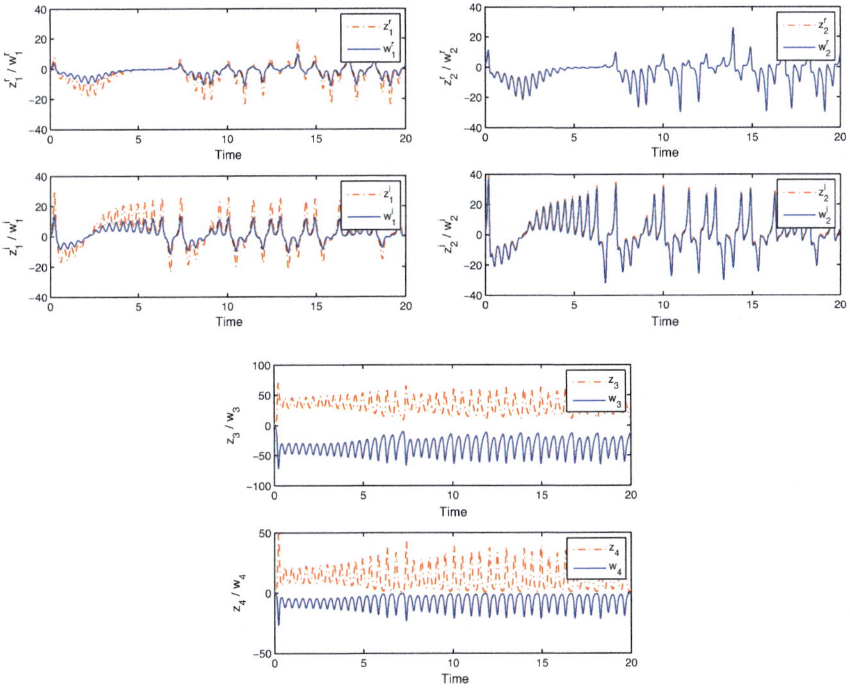

Fig. 4.7 Hyperchaotic MFPS of the hyperchaotic complex Lorenz system with the controller (4.3.11) and the update laws of the estimation (4.3.12)

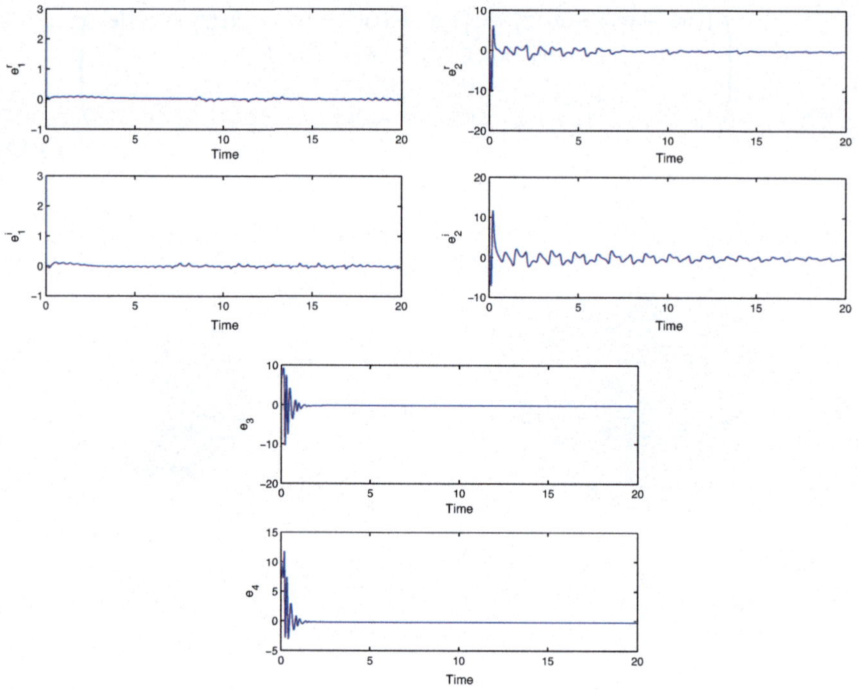

Fig. 4.8 Error dynamics of the drive system (4.3.11) and response system (4.3.12)

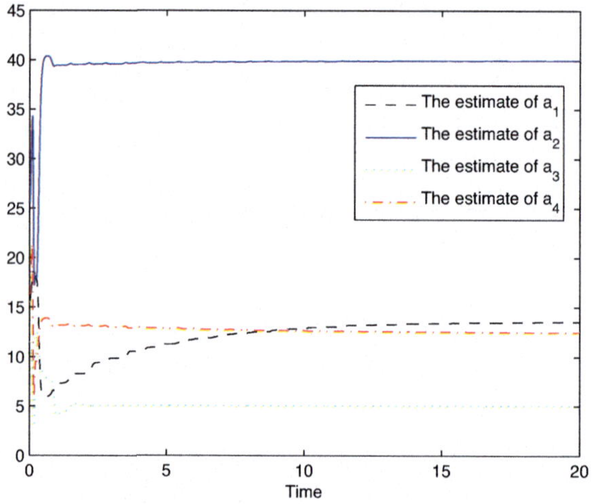

Fig. 4.9 The estimates of unknown parameters a_1, a_2, a_3 and a_4

In number simulation, we take $h_1 = 0.5$, $h_2 = 1$, $h_3 = -1$, $h_4 = -0.5$ and the external disturbances are chosen as $d_{11} = \sin(0.25\pi t) + j\cos(\pi t)$, $d_{12} = 0.3\sin(\pi t) + j0.5\cos(\pi t)$, $d_{13} = 0.2\sin(0.5\pi t)$, $d_{14} = 0.5\cos(0.5\pi t)$ and $d_{21} =$

$0.2cos(1.5\pi t) + j0.5\sin(2\pi t)$, $d_{22} = 0.5\sin(0.5\pi t) + j0.3\sin(\pi t)$, $d_{23} = 0.8\sin(1.5\pi t)$, $d_{24} = 0.3\cos(0.3\pi t)$, then we get the bound as $\rho_{11} = 0.5$, $\rho_{12} = 0.8$, $\rho_{13} = 0.2$,$\rho_{14} = 0.25$ and $\rho_{21} = 0.7$, $\rho_{22} = 0.8$, $\rho_{23} = 0.8$, $\rho_{24} = 0.3$. The unknown parameters are taken as $a_1 = 14, a_2 = 40, a_3 = 5$ and $a_4 = 13$, the hyperchaotic complex Lorenz system has hyperchaotic attractors see Fig. 4.6. The initial states of the drive system (4.3.9) and response system (4.3.10) are selected as $z_1(0) = 1 + 2j$, $z_2(0) = 3 + 4j$, $z_3(0) = 5$, $z_4(0) = 6$ and $w_1(0) = 5 + 4j$, $w_2(0) = 2 + 3j$, $w_3(0) = 6$, $w_4(0) = 1$, separately; the estimations of the parameters have the initial values $\hat{a}_1(0) = 20$, $\hat{a}_2(0) = 15$, $\hat{a}_3(0) = 20$ and $\hat{a}_4(0) = 15$. Also we choose $k = 1.5$ and $\xi = 0.1$.

In this set of numerical simulations, the controller vector (4.3.11) is assumed to be initiated at at $t = 0s$, and the corresponding simulation outcomes are presented in Figs. 4.7, 4.8 and 4.9. Figure 4.7 shows clearly that the response system (4.3.10) achieves asymptotic synchronization with the drive system (4.3.9) in accordance with the MFPS criterion. The real and imaginary parts of the MFPS errors are respectively given in Fig. 4.8, from which it can be concluded that each error signal converges asymptotically to a value close to zero. Additionally, Fig. 4.9 demonstrates that the estimates of the unknown parameters parameters a_1, a_2, a_3 and a_4 converge to 14, 40, 5 and 13, respectively, as $t \to \infty$.

Chapter 5
Adaptive Modified Function Projective Synchronization of Chaotic Complex Systems

Modified function projective synchronization (MFPS), a synchronization mode that enables drive-response systems to achieve synchronization with respect to a predefined scale function matrix, plays a crucial role in the field of secure communications [109–112]. Adaptive control [113–115] emerges as an effective approach to accomplish MFPS for chaotic systems. This chapter focuses on the exploration of adaptive MFPS and is structured into two sections. Section 5.1 presents the adaptive MFPS scheme for general uncertain chaotic complex systems, while Sect. 5.2 investigates the adaptive hybrid function projective synchronization problem for general chaotic complex systems with mismatched orders.

5.1 Adaptive MFPS of General Uncertain Chaotic Complex Systems

5.1.1 The Definition of MFPS and Problem Formulation

First, the definition of modified function projective synchronization (MFPS) is presented as follows. We then consider two general chaotic complex systems

$$\begin{cases} \dot{\mathbf{z}}(t) &= H(\mathbf{z}, t), \\ \dot{\mathbf{w}}(t) &= L(\mathbf{w}, t), \end{cases} \tag{5.1.1}$$

where $\mathbf{z} = [z_1, z_2, \cdots, z_n]^T$ and $\mathbf{w} = [w_1, w_2, \cdots, w_n]^T$ are the state complex vectors.

Definition 5.1 If there exists a function matrix $\mathbf{D(t)}$=diag$\{d_1(t), d_2(t), \cdots, d_n(t)\}$ such that $\lim_{t\to\infty} \|\mathbf{w}(t) - \mathbf{D}(t)\mathbf{z}(t)\| = 0$, then we regard that $\mathbf{z}$ synchronizes $\mathbf{w}$ up to

S. Liu et al., *Chaos in Complex Dynamical System: Theory and Application*,
https://doi.org/10.1007/978-981-95-7762-0_5

a scaling function matrix $\mathbf{D}(t)$, such synchronization is called as modified function projective synchronization (MFPS).

Remark 5.1 CS, AS, PS, MPS and FPS are the special cases of MFPS, where $d_1(t) = d_2(t) = \cdots = d_n(t) = 1$, $d_1(t) = d_2(t) = \cdots = d_n(t) = -1$, $d_1(t) = d_2(t) = \cdots = d_n(t) = \alpha$ (α is a constant), $d_1(t) = \alpha_1,\ d_2(t) = \alpha_2,\ \cdots,\ d_n(t) = \alpha_n$ and $d_1(t) = d_2(t) = \cdots = d_n(t) = \phi(t)$, respectively. Specially, when $d_1(t) = d_2(t) = \cdots = d_n(t) = 0$, the synchronization problem will be turned into a control problem of chaotic complex systems.

Remark 5.2 All the dynamical states from two synchronized systems in the sense of MFPS can be flexed by the scale function, since there are many scaling function factors in the scaling function matrix $\mathbf{D}(t)$.

We consider the general drive chaotic complex system described by the following form

$$\dot{\mathbf{z}}(t) = H(\mathbf{z}, t) = \mathbf{F}(\mathbf{z})\mathbf{A} + \mathbf{f}(\mathbf{z}), \tag{5.1.2}$$

where $\mathbf{z} = \mathbf{z}^r + j\mathbf{z}^i$, $\mathbf{z}^r = (u_1,\ u_3,\ \cdots,\ u_{2n-1})$, $\mathbf{z}^i = (u_2,\ u_4,\ \cdots,\ u_{2n})^T$. $\mathbf{f} = (f_1,\ f_2,\ \cdots,\ f_n)$ is a vector of complex nonlinear functions, $\mathbf{F}(\mathbf{z})$ is an $n \times n$ complex matrix and the elements of this matrix are complex state variables, $\mathbf{A}$ is an $n \times 1$ real (or complex) vector of system parameters. On the other hand, the response chaotic complex system is assumed as

$$\dot{\mathbf{w}}(t) = L(\mathbf{w}, t) = \mathbf{G}(\mathbf{w})\mathbf{B} + \mathbf{g}(\mathbf{w}) + \mathbf{v}, \tag{5.1.3}$$

where $\mathbf{w} = \mathbf{w}^r + j\mathbf{w}^i$, $\mathbf{w}^r = (u'_1,\ u'_3,\ \cdots,\ u'_{2n-1})$ and $\mathbf{w}^i = (u'_2,\ u'_4,\ \cdots,\ u'_{2n})^T$. $\mathbf{g} = (g_1,\ g_2,\ \cdots,\ g_n)$ is a vector of complex nonlinear functions, $\mathbf{G}(\mathbf{w})$ is an $n \times n$ complex matrix and its elements are state complex variables, $\mathbf{B}$ is an $n \times 1$ real (or complex) vector of system parameters and the controller $\mathbf{v} = \mathbf{v}^r + j\mathbf{v}^i$ needs to be designed, where $\mathbf{v} = (v_1,\ v_2,\ \cdots,\ v_n)$.

Let $\mathbf{e}(t) = \mathbf{w}(t) - \mathbf{D}(t)\mathbf{z}(t)$, where $\mathbf{e}(t)$ denotes the MFPS error vector. The core objective is to design the controller $\mathbf{v}$ such that the synchronization error vector $\mathbf{e}(t)$ converges to zero as $t \to \infty$, i.e.

$$\begin{aligned}
\lim_{t\to\infty} \|\mathbf{e}^r(t)\|^2 &= \lim_{t\to\infty} \|\mathbf{w}^r(t,\ \mathbf{w_0}^r) - \mathbf{D}(t)\mathbf{z}^r(t,\ \mathbf{z_0}^r)\|^2 \\
&= \lim_{t\to\infty} \left(\sum_{k=1}^{n} \|u'_{2k-1}(t,\ u'_{2k-1}(0) \right) - d_k(t)u_{2k-1}(t,\ u_{2k-1}(0))\|)^2 = 0, \\
\lim_{t\to\infty} \|\mathbf{e}^i(t)\|^2 &= \lim_{t\to\infty} \|\mathbf{w}^i(t,\ \mathbf{w_0}^i) - \mathbf{D}(t)\mathbf{z}^i(t,\ \mathbf{z_0}^i)\|^2 \\
&= \lim_{t\to\infty} \left(\sum_{k=1}^{n} \|u'_{2k}(t,\ u'_{2k}(0) \right) - d_k(t)u_{2k}(t,\ u_{2k}(0))\|)^2 = 0,
\end{aligned}$$

then

$$\lim_{t\to\infty}\|\mathbf{e}(t)\| = (\lim_{t\to\infty}\|\mathbf{e}^r(t)\|^2 + \lim_{t\to\infty}\|\mathbf{e}^i(t)\|^2)^{\frac{1}{2}} = 0,$$

where $\mathbf{z_0}^r = [u_1(0), u_3(0), \cdots, u_{2n-1}(0)]^T$ and $\mathbf{z_0}^i = [u_2(0), u_4(0), \cdots, u_{2n}(0)]^T$ are initial values of real and imaginary parts of the complex state variables from the complex system (5.1.2), while $\mathbf{w_0}^r = [u_1'(0), u_3'(0), \cdots, u_{2n-1}'(0)]^T$ and $\mathbf{w_0}^i = [u_2'(0), u_4'(0), \cdots, u_{2n}'(0)]^T$ are ones of the complex system (5.1.3).

5.1.2 Controller Design of Adaptive MFPS

Theorem 5.1 *If the nonlinear controller is designed as*

$$\begin{aligned}\mathbf{v} &= \mathbf{v}^r + j\mathbf{v}^i\\ &= \mathbf{D}(t)\mathbf{F}(\mathbf{z})\hat{\mathbf{A}} + \mathbf{D}(t)\mathbf{f}(\mathbf{z}) + \dot{\mathbf{D}}(t)\mathbf{z}(t) - \mathbf{G}(\mathbf{w})\hat{\mathbf{B}} - \mathbf{g}(\mathbf{w}) - K\mathbf{e}\\ &= \mathbf{D}(t)\mathbf{F}^r(u_1, u_3, \cdots, u_{2n-1})\hat{\mathbf{A}} + \mathbf{D}(t)\mathbf{f}^r(u_1, u_2, \cdots, u_n)\\ &\quad + \dot{\mathbf{D}}(t)\mathbf{u}_{2n-1}(t) - \mathbf{G}^r(u_1', u_2', \cdots, u_{2n-1}')\hat{\mathbf{B}} - \mathbf{g}^r(u_1', u_2', \cdots, u_n')\\ &\quad - K\mathbf{e}^r + j[\mathbf{D(t)}\mathbf{F}^i(u_2, u_4, \cdots, u_{2n})\hat{\mathbf{A}} + \mathbf{D}(t)\mathbf{f}^i(u_1, u_2, \cdots, u_n)\\ &\quad + \dot{\mathbf{D}}(t)\mathbf{u}_{2n}(t) - \mathbf{G}^i(u_2', u_4', \cdots, u_{2n}')\hat{\mathbf{B}} - \mathbf{g}^i(u_1', u_2', \cdots, u_n') - K\mathbf{e}^i],\end{aligned} \tag{5.1.4}$$

the adaptive laws of parameters are set as

$$\begin{cases}\dot{\hat{\mathbf{A}}} = [-(\mathbf{F}^r(u_1, u_3, \cdots, u_{2n-1}))^T\mathbf{D(t)}, -(\mathbf{F}^i(u_2, u_4, \cdots, u_{2n}))^T\mathbf{D(t)}]\begin{bmatrix}\mathbf{e}^r\\ \mathbf{e}^i\end{bmatrix},\\ \dot{\hat{\mathbf{B}}} = [(\mathbf{G}^r(u_1', u_3', \cdots, u_{2n-1}'))^T, (\mathbf{G}^i(u_2', u_4', \cdots, u_{2n}'))^T]\begin{bmatrix}\mathbf{e}^r\\ \mathbf{e}^i\end{bmatrix},\end{cases} \tag{5.1.5}$$

and updated laws of control coefficients are selected as

$$\dot{k}_l = \xi_l[(e_l^r)^2 + (e_l^i)^2](l = 1, 2, 3), \tag{5.1.6}$$

then MFPS between the response system (5.1.3) *and the drive system* (5.1.2) *will occur with desired scaling function matrix* $\mathbf{D}(t)$=*diag*$(d_1(t), d_2(t), \cdots, d_n(t))$ *asymptotically, where* $K = diag\{k_1, k_2, k_3\}$ *is a positive definitional matrix, and the parameters of matrixes* $\hat{\mathbf{A}}$ *and* $\hat{\mathbf{B}}$ *are parameters estimation of the unknown matrixes* $\mathbf{A}$ *and* $\mathbf{B}$ *, respectively.*

Proof From chaotic complex systems (5.1.2) and (5.1.3), we get the error dynamical system as follows

$$\begin{aligned}\dot{\mathbf{e}} &= \dot{\mathbf{e}}^r + j\dot{\mathbf{e}}^i\\ &= \mathbf{G}^r(u_1', u_2', \cdots, u_{2n-1}')\mathbf{B} + \mathbf{g}^r(u_1', u_2', \cdots, u_n') - \mathbf{D}(t)\mathbf{F}^r(u_1, u_3, \cdots, u_{2n-1})\mathbf{A}\\ &\quad - \mathbf{D}(t)\mathbf{f}^r(u_1, u_2, \cdots, u_n) - \dot{\mathbf{D}}(t)u_{2n-1}(t) + \mathbf{v}^r + j[\mathbf{G}^i(u_2', u_4', \cdots, u_{2n}')\mathbf{B}\end{aligned}$$

$$+\mathbf{g}^i(u_1', u_2', \cdots, u_n') - \mathbf{D}(t)\mathbf{F}^i(u_2, u_4, \cdots, u_{2n})\mathbf{A}$$
$$-\mathbf{D}(t)\mathbf{f}^i(u_1, u_2, \cdots, u_n) - \dot{\mathbf{D}}(t)u_{2n}(t) + \mathbf{v}^i]. \tag{5.1.7}$$

Hence, upon substituting the controller (5.1.4) into the error system (5.1.7) and decomposing these equations into real and imaginary parts, the error system can be re-expressed as

$$\begin{cases} \dot{\mathbf{e}}^r = \mathbf{D}(t)\mathbf{F}^r(u_1, u_3, \cdots, u_{2n-1})(\hat{\mathbf{A}} - \mathbf{A}) - \mathbf{G}^r(u_1', u_2', \cdots, u_{2n-1}')(\hat{\mathbf{B}} - \mathbf{B}) - \mathbf{K}\mathbf{e}^r, \\ \dot{\mathbf{e}}^i = \mathbf{D}(t)\mathbf{F}^i(u_1, u_3, \cdots, u_{2n-1})(\hat{\mathbf{A}} - \mathbf{A}) - \mathbf{G}^i(u_1', u_2', \cdots, u_{2n-1}')(\hat{\mathbf{B}} - \mathbf{B}) - \mathbf{K}\mathbf{e}^i. \end{cases} \tag{5.1.8}$$

We define the parameter error matrices as $\tilde{\mathbf{A}} = \hat{\mathbf{A}} - \mathbf{A}$ and $\tilde{\mathbf{B}} = \hat{\mathbf{B}} - \mathbf{B}$. Consider a Lyapunov function candidate as

$$V(\mathbf{e}^r, \mathbf{e}^i, \tilde{\mathbf{A}}, \tilde{\mathbf{B}}, k_l) = \frac{1}{2}[(\mathbf{e}^r)^T\mathbf{e}^r + (\mathbf{e}^i)^T\mathbf{e}^i + \tilde{\mathbf{A}}^T\tilde{\mathbf{A}} + \tilde{\mathbf{B}}^T\tilde{\mathbf{B}} + \sum_{l=1}^{3}\frac{1}{\xi_l}(k_l - k)^2],$$

where k is a positive constant. On the basis of the controller (5.1.4), the adaptive parameters laws (5.1.5) and updated laws of control coefficients (5.1.6), the time derivative of V along the trajectory of the error dynamical system (5.1.8) is given as follows

$$\begin{aligned} \dot{V}(t) &= (\dot{\mathbf{e}}^r)^T\mathbf{e}^r + (\dot{\mathbf{e}}^i)^T\mathbf{e}^i + \tilde{\mathbf{A}}^T\dot{\tilde{\mathbf{A}}} + \tilde{\mathbf{B}}^T\dot{\tilde{\mathbf{B}}} + \sum_{l=1}^{3}\frac{1}{\xi_l}(k_l - k)\dot{k}_l \\ &= [\mathbf{D}(t)\mathbf{F}^r(u_1, u_3, \cdots, u_{2n-1})(\hat{\mathbf{A}} - \mathbf{A}) - \mathbf{G}^r(u_1', u_2', \cdots, u_{2n-1}')(\hat{\mathbf{B}} - \mathbf{B}) - \mathbf{K}\mathbf{e}^r]^T\mathbf{e}^r \\ &\quad + [\mathbf{D}(t)\mathbf{F}^i(u_1, u_3, \cdots, u_{2n-1})(\hat{\mathbf{A}} - \mathbf{A}) - \mathbf{G}^i(u_1', u_2', \cdots, u_{2n-1}')(\hat{\mathbf{B}} - \mathbf{B}) - \mathbf{K}\mathbf{e}^i]^T\mathbf{e}^i \\ &\quad + \tilde{\mathbf{A}}^T[-(\mathbf{F}^r(u_1, u_3, \cdots, u_{2n-1}))^T\mathbf{D(t)}\mathbf{e}^r - (\mathbf{F}^i(u_2, u_4, \cdots, u_{2n}))^T\mathbf{D(t)}\mathbf{e}^i] \\ &\quad + \tilde{\mathbf{B}}^T[(\mathbf{G}^r(u_1', u_3', \cdots, u_{2n-1}'))^T\mathbf{e}^r + (\mathbf{G}^i(u_2', u_4', \cdots, u_{2n}'))^T\mathbf{e}^i] \\ &= -k[(\mathbf{e}^r)^T\mathbf{e}^r + (\mathbf{e}^i)^T\mathbf{e}^i] \le 0. \end{aligned}$$

Since V is positive definite and $\dot{V}$ is negative semi-definite, we cannot immediately obtain that the origin of error system (5.1.8) is asymptotically stable. In fact, since $\dot{V} \le 0$, then $\mathbf{e}^r$, $\mathbf{e}^i$, $\tilde{\mathbf{A}}$, $\tilde{\mathbf{B}} \in L_\infty$. From the fact that

$$\int_0^t \|\mathbf{e}^r\| + \|\mathbf{e}^i\| = -\frac{1}{k}(V(t) - V(0)) \le \frac{V(0)}{k},$$

we get $\mathbf{e}^r$, $\mathbf{e}^i \in L_2$. From error system (5.1.8), we obtain $\dot{\mathbf{e}}^r$, $\dot{\mathbf{e}}^i \in L_\infty$. According to the Barbalat's Lemma [116, 117], we have

$$\lim_{t\to\infty} \mathbf{e}^r + \mathbf{e}^i = 0,$$

namely,

$$
\begin{aligned}
&\lim_{t\to\infty} \|\mathbf{e}^r(t)\|^2 + \lim_{t\to\infty} \|\mathbf{e}^i(t)\|^2 \\
&= \lim_{t\to\infty} \|\mathbf{w}^r(t, \mathbf{w_0}^r) - \mathbf{D}(t)\mathbf{z}^r(t, \mathbf{z_0}^r)\|^2 + \lim_{t\to\infty} \|\mathbf{w}^i(t, \mathbf{w_0}^i) - \mathbf{D}(t)\mathbf{z}^i(t, \mathbf{z_0}^i)\|^2 \\
&= \lim_{t\to\infty} \sum_{k=1}^{m} \|u'_{2k-1}(t, u'_{2k-1}(0)) - d_k(t)u_{2k-1}(t, u_{2k-1}(0))\|^2 \\
&\quad + \lim_{t\to\infty} \sum_{k=1}^{m} \|u'_{2k}(t, u'_{2k}(0)) - d_k(t)u_{2k}(t, u_{2k}(0))\|^2 \\
&= 0.
\end{aligned}
$$

It follows that the general response chaotic complex system (5.1.3) realizes asymptotic MFPS with the general drive chaotic complex system (5.1.2). This concludes the proof. □

Corollary 5.1 *If the structures of general chaotic complex systems* (5.1.2) *and* (5.1.3) *are identical, i.e.,* $\mathbf{F}(\mathbf{z}) = \mathbf{G}(\mathbf{w})$ *and* $\mathbf{f}(\mathbf{z}) = \mathbf{g}(\mathbf{w})$, *then the controller is designed as*

$$
\begin{aligned}
\mathbf{v} &= [\mathbf{D}(t)F(\mathbf{z}) - \mathbf{F}(\mathbf{w})]\hat{\mathbf{A}} + \mathbf{D}(t)\mathbf{f}(\mathbf{z}) - \mathbf{f}(\mathbf{w}) + \dot{\mathbf{D}}(t)\mathbf{z}(t) - K\mathbf{e} \\
&= [\mathbf{D}(t)\mathbf{F}^r(u_1, u_3, \cdots, u_{2n-1}) - \mathbf{F}^r(u'_1, u'_2, \cdots, u'_{2n-1})]\hat{\mathbf{A}} + \mathbf{D}(t)\mathbf{f}^r(u_1, u_2, \\
&\quad \cdots, u_n) - \mathbf{f}^r(u'_1, u'_2, \cdots, u'_n + \dot{\mathbf{D}}(t)u_{2n-1}(t) - K\mathbf{e}^r \\
&\quad + j[(\mathbf{D(t)}\mathbf{F}^i(u_2, u_4, \cdots, u_{2n}) - \mathbf{F}^i(u'_2, u'_4, \cdots, u'_{2n}))\hat{\mathbf{A}} + \mathbf{D}(t)\mathbf{f}^i(u_1, u_2, \\
&\quad \cdots, u_n) - \mathbf{f}^i(u'_1, u'_2, \cdots, u'_n) + \dot{\mathbf{D}}(t)u_{2n}(t) - K\mathbf{e}^i],
\end{aligned} \tag{5.1.9}
$$

where the control coefficients matrix $K = diag\{k_1, k_2, k_3\}$ *is updated according to system* (5.1.6) *and the parameters adaptive laws are chosen as*

$$
\begin{aligned}
\dot{\hat{\mathbf{A}}} &= [\mathbf{F}^r(u'_1, u'_3, \cdots, u'_{2n-1}) - \mathbf{D}(t)\mathbf{F}^r(u_1, u_3, \cdots, u_{2n-1})]^T\mathbf{e}^r \\
&\quad + [\mathbf{F}^i(u'_2, u'_4, \cdots, u'_{2n}) - \mathbf{D}(t)\mathbf{F}^i(u_2, u_4, \cdots, u_{2n})]^T\mathbf{e}^i.
\end{aligned} \tag{5.1.10}
$$

Thus, MFPS of two identical general chaotic complex systems is also achieved.

Remark 5.3 If some parameters from parameters matrix **A** or **B** are unknown, Theorem 5.1 and Corollary 5.1 can also be applied to realize MFPS of chaotic complex systems with partly unknown parameters. Specially, when parameters matrixs **A** and **B** are known, then Theorem 5.1 and Corollary 5.1 can also be applied to achieve MFPS of chaotic complex systems with certain parameters.

Remark 5.4 Obviously, the control coefficient matrix K can be chosen as any positive matrix to derive the controller **v** in Theorem 5.1. Moreover, the control coefficients can be automatically adapted to updated laws by the adaptive feedback control method. Hence, the adaptive control coefficients are more efficient than the fixed one presented in some existing works.

Remark 5.5 In Theorem 5.1, it is obvious that the real and imaginary parts of function $\mathbf{F}(\cdot)$ can be separated completely and the functions $\mathbf{F}^r(\cdot)$ and $\mathbf{F}^i(\cdot)$ are denoted as functions of the variables $u_1, u_3, \cdots, u_{2n-1}$ and $u_2, u_4, \cdots, u_{2n}$, respectively. The functions $\mathbf{f}^r(\cdot)$ and $\mathbf{f}^i(\cdot)$ are denoted with the variables $u_1, u_2, \cdots, u_n$, which are determined by the special complex systems, that is, the real and imaginary parts of function $\mathbf{f}(\cdot)$ are not separated completely. The same cases exist in the functions $\mathbf{G}^r(\cdot)$, $\mathbf{G}^i(\cdot)$ and $\mathbf{g}^r(\cdot)$, $\mathbf{g}^i(\cdot)$.

Remark 5.6 If $\mathbf{A}$ is a complex vector, then $\mathbf{A}$ can be rewritten as $\mathbf{A} = \mathbf{A}^r + j\mathbf{A}^i$. So the general drive chaotic complex system (5.1.2) becomes

$$\dot{\mathbf{z}} = \mathbf{F}(\mathbf{z})(\mathbf{A}^r + j\mathbf{A}^i) + \mathbf{f}(\mathbf{z}), \tag{5.1.11}$$

which can also be rewritten as

$$\dot{\mathbf{z}}_1 = \mathbf{F}(\mathbf{z})\mathbf{A}^r + \mathbf{f}_1(\mathbf{z}) \tag{5.1.12}$$

and

$$\dot{\mathbf{z}}_2 = \mathbf{F}(\mathbf{z})\mathbf{A}^i - j\mathbf{f}_2(\mathbf{z}), \tag{5.1.13}$$

where $\mathbf{f}_1(\mathbf{z}) + \mathbf{f}_2(\mathbf{z}) = \mathbf{f}(\mathbf{z})$. Similarly, when $\mathbf{B}$ is a complex vector, then response chaotic complex system (5.1.3) also become

$$\dot{\mathbf{w}} = \mathbf{G}(\mathbf{w})(\mathbf{B}^r + j\mathbf{B}^i) + \mathbf{g}(\mathbf{w}) + \mathbf{v}', \tag{5.1.14}$$

then

$$\dot{\mathbf{w}}_1 = \mathbf{G}(\mathbf{w})\mathbf{B}^r + \mathbf{g}_1(\mathbf{w}) + \mathbf{v}'' \tag{5.1.15}$$

and

$$\dot{\mathbf{w}}_2 = \mathbf{G}(\mathbf{w})\mathbf{B}^i - j\mathbf{g}_2(\mathbf{w}) + \mathbf{v}''', \tag{5.1.16}$$

where $\mathbf{g}_1(\mathbf{w}) + \mathbf{g}_2(\mathbf{w}) = \mathbf{g}(\mathbf{w})$. The structures of general chaotic complex subsystems (5.1.12), (5.1.13), (5.1.15) and (5.1.16) are the similar with ones of the general chaotic complex systems (5.1.2) and (5.1.3), respectively. Then the design of the controller and adaptive laws of parameters to achieve MFPS between chaotic complex subsystems (5.1.12) and (5.1.15)(or (5.1.13) and (5.1.16)) is the same (or similar) as one of general chaotic complex systems (5.1.2) and (5.1.3). If the chaotic complex subsystems (5.1.12) and (5.1.15), and the chaotic complex subsystems (5.1.13) and (5.1.16) realize MFPS respectively, the general chaotic complex systems (5.1.11) and (5.1.14) also accomplish MFPS. According to Theorem 5.1, the controllers $\mathbf{v}''$, $\mathbf{v}'''$ and the estimates $\hat{\mathbf{A}}^r$, $\hat{\mathbf{B}}^r$, $\hat{\mathbf{A}}^i$, $\hat{\mathbf{B}}^i$ are designed to realize MFPS of the chaotic complex subsystems (5.1.12) and (5.1.15), and the chaotic complex subsystems (5.1.13) and (5.1.16), respectively. Hence, the general chaotic complex systems (5.1.11) and (5.1.14) achieve MFPS, and the controller $\mathbf{v}' = \mathbf{v}'' + j\mathbf{v}'''$ and the estimates $\hat{\mathbf{A}} = \hat{\mathbf{A}}^r + j\hat{\mathbf{A}}^i$, $\hat{\mathbf{B}} = \hat{\mathbf{B}}^r + j\hat{\mathbf{B}}^i$ are obtained.

According to Remark 5.6, the controller design strategy for complex parameter vectors $\mathbf{A}$ and $\mathbf{B}$ bears resemblance to that applicable to real parameter vectors. To streamline the subsequent analysis, we hereby choose chaotic complex systems with real parameter vectors as cases for the verification of Theorem 5.1.

5.1.3 Simulation Examples

Herein, we adopt the MFPS of two chaotic complex systems (both different and identical) to validate the effectiveness of the proposed adaptive MFPS schemes.

I. Adaptive MFPS of Two Different Chaotic Complex Systems with Fully Uncertain Parameters

To characterize the MFPS behavior between two chaotic complex systems, we designate the chaotic complex system discussed in Ref.[118] as the drive system, which is expressed as follows

$$\begin{cases} \dot{z}_1 = a_1(z_1 - z_2), \\ \dot{z}_2 = -a_2 z_2 + z_1 z_3, \\ \dot{z}_3 = -a_3 z_3 + \frac{1}{2}(\bar{z}_1 z_2 + z_1 \bar{z}_2). \end{cases} \tag{5.1.17}$$

Furthermore, we employ another chaotic complex system presented by Mahmoud et al. [119]; this system, combined with the controller, serves as the response system, which is expressed as follows

$$\begin{cases} \dot{w}_1 = b_1 w_1 - w_2 w_3 + v_1^r + j v_1^i, \\ \dot{w}_2 = -b_2 w_2 + w_1 w_3 + v_2^r + j v_2^i, \\ \dot{w}_3 = -b_3 w_3 + \frac{1}{2}(\bar{w}_1 w_2 + w_1 \bar{w}_2) + v_3. \end{cases} \tag{5.1.18}$$

First, we rearrange the chaotic complex systems (5.1.17) and (5.1.18) into the forms corresponding to systems (5.1.2) and (5.1.3), respectively, as follows

$$\dot{\mathbf{z}} = \mathbf{F}(\mathbf{z})\mathbf{A} + \mathbf{f}(\mathbf{z}), \tag{5.1.19}$$

where

$$\mathbf{F}(\mathbf{z}) = \begin{pmatrix} z_1 - z_2 & 0 & 0 \\ 0 & -z_2 & 0 \\ 0 & 0 & -z_3 \end{pmatrix}, \ \mathbf{f}(\mathbf{z}) = \begin{pmatrix} 0 \\ z_1 z_3 \\ \frac{1}{2}(\bar{z}_1 z_2 + z_1 \bar{z}_2) \end{pmatrix}.$$

Similarly,

$$\mathbf{G(w)} = \begin{pmatrix} w_1 & 0 & 0 \\ 0 & -w_2 & 0 \\ 0 & 0 & -w_3 \end{pmatrix}, \ \mathbf{g(w)} = \begin{pmatrix} -w_2w_3 \\ w_1w_3 \\ \frac{1}{2}(\bar{w}_1w_2 + w_1\bar{w}_2) \end{pmatrix},$$

where $\mathbf{z} = (z_1, z_2, z_3)^T$ and $\mathbf{w} = (w_1, w_2, w_3)^T$ are the state vectors, $\mathbf{A} = (a_1, a_2, a_3)^T$ and $\mathbf{B} = (b_1, b_2, b_3)^T$ are unknown parameter matrixes, v_1^r, v_1^i, v_2^r, v_2^i and v_3 are five control inputs, which are needed to be designed.

According to Theorem 5.1, the controller is designed as

$$\begin{aligned}
\mathbf{v} &= \mathbf{D(t)F(z)\hat{A}} + \mathbf{D(t)f(z)} + \mathbf{D}(\dot{t})z(t) - \mathbf{G(w)\hat{B}} - \mathbf{g(w)} - K\mathbf{e} \\
&= \begin{pmatrix} -\hat{b}_1w_1 + w_2w_3 + \hat{a}_1d_1(t)(z_1 - z_2) + \dot{d}_1(t)z_1 - k_1e_1 \\ \hat{b}_2w_2 - w_1w_3 - d_2(t)(\hat{a}_2z_2 - z_1z_3) + \dot{d}_2(t)z_2 - k_2e_2 \\ \hat{b}_3w_3 - \frac{1}{2}(\bar{w}_1w_2 + w_1\bar{w}_2) - d_3(t)[\hat{a}_3z_3 - \frac{1}{2}(\bar{z}_1z_2 + z_1\bar{z}_2)] + \dot{d}_3(t)z_3 - k_3e_3 \end{pmatrix} \\
&= \begin{pmatrix} -\hat{b}_1u_1' + u_3'u_5' + d_1(t)(u_1 - u_3)\hat{a}_1 + \dot{d}_1(t)u_1 - k_1e_1^r \\ \hat{b}_2u_3' - u_1'u_5' + d_2(t)u_1u_5 - d_2(t)u_3\hat{a}_2 + \dot{d}_2(t)u_3 - k_2e_2^r \\ \hat{b}_3u_5' - (u_1'u_3' + u_2'u_4') - d_3(t)u_5\hat{a}_3 + d_3(t)(u_1u_3 + u_2u_4) + \dot{d}_3(t)u_5 - k_3e_3 \end{pmatrix} \\
&\quad + j\begin{pmatrix} -\hat{b}_1u_2' + u_4'u_5' + d_1(t)(u_2 - u_4)\hat{a}_1 + \dot{d}_1(t)u_2 - k_1e_1^i \\ \hat{b}_2u_4' - u_2'u_5' + d_2(t)u_2u_5 - d_2(t)u_4\hat{a}_2 + \dot{d}_2(t)u_4 - k_2e_2^i \\ 0 \end{pmatrix},
\end{aligned} \tag{5.1.20}$$

adaptive laws of parameters are taken as

$$\begin{aligned}
\dot{\hat{\mathbf{A}}} &= \left[-(\mathbf{F}^r)^T\mathbf{D}(t) \ -(\mathbf{F}^i)^T\mathbf{D}(t) \right] \begin{pmatrix} \mathbf{e}^r \\ \mathbf{e}^i \end{pmatrix} = -(\mathbf{F}^r)^T\mathbf{D}(t)\mathbf{e}^r - (\mathbf{F}^i)^T\mathbf{D}(t)\mathbf{e}^i \\
&= \begin{pmatrix} (u_3 - u_1)e_1^r - u_1e_2^r + (u_4 - u_2)e_1^i - u_1e_2^i \\ (u_1 + u_3)e_2^r + (u_2 + u_4)e_2^i \\ -u_5e_3 \end{pmatrix},
\end{aligned} \tag{5.1.21}$$

$$\dot{\hat{\mathbf{B}}} = \left[(\mathbf{G}^r)^T \ (\mathbf{G}^i)^T \right] \begin{pmatrix} \mathbf{e}^r \\ \mathbf{e}^i \end{pmatrix} = (\mathbf{G}^r)^T\mathbf{e}^r + (\mathbf{G}^i)^T\mathbf{e}^i = \begin{pmatrix} u_1'e_1^r + u_2'e_1^i \\ -u_3'e_2^r - u_4'e_2^i \\ -u_5'e_3 \end{pmatrix}, \tag{5.1.22}$$

and updated laws of control coefficients are taken as

$$\dot{k}_l = (e_l^r)^2 + (e_l^i)^2 (l = 1, 2, 3). \tag{5.1.23}$$

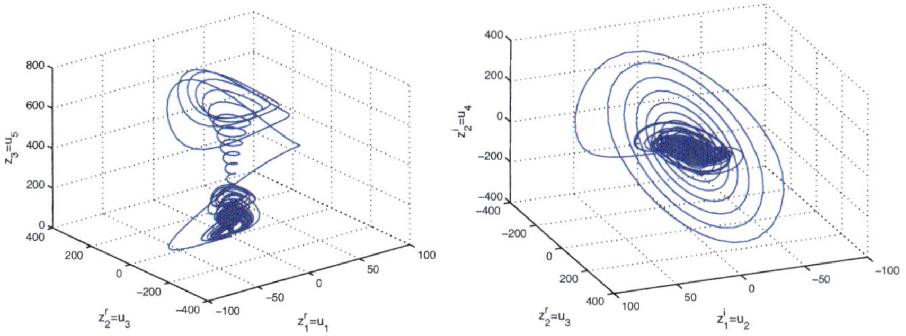

Fig. 5.1 Chaotic attractors of the chaotic complex system (5.1.17)

To demonstrate the effectiveness of the proposed adaptive MFPS scheme, numerical simulations are performed on the chaotic complex systems (5.1.17) and (5.1.18). The Runge-Kutta integration method is utilized for the numerical solution of the aforementioned systems. The true values of the unknown parameters are specified as $a_1 = 20$, $a_2 = 28$, $a_3 = 2$ and $b_1 = 9.5$, $b_2 = 19$, $b_3 = 2$, and the chaotic attractors of systems (5.1.17) and (5.1.18) under these parameter settings are depicted in Figs. 5.1 and 5.2. The initial state vectors of the drive system and the response system are chosen as $(z_1(0), z_2(0), z_3(0))^T = (1 + 2j, 3 + 4j, 5)^T$ and $(w_1(0), w_2(0), w_3(0))^T = (4 + j, -5 + 3j, 2)^T$, respectively. The initial estimates of the parameter matrixes $\hat{\mathbf{A}}$ and $\hat{\mathbf{B}}$ are set to $(a_1(0),\ a_2(0),\ a_3(0))^T = (10,\ 10,\ 10)^T$, $(b_1(0), b_2(0), b_3(0))^T = (5, 5, 5)^T$; the scaling function matrix is defined as $\mathbf{D}(t) = diag\{2 + 0.2\sin(\frac{\pi}{10}t),\ 2 + 0.3\sin(\frac{\pi}{10}t),\ 2 + 0.5\sin(\frac{\pi}{10}t)\}$; and the initial values of the control coefficients are assigned as $k_1(0) = 1$,$k_2(0) = 1$,$k_3(0) = 1$. Consequently, the initial error vector is determined to be $(e_1(0),\ e_2(0),\ e_3(0))^T = (2 - 3j,\ -1 - 5j,\ -8)^T$. The simulation results of adaptive MFPS for systems (5.1.17) and (5.1.18), implemented via the controller (5.1.20), parameter adaptive laws (5.1.21)–(5.1.22) and control coefficient update laws (5.1.23), are illustrated in Figs. 5.3, 5.4 and 5.5. In particular,Figs. 5.3 and 5.4 display the MFPS state time response of chaotic complex systems (5.1.17) and (5.1.18) and MFPS error dynamics between state variables from chaotic complex systems (5.1.17) and (5.1.18), respectively. These errors converge asymptotically to zero. As shown in Fig. 5.5, the parameter estimates $\hat{a}_1$, $\hat{a}_2$, $\hat{a}_3$ and $\hat{b}_1$, $\hat{b}_2$, $\hat{b}_3$ converge to their preset true values as $t \to \infty$, and the control coefficients k_l $(l = 1,\ 2,\ 3)$ converge to stable constant values over time.

II. Adaptive MFPS of Two Identical Chaotic Complex Systems with Fully Unknown Parameters

The chaotic complex system (5.1.17) with fully unknown parameters are taken as drive and response systems to illustrate the adaptive MFPS of two identical chaotic complex systems. The response system with the controller is described as

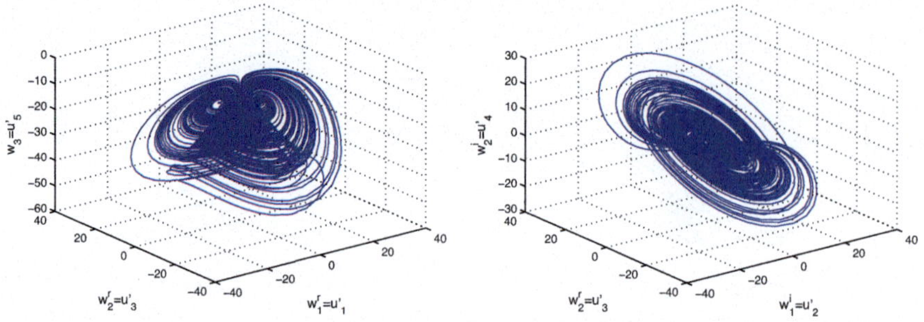

Fig. 5.2 Chaotic attractors of the chaotic complex system (5.1.18)

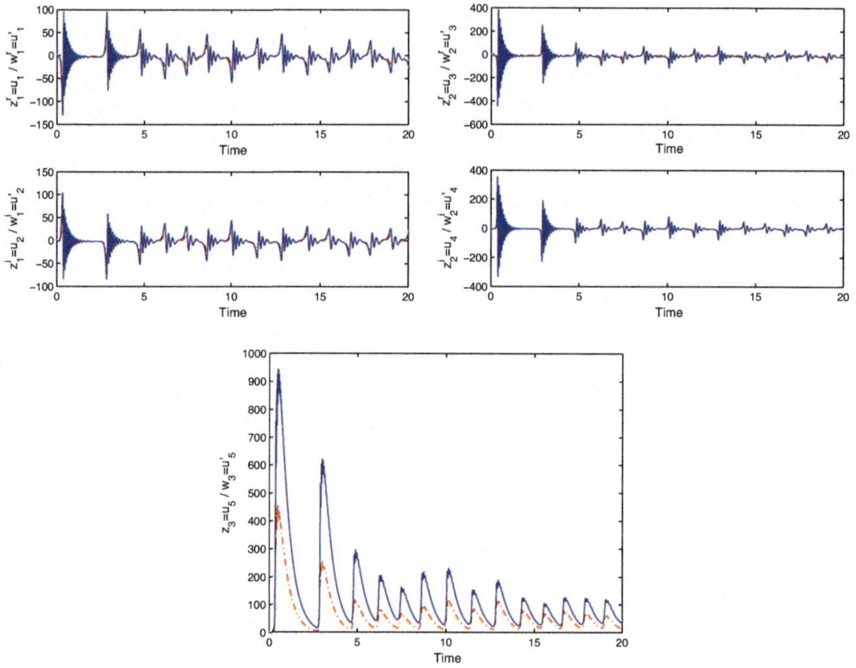

Fig. 5.3 The adaptive MFPS of two different chaotic complex systems (5.1.17) and (5.1.18) with the controller (5.1.20), the adaptive laws (5.1.21) and (5.1.22), and the update laws (5.1.23)

$$\begin{cases} \dot{w}_1 = a_1(w_1 - w_2) + v_1, \\ \dot{w}_2 = -a_2 w_2 + w_1 w_3 + v_2, \\ \dot{w}_3 = -a_3 w_3 + \frac{1}{2}(\bar{z}_1 z_2 + z_1 \bar{z}_2) + v_3, \end{cases} \tag{5.1.24}$$

where a_1, a_2, a_3 are unknown parameters, v_1, v_2, v_3 are controller inputs needed to be designed.

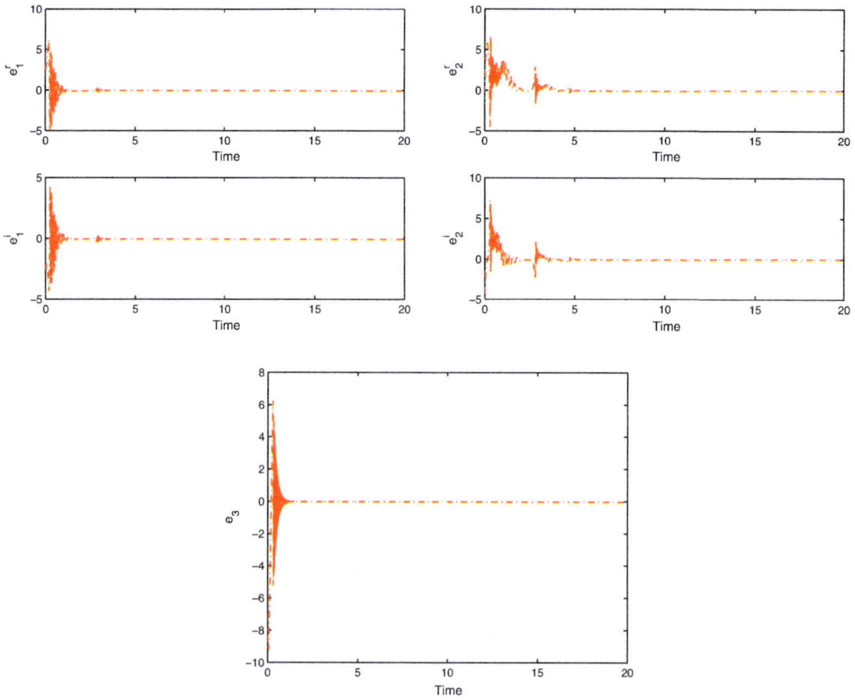

Fig. 5.4 The error dynamics of MFPS of two different chaotic complex systems (5.1.17) and (5.1.18)

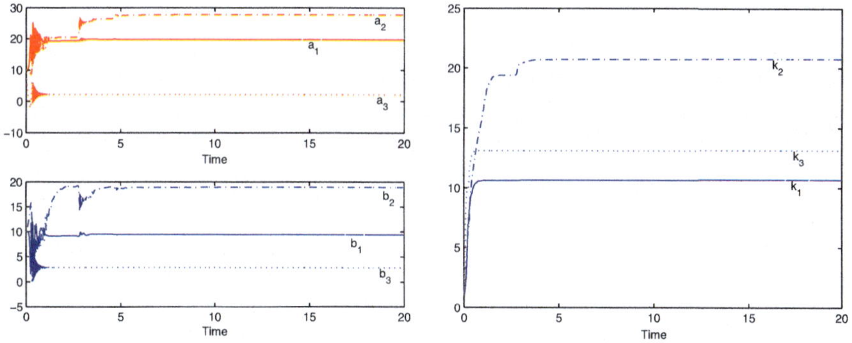

(a) Parameters $\hat{a}_1, \hat{a}_2, \hat{a}_3$ and $\hat{b}_1, \hat{b}_2, \hat{b}_3$ (b) Control coefficients k_1, k_2 and k_3

Fig. 5.5 Time evolution of parameters and control coefficients

Then, based on Eqs. (5.1.9) and (5.1.10) in Corollary 5.1, the controller is designed as

$$\begin{aligned}
\mathbf{v} &= [\mathbf{D}(t)F(\mathbf{z}) - \mathbf{F}(\mathbf{w})]\hat{\mathbf{A}} + \mathbf{D}(t)\mathbf{f}(\mathbf{z}) - \mathbf{f}(\mathbf{w}) + \dot{\mathbf{D}}(t)\mathbf{z}(t) - K\mathbf{e} \\
&= \begin{pmatrix} [d_1(t)(z_1 - z_2) - (w_1 - w_2)]\hat{a}_1 + \dot{d}_1(t)z_1 - k_1 e_1 \\ (-d_2(t)z_2 + w_2)\hat{a}_2 + d_2(t)z_1 z_3 - w_1 w_3 + \dot{d}_2(t)z_2 - k_2 e_2 \\ (-d_3(t)z_3 + w_3)\hat{a}_3 + \frac{1}{2}d_3(t)(\bar{z}_1 z_2 + z_1\bar{z}_2) - \frac{1}{2}(\bar{w}_1 w_2 + w_1\bar{w}_2) + \dot{d}_3(t)z_3 - k_3 e_3 \end{pmatrix} \\
&= \begin{pmatrix} [d_1(t)(u_1 - u_3) - (u_1' - u_3')]\hat{a}_1 + \dot{d}_1(t)u_1 - k_1 e_1^r \\ (-d_2(t)u_3 + u_3')\hat{a}_2 + d_2(t)u_1 u_5 - u_1' u_5' + \dot{d}_2(t)u_3 - k_2 e_2^r \\ (-d_3(t)u_5 + u_5')\hat{a}_3 + d_3(t)(u_1 u_3 + u_2 u_4) - (u_1' u_3' + u_2' u_4') + \dot{d}_3(t)u_5 - k_3 e_3 \end{pmatrix} \\
&\quad + j\begin{pmatrix} [d_1(t)(u_2 - u_4) - (u_2' - u_4')]\hat{a}_1 + \dot{d}_1(t)u_2 - k_1 e_1^i \\ (-d_2(t)u_4 + u_4')\hat{a}_2 + d_2(t)u_2 u_5 - u_2' u_5' + \dot{d}_2(t)u_4 - k_2 e_2^i \\ 0 \end{pmatrix},
\end{aligned} \tag{5.1.25}$$

the adaptive laws of parameters are given as

$$\begin{aligned}
\dot{\hat{\mathbf{A}}} &= [\mathbf{F}^r(u_1', u_3', \cdots, u_{2n-1}') - \mathbf{D}(t)\mathbf{F}^r(u_1, u_3, \cdots, u_{2n-1})]^T \mathbf{e}^r \\
&\quad + [\mathbf{F}^i(u_2', u_4', \cdots, u_{2n}') - \mathbf{D}(t)\mathbf{F}^i(u_2, u_4, \cdots, u_{2n})]^T \mathbf{e}^i \\
&= \begin{pmatrix} (u_1' - u_3' - (u_1 - u_3)d_1(t))e_1^r + (u_2' - u_4' - (u_2 - u_4)d_1(t))e_1^i \\ (-u_3' + d_2(t)u_3)e_2^r - (u_4' - d_2(t)u_4)e_2^i \\ (-u_5' + d_3(t)u_5)e_3 \end{pmatrix}
\end{aligned} \tag{5.1.26}$$

and updated laws of control coefficients are selected as

$$\dot{k}_l = (e_l^r)^2 + (e_l^i)^2 (l = 1, 2, 3). \tag{5.1.27}$$

The true parameter values are selected as $a_1 = 20$, $a_2 = 28$, $a_3 = 2$. The initial conditions for both the drive and response systems, the initial estimates of the parameters in matrices $\hat{\mathbf{A}}$ and $\hat{\mathbf{B}}$, the scaling function matrix $\mathbf{D}(t)$, and the initial values of the control coefficients k_1, k_2, k_3 are chosen as follows: $(z_1(0), z_2(0), z_3(0))^T = (2 + j, 4 - 3j, 2)^T$, $(w_1(0), w_2(0), w_3(0))^T = (1 + 2j, 3 + 4j, 5)^T$, $(a_1(0), a_2(0), a_3(0))^T = (15, 20, 5)^T$, $\mathbf{D}(t) = diag\{1 + 0.5\sin(\frac{\pi}{20}t), 1 + 0.7\sin(\frac{\pi}{20}t), 1 + 0.9\sin(\frac{\pi}{20}t)\}$, and $k_1(0) = 2$, $k_2(0) = 1$, $k_3(0) = 1.5$. Hence, the initial value of the error system is $(e_1(0), e_2(0), e_3(0))^T = (-1 + j, -1 + 7j, 3)^T$. The adaptive MFPS between systems (5.1.17) and (5.1.24), achieved using the controller (5.1.25) and the parameter adaptive laws (5.1.26), is illustrated in Figs. 5.6–5.8. Figure 5.6 displays the state trajectories of the MFPS for the chaotic complex systems (5.1.17) and (5.1.24). Figure 5.7 shows the MFPS error dynamics between the state variables of the two systems, which converge asymptotically to zero. Figure 5.8 demonstrates that the estimates $\hat{a}_1$, $\hat{a}_2$, and $\hat{a}_3$ of the unknown parameters converge to the true values $a_1 = 20$, $a_2 = 28$, and $a_3 = 2$, respectively, as $t \to \infty$. Furthermore, Fig. 5.8 also indicates that the control coefficients k_l $(l = 1, 2, 3)$ each converge to a constant value as t approaches infinity.

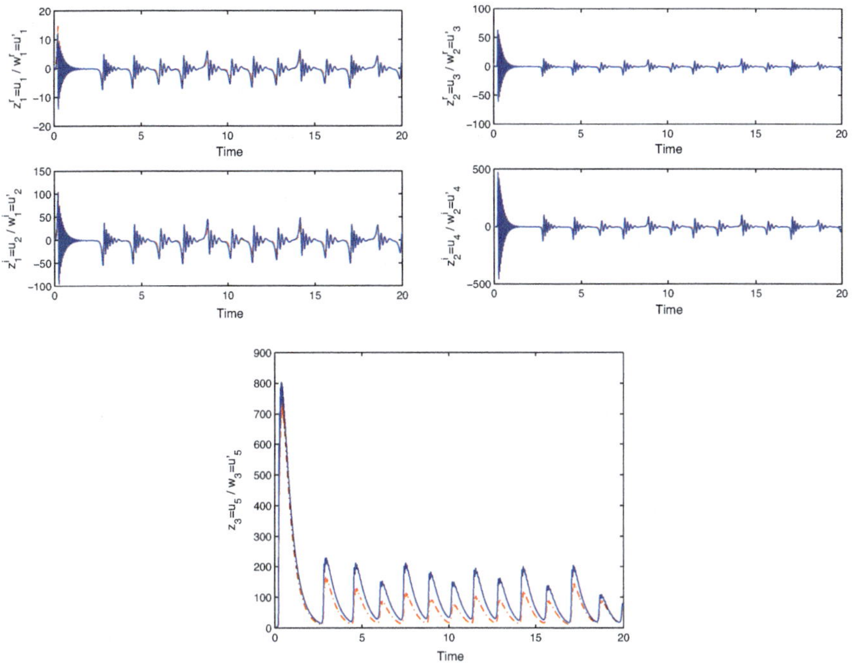

Fig. 5.6 The adaptive MFPS of two identical chaotic complex systems (5.1.17) and (5.1.24) with the controller (5.1.25), the adaptive laws (5.1.26) and the update laws (5.1.27)

5.2 Adaptive Hybrid Function Projective Synchronization of General Chaotic Complex Systems with Different Orders

5.2.1 Problem Formulations

Consider the general drive chaotic complex system in the form of

$$\dot{\mathbf{z}}(t) = \mathbf{F}(\mathbf{z})\mathbf{A} + \mathbf{f}(\mathbf{z}), \tag{5.2.1}$$

here, $\mathbf{z} = \mathbf{z}^r + j\mathbf{z}^i \in \mathbb{C}^m$ represents the complex state vector. The term $\mathbf{f} = (f_1, f_2, \cdots, f_m)$ denotes a vector of complex nonlinear functions, while $\mathbf{F}(\mathbf{z}) : \mathbb{C}^m \to \mathbb{C}^{m \times k}$ is a complex matrix whose elements are functions of the complex state variables. The vector $\mathbf{A}$, which is a $k \times 1$ matrix, contains the real (or complex) parameters of the system. The superscripts r and i indicate the real and imaginary parts of the complex state vector $\mathbf{z}$, respectively. On the other hand, the response chaotic complex system is described by

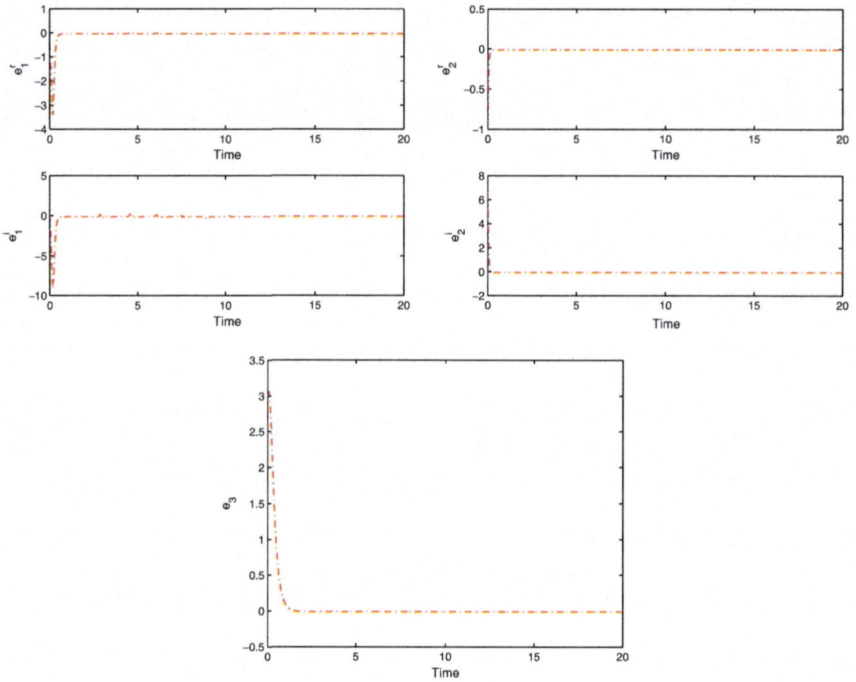

Fig. 5.7 The error dynamics of MFPS of two identical chaotic complex systems (5.1.17) and (5.1.24)

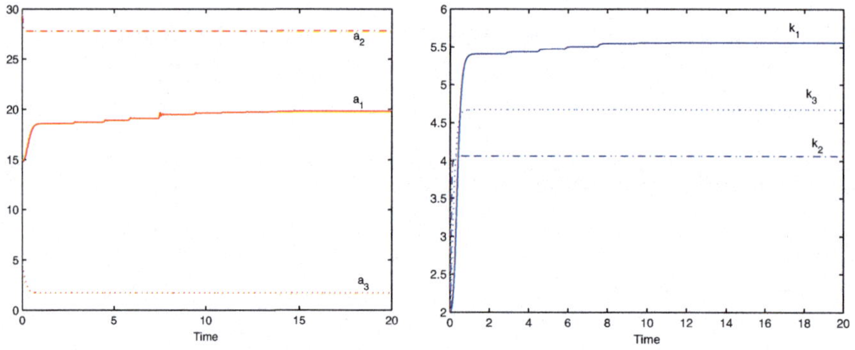

(a) Parameters $\hat{a}_1, \hat{a}_2, \hat{a}_3$

(b) Control coefficients k_1, k_2 and k_3

Fig. 5.8 Time evolution of parameters and control coefficients

$$\dot{\mathbf{w}}(t) = \mathbf{G}(\mathbf{w})\mathbf{B} + \mathbf{g}(\mathbf{w}) + \mathbf{u}, \tag{5.2.2}$$

here, $\mathbf{w} = \mathbf{w}^r + j\mathbf{w}^i \in \mathbb{C}^n$ denotes the complex state vector. The function $\mathbf{g} : \mathbb{C}^n \to \mathbb{C}^n$ is a vector of complex nonlinear functions depending on w, and $\mathbf{G}(\mathbf{w})$ is a $n \times l$ complex matrix whose elements are functions of the complex state variables. The vector $\mathbf{B}$, which is $l \times 1$, contains the real (or complex) system parameters. The controller vector $\mathbf{u} = \mathbf{u}^r + j\mathbf{u}^i$, with $\mathbf{u} = (u_1, u_2, \cdots, u_n)$, remains to be designed. Under the condition where the order $n = m$, $k = l$, and the functions satisfy $\mathbf{f}(\cdot) = \mathbf{g}(\cdot)$, $\mathbf{F}(\cdot) = \mathbf{G}(\cdot)$, the response chaotic complex system becomes identical to the drive chaotic complex system. In this specific case, the problem reduces to the HFPS of chaotic complex systems, which has been previously studied. When $m > n$ (which inherently implies $\mathbf{f}(\cdot) \neq \mathbf{g}(\cdot)$ and $\mathbf{F}(\cdot) \neq \mathbf{G}(\cdot)$, meaning the order of the response chaotic complex system is lower than that of the drive system, HFPS can only be realized in a reduced-order manner. In essence, reduced-order HFPS involves controlling the response chaotic complex system (5.2.2) to behave as a projection of the drive chaotic complex system under the HFPS framework. To analyze this, the drive chaotic complex system can be conceptually divided into two parts. The first part is the projected chaotic complex subsystem:

$$\dot{\mathbf{z}_p} = \mathbf{F}_p(\mathbf{z})\mathbf{A} + \mathbf{f}_p(\mathbf{z}), \tag{5.2.3}$$

the rest chaotic complex subsystem:

$$\dot{\mathbf{z}_q} = \mathbf{F}_q(\mathbf{z})\mathbf{A} + \mathbf{f}_q(\mathbf{z}), \tag{5.2.4}$$

where $\mathbf{z}_p = [z_1, z_2, \cdots, z_n]^T$, $\mathbf{z}_q = [z_{n+1}, z_{n+2}, \cdots, z_m]^T$, $\mathbf{f}_p : \mathbb{C}^m \to \mathbb{C}^n$, $\mathbf{f}_q : \mathbb{C}^m \to \mathbb{C}^u$, $F_p : \mathbb{C}^m \to \mathbb{C}^{n\times k}$, $F_q : \mathbb{C}^m \to \mathbb{C}^{u\times k}$ and u satisfies $n + u = m$. When $m < n$, indicating that the order of the response chaotic complex system is higher than that of the drive system, HFPS can only be achieved in an increased-order manner. In this case, auxiliary complex state vectors, which are functions of the complex state z, are incorporated into the drive chaotic complex system. By setting $q = n - m$ and defining $z_{m+1} = \phi_1(z), z_{m+2} = \phi_2(z), \cdots, z_{m+q} = z_n = \phi_q(z)$, a new n-dimensional complex state vector $\mathbf{z}_e = (z_1, z_2, \cdots, z_m, z_{m+1}, \cdots, z_n)$ is obtained. Consequently, the new drive chaotic complex system can be expressed as

$$\dot{\mathbf{z}_e} = \mathbf{F}_e(\mathbf{z})\mathbf{A} + \mathbf{f}_e(\mathbf{z}_e), \tag{5.2.5}$$

where $\mathbf{z}_e = [z_1, z_2, \cdots, z_n]^T$, $\mathbf{f}_e : \mathbb{C}^m \to \mathbb{C}^n$, $F_e : \mathbb{C}^m \to \mathbb{C}^{n\times k}$ and $\mathbf{A} = (a_1, a_2, \cdots, a_\kappa)^T$ is a real (or complex) uncertain parameter vector. The complex vector error is defined as $\mathbf{e}(t) = \mathbf{w}(t) - \mathbf{D}(t)\mathbf{z}_v(t)$, where $v = p$ when $m > n$, and $v = e$ when $m < n$. Here, $\mathbf{D}(t)$ is an nth-order diagonal function matrix, expressed as $\mathbf{D}(t) = diag(d_1(t), d_2(t), \cdots, d_n(t))$. This matrix $\mathbf{D}(t)$ is referred to as the scaling function matrix, and its elements $d_1(t), d_2(t), \cdots, d_n(t)$ are known as the scaling function factors, all of which are bounded. The fundamental objective of achieving HFPS is to design a controller $\mathbf{U}$ that enables the response chaotic complex system

(5.2.2) to asymptotically synchronize with either the projection system (5.2.3) or the extended system (5.2.5) of the drive chaotic complex system (5.2.1), in the sense of HFPS and under nonidentical initial conditions. Specifically, the complex error vector $\mathbf{e} = \mathbf{e}^r + j\mathbf{e}^i = \mathbf{w} - \mathbf{D}(t)\mathbf{z}_v$ must satisfy

$$\lim_{t\to\infty} \|\mathbf{e}^r(t)\|^2 = \lim_{t\to\infty} \|\mathbf{w}^r(t, \mathbf{w_0}^r) - \mathbf{D}(t)\mathbf{z}_v^r(t, \mathbf{z_0}^r)\|^2 = 0,$$

and

$$\lim_{t\to\infty} \|\mathbf{e}^i(t)\|^2 = \lim_{t\to\infty} \|\mathbf{w}^i(t, \mathbf{w_0}^i) - \mathbf{D}(t)\mathbf{z}_v^i(t, \mathbf{z_0}^i)\|^2 = 0,$$

then

$$\lim_{t\to\infty} \|\mathbf{e}(t)\| = \lim_{t\to\infty} (\|\mathbf{e}^r(t)\|^2 + \|\mathbf{e}^i(t)\|^2)^{\frac{1}{2}} = 0,$$

here, $\mathbf{z_0}^r$ and $\mathbf{z_0}^i$ represent the initial values of the complex state variables for system (5.2.3) or (5.2.5), while $\mathbf{w_0}^r$ and $\mathbf{w_0}^i$ denote the corresponding initial values for system (5.2.2).

5.2.2 *Adaptive Reduced Order HFPS of General Chaotic Complex Systems with Different Orders*

I. Controller Design of Adaptive Reduced Order HFPS

Based on adaptive control and parameter identification techniques, the main results concerning HFPS are presented in this section.

Theorem 5.2 *If the adaptive nonlinear controller is designed as*

$$\begin{aligned}
\mathbf{U} = \mathbf{U}^r + j\mathbf{U}^i &= \mathbf{D}(t)\mathbf{F}_v(\mathbf{z})\hat{\mathbf{A}} + \mathbf{D}(t)\mathbf{f}_v(\mathbf{z}) + \dot{\mathbf{D}}(t)\mathbf{z}_v - \mathbf{G}(\mathbf{w})\hat{\mathbf{B}} - \mathbf{g}(\mathbf{w}) - \mathbf{K}\mathbf{e} \\
&= \mathbf{D}(t)\mathbf{F}_v^r(z_1^r, z_2^r, \cdots, z_n^r)\hat{\mathbf{A}} + \mathbf{D}(t)\mathbf{f}_v^r(z_1^r, z_1^i, \cdots, z_n^r, z_n^i) + \dot{\mathbf{D}}(t)\mathbf{z}_v^r \\
&\quad -\mathbf{G}^r(w_1^r, w_2^r, \cdots, w_n^r)\hat{\mathbf{B}} - \mathbf{g}^r(w_1^r, w_1^i, \cdots, w_n^r, w_n^i) - \mathbf{K}\mathbf{e}^r + j[\mathbf{D}(t)\mathbf{F}_v^i(z_1^i, z_2^i, \cdots, z_n^i)\hat{\mathbf{A}} \\
&\quad + \mathbf{D}(t)\mathbf{f}_v^i(z_1^r, z_1^i, \cdots, z_n^r, z_n^i) + \dot{\mathbf{D}}(t)\mathbf{z}_v^i - \mathbf{G}^i(w_1^i, w_2^i, \cdots, w_n^i)\hat{\mathbf{B}} \\
&\quad -\mathbf{g}^i(w_1^r, w_1^i, \cdots, w_n^r, w_n^i) - \mathbf{K}\mathbf{e}^i],
\end{aligned} \tag{5.2.6}$$

adaptive laws of parameters are taken as

$$\begin{cases}
\dot{\hat{\mathbf{A}}} = [-(\mathbf{F}_v^r(z_1^r, z_2^r, \cdots, z_n^r))^T\mathbf{D}(t), -(\mathbf{F}_v^i(z_1^i, z_2^i, \cdots, z_n^i))^T\mathbf{D}(t)]\begin{bmatrix}\mathbf{e}^r \\ \mathbf{e}^i\end{bmatrix}, \\
\dot{\hat{\mathbf{B}}} = [(\mathbf{G}^r(w_1^r, w_2^r, \cdots, w_n^r))^T, (\mathbf{G}^i(w_1^i, w_2^i, \cdots, w_n^i))^T]\begin{bmatrix}\mathbf{e}^r \\ \mathbf{e}^i\end{bmatrix}
\end{cases} \tag{5.2.7}$$

and updated laws of control coefficients are selected as

$$\dot{k}_\mu = \xi_\mu[(e_\mu^r)^2 + (e_\mu^i)^2](\mu = 1, 2, \cdots, n), \tag{5.2.8}$$

then **(i)** *when* $m > n$*, then* $v = p$*, namely, the response chaotic complex system* (5.2.2) *synchronizes the projection system* (5.2.3) *of drive chaotic complex system* (5.2.1) *asymptotically in the sense of HFPS;* **(ii)** *when* $m < n$*, then* $v = e$*, namely, the response chaotic complex system* (5.2.2) *synchronizes the extended drive chaotic complex system* (5.2.5) *asymptotically in the sense of HFPS; where* $K = diag\{k_1, k_2, \cdots, k_n\}$ *is a positive definitional matrix,* $\xi_\mu > 0$ *is constant, and the parameter vectors* $\hat{\mathbf{A}}$ *and* $\hat{\mathbf{B}}$ *are, respectively, estimation of the unknown parameter vectors* $\mathbf{A}$ *and* $\mathbf{B}$.

Proof From the chaotic complex systems described by (5.2.2) and (5.2.3) (or alternatively (5.2.5)), the error dynamical system is derived as follows:

$$\begin{aligned}
\dot{\mathbf{e}} &= \dot{\mathbf{w}} - \dot{\mathbf{D}}(t)\mathbf{z} - \mathbf{D}(t)\dot{\mathbf{z}} = \mathbf{G}(\mathbf{w})\mathbf{B} + \mathbf{g}(\mathbf{w}) - \mathbf{D}(t)\mathbf{F}_v(\mathbf{z})\mathbf{A} - \mathbf{D}(t)f_v(\mathbf{z}) + \mathbf{U} \\
&= \mathbf{G}^r(w_1^r, w_2^r, \cdots, w_n^r)\mathbf{B} + \mathbf{g}^r(w_1^r, w_1^i, \cdots, w_n^r, w_n^i) - \\
&\quad \mathbf{D}(t)\mathbf{F}_v^r(z_1^r, z_2^r, \cdots, z_n^r)\mathbf{A} - \mathbf{D}(t)\mathbf{f}_v^r(u_1, u_2, \cdots, u_n) + \mathbf{U}^r \\
&\quad + j[\mathbf{G}^i(w_1^i, w_2^i, \cdots, w_n^i)\mathbf{B} + \mathbf{g}^i(w_1^r, w_1^i, \cdots, w_n^r, w_n^i) \\
&\quad - \mathbf{D}(t)\mathbf{F}_v^i(z_1^i, z_2^i, \cdots, z_n^i)\mathbf{A} - \mathbf{D}(t)\mathbf{f}_v^i(u_1, u_2, \cdots, u_n) + \mathbf{U}^i].
\end{aligned} \tag{5.2.9}$$

By substituting systems (5.2.2), (5.2.3)(or (5.2.5)), and (5.2.6) into the error system (5.2.9), and subsequently separating the real and imaginary parts of the resulting equations, the error system (5.2.9) can be expressed as

$$\begin{aligned}
\dot{\mathbf{e}}^r &= \mathbf{G}^r(w_1^r, w_2^r, \cdots, w_n^r)(\mathbf{B} - \hat{\mathbf{B}}) - \mathbf{D}(t)\mathbf{F}_v^r(z_1^r, z_2^r, \cdots, z_n^r)(\mathbf{A} - \hat{\mathbf{A}}) - \mathbf{K}\mathbf{e}^r, \\
\dot{\mathbf{e}}^i &= \mathbf{G}^i(w_1^i, w_2^i, \cdots, w_n^i)(\mathbf{B} - \hat{\mathbf{B}}) - \mathbf{D}(t)\mathbf{F}_v^i(z_1^i, z_2^i, \cdots, z_n^i)(\mathbf{A} - \hat{\mathbf{A}}) - \mathbf{K}\mathbf{e}^i.
\end{aligned} \tag{5.2.10}$$

The parameter error vectors are defined as $\tilde{\mathbf{A}} = \hat{\mathbf{A}} - \mathbf{A}$ and $\tilde{\mathbf{B}} = \hat{\mathbf{B}} - \mathbf{B}$. Consider the following candidate Lyapunov function:

$$V(\mathbf{e}^r, \mathbf{e}^i, \tilde{\mathbf{A}}, \tilde{\mathbf{B}}) = \frac{1}{2}[(\mathbf{e}^r)^T\mathbf{e}^r + (\mathbf{e}^i)^T\mathbf{e}^i + \tilde{\mathbf{A}}^T\tilde{\mathbf{A}} + \tilde{\mathbf{B}}^T\tilde{\mathbf{B}} + \sum_{\mu=1}^{n}\frac{1}{\xi_\mu}(k_\mu - k)^2] > 0, \ (k > 0),$$

the time derivative of V along the trajectories of the error system (5.2.9) is given by

$$\begin{aligned}
\dot{V}(t) &= (\dot{\mathbf{e}}^r)^T\mathbf{e}^r + (\dot{\mathbf{e}}^i)^T\mathbf{e}^i + \tilde{\mathbf{A}}^T\dot{\hat{\mathbf{A}}} + \tilde{\mathbf{B}}^T\dot{\hat{\mathbf{B}}} + \sum_{\mu=1}^{n}\frac{1}{\xi_\mu}(k_\mu - k)\dot{k}_\mu \\
&= [\mathbf{G}^r(w_1^r, w_2^r, \cdots, w_n^r)(\mathbf{B} - \hat{\mathbf{B}}) - \mathbf{D}(t)\mathbf{F}_v^r(z_1^r, z_2^r, \cdots, z_n^r)(\mathbf{A} - \hat{\mathbf{A}}) - \mathbf{K}\mathbf{e}^r]^T\mathbf{e}^r \\
&\quad + [\mathbf{G}^i(w_1^i, w_2^i, \cdots, w_n^i)(\mathbf{B} - \hat{\mathbf{B}}) - \mathbf{D}(t)\mathbf{F}_v^i(z_1^i, z_2^i, \cdots, z_n^i)(\mathbf{A} - \hat{\mathbf{A}}) - \mathbf{K}\mathbf{e}^i]^T\mathbf{e}^i \\
&\quad - \tilde{\mathbf{A}}^T[(\mathbf{F}_v^r(z_1^r, z_2^r, \cdots, z_n^r))^T\mathbf{D}(t)\mathbf{e}^r + (\mathbf{F}_v^i(z_1^i, z_2^i, \cdots, z_n^i))^T\mathbf{D}(t)\mathbf{e}^i] \\
&\quad + \tilde{\mathbf{B}}^T[(\mathbf{G}^r(w_1^r, w_2^r, \cdots, w_n^r))^T\mathbf{e}^r + (\mathbf{G}^i(w_1^i, w_2^i, \cdots, w_n^i))^T]\mathbf{e}^i \\
&\quad + \sum_{\mu=1}^{n}(k_\mu - k)[(e_\mu^r)^2 + (e_\mu^i)^2] \\
&= -k[(\mathbf{e}^r)^T\mathbf{e}^r + (\mathbf{e}^i)^T\mathbf{e}^i] \le 0.
\end{aligned}$$

Since $V(t)$ is positive definite, and $\dot{V}(t)$ is negative semi-definite, we cannot immediately conclude that the origin of error system (5.2.9) is asymptotically stable. Specifically, given that $V(t) > 0$ and $\dot{V}(t) \le 0$, it follows that $V(t)$ is bounded. Consequently, its arguments-namely e_μ^r, e_μ^i, $\tilde{a}_\mu$ and $\tilde{b}_\mu$ for $\mu = 1, 2, \cdots, n$- must be bounded. This can be compactly expressed as $e^r_\mu \in L_\infty$, $e^i_\mu \in L_\infty$, $\tilde{a}_\mu \in L_\infty$ and $\tilde{b}_\mu \in L_\infty$ for $\mu = 1, 2, \cdots, n$. It should be noted that

$$\int_0^t \dot{V}(\tau)d_\tau = V(t) - V(0).$$

Since V is non-increasing and positive definite, $V(0) - V(t) \leq V(0)$, i.e.

$$-\int_0^t \dot{V}(\tau)d_\tau \leq V(0).$$

And $\dot{V}(t) = -k[(\mathbf{e}^r)^T\mathbf{e}^r + (\mathbf{e}^i)^T\mathbf{e}^i](k > 0)$, then

$$k\int_0^t [(\mathbf{e}^r(\tau))^T\mathbf{e}^r(\tau) + (\mathbf{e}^i(\tau))^T\mathbf{e}^i(\tau)]d_\tau \leq V(0),$$

which is equivalent to

$$\int_0^t (\|\mathbf{e}^r(\tau)\|^2 + \|\mathbf{e}^i(\tau)\|^2)d_\tau \leq V(0)/k < \infty.$$

Hence, we have $\int_0^t \|\mathbf{e}^r(\tau)\|^2 d_\tau < \infty$ and $\int_0^t \|\mathbf{e}^i(\tau)\|^2 d_\tau < \infty$, which simplify to $\sqrt{\int_0^t \|\mathbf{e}^r(\tau)\|^2 d_\tau} < \infty$ and $\sqrt{\int_0^t \|\mathbf{e}^i(\tau)\|^2 d_\tau} < \infty$, which are just $\|\mathbf{e}^r(t)\|_{L_2} < \infty$ and $\|\mathbf{e}^i(t)\|_{L_2} < \infty$, namely, $e^r_\mu \in L_2$ and $e^i_\mu \in L_2(\mu = 1, 2, \cdots, n)$. From the error dynamics (10) with bounded scaling function factor $d_\mu(t)$ and e^r_μ, e^i_μ, $\tilde{a}_\mu$, $\tilde{b}_\mu \in L_\infty(\mu = 1, 2, \cdots, n)$, we get $\dot{e}^r_\mu \in L_2$ and $\dot{e}^i_\mu \in L_2$. According to the Barbalat's lemma, we have $e^r_\mu \to 0$ and $e^i_\mu \to 0$ as $t \to \infty$, that is

$$\begin{aligned}
\lim_{t\to\infty} \|\mathbf{e}^r(t)\|^2 + \|\mathbf{e}^i(t)\|^2 &= \lim_{t\to\infty}\left[\left(\sum_{\mu=1}^{n} \|e^r_\mu(t,\, e_\mu(0)\right)\|)^2 + \left(\sum_{\mu=1}^{n} \|e^i_\mu(t,\, e_\mu(0)\right)\|)^2\right] \\
&= \lim_{t\to\infty}\left[\left(\sum_{\mu=1}^{n} \|w^r_\mu(t,\, w^r_\mu(0)\right) - d_\mu z^r_\mu(t,\, z^r_\mu(0))\|)^2\right. \\
&\quad \left. + \left(\sum_{\mu=1}^{n} \|w^i_\mu(t,\, w^i_\mu(0)\right) - d_\mu z^i_\mu(t,\, z^i_\mu(0))\|)^2\right] = 0,
\end{aligned}$$

then, we obtain

$$\lim_{t\to\infty} \|\mathbf{e}(t)\| = \lim_{t\to\infty} \sqrt{\|\mathbf{e}^r(t)\|^2 + \|\mathbf{e}^i(t)\|^2} = 0.$$

Therefore, with the designed controller (5.2.6), the parameter estimation update laws (5.2.7), and the control coefficient update laws (5.2.8), the response complex system (5.2.2) asymptotically synchronizes with the projection drive system (5.2.3) (or the extended drive system (5.2.5)). This completes the proof. □

Remark 5.7 From Theorem 5.2, it is clear that the functions $\mathbf{F}_v^r(\cdot)$ and $\mathbf{F}_v^i(\cdot)$ denote as functions of the variables $z_1^r,\ z_2^r,\ \cdots,\ z_n^r$ and $z_1^i,\ z_2^i,\ \cdots,\ z_n^i$, respectively. The functions $\mathbf{f}^r(\cdot)$ and $\mathbf{f}^i(\cdot)$ are denoted with the variables $z_1^r,\ z_1^i,\ \cdots,\ z_n^r,\ z_n^i$, which are determined by the special chaotic complex systems, that is, the real and imaginary parts of functions $\mathbf{f}^r(\cdot)$ and $\mathbf{f}^i(\cdot)$ aren't separated completely. The same cases exist in the functions $\mathbf{G}^r(\cdot)$, $\mathbf{G}^i(\cdot)$ and $\mathbf{g}^r(\cdot)$, $\mathbf{g}^i(\cdot)$.

Remark 5.8 If $\mathbf{A}$ is a complex parameter vector, it can be expressed as $\mathbf{A} = \mathbf{A}^r + j\mathbf{A}^i$. In this case, the complex system (5.2.1) becomes:

$$\dot{\mathbf{z}} = \mathbf{F}(\mathbf{z})(\mathbf{A}^r + j\mathbf{A}^i) + \mathbf{f}(\mathbf{z}). \tag{5.2.11}$$

This expression can also be reformulated as follows:

$$\dot{\mathbf{z}}_1 = \mathbf{F}(\mathbf{z})\mathbf{A}^r + \mathbf{f}_1(\mathbf{z}) \tag{5.2.12}$$

and

$$\dot{\mathbf{z}}_2 = \mathbf{F}(\mathbf{z})\mathbf{A}^i - j\mathbf{f}_2(\mathbf{z}), \tag{5.2.13}$$

where $\mathbf{z}_1 + j\mathbf{z}_2 = \mathbf{z}$ and $\mathbf{f}_1(\mathbf{z}) + \mathbf{f}_2(\mathbf{z}) = \mathbf{f}(\mathbf{z})$. Similarly, the chaotic complex system (5.2.3) (or (5.2.5)) can be expressed as

$$\dot{\mathbf{z}}_v = \mathbf{F}_v(\mathbf{z})(\mathbf{A}^r + j\mathbf{A}^i) + \mathbf{f}_v(\mathbf{z}), \tag{5.2.14}$$

where $v = p$ or $v = e$, system (5.2.14) can be reformulated as follows:

$$\dot{\mathbf{z}}_{v1} = \mathbf{F}_v(\mathbf{z})\mathbf{A}^r + \mathbf{f}_{v1}(\mathbf{z}) \tag{5.2.15}$$

and

$$\dot{\mathbf{z}}_{v2} = \mathbf{F}_v(\mathbf{z})\mathbf{A}^i - j\mathbf{f}_{v2}\mathbf{z}), \tag{5.2.16}$$

where $\mathbf{z}_v = \mathbf{z}_{v1} + j\mathbf{z}_{v2}$ and $\mathbf{f}_{v1}(\mathbf{z}) + \mathbf{f}_{v2}(\mathbf{z}) = \mathbf{f}_v(\mathbf{z})$. Similarly, if $\mathbf{B}$ is a complex vector, the chaotic complex system (5.2.2) becomes

$$\dot{\mathbf{w}} = \mathbf{G}(\mathbf{w})(\mathbf{B}^r + j\mathbf{B}^i) + \mathbf{g}(\mathbf{w}) + \mathbf{U}', \tag{5.2.17}$$

then

$$\dot{\mathbf{w}}_1 = \mathbf{G}(\mathbf{w})\mathbf{B}^r + \mathbf{g}_1(\mathbf{w}) + \mathbf{U}'' \tag{5.2.18}$$

and

$$\dot{\mathbf{w}}_2 = \mathbf{G}(\mathbf{w})\mathbf{B}^i - j\mathbf{g}_2(\mathbf{w}) + \mathbf{U}''', \tag{5.2.19}$$

where $\mathbf{w} = \mathbf{w}_1 + j\mathbf{w}_2$ and $\mathbf{g}_1(\mathbf{z}) + \mathbf{g}_2(\mathbf{z}) = \mathbf{g}(\mathbf{z})$. The structures of the complex systems (5.2.14), (5.2.15), (5.2.17), and (5.2.18) are analogous to those of the complex systems (5.2.3) and (5.2.2), respectively. Hence, the design of the controller to achieve HFPS between systems (5.2.15) and (5.2.18) (or between (5.2.16) and (5.2.19)) follows a similar approach to that used for systems (5.2.3) and (5.2.2). If systems (5.2.15) and (5.2.18), as well as systems (5.2.16) and (5.2.19), achieve HFPS individually, then the switching system (5.2.14) and system (5.2.17) also accomplish HFPS. By designing the controller $\mathbf{U}''$ and the parameter estimates $\hat{\mathbf{A}}^r$, $\hat{\mathbf{B}}^r$ in accordance with Theorem 5.2, HFPS between the systems (5.2.15) and (5.2.18) is realized. Similarly, the controller $\mathbf{U}'''$ and the estimates $\hat{\mathbf{A}}^i$, $\hat{\mathbf{B}}^i$ can be designed following the same methodology. The overall controller $\mathbf{U}' = \mathbf{U}'' + j\mathbf{U}'''$ and the parameter estimates $\hat{\mathbf{A}} = \hat{\mathbf{A}}^r + j\hat{\mathbf{A}}^i$, $\hat{\mathbf{B}} = \hat{\mathbf{B}}^r + j\hat{\mathbf{B}}^i$ are thereby obtained.

According to Remark 5.8, the design procedure for the controller remains similar regardless of whether the parameter vectors **A** and **B** are complex or real. Therefore, for simplicity, the subsequent sections will use chaotic complex systems with real parameter vectors to validate the performance of the designed controller.

II. Application of the Adaptive Reduced Order Controller

Assuming the hyper-chaotic complex Lorenz system [120] acts as the drive system for the chaotic complex Chen system, we note that the drive system is 4-dimensional which is higher than the 3-dimensional response system. Therefore, the case of $v = p$ in Theorem 5.2 must be considered.

(1) Taking the projection $z_1 - z_2 - z_3$ of the hyper-chaotic complex Lorenz system

The hyper-chaotic complex Lorenz system is described by

$$\begin{cases} \dot{z}_1 = a_1(z_2 - z_1) + z_4(1+i), \\ \dot{z}_2 = a_3 z_1 - z_2 - z_1 z_3, \\ \dot{z}_3 = \frac{1}{2}(\bar{z}_1 z_2 + z_1 \bar{z}_2) - a_2 z_3, \\ \dot{z}_4 = \frac{1}{2}(\bar{z}_1 z_2 + z_1 \bar{z}_2) - a_4 z_4, \end{cases} \tag{5.2.20}$$

where a_1, a_2, and a_3 are positive real parameters that influence the system's dynamics, $z_1 = z_1^r + jz_1^i$ and $z_2 = z_2^r + jz_2^i$ are complex variables, and z_3 and z_4 are real variables, with a_4 serving as a control parameter. When $a_1 = 14$, $a_2 = 5$, $a_3 = 45$ and $a_4 = 5.5$, the hyper-chaotic complex Lorenz system exhibits two positive Lyapunov exponents, indicating the presence of two chaotic attractors and hyper-chaotic behavior (see Ref.[120]).

The chaotic complex Chen system is introduced by

$$\begin{cases} \dot{w}_1 = b_1(w_2 - w_1), \\ \dot{w}_2 = b_3(w_1 + w_2) - b_1 w_1 - w_1 w_3, \\ \dot{w}_3 = \frac{1}{2}(\bar{w}_1 w_2 + w_1 \bar{w}_2) - b_2 w_3, \end{cases} \tag{5.2.21}$$

where b_1, b_2 and b_3 are positive real parameters that govern the dynamics of system(5.2.21), $w_1 = w_1^r + jw_1^i$ and $w_2 = w_2^r + jw_2^i$ are complex variables, and w_3 is a real variable. When $b_1 = 42$, $b_2 = 4$ and $b_3 = 26$, the chaotic complex Chen system possesses a single positive Lyapunov exponent, indicating the presence of one chaotic attractor.

To investigate the reduced-order HFPS behavior between the hyperchaotic complex Lorenz system and the chaotic complex Chen system, we take the $z_1 - z_2 - z_3$ projection of the hyperchaotic complex Lorenz system as the drive system. Thus, the drive system corresponds to the projected subsystem of the hyperchaotic complex Lorenz system (5.2.20), and can be expressed in the following form:

$$\begin{pmatrix} \dot{z}_1 \\ \dot{z}_2 \\ \dot{z}_3 \end{pmatrix} = \begin{pmatrix} z_2 - z_1 & 0 & 0 \\ 0 & 0 & z_1 \\ 0 & -z_3 & 0 \end{pmatrix} \begin{pmatrix} a_1 \\ a_2 \\ a_3 \end{pmatrix} + \begin{pmatrix} z_4(1+i) \\ -z_2 - z_1 z_3 \\ \frac{1}{2}(\bar{z}_1 z_2 + z_1 \bar{z}_2) \end{pmatrix}. \tag{5.2.22}$$

We also assume that the complex Chen system (5.2.21) serves as the response system, which can be expressed in the following form:

$$\begin{pmatrix} \dot{w}_1 \\ \dot{w}_2 \\ \dot{w}_3 \end{pmatrix} = \begin{pmatrix} w_2 - w_1 & 0 & 0 \\ -w_1 & 0 & w_1 + w_2 \\ 0 & -w_3 & 0 \end{pmatrix} \begin{pmatrix} b_1 \\ b_2 \\ b_3 \end{pmatrix} + \begin{pmatrix} 0 \\ -w_1 w_3 \\ \frac{1}{2}(\bar{w}_1 w_2 + w_1 \bar{w}_2) \end{pmatrix} \tag{5.2.23}$$

$$+ \begin{pmatrix} u_1 \\ u_2 \\ u_3 \end{pmatrix}, \tag{5.2.24}$$

where u_1, u_2 and u_3 are three control inputs needing to be determined and all parameters a_1, a_2, a_3, b_1, b_2 and b_3 are unknown real parameters.

Then, according to the Theorem 5.2, we design the controller as

$$\begin{cases} u_1 = \dot{d}_1(t)z_1 + d_1(t)\hat{a}_1(z_2 - z_1) - k_1 e_1 + d_1(t) z_4(1+i) - \hat{b}_1(w_2 - w_1), \\ u_2 = \dot{d}_2(t)z_2 + d_2(t)(\hat{a}_3 z_1 - z_2 - z_1 z_3) - \hat{b}_3(w_1 + w_2) + \hat{b}_1 w_1 + w_1 w_3 - k_2 e_2, \\ u_3 = \dot{d}_3(t)z_3 - d_3(t)(\hat{a}_2 z_3 - \frac{1}{2}(\bar{z}_1 z_2 + z_1 \bar{z}_2)) - \frac{1}{2}(\bar{w}_1 w_2 + w_1 \bar{w}_2) + \hat{b}_2 w_3 - k_3 e_3, \end{cases} \tag{5.2.25}$$

the estimates $\hat{a}_1$, $\hat{a}_2$, $\hat{a}_3$ and $\hat{b}_1$, $\hat{b}_2$, $\hat{b}_3$ obey the update rules

$$\begin{cases} \dot{\hat{a}}_1 = -(z_2^r - z_1^r)d_1(t)e_1^r - (z_2^i - z_1^i)d_1(t)e_1^i, \\ \dot{\hat{a}}_2 = d_3(t)z_3 e_3, \\ \dot{\hat{a}}_3 = -z_1^r d_2(t)e_2^r - z_1^i d_2(t)e_2^i, \end{cases} \tag{5.2.26}$$

and

$$\begin{cases} \dot{\hat{b}}_1 = (w_2^r - w_1^r)e_1^r - w_1^r e_2^r + (w_2^i - w_1^i)e_1^i - w_1^i e_2^i, \\ \dot{\hat{b}}_2 = -w_3 e_3, \\ \dot{\hat{b}}_3 = (w_1^r + w_2^r)e_2^r + (w_1^i + w_2^i)e_2^i, \end{cases} \tag{5.2.27}$$

and the updated laws of control coefficients are

$$\dot{k}_\mu = (e_\mu^r)^2 + (e_\mu^i)^2,\ \mu = 1,\ 2,\ 3.$$

To verify and demonstrate the effectiveness of the proposed reduced-order scheme, we present simulation results for the adaptive HFPS between the $z_1 - z_2 - z_3$ projection of the hyper-chaotic complex Lorenz system and the chaotic complex Chen system. The fourth-order Runge–Kutta method is employed for numerical integration. The initial conditions are chosen as$(z_1(0), z_2(0), z_3(0), z_4(0))^T = (1 - 0.5j, -2 + j, 3, 2)^T$ and $(w_1(0), w_2(0), w_3(0))^T = (-3 + j, 2 + 0.5j, 1)^T$, with the time-varying scaling matrix defined as $\mathbf{D}(t) = diag\{1 - 0.2\sin(\frac{\pi}{10}t), -1 + 0.3\sin(\frac{\pi}{10}t), 2 - 0.5\sin(\frac{\pi}{10}t)\}$. Consequently, the error system starts from $e_1(0) = -5 + 2j,\ e_2(0) = 6 - 1.5j,\ e_3(0) = -5$. Additionally, the initial values of the adaptive gains are set as $k_\mu(\mu = 1,\ 2,\ 3)$ are $k_1(0) = 4,\ k_2(0) = 4$ and $k_3(0) = 2$.. The "unknown" parameters are set to $a_1 = 14,\ a_2 = 5,\ a_3 = 45,\ a_4 = 5.5$ and $b_1 = 42,\ b_2 = 4,\ b_3 = 26$. The adaptive HFPS between systems (5.2.22) and (5.2.23), achieved through the controller (5.2.25) together with the parameter update laws (5.2.26) and (5.2.27), is illustrated in Figs. 5.9 and 5.10, using the initial parameter estimates $\hat{a}_1(0) = 5,\ \hat{a}_2(0) = 5,\ \hat{a}_3(0) = 5,\ \hat{a}_4(0) = 5$ and $\hat{b}_1(0) = 15,\ \hat{b}_2(0) = 15,\ \hat{b}_3(0) = 15$. As shown in Fig. 5.9, the synchronization errors converge rapidly to zero under the action of the controller (5.2.25) and the adaptation laws (5.2.26) and (5.2.27), demonstrating successful HFPS. Figure 5.10 shows that the parameter estimates $\hat{a}_1,\ \hat{a}_2,\ \hat{a}_3,\ \hat{a}_4$ and $\hat{b}_1,\ \hat{b}_2,\ \hat{b}_3$ converge asymptotically to the true values $a_1 = 14,\ a_2 = 5,\ a_3 = 45,\ a_4 = 5.5$ and $b_1 = 42,\ b_2 = 4,\ b_3 = 26$ as $t \to \infty$, respectively. Furthermore, Fig. 5.10 illustrates that the adaptive control gains k_1, k_2, and k_3 each converge to a constant value as $t \to \infty$. These results confirm that both reduced-order HFPS and accurate parameter identification have been successfully achieved using the controller (5.2.25) together with the parameter update laws (5.2.26) and (5.2.27).

(2) Taking the projection $z_1 - z_2 - z_4$ of the hyper-chaotic complex Lorenz system

The $z_1 - z_2 - z_4$ projection of the hyper-chaotic complex Lorenz system is taken as the drive system; that is, the drive system corresponds to the projected subsystem of the hyper-chaotic complex Lorenz system (5.2.20) and is described as

$$\begin{pmatrix} \dot{z}_1 \\ \dot{z}_2 \\ \dot{z}_4 \end{pmatrix} = \begin{pmatrix} z_2 - z_1 & 0 & 0 \\ 0 & z_1 & 0 \\ 0 & 0 & -z_4 \end{pmatrix} \begin{pmatrix} a_1 \\ a_3 \\ a_4 \end{pmatrix} + \begin{pmatrix} z_4(1+i) \\ -z_2 - z_1 z_3 \\ \frac{1}{2}(\bar{z}_1 z_2 + z_1 \bar{z}_2) \end{pmatrix}, \tag{5.2.28}$$

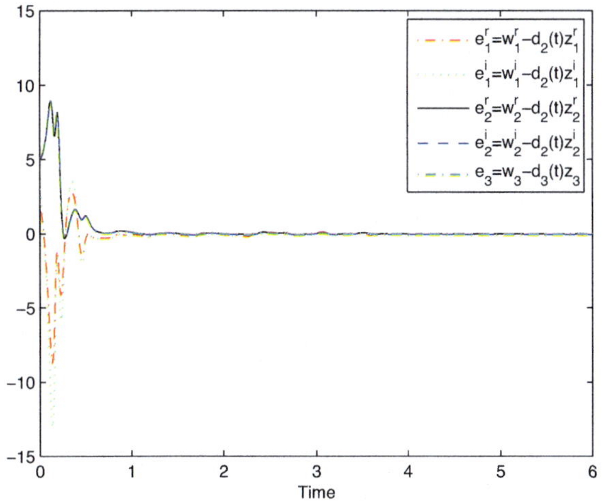

Fig. 5.9 The graph of the errors between the projection $z_1 - z_2 - z_3$ of the hyper-chaotic complex Lorenz system and chaotic complex Chen system

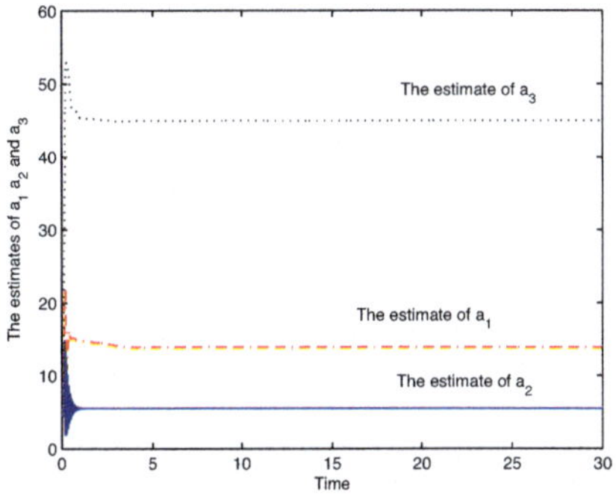

(a) Parameters $\hat{a}_1$, $\hat{a}_2$ and $\hat{a}_3$.

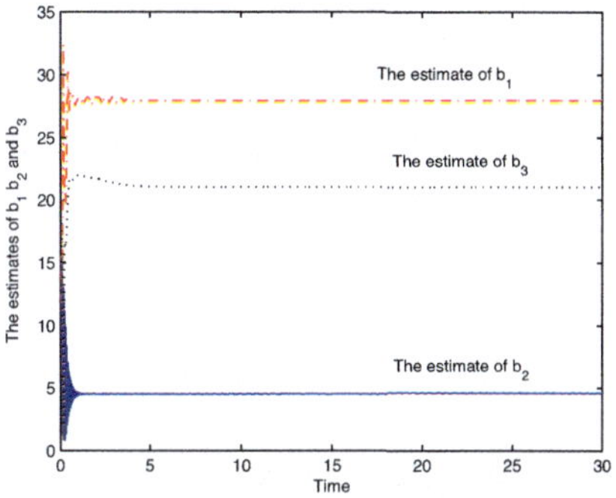

(b) Parameters $\hat{b}_1$, $\hat{b}_2$ and $\hat{b}_3$.

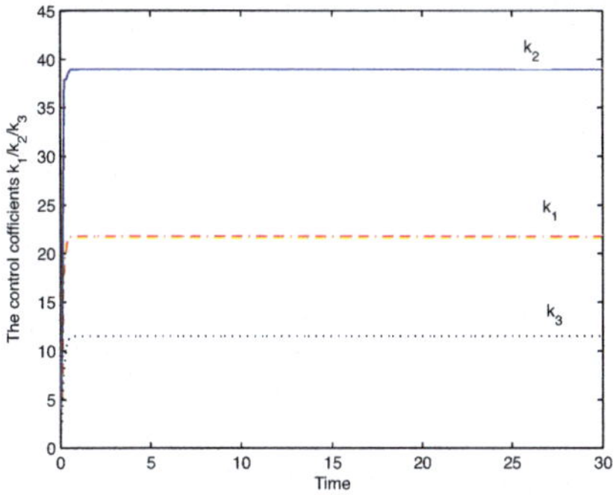

(c) Control coefficients k_1, k_2, k_3.

Fig. 5.10 Time evolution of parameters and control coefficients

and the response complex Chen system (5.2.21) can be described in the same way as the form of system (5.2.23). Similarly, according to the Theorem 5.2, we design the controller as

$$\begin{cases} u_1 = \dot{d}_1(t)z_1 + d_1(t)\hat{a}_1(z_2 - z_1) - k_1e_1 + d_1(t)z_4(1+i) - \hat{b}_1(w_2 - w_1), \\ u_2 = \dot{d}_2(t)z_2 + d_2(t)(\hat{a}_3z_1 - z_2 - z_1z_3) - k_2e_2 - \hat{b}_3(w_1 + w_2) + \hat{b}_1w_1 + w_1w_3, \\ u_4 = \dot{d}_3(t)z_4 - d_3(t)[\hat{a}_4z_4 - \frac{1}{2}(\bar{z}_1z_2 + z_1\bar{z}_2)] - \frac{1}{2}(\bar{w}_1w_2 + w_1\bar{w}_2) + \hat{b}_2w_3 - k_3e_3, \end{cases} \tag{5.2.29}$$

the estimates $\hat{a}_1$, $\hat{a}_2$, $\hat{a}_3$ and $\hat{b}_1$, $\hat{b}_2$, $\hat{b}_3$ obey the update rules

$$\begin{cases} \dot{\hat{a}}_1 = -(z_2^r - z_1^r)d_1(t)e_1^r - (z_2^i - z_1^i)d_1(t)e_1^i, \\ \dot{\hat{a}}_3 = -z_1^r d_2(t)e_2^r - z_1^i d_2(t)e_2^i, \\ \dot{\hat{a}}_4 = d_3(t)z_4e_4, \end{cases} \tag{5.2.30}$$

and

$$\begin{cases} \dot{\hat{b}}_1 = (w_2^r - w_1^r)e_1^r - w_1^r e_2^r + (w_2^i - w_1^i)e_1^i - w_1^i e_2^i, \\ \dot{\hat{b}}_2 = -w_3e_3, \\ \dot{\hat{b}}_3 = (w_1^r + w_2^r)e_2^r + (w_1^i + w_2^i)e_2^i, \end{cases} \tag{5.2.31}$$

and the updated laws of control coefficients are

$$\dot{k}_\mu = (e_\mu^r)^2 + (e_\mu^i)^2, \ \mu = 1, 2, 3.$$

We present simulation results for the adaptive HFPS between the $z_1 - z_2 - z_4$ projection of the hyper-chaotic complex Lorenz system and the chaotic complex Chen system. The initial condition $(z_1(0), z_2(0), z_3(0), z_4(0))^T$, $(w_1(0), w_2(0), w_3(0))^T$ and $(k_1(0), k_2(0), k_3(0))^T$ are taken as the same vector values as Sect. 5.2.2, and the matrix scaling factor $\mathbf{D}(t) = diag\{1 - 0.2\sin(\frac{\pi}{10}t), -1 + 0.3\sin(\frac{\pi}{10}t), 2 - 0.5\sin(\frac{\pi}{10}t)\}$ is also considered. Hence the error system has also same initial values as Section 5.2.2. The "unknown" parameters are also selected as $a_1 = 14$, $a_2 = 5$, $a_3 = 45$, $a_4 = 5.5$ and $b_1 = 42$, $b_2 = 4$, $b_3 = 26$. The adaptive HFPS of the systems (5.2.28) and (5.2.23) via adaptive controller (5.2.29) and parameter vector update laws (5.2.30) and (5.2.31) with the initial estimated parameters $\hat{a}_1(0) = 5$, $\hat{a}_2(0) = 5$, $\hat{a}_4(0) = 5$ and $\hat{b}_1(0) = 15$, $\hat{b}_2(0) = 15$, $\hat{b}_3(0) = 15$ are shown in Figs. 5.11-5.12. Figure 5.11 displays the HFPS error state response between systems (5.2.28) and (5.2.23). Figure 5.12 shows that the estimates $\hat{a}_1$, $\hat{a}_3$, $\hat{a}_4$ and $\hat{b}_1$, $\hat{b}_2$, $\hat{b}_3$ of the unknown parameters converges to $a_1 = 14$, $a_2 = 5$, $a_3 = 45$, $a_4 = 5.5$ and $b_1 = 42$, $b_2 = 4$, $b_3 = 26$ as $t \to \infty$, respectively. Figure 5.12 shows that the control gains k_1, k_2 and k_3 each converge to a constant value as $t \to \infty$. This confirms that reduced-order HFPS and parameter identification have been successfully achieved using the controller (5.2.29) together with the parameter update laws (5.2.30) and (5.2.31).

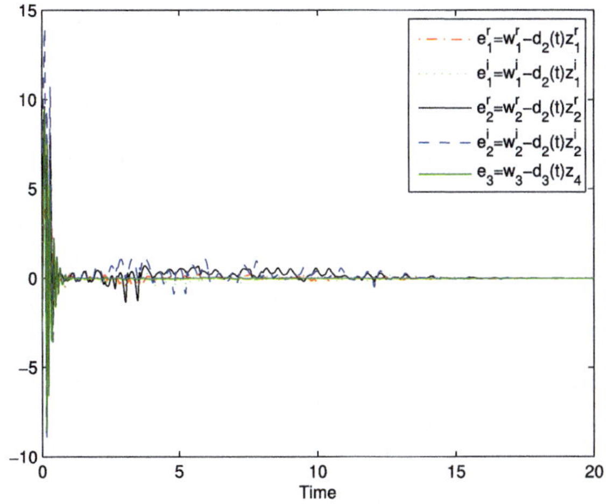

Fig. 5.11 The graph of the errors between the projection $z_1 - z_2 - z_4$ of the hyper-chaotic complex Lorenz system and chaotic complex Chen system

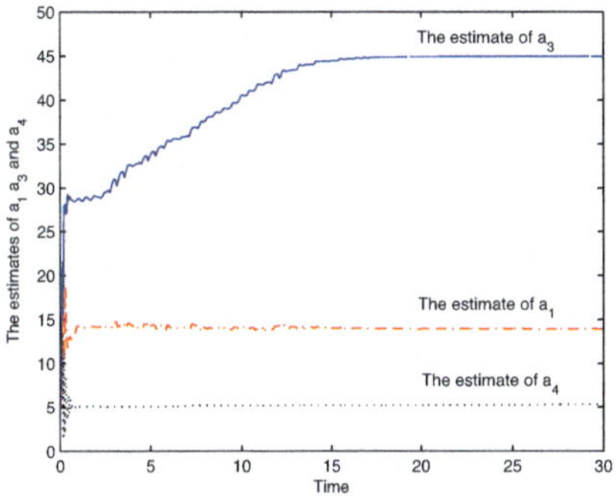

(a) Parameters $\hat{a}_1$, $\hat{a}_3$ and $\hat{a}_4$.

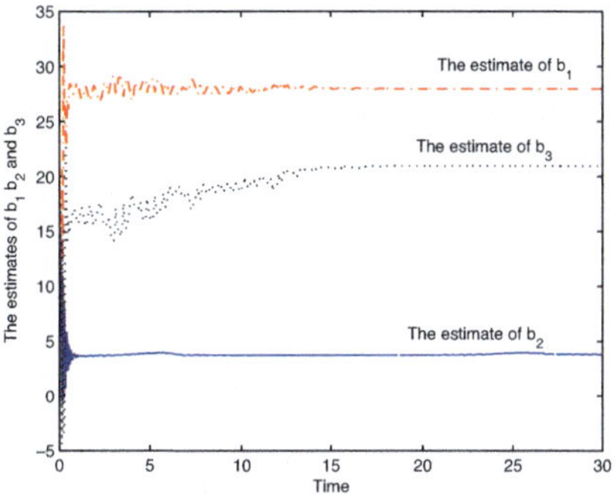

(b) Parameters $\hat{b}_1$, $\hat{b}_2$ and $\hat{b}_3$.

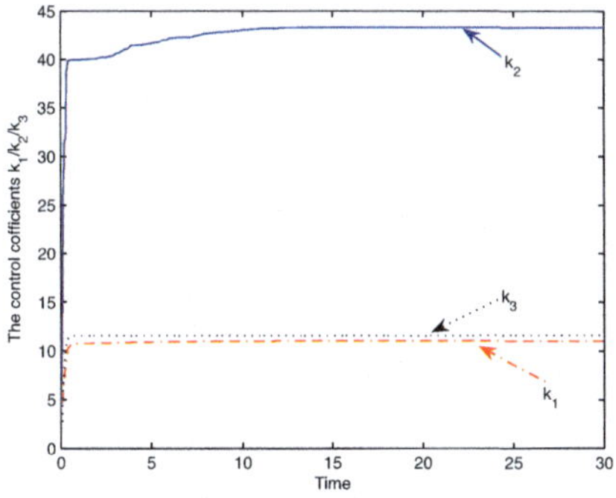

(c) Control coefficients k_1, k_2, k_3.

Fig. 5.12 Time evolution of parameters and control coefficients

5.2.3 *Adaptive Added Order HFPS of General Chaotic Complex Systems with Different Orders*

I. Controller Design of Adaptive Added Order HFPS

Let the complex Lorenz system drive the complex hyperchaotic Lü system. Clearly, the order of the drive system (the complex Lorenz system) is 3, which is less than the order 4 of the response system (the complex hyper-chaotic Lü system). Therefore, we must consider the case $v = e$ in Theorem 5.2.

II. Application of the Adaptive Added Order Controller

(1) Adaptive HFPS of system $z_4 = \phi(z) = 2z_3$

The complex Lorenz system is described as

$$\begin{cases} \dot{z}_1 = a_1(z_2 - z_1), \\ \dot{z}_2 = a_3 z_1 - z_2 - z_1 z_3, \\ \dot{z}_3 = \frac{1}{2}(\bar{z}_1 z_2 + z_1 \bar{z}_2) - a_2 z_3, \end{cases} \tag{5.2.32}$$

where $a_1,\ a_2$ and a_3 are positive real parameters, $z_1 = z_1^r + jz_1^i$ and $z_2 = z_2^r + jz_2^i$ are complex variables, and z_3 is a real variable. When $a_1 = 14,\ a_2 = 3.7$ and $a_3 = 35$, the complex Lorenz system possesses a single positive Lyapunov exponent, indicating the presence of one chaotic attractor.

The response system is given by the hyper-chaotic complex Lü system as

$$\begin{cases} \dot{w}_1 = b_1(w_2 - w_1) + w_4 + u_1, \\ \dot{w}_2 = w_4 + b_3 w_2 - w_1 w_3 + u_2, \\ \dot{w}_3 = \frac{1}{2}(\bar{w}_1 w_2 + w_1 \bar{w}_2) - b_2 w_3 + u_3, \\ \dot{w}_4 = \frac{1}{2}(\bar{w}_1 w_2 + w_1 \bar{w}_2) - b_4 w_4 + u_4, \end{cases} \tag{5.2.33}$$

where $b_1,\ b_2,\ b_3$ and b_4 are positive real parameters, $w_1 = w_1^r + jw_1^i$ and $w_2 = w_2^r + jw_2^i$ are complex variables, and w_3 and w_4 are real variables. When $b_1 = 42,\ b_2 = 6,\ b_3 = 25$ and $b_4 = 5$, the hyper-chaotic complex Lü system exhibits two positive Lyapunov exponents, indicating the presence of two chaotic attractors, hyper-chaotic behavior, and richer dynamical properties (see Ref.[121]).

Taking $z_4 = \phi(z) = 2z_3$, we obtain $\dot{z}_4 = \bar{z}_1 z_2 + z_1 \bar{z}_2 - 2a_2 z_3$. Based on Theorem 5.2, the controller is designed as

$$\begin{cases} u_1 = d_1(t)(z_2 - z_1)\hat{a}_1 + \dot{d}_1(t) z_1 - \hat{b}_1(w_2 - w_1) - w_4 - k_1 e_1, \\ u_2 = d_2(t)(z_1 \hat{a}_3 - z_2 - z_1 z_3) + \dot{d}_2(t) z_2 - w_4 - \hat{b}_3 w_2 + w_1 w_3 - k_2 e_2, \\ u_3 = d_3(t)[\frac{1}{2}(\bar{z}_1 z_2 + z_1 \bar{z}_2) - z_3 \hat{a}_2] + \dot{d}_3(t) z_3 - \frac{1}{2}(\bar{w}_1 w_2 + w_1 \bar{w}_2) + \hat{b}_2 w_3 - k_3 e_3, \\ u_4 = d_4(t)[\bar{z}_1 z_2 + z_1 \bar{z}_2 - 2\hat{a}_2 z_3] + \dot{d}_4(t) z_4 + w_4 \hat{b}_4 - \frac{1}{2}(\bar{w}_1 w_2 + w_1 \bar{w}_2) - k_4 e_4, \end{cases} \tag{5.2.34}$$

the estimates of $a_1,\ a_2,\ a_3$ and $b_1,\ b_2,\ b_3,\ b_4$ are obtained as follows

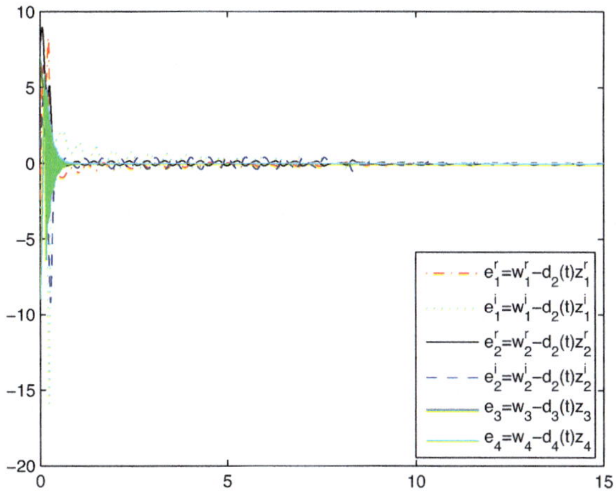

Fig. 5.13 The graph of the errors between the added-order complex Lorenz system and the hyper-chaotic chaotic complex Lü system

$$\begin{cases} \dot{\hat{a}}_1 = -d_1(t)(z_2^r - z_1^r)e_1^r - d_1(t)(z_2^i - z_1^i)e_1^i, \\ \dot{\hat{a}}_2 = d_3(t)z_3e_3 + 2d_4(t)z_3e_4, \\ \dot{\hat{a}}_3 = -d_2(t)z_1^r e_2^r - d_2(t)z_1^i e_2^i, \end{cases} \tag{5.2.35}$$

and

$$\begin{cases} \dot{\hat{b}}_1 = (w_2^r - w_1^r)e_1^r + (w_2^i - w_1^i)e_1^i, \\ \dot{\hat{b}}_2 = -w_3e_3, \\ \dot{\hat{b}}_3 = w_2^r e_2^r + w_2^i e_2^i, \\ \dot{\hat{b}}_4 = -w_4e_4, \end{cases} \tag{5.2.36}$$

and the updated laws of control coefficients are

$$\dot{k}_\mu = 2(e_\mu^r)^2 + (e_\mu^i)^2, \ \mu = 1, 2, 3, 4.$$

In these numerical simulations, the "unknown" parameters of the complex Lorenz system are set to $a_1 = 14$, $a_2 = 3.7$, and $a_3 = 35$, and those of the hyper-chaotic complex Lü system are set to $b_1 = 1$, $b_2 = 26$, $b_3 = 0.7$, and $b_4 = 1.5$ to ensure chaotic behavior. The initial states of the drive and response systems are given by $z_1(0) = 2 + 3j$, $z_2(0) = -5 + 3j$, $z_3(0) = 2$, with $z_4(0) = 4$, and $w_1(0) = 1 + 2j$, $w_2(0) = 3 + 4j$, $w_3(0) = 5$, $w_4(0) = -1$, respectively. The estimation of parameters have initial conditions $\hat{a}_1(0) = 10$, $\hat{a}_2(0) = 10$, $\hat{a}_3(0) = 10$, $\hat{b}_1(0) = 5$, $\hat{b}_2(0) = 5$, $\hat{b}_3(0) = 5$, $\hat{b}_4(0) = 5$ and the function scaling matrix $\mathbf{D}(t)$ is chosen as $\mathbf{D}(t) = diag\{2 - 0.5\sin(\frac{\pi}{10}t),\ 1 + 0.2\sin(\frac{\pi}{10}t),\ -1 - 0.8\sin(\frac{\pi}{10}t),\ -2 - 0.5\sin(\frac{\pi}{10}t)\}$.

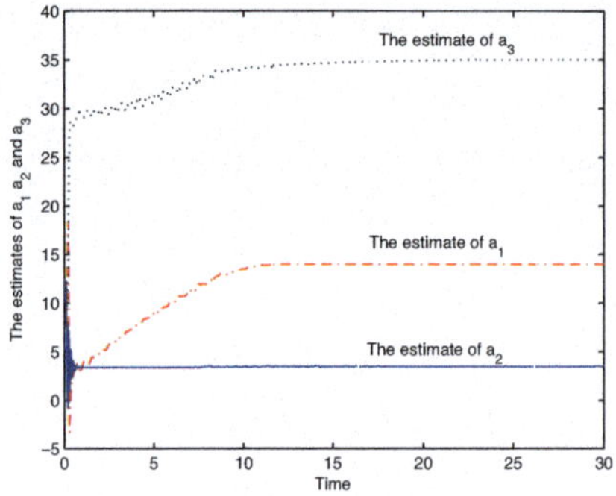

(a) Parameters $\hat{a}_1$, $\hat{a}_2$ and $\hat{a}_3$.

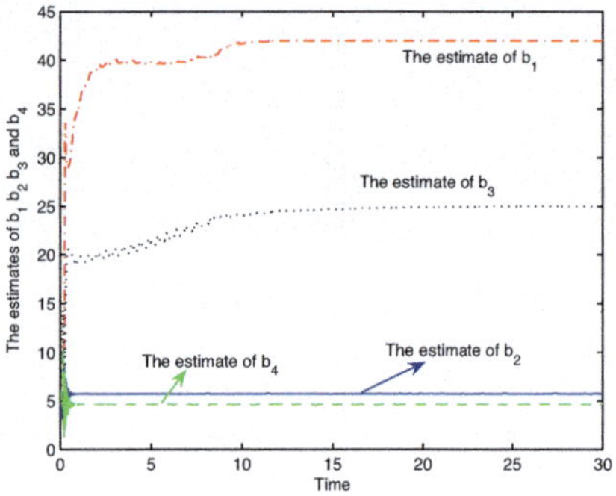

(b) Parameters $\hat{b}_1$, $\hat{b}_2$, $\hat{b}_3$ and $\hat{b}_4$.

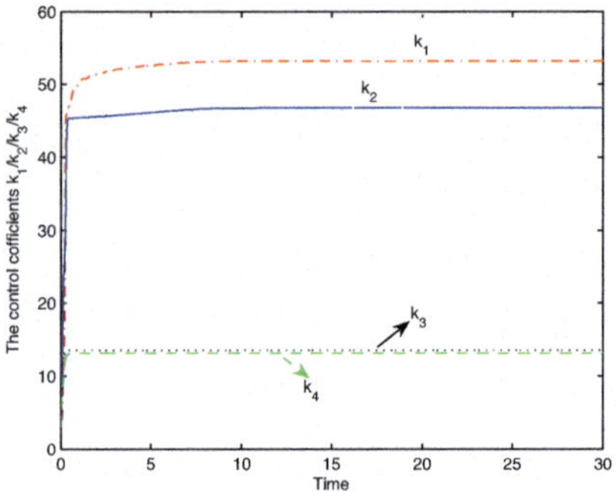

(c) Control coefficients.

Fig. 5.14 Time evolution of parameters and control coefficients

The error system is initialized with the values $e_1(0) = -3 - 4j$, $e_2(0) = 8 + j$, $e_3(0) = 7$, $e_4(0) = -9$. Moreover, the initial values $k_\mu(\mu = 1, 2, 3, 4)$ are $k_1(0) = 3$, $k_2(0) = 3$, $k_3(0) = 3$ and $k_4(0) = 3$. The simulation results are presented in Figs. 5.13-5.14. Figure 5.13 displays the error dynamics between the extended complex Lorenz system and the hyper-chaotic complex Lü system. As the errors converge to zero (as observed in Fig. 5.13), the added-order HFPS between the two uncertain chaotic complex systems is successfully achieved. Furthermore, Fig. 5.14 illustrates how the estimated parameters $\hat{a}_1$, $\hat{a}_2$, $\hat{a}_3$, $\hat{b}_1$, $\hat{b}_2$, $\hat{b}_3$ and $\hat{b}_4$ converge to constant values as time $t \to \infty$. The time evolution of the adaptive control coefficients $k_\mu(\mu = 1, 2, 3, 4)$ is also shown in Fig. 5.14.

(2) Adaptive HFPS of system $z_4 = \phi(z) = z_3^2$

We define $z_4 = \phi(z) = z_3^2$, which leads to $\dot{z}_4 = 2z_3(\frac{1}{2}(\bar{z}_1 z_2 + z_1 \bar{z}_2) - a_2 z_3)$. Following Theorem 5.2, the controller is designed analogously as follows:

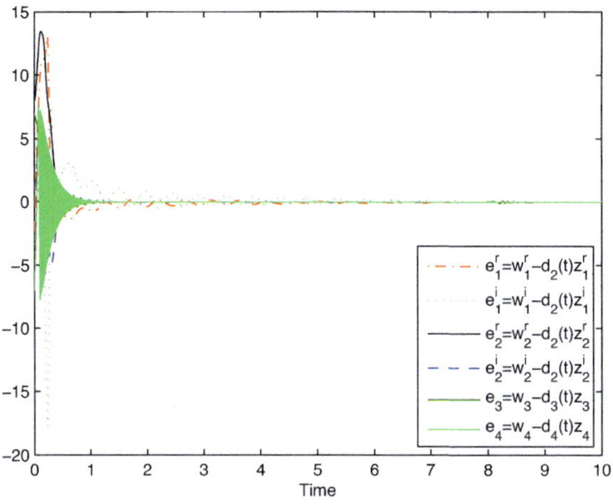

Fig. 5.15 The graph of the errors between the added-order complex Lorenz system and the hyper-chaotic chaotic complex Lü system

$$\begin{cases} u_1 = d_1(z_2 - z_1)\hat{a}_1 - \hat{b}_1(w_2 - w_1) - w_4 - k_1 e_1, \\ u_2 = d_2 z_1 \hat{a}_3 - d_2(z_2 + z_1 z_3) - w_4 - \hat{b}_3 w_2 + w_1 w_3 - k_2 e_2, \\ u_3 = d_3(t)[\frac{1}{2}(\bar{z}_1 z_2 + z_1 \bar{z}_2) - z_3 \hat{a}_2] + \dot{d}_3(t) z_3 - \frac{1}{2}(\bar{w}_1 w_2 + w_1 \bar{w}_2) + \hat{b}_2 w_3 - k_3 e_3, \\ u_4 = w_4 \hat{b}_4 + d_4(t)[z_3(\bar{z}_1 z_2 + z_1 \bar{z}_2) - 2\hat{a}_2 z_3^2] + \dot{d}_4(t) z_4 - \frac{1}{2}(\bar{w}_1 w_2 + w_1 \bar{w}_2) - k_4 e_4, \end{cases} \tag{5.2.37}$$

the estimates of $a_1,\ a_2,\ a_3$ and $b_1,\ b_2,\ b_3,\ b_4$ are given as

$$\begin{cases} \dot{\hat{a}}_1 = -d_1(z_2^r - z_1^r)e_1^r - d_1(z_2^i - z_1^i)e_1^i, \\ \dot{\hat{a}}_2 = d_3 z_3 e_3 + 2 d_4 z_3^2 e_4, \\ \dot{\hat{a}}_3 = -d_2 z_1^r e_2^r - d_2 z_1^i e_2^i, \end{cases} \tag{5.2.38}$$

and

$$\begin{cases} \dot{\hat{b}}_1 = (w_2^r - w_1^r)e_1^r + (w_2^i - w_1^i)e_1^i, \\ \dot{\hat{b}}_2 = -w_3 e_3, \\ \dot{\hat{b}}_3 = w_2^r e_2^r + w_2^i e_2^i, \\ \dot{\hat{b}}_4 = -w_4 e_4 \end{cases} \tag{5.2.39}$$

and the updated laws of control coefficients are

$$\dot{k}_\mu = 0.5(e_\mu^r)^2 + (e_\mu^i)^2,\ \mu = 1,\ 2,\ 3,\ 4.$$

In these numerical simulations, we also set the "unknown" parameters of the complex Lorenz system to $a_1 = 10$, $a_2 = 28$, and $a_3 = 8/3$, and those of the hyper-chaotic complex Lü system to $b_1 = 1$, $b_2 = 26$, $b_3 = 0.7$, and $b_4 = 1.5$, to ensure chaotic behavior.

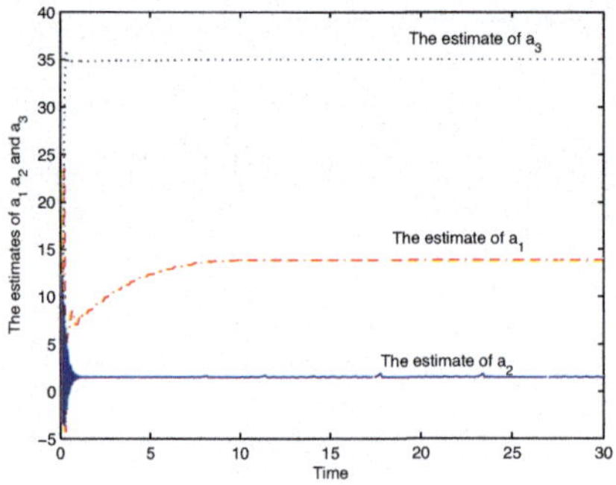

(a) Parameters $\hat{a}_1$, $\hat{a}_2$ and $\hat{a}_3$.

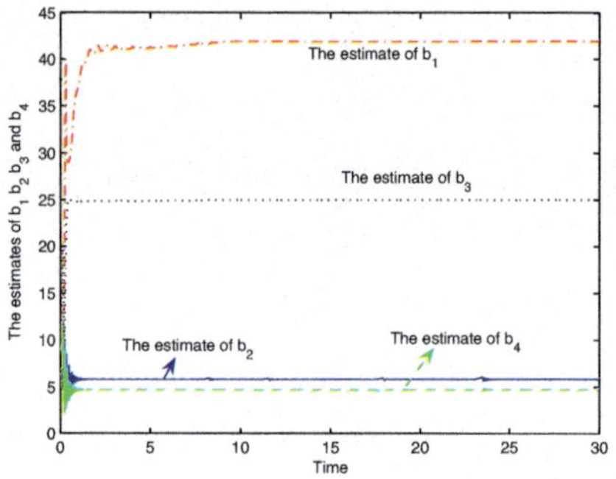

(b) Parameters $\hat{b}_1, \hat{b}_2, \hat{b}_3, \hat{b}_4$.

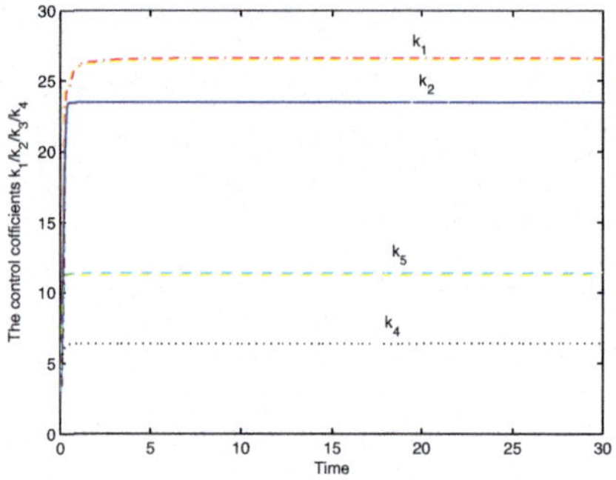

(c) Control coefficients.

Fig. 5.16 Time evolution of parameters and control coefficients

The initial values $(z_1(0), z_2(0), z_3(0))^T$, $(w_1(0), w_2(0), w_3(0))^T$ and $(k_1(0), k_2(0), k_3(0))^T$ are assigned the same vector values as in Sect. 5.2.3. Consequently, the initial value of variable z_4 is $z_4(0) = 8$, and the scaling function matrix is chosen as $\mathbf{D}(t) = diag\{1 - 0.2\sin(\frac{\pi}{10}t),\ -1 + 0.3\sin(\frac{\pi}{10}t),\ ,\ 2 - 0.5\sin(\frac{\pi}{10}t)\}$. Accordingly, the error system has initial values $e_1(0) = -3 - 4j$, $e_2(0) = 8 + j$, $e_3(0) = 7$, $e_4(0) = -17$. The simulation results are presented in Figs. 5.15 and 5.16. Figure 5.15 displays the error dynamics between the extended complex Lorenz system and the hyper-chaotic complex Lü system. As the error variables converge to zero (see Fig. 5.15), added-order HFPS between the two uncertain chaotic complex systems is achieved. Figure 5.16 illustrates that the parameter estimates $\hat{a}_1, \hat{a}_2, \hat{a}_3, \hat{b}_1, \hat{b}_2, \hat{b}_3$, and $\hat{b}_4$ each approach constant values as $t \to \infty$. The time evolution of the adaptive control coefficients k_1, k_2, k_3 and k_4 is shown in Fig. 5.16.

Chapter 6
Complex Modified Projective Synchronization and Parameters Identification for Uncertain Chaotic Complex Systems

This chapter proposes the complex modified projective synchronization (CMPS) and investigates parameters identification for uncertain chaotic complex systems. Section 6.1 presents CMPS and parameters identification for uncertain chaotic complex systems. Section 6.2 shows modified projective synchronization (MPS) with complex scaling factors of uncertain real chaos and complex chaos.

6.1 CMPS and Parameters Identification for Uncertain Chaotic Complex Systems

6.1.1 Mathematical Model and Problem Descriptions

We analyze an n-dimensional chaotic complex nonlinear system of the form:

$$\dot{\mathbf{x}} = \mathbf{F}(\mathbf{x})\mathbf{A} + \mathbf{f}(\mathbf{x}) + \mathbf{v} + \mathbf{d}_1, \tag{6.1.1}$$

where $\mathbf{x} = (x_1, x_2, \ldots, x_n)^{\mathrm{T}} = \mathbf{x}^r + j\mathbf{x}^i$ is a complex state vector. Set $x_1 = u_1 + ju_2$, $x_2 = u_3 + ju_4, \ldots, x_n = u_{2n-1} + ju_{2n}$, $\mathbf{x}^r = (u_1, u_3, \ldots, u_{2n-1})^{\mathrm{T}}$, and $\mathbf{x}^i = (u_2, u_4, \ldots, u_{2n})^{\mathrm{T}}$. $\mathbf{F}(\mathbf{x})$ is a $n \times m$ complex matrix and its elements are functions of complex state variables, and $\mathbf{A} = (a_1, a_2, \ldots, a_m)^{\mathrm{T}}$ is a real (or complex) vector of unknown parameters, and $\mathbf{f} = (f_1, f_2, \ldots, f_n)^{\mathrm{T}}$ is a vector of complex nonlinear function. $\mathbf{d}_1 = (d_{11}, d_{12}, \ldots, d_{1n})^{\mathrm{T}}$ represents the bounded external disturbances, and $| d_{1l} | < \rho_1$ $(l = 1, 2, \ldots, n)$, where ρ_1 is a positive constant.

Consider a class of drive chaotic complex systems governed by $\mathbf{z} = (z_1, z_2, \ldots, z_n)^{\mathrm{T}}$, which satisfies

$$\dot{\mathbf{z}} = \mathbf{g}(\mathbf{z}) + \mathbf{d}_2, \tag{6.1.2}$$

S. Liu et al., *Chaos in Complex Dynamical System: Theory and Application*,
https://doi.org/10.1007/978-981-95-7762-0_6

where $z_1 = u_1' + ju_2'$, ..., $z_n = u_{2n-1}' + ju_{2n}'$, $\mathbf{z}^r = (u_1', u_3', \ldots, u_{2n-1}')^{\mathrm{T}}$, $\mathbf{z}^i = (u_2', u_4', \ldots, u_{2n}')^{\mathrm{T}}$. $\mathbf{g} = (g_1, g_2, \ldots, g_n)^{\mathrm{T}}$ is a vector of complex nonlinear function. $\mathbf{d}_2 = (d_{21}, d_{22}, \ldots, d_{2n})^{\mathrm{T}}$ is the bounded external disturbances, and $| d_{2l} | < \rho_2$ $(l = 1, 2, \ldots, n)$, where ρ_2 is a positive constant. These models contain hyperchaotic systems.

According to MPS with real scaling factors [122–124], the definition of CMPS is introduced.

Definition 6.1 For the response system (6.1.1) and drive system (6.1.2), if there exists a complex constant matrix $\mathbf{H} = diag\{h_1 + jh_2, h_3 + jh_4, \ldots, h_{2n-1} + jh_{2n}\}$ such that

$$
\begin{aligned}
\lim_{t\to\infty} \|\mathbf{e}(t)\|^2 &= \lim_{t\to\infty} \|\mathbf{x(t)} - \mathbf{Hz(t)}\|^2 \\
&= \lim_{t\to\infty} (\|\mathbf{x}(t)^r - \mathbf{H}^r\mathbf{z}(t)^r + \mathbf{H}^i\mathbf{z}(t)^i\|^2 + \|\mathbf{x}(t)^i - \mathbf{H}^r\mathbf{z}(t)^i - \mathbf{H}^i\mathbf{z}(t)^r\|^2) \\
&= 0,
\end{aligned} \tag{6.1.3}
$$

where $\mathbf{H}^r = diag\{h_1, h_3, \ldots, h_{2n-1}\}$ and $\mathbf{H}^i = diag\{h_2, h_4, \ldots, h_{2n}\}$, then (6.1.1) and (6.1.2) achieve complex modified projective synchronization (CMPS). $\mathbf{H}$ is the scaling matrix, $h_1 + jh_2, h_3 + jh_4, \ldots, h_{2n-1} + jh_{2n}$ are scaling factors.

Remark 6.1 If $h_2 = h_4 = \cdots = h_{2n} = 0$, it is called as modified projective synchronization (MPS). The full state hybrid projective synchronization (FSHPS) with real scaling factors in [127] is MPS for all state variables in fact.

We aim to design a controller $\mathbf{v}$ such that the drive system (6.1.2) and the response system (6.1.1) can reach CMPS. The designed controller is $\mathbf{v} = \mathbf{v}^r + j\mathbf{v}^i$, $\mathbf{v}^r = (v_1, v_3, \ldots, v_{2n-1})^{\mathrm{T}}$ and $\mathbf{v}^i = (v_2, v_4, \ldots, v_{2n})^{\mathrm{T}}$.

6.1.2 *Adaptive CMPS Schemes*

(I) The drive signal without unknown parameters.

For simplicity, we first consider the case where $\mathbf{A}$ is a real vector.

Theorem 6.1 *If the complex controller is designed as*

$$
\begin{aligned}
\mathbf{v} &= -\mathbf{F}(\mathbf{x})\hat{\mathbf{A}} - \mathbf{f}(\mathbf{x}) + \mathbf{Hg}(\mathbf{z}) + \mathbf{Ke} \\
&= -\mathbf{F}(\mathbf{x})^r\hat{\mathbf{A}} - \mathbf{f}(\mathbf{x})^r + \mathbf{H}^r\mathbf{g}(\mathbf{z})^r - \mathbf{H}^i\mathbf{g}(\mathbf{z})^i + \mathbf{Ke}^r + j[-\mathbf{F}(\mathbf{x})^i\hat{\mathbf{A}} - \mathbf{f}(\mathbf{x})^i \\
&\quad + \mathbf{H}^r\mathbf{g}(\mathbf{z})^i + \mathbf{H}^i\mathbf{g}(\mathbf{z})^r + \mathbf{Ke}^i],
\end{aligned} \tag{6.1.4}
$$

where $\mathbf{K} = diag(k_1, k_2, \ldots, k_n)$ *represents the real control strength matrix,* $\hat{\mathbf{A}}$ *denotes the estimated value of the unknown parameter vector* $\mathbf{A}$, *and the adaptive laws are selected as*

$$\begin{cases} \dot{\hat{\mathbf{A}}} = \eta(\mathbf{F}^{\mathrm{T}}(\mathbf{x})^r, \mathbf{F}^{\mathrm{T}}(\mathbf{x})^i) \begin{bmatrix} \mathbf{e}^r \\ \mathbf{e}^i \end{bmatrix}, \\ \dot{k}_l = -\gamma_l[e_l^r(t)^2 + e_l^i(t)^2], l = 1, 2, \ldots, n, \end{cases} \tag{6.1.5}$$

where $\eta = diag(\eta_1, \eta_2, \ldots, \eta_n)$ *and* $\gamma = diag(\gamma_1, \gamma_2, \ldots, \gamma_n)$ *are corresponding convergence factor matrixes and their elements are positive real constants, then the response system* (6.1.1) *can synchronize the drive system* (6.1.2) *in the sense of CMPS asymptotically with an arbitrarily small error bound.*

Proof Based on systems (6.1.1), (6.1.2) and (6.1.4), we have

$$\begin{aligned} \dot{\mathbf{e}} &= \dot{\mathbf{x}} - H\dot{\mathbf{z}} = \mathbf{F}(\mathbf{x})\mathbf{A} + \mathbf{f}(\mathbf{x}) - \mathbf{F}(\mathbf{x})\hat{\mathbf{A}} - \mathbf{f}(\mathbf{x}) + \mathbf{Hg}(\mathbf{z}) + \mathbf{Ke} - \mathbf{Hg}(\mathbf{z}) + \mathbf{d}_1 - \mathbf{Hd}_2 \\ &= \mathbf{F}(\mathbf{x})\tilde{\mathbf{A}} + \mathbf{Ke} + \Delta\mathbf{d}, \end{aligned} \tag{6.1.6}$$

where $\tilde{\mathbf{A}} = \mathbf{A} - \hat{\mathbf{A}} = (\tilde{a}_1, \tilde{a}_2, \ldots, \tilde{a}_m)^{\mathrm{T}}$ are the errors between the true values and estimated values of unknown parameters. $\Delta\mathbf{d} = \mathbf{d}_1 - \mathbf{Hd}_2$ is the difference between $\mathbf{d}_1$ and $\mathbf{d}_2$ with scaling matrix $\mathbf{H}$, and $\|\Delta\mathbf{d}\| \le \sqrt{n}(\rho_1 + h_{max}\rho_2)$ where $h_{max} = max\{| h_1 + jh_2 |, | h_3 + jh_4 |, \ldots, | h_{2n-1} + jh_{2n} |\}$. Therefore, we have $\dot{\mathbf{e}}^r = \mathbf{F}^r\tilde{\mathbf{A}} + \mathbf{Ke}^r + \Delta\mathbf{d}^r$, $\dot{\mathbf{e}}^i = \mathbf{F}^i\tilde{\mathbf{A}} + \mathbf{Ke}^i + \Delta\mathbf{d}^i$.

The following nonnegative Lyapunov function is defined as:

$$V(e,t) = \frac{1}{2}\sum_{l=1}^{n}[e_l^r(t)^2 + e_l^i(t)^2] + \frac{1}{2}\sum_{l=1}^{n}\frac{1}{\gamma_l}(k_l + L)^2 + \frac{1}{2}\sum_{l=1}^{m}\frac{1}{\eta_l}\tilde{a}_l^2,$$

where L is an arbitrary large positive constant. Hence, we obtain that

$$\begin{aligned} \dot{V} &= (\dot{\mathbf{e}}^r)^{\mathrm{T}}\mathbf{e}^r + (\dot{\mathbf{e}}^i)^{\mathrm{T}}\mathbf{e}^i - \sum_{l=1}^{n}(k_l + L)[e_l^r(t)^2 + e_l^i(t)^2] - \sum_{l-1}^{m}\tilde{a}_l[(F^{\mathrm{T}})_l^r\mathbf{e}^r + (F^{\mathrm{T}})_l^i\mathbf{e}^i] \\ &= (\mathbf{F}^r\tilde{\mathbf{A}} + \mathbf{Ke}^r + \Delta\mathbf{d}^r)^{\mathrm{T}}\mathbf{e}^r + (\mathbf{F}^i\tilde{\mathbf{A}} + \mathbf{Ke}^i + \Delta\mathbf{d}^i)^{\mathrm{T}}\mathbf{e}^i - \sum_{l=1}^{n}k_l[e_l^r(t)^2 + e_l^i(t)^2] \\ &\quad - L\|\mathbf{e}\|^2 - \sum_{l=1}^{m}\tilde{a}_l((F^{\mathrm{T}})_l^r\mathbf{e}^r + (F^{\mathrm{T}})_l^i\mathbf{e}^i) \\ &= -L\|\mathbf{e}\|^2 + (\Delta\mathbf{d}^r)^{\mathrm{T}}\mathbf{e}^r + (\Delta\mathbf{d}^i)^{\mathrm{T}}\mathbf{e}^i. \end{aligned}$$

If $\|\mathbf{e}\|^2 = \delta$ is taken, then we have

$$(\Delta\mathbf{d}^r)^{\mathrm{T}}\mathbf{e}^r + (\Delta\mathbf{d}^i)^{\mathrm{T}}\mathbf{e}^i \le \sqrt{n}(\rho_1 + h_{max}\rho_2)(\|\mathbf{e}^r\| + \|\mathbf{e}^i\|) \le (\rho_1 + h_{max}\rho_2)\sqrt{2n\delta} < C\delta,$$

where C is a positive real constant satisfying $C > \sqrt{2n/\delta}(\rho_1 + h_{max}\rho_2)$. Since L is an arbitrary large positive constant, we choose $L > C$ and get $\dot{V} \le -(L - C)\|\mathbf{e}\|^2 < 0$.

As $\|\mathbf{e}\|^2 = \delta$ is an invariant set, and according to LaSalle invariant principle, then the error resides in a ball with the radius $2n(\rho_1 + h_{max}\rho_2)^2/C^2$. The appropriate C can be chosen to make the error bound as small as possible. Therefore, the response system (6.1.1) will synchronize (6.1.2) in the sense of CMPS asymptotically with an arbitrarily small error bound. The proof is completed. □

If $\mathbf{A} = \mathbf{A}^r + j\mathbf{A}^i$ is a complex vector, then the complex system (6.1.1) takes the form

$$\dot{\mathbf{x}} = \mathbf{F}(\mathbf{x})\mathbf{A}^r + \mathbf{j}\mathbf{F}(\mathbf{x})\mathbf{A}^i + \mathbf{f}(\mathbf{x}) + \mathbf{v} = \mathbf{F}(\mathbf{x})\mathbf{A}^r + \mathbf{R}(\mathbf{x})\mathbf{A}^i + \mathbf{f}(\mathbf{x}) + \mathbf{v}, \quad (6.1.7)$$

where $\mathbf{R}(\mathbf{x}) = \mathbf{j}\mathbf{F}(\mathbf{x})$ is a new $n \times m$ complex matrix. Therefore, we obtain the following result.

Theorem 6.2 *If the adaptive controller is designed as*

$$\begin{aligned} \mathbf{v} &= -\mathbf{F}(\mathbf{x})\hat{\mathbf{A}}^r - \mathbf{R}(\mathbf{x})\hat{\mathbf{A}}^i - \mathbf{f}(\mathbf{x}) + \mathbf{H}\mathbf{g}(\mathbf{z}) + \mathbf{K}\mathbf{e} \\ &= -\mathbf{F}^r\hat{\mathbf{A}}^r - \mathbf{R}^r\hat{\mathbf{A}}^i - \mathbf{f}^r + \mathbf{H}^r\mathbf{g}(\mathbf{z})^r - \mathbf{H}^i\mathbf{g}(\mathbf{z})^i + \mathbf{K}\mathbf{e}^r + j[-\mathbf{F}^i\hat{\mathbf{A}}^r - \mathbf{R}^i\hat{\mathbf{A}}^i - \mathbf{f}^i + \mathbf{H}^r\mathbf{g}(\mathbf{z})^i \\ &\quad + \mathbf{H}^i\mathbf{g}(\mathbf{z})^r + \mathbf{K}\mathbf{e}^i], \end{aligned}$$

and the adaptive laws are selected as

$$\begin{cases} \dot{\hat{\mathbf{A}}}^r = \eta(\mathbf{F}^{\mathrm{T}}(\mathbf{x})^r, \mathbf{F}^{\mathrm{T}}(\mathbf{x})^i)\begin{bmatrix}\mathbf{e}^r\\ \mathbf{e}^i\end{bmatrix}, \quad \dot{\hat{\mathbf{A}}}^i = \eta'(\mathbf{R}^{\mathrm{T}}(\mathbf{x})^r, \mathbf{R}^{\mathrm{T}}(\mathbf{x})^i)\begin{bmatrix}\mathbf{e}^r\\ \mathbf{e}^i\end{bmatrix}, \\ \dot{k}_l = -\gamma_l[e_l^r(t)^2 + e_l^i(t)^2], l = 1, 2, \ldots, n, \end{cases}$$

where $\eta' = diag(\eta_1', \eta_2', \ldots, \eta_n')$ is a convergence factor matrix and its elements are positive real constants, then the response system (6.1.7) *will synchronize the drive system* (6.1.2) *in the sense of CMPS asymptotically with an arbitrarily small error bound.*

Proof It is similar to the proof in Theorem 6.1 and thus is omitted. □

(II) The drive signal with unknown parameters.

When the reference signal has unknown parameters, the drive system is described by

$$\dot{\mathbf{z}} = \mathbf{G}(\mathbf{z})\mathbf{B} + \mathbf{p}(\mathbf{z}) + \mathbf{d}_2, \quad (6.1.8)$$

where $\mathbf{B} = (b_1, b_2, \ldots, b_s)^{\mathrm{T}}$ is an $s \times 1$ real (or complex) vector of unknown parameters, and $\mathbf{G}(\mathbf{z})$ is an $n \times s$ complex matrix and its elements are functions of complex state variables. We still realize the CMPS between systems (6.1.1) (or (6.1.7)) and (6.1.8).

If both **A** and **B** are real vectors, then we can obtain

Theorem 6.3 *If the adaptive controller is designed as*

$$\begin{aligned}\mathbf{v} &= -\mathbf{F}(\mathbf{x})\hat{\mathbf{A}} - \mathbf{f}(\mathbf{x}) + \mathbf{H}\mathbf{G}(\mathbf{z})\hat{\mathbf{B}} + \mathbf{H}\mathbf{p}(\mathbf{z}) + \mathbf{K}\mathbf{e} \\ &= -\mathbf{F}(\mathbf{x})^r\hat{\mathbf{A}} - \mathbf{f}(\mathbf{x})^r + (\mathbf{H}^r\mathbf{G}(\mathbf{z})^r - \mathbf{H}^i\mathbf{G}(\mathbf{z})^i)\hat{\mathbf{B}} + (\mathbf{H}^r\mathbf{p}(\mathbf{z})^r - \mathbf{H}^i\mathbf{p}(\mathbf{z})^i) + \mathbf{K}\mathbf{e}^r \\ &+ j[-\mathbf{F}(\mathbf{x})^i\hat{\mathbf{A}} - \mathbf{f}(\mathbf{x})^i + (\mathbf{H}^r\mathbf{G}(\mathbf{z})^i + \mathbf{H}^i\mathbf{G}(\mathbf{z})^r)\hat{\mathbf{B}} + (\mathbf{H}^r\mathbf{p}(\mathbf{z})^i + \mathbf{H}^i\mathbf{p}(\mathbf{z})^r) + \mathbf{K}\mathbf{e}^i],\end{aligned} \tag{6.1.9}$$

and the adaptive laws are selected as

$$\begin{cases}\dot{\hat{\mathbf{A}}} = \eta(\mathbf{F}^{\mathrm{T}}(\mathbf{x})^r, \mathbf{F}^{\mathrm{T}}(\mathbf{x})^i)\begin{bmatrix}\mathbf{e}^r \\ \mathbf{e}^i\end{bmatrix}, \\ \dot{\hat{\mathbf{B}}} = -\tau((\mathbf{H}^r\mathbf{G}(\mathbf{z})^r - \mathbf{H}^i\mathbf{G}(\mathbf{z})^i)^{\mathrm{T}}, (\mathbf{H}^r\mathbf{G}(\mathbf{z})^i + \mathbf{H}^i\mathbf{G}(\mathbf{z})^r)^{\mathrm{T}})\begin{bmatrix}\mathbf{e}^r \\ \mathbf{e}^i\end{bmatrix}, \\ \dot{k}_l = -\gamma_l[e_l^r(t)^2 + e_l^i(t)^2], l = 1, 2, \ldots, n,\end{cases}$$

where $\tau = diag(\tau_1, \tau_2, \ldots, \tau_n)$ is convergence factor matrix and its elements are positive real constants, then the response system (6.1.1) *will synchronize* (6.1.8) *in the sense of CMPS asymptotically with an arbitrarily small error bound.*

Proof On the basis of (6.1.1), (6.1.8) and (6.1.9), we get

$$\begin{aligned}\dot{\mathbf{e}} &= \mathbf{F}(\mathbf{x})\mathbf{A} + \mathbf{f}(\mathbf{x}) - \mathbf{F}(\mathbf{x})\hat{\mathbf{A}} - \mathbf{f}(\mathbf{x}) + \mathbf{H}\mathbf{G}(\mathbf{z})\hat{\mathbf{B}} + \mathbf{H}\mathbf{p}(\mathbf{z}) + \mathbf{K}\mathbf{e} - \mathbf{H}\mathbf{G}(\mathbf{z})\mathbf{B} - \mathbf{H}\mathbf{p}(\mathbf{z}) + \mathbf{d}_1 - \mathbf{H}\mathbf{d}_2 \\ &= \mathbf{F}(\mathbf{x})\tilde{\mathbf{A}} - \mathbf{H}\mathbf{G}(\mathbf{z})\tilde{\mathbf{B}} + \mathbf{K}\mathbf{e} + \Delta\mathbf{d},\end{aligned}$$

where $\tilde{\mathbf{B}} = \mathbf{B} - \hat{\mathbf{B}} = (\tilde{b}_1, \tilde{b}_2, \ldots, \tilde{b}_s)^{\mathrm{T}}$ are the errors between the true values and estimated values of unknown parameters in drive system. Setting $\phi\mathbf{1}(\mathbf{z}) = \mathbf{H}\mathbf{G}(\mathbf{z})$, we get $\phi\mathbf{1}(\mathbf{z})^r = \mathbf{H}^r\mathbf{G}(\mathbf{z})^r - \mathbf{H}^i\mathbf{G}(\mathbf{z})^i$, and $\phi\mathbf{1}(\mathbf{z})^i = \mathbf{H}^r\mathbf{G}(\mathbf{z})^i + \mathbf{H}^r\mathbf{G}(\mathbf{z})^i$. Hence, we obtain $\dot{\mathbf{e}}^r = \mathbf{F}(\mathbf{x})^r\tilde{\mathbf{A}} - \phi\mathbf{1}(\mathbf{z})^r\tilde{\mathbf{B}} + \mathbf{K}\mathbf{e}^r + \Delta\mathbf{d}^r$, $\dot{\mathbf{e}}^i = \mathbf{F}(\mathbf{x})^i\tilde{\mathbf{A}} - \phi\mathbf{1}(\mathbf{z})^i\tilde{\mathbf{B}} + \mathbf{K}\mathbf{e}^i + \Delta\mathbf{d}^i$. Consider the following nonnegative Lyapunov function:

$$V(e,t) = \frac{1}{2}\sum_{l=1}^{n}[e_l^r(t)^2 + e_l^i(t)^2] + \frac{1}{2}\sum_{l=1}^{n}\frac{1}{\gamma_l}(k_l + L)^2 + \frac{1}{2}\sum_{l=1}^{m}\frac{1}{\eta_l}\tilde{a}_l^2 + \frac{1}{2}\sum_{q=1}^{s}\frac{1}{\tau_l}\tilde{b}_q^2,$$

where L is an arbitrary large positive constant. Then

$$\begin{aligned}\dot{V} &= (\dot{\mathbf{e}}^r)^{\mathrm{T}}\mathbf{e}^r + (\dot{\mathbf{e}}^i)^{\mathrm{T}}\mathbf{e}^i - \sum_{l=1}^{n}(k_l + L)[e_l^r(t)^2 + e_l^i(t)^2] - \sum_{l=1}^{m}\tilde{a}_l[(F^{\mathrm{T}})_l^r\mathbf{e}^r + (F^{\mathrm{T}})_l^i\mathbf{e}^i] \\ &\quad + \sum_{q=1}^{s}\tilde{b}_q[(\phi 1^{\mathrm{T}})_q^r\mathbf{e}^r + (\phi 1^{\mathrm{T}})_q^i\mathbf{e}^i] \\ &= (\mathbf{F}^r\tilde{\mathbf{A}} - \phi\mathbf{1}^r\tilde{\mathbf{B}} + \mathbf{K}\mathbf{e}^r + \Delta\mathbf{d}^r)^{\mathrm{T}}\mathbf{e}^r + (\mathbf{F}^i\tilde{\mathbf{A}} - \phi\mathbf{1}^i\tilde{\mathbf{B}} + \mathbf{K}\mathbf{e}^i + \Delta\mathbf{d}^i)^{\mathrm{T}}\mathbf{e}^i - \sum_{l=1}^{n}k_l[e_l^r(t)^2 + e_l^i(t)^2] \\ &\quad - L\|\mathbf{e}\|^2 - \sum_{l=1}^{m}\tilde{a}_l[(F^{\mathrm{T}})\mathbf{e}^r + (F^{\mathrm{T}})_l^i\mathbf{e}^i] + \sum_{q=1}^{s}\tilde{b}_q[(\phi 1^{\mathrm{T}})_q^r\mathbf{e}^r + (\phi 1^{\mathrm{T}})_q^i\mathbf{e}^i] \\ &= -L\|\mathbf{e}\|^2 + (\Delta\mathbf{d}^r)^{\mathrm{T}}\mathbf{e}^r + (\Delta\mathbf{d}^i)^{\mathrm{T}}\mathbf{e}^i.\end{aligned}$$

Similarly to Theorem 6.1, we realize CMPS between the drive system (6.1.8) and the response system (6.1.1). The proof is completed. □

When **A** and **B** are complex vectors, system (6.1.10) is as the response system, the drive system (6.1.2) is described as

$$\dot{\mathbf{z}} = \mathbf{G}(\mathbf{z})\mathbf{B}^r + \mathbf{j}\mathbf{G}(\mathbf{z})\mathbf{B}^i + \mathbf{p}(\mathbf{z}, t) = \mathbf{G}(\mathbf{z})\mathbf{B}^r + \mathbf{Q}(\mathbf{z})\mathbf{B}^i + \mathbf{p}(\mathbf{z}, t), \quad (6.1.10)$$

where $\mathbf{Q}(\mathbf{z}) = \mathbf{j}\mathbf{G}(\mathbf{z})$ is a new $n \times s$ complex matrix.

Theorem 6.4 *If the adaptive controller is designed as*

$$\begin{aligned}\mathbf{v} &= -\mathbf{F}(\mathbf{x})\hat{\mathbf{A}}^r - \mathbf{R}(\mathbf{x})\hat{\mathbf{A}}^i - \mathbf{f}(\mathbf{x}) + \mathbf{H}\mathbf{G}(\mathbf{z})\hat{\mathbf{B}}^r + \mathbf{H}\mathbf{Q}(\mathbf{z})\hat{\mathbf{B}}^i + \mathbf{H}\mathbf{p}(\mathbf{z}) + \mathbf{K}\mathbf{e} \\ &= -\mathbf{F}(\mathbf{x})^r\hat{\mathbf{A}}^r - \mathbf{R}(\mathbf{x})^r\hat{\mathbf{A}}^i - \mathbf{f}(\mathbf{x})^r + (\mathbf{H}^r\mathbf{G}(\mathbf{z})^r - \mathbf{H}^i\mathbf{G}(\mathbf{z})^i)\hat{\mathbf{B}}^r + (\mathbf{H}^r\mathbf{Q}(\mathbf{z})^r - \mathbf{H}^i\mathbf{Q}(\mathbf{z})^i)\hat{\mathbf{B}}^i \\ &\quad + (\mathbf{H}^r\mathbf{p}(\mathbf{z})^r - \mathbf{H}^i\mathbf{p}(\mathbf{z})^i) + \mathbf{K}\mathbf{e}^r + j[-\mathbf{F}(\mathbf{x})^i\hat{\mathbf{A}}^r - \mathbf{R}(\mathbf{x})^i\hat{\mathbf{A}}^i - \mathbf{f}(\mathbf{x})^i + (\mathbf{H}^i\mathbf{G}(\mathbf{z})^r + \mathbf{H}^r\mathbf{G}(\mathbf{z})^i)\hat{\mathbf{B}}^r \\ &\quad + (\mathbf{H}^i\mathbf{Q}(\mathbf{z})^r + \mathbf{H}^r\mathbf{Q}(\mathbf{z})^i)\hat{\mathbf{B}}^i + (\mathbf{H}^r\mathbf{p}(\mathbf{z})^i + \mathbf{H}^r\mathbf{p}(\mathbf{z})^i) + \mathbf{K}\mathbf{e}^i],\end{aligned}$$

and the adaptive laws are selected as

$$\begin{cases} \dot{\hat{\mathbf{A}}}^r = \eta(\mathbf{F}^{\mathrm{T}}(\mathbf{x})^r, \mathbf{F}^{\mathrm{T}}(\mathbf{x})^i)\begin{bmatrix}\mathbf{e}^r \\ \mathbf{e}^i\end{bmatrix}, \quad \dot{\hat{\mathbf{A}}}^i = \eta'(\mathbf{R}^{\mathrm{T}}(\mathbf{x})^r, \mathbf{R}^{\mathrm{T}}(\mathbf{x})^i)\begin{bmatrix}\mathbf{e}^r \\ \mathbf{e}^i\end{bmatrix}, \\ \dot{\hat{\mathbf{B}}}^r = -\tau((\mathbf{H}^r\mathbf{G}(\mathbf{z})^r - \mathbf{H}^i\mathbf{G}(\mathbf{z})^i)^{\mathrm{T}}, (\mathbf{H}^r\mathbf{G}(\mathbf{z})^i + \mathbf{H}^i\mathbf{G}(\mathbf{z})^r)^{\mathrm{T}})\begin{bmatrix}\mathbf{e}^r \\ \mathbf{e}^i\end{bmatrix}, \\ \dot{\hat{\mathbf{B}}}^i = -\tau'((\mathbf{H}^r\mathbf{Q}(\mathbf{z})^r - \mathbf{H}^i\mathbf{Q}(\mathbf{z})^i)^{\mathrm{T}}, (\mathbf{H}^r\mathbf{Q}(\mathbf{z})^i + \mathbf{H}^i\mathbf{Q}(\mathbf{z})^r)^{\mathrm{T}})\begin{bmatrix}\mathbf{e}^r \\ \mathbf{e}^i\end{bmatrix}, \\ \dot{k}_l = -\gamma_l[e_l^r(t)^2 + e_l^i(t)^2], l = 1, 2, \ldots, n, \end{cases}$$

where $\tau' = diag(\tau_1', \tau_2', \ldots, \tau_n')$ *is a convergence factor matrix and its elements are positive real constants. Then system* (6.1.7) *will synchronize the drive system* (6.1.10) *in the sense of CMPS asymptotically with an arbitrarily small error bound.*

Proof It is similar to the proof in Theorem 6.3 and thus is omitted. □

If **A** is a real vector, and **B** is a complex vector, we have

Theorem 6.5 *If the adaptive controller is designed as*

$$\begin{aligned}\mathbf{v} &= -\mathbf{F}(\mathbf{x})\hat{\mathbf{A}} - \mathbf{f}(\mathbf{x}) + \mathbf{H}\mathbf{G}(\mathbf{z})\hat{\mathbf{B}}^r + \mathbf{H}\mathbf{Q}(\mathbf{z})\hat{\mathbf{B}}^i + \mathbf{H}\mathbf{p}(\mathbf{z}) + \mathbf{K}\mathbf{e} \\ &= -\mathbf{F}(\mathbf{x})^r\hat{\mathbf{A}} - \mathbf{f}(\mathbf{x})^r + (\mathbf{H}^r\mathbf{G}(\mathbf{z})^r - \mathbf{H}^i\mathbf{G}(\mathbf{z})^i)\hat{\mathbf{B}}^r + (\mathbf{H}^r\mathbf{Q}(\mathbf{z})^r - \mathbf{H}^i\mathbf{Q}(\mathbf{z})^i)\hat{\mathbf{B}}^i \\ &\quad + (\mathbf{H}^r\mathbf{p}(\mathbf{z})^r - \mathbf{H}^i\mathbf{p}(\mathbf{z})^i) + \mathbf{K}\mathbf{e}^r \\ &\quad + j[-\mathbf{F}(\mathbf{x})^i\hat{\mathbf{A}} - \mathbf{f}(\mathbf{x})^i + (\mathbf{H}^i\mathbf{G}(\mathbf{z})^r + \mathbf{H}^r\mathbf{G}(\mathbf{z})^i)\hat{\mathbf{B}}^r + (\mathbf{H}^i\mathbf{Q}(\mathbf{z})^r + \mathbf{H}^r\mathbf{Q}(\mathbf{z})^i)\hat{\mathbf{B}}^i \\ &\quad + (\mathbf{H}^r\mathbf{p}(\mathbf{z})^i + \mathbf{H}^r\mathbf{p}(\mathbf{z})^i) + \mathbf{K}\mathbf{e}^i],\end{aligned}$$

and the adaptive laws are selected as

$$\begin{cases} \dot{\hat{\mathbf{A}}} &= \eta(\mathbf{F}^{\mathrm{T}}(\mathbf{x})^r, \mathbf{F}^{\mathrm{T}}(\mathbf{x})^i)\begin{bmatrix}\mathbf{e}^r \\ \mathbf{e}^i\end{bmatrix}, \\ \dot{\hat{\mathbf{B}}}^r &= -\tau((\mathbf{H}^r\mathbf{G}(\mathbf{z})^r - \mathbf{H}^i\mathbf{G}(\mathbf{z})^i)^{\mathrm{T}}, (\mathbf{H}^r\mathbf{G}(\mathbf{z})^i + \mathbf{H}^i\mathbf{G}(\mathbf{z})^r)^{\mathrm{T}})\begin{bmatrix}\mathbf{e}^r \\ \mathbf{e}^i\end{bmatrix}, \\ \dot{\hat{\mathbf{B}}}^i &= -\tau'((\mathbf{H}^r\mathbf{Q}(\mathbf{z})^r - \mathbf{H}^i\mathbf{Q}(\mathbf{z})^i)^{\mathrm{T}}, (\mathbf{H}^r\mathbf{Q}(\mathbf{z})^i + \mathbf{H}^i\mathbf{Q}(\mathbf{z})^r)^{\mathrm{T}})\begin{bmatrix}\mathbf{e}^r \\ \mathbf{e}^i\end{bmatrix}, \\ \dot{k}_l &= -\gamma_l[e_l^r(t)^2 + e_l^i(t)^2], l = 1, 2, \ldots, n, \end{cases}$$

then system (6.1.1) *will synchronize system* (6.1.10) *in the sense of CMPS asymptotically with an arbitrarily small error bound.*

Proof It is similar to the proof in Theorem 6.3 and thus is omitted. □

If **A** is a complex vector, and **B** is a real vector, we get

Theorem 6.6 *If the adaptive controller is designed as*

$$\begin{aligned} \mathbf{v} &= -\mathbf{F}(\mathbf{x})\hat{\mathbf{A}}^r - \mathbf{R}(\mathbf{x})\hat{\mathbf{A}}^i - \mathbf{f}(\mathbf{x}) + \mathbf{H}\mathbf{G}(\mathbf{z})\hat{\mathbf{B}} + \mathbf{H}\mathbf{p}(\mathbf{z}) + \mathbf{K}\mathbf{e} \\ &= -\mathbf{F}(\mathbf{x})^r\hat{\mathbf{A}}^r - \mathbf{R}(\mathbf{x})^r\hat{\mathbf{A}}^i - \mathbf{f}(\mathbf{x})^r + (\mathbf{H}^r\mathbf{G}(\mathbf{z})^r - \mathbf{H}^i\mathbf{G}(\mathbf{z})^i)\hat{\mathbf{B}} + (\mathbf{H}^r\mathbf{p}(\mathbf{z})^r - \mathbf{H}^i\mathbf{p}(\mathbf{z})^i) + \mathbf{K}\mathbf{e}^r \\ &\quad + j[-\mathbf{F}(\mathbf{x})^i\hat{\mathbf{A}}^r - \mathbf{R}(\mathbf{x})^i\hat{\mathbf{A}}^i - \mathbf{f}(\mathbf{x})^i + (\mathbf{H}^i\mathbf{G}(\mathbf{z})^r + \mathbf{H}^r\mathbf{G}(\mathbf{z})^i)\hat{\mathbf{B}} + (\mathbf{H}^r\mathbf{p}(\mathbf{z})^i + \mathbf{H}^r\mathbf{p}(\mathbf{z})^i) + \mathbf{K}\mathbf{e}^i], \end{aligned}$$

and the adaptive laws are selected as

$$\begin{cases} \dot{\hat{\mathbf{A}}}^r = \eta(\mathbf{F}^{\mathrm{T}}(\mathbf{x})^r, \mathbf{F}^{\mathrm{T}}(\mathbf{x})^i)\begin{bmatrix}\mathbf{e}^r \\ \mathbf{e}^i\end{bmatrix}, \quad \dot{\hat{\mathbf{A}}}^i = \eta'(\mathbf{R}^{\mathrm{T}}(\mathbf{x})^r, \mathbf{R}^{\mathrm{T}}(\mathbf{x})^i)\begin{bmatrix}\mathbf{e}^r \\ \mathbf{e}^i\end{bmatrix}, \\ \dot{\hat{\mathbf{B}}} \;= -\tau((\mathbf{H}^r\mathbf{G}(\mathbf{z})^r - \mathbf{H}^i\mathbf{G}(\mathbf{z})^i)^{\mathrm{T}}, (\mathbf{H}^r\mathbf{G}(\mathbf{z})^i + \mathbf{H}^i\mathbf{G}(\mathbf{z})^r)^{\mathrm{T}})\begin{bmatrix}\mathbf{e}^r \\ \mathbf{e}^i\end{bmatrix}, \\ \dot{k}_l \;= -\gamma_l[e_l^r(t)^2 + e_l^i(t)^2], l = 1, 2, \ldots, n, \end{cases}$$

then system (6.1.7) *synchronizes system* (6.1.8) *in the sense of CMPS asymptotically with an arbitrarily small error bound.*

Proof It is similar to the proof in Theorem 6.3 and thus is omitted. □

In a word, we have realized CMPS in the presence of unknown parameters and bounded disturbances, eliminating the need for compensators. The resulting framework offers a simpler and more general solution compared to the controller presented in [125].

6.1.3 The Sufficient Condition and Necessary Condition of Identification in CMPS

The identification of unknown parameters relies on a condition of persistency of excitation (PE) and linear independence (LI) tailored for nonlinear systems. First, let us recall definitions of PE and LI in real field.

Definition 6.2 (PE) The locally integrable real function $\phi 2 : R_{\geq 0} \rightarrow R^{n\times m}$ is said to be persistently exciting if there exists $\mu > 0$ and $T > 0$ such that $\int_t^{t+T} \phi 2(x)\phi 2(x)^{\mathrm{T}} dx \geq \mu \mathbf{I}_{n\times n}, \forall t \in R_{\geq 0}$.

Definition 6.3 (LT) The set of real function $\{\varphi_k(x), k = 1, 2, \ldots, n\}$ is called as linear independence in the set S if and only if $\sum_{k=1}^{n} d_k \varphi_k(x) = 0, \forall x \in S, d_k \in R$ under the condition that $d_k = 0, k = 1, 2, \ldots, n$.

Since the disturbances are confined to a small region, we neglect their influence to simplify the analysis of identification conditions in the complex domain.

(I) The drive signal without unknown parameters.

Based on (6.1.6), it follows that $\dot{\mathbf{e}} = \mathbf{F(x)}(\mathbf{A} - \hat{\mathbf{A}}) + \mathbf{Ke}$. Following Theorems 6.1 and 6.2, we conclude that $\mathbf{e} \rightarrow 0, \dot{\mathbf{e}} \rightarrow 0$. Consequently, $\mathbf{F(x)}(\mathbf{A} - \hat{\mathbf{A}}) \rightarrow 0$.

If $\mathbf{A}$ is a real vector, then we obtain $\mathbf{F(x)}^r(\mathbf{A} - \hat{\mathbf{A}}) \rightarrow 0,\ \mathbf{F(x)}^i(\mathbf{A} - \hat{\mathbf{A}}) \rightarrow 0$. Let $\phi 2(x) = (\mathbf{F(x)}^r, \mathbf{F(x)}^i)$ be an $n \times 2m$ matrix. When $n \geq m$, we derive the sufficient condition in Theorem 6.1 for the convergence of unknown parameters to their true values: $\phi 2(x)$ must be persistently exciting. When $n < m$, the necessary condition is that $\phi 2(x)$ is persistently exciting, i.e., there exist $\mu > 0$ and $T > 0$ such that

$$\int_t^{t+T} (\mathbf{F(x)}^r, \mathbf{F(x)}^i) \begin{bmatrix} \mathbf{F}^{\mathrm{T}}(\mathbf{x})^r \\ \mathbf{F}^{\mathrm{T}}(\mathbf{x})^i \end{bmatrix} dx \geq \mu \mathbf{I}_{n\times n}, \forall t \in R_{\geq 0}.$$

From the point of LI, we obtain $\sum_{l=1}^{m} F_{kl}(x)^r \tilde{a}_l = 0$ and $\sum_{l=1}^{m} F_{kl}(x)^i \tilde{a}_l = 0$ $(k = 1, 2, \ldots, n, l = 1, 2, \ldots, m)$. If there exists one k to ensure one of $\{F_{kl}(x)^r, l = 1, 2, \ldots, m\}$ and $\{F_{kl}(x)^i, l = 1, 2, \ldots, m\}$ is linearly independent, then corresponding $a_l - \hat{a}_l \rightarrow 0$.

If $\mathbf{A}$ is a complex vector, then $\mathbf{F(x)}(\mathbf{A}^r - \hat{\mathbf{A}}^r) + \mathbf{R(x)}(\mathbf{A}^i - \hat{\mathbf{A}}^i) \rightarrow 0$ can be obtained. Separating the real part from the imaginary part, we have $\mathbf{F(x)}^r(\mathbf{A}^r - \hat{\mathbf{A}}^r) + \mathbf{R(x)}^r(\mathbf{A}^i - \hat{\mathbf{A}}^i) \rightarrow 0$ and $\mathbf{F(x)}^i(\mathbf{A}^r - \hat{\mathbf{A}}^r) + \mathbf{R(x)}^i(\mathbf{A}^i - \hat{\mathbf{A}}^i) \rightarrow 0$. Hence, we choose

$$\phi 2(x) = \begin{bmatrix} \mathbf{F(x)}^r & \mathbf{R(x)}^r \\ \mathbf{F(x)}^i & \mathbf{R(x)}^i \end{bmatrix},$$

which is a $2n \times 2m$ matrix. When $n \geq m$, we derive the sufficient condition in Theorem 6.2 for the convergence of unknown parameters to their true values: $\phi 2(x)$ must be persistently exciting. When $n < m$, we obtain the necessary condition is that $\phi 2(x)$ is persistently exciting. It means that if there exists one k to ensure $\{F_{kl}(x)^r, l = 1, 2, \ldots, m, R_{kl}(x)^r, l = 1, 2, \ldots, m\}$ linearly independent or $\{F_{kl}(x)^i, l = 1, 2, \ldots, m, R_{kl}(x)^i, l = 1, 2, \ldots, m\}$ linearly independent, then corresponding $a_l^r - \hat{a}_l^r \rightarrow 0$ and $a_l^i - \hat{a}_l^i \rightarrow 0$.

(II) The drive signal with unknown parameters.

If the drive system has unknown parameters, then

$$\mathbf{F}(\mathbf{x})(\mathbf{A}-\hat{\mathbf{A}})-\mathbf{H}\mathbf{G}(\mathbf{z})(\mathbf{B}-\hat{\mathbf{B}})\to 0. \tag{6.1.11}$$

When **A** and **B** are real vectors, then separating the real and imaginary parts yields

$$\begin{aligned}\mathbf{F}(\mathbf{x})^r(\mathbf{A}-\hat{\mathbf{A}})-(\mathbf{H}^r\mathbf{G}(\mathbf{z})^r-\mathbf{H}^i\mathbf{G}(\mathbf{z})^i)(\mathbf{B}-\hat{\mathbf{B}})\to 0,\\ \mathbf{F}(\mathbf{x})^i(\mathbf{A}-\hat{\mathbf{A}})-(\mathbf{H}^r\mathbf{G}(\mathbf{z})^i+\mathbf{H}^r\mathbf{G}(\mathbf{z})^i)(\mathbf{B}-\hat{\mathbf{B}})\to 0.\end{aligned}$$

In fact, we have designed a controller to make $\mathbf{x}\to\mathbf{Hz}$, and thus

$$\phi 2(z)=\begin{bmatrix}\mathbf{F}(\mathbf{Hz})^r & -(\mathbf{H}^r\mathbf{G}(\mathbf{z})^r-\mathbf{H}^i\mathbf{G}(\mathbf{z})^i)\\ \mathbf{F}(\mathbf{Hz})^i & -(\mathbf{H}^r\mathbf{G}(\mathbf{z})^i+\mathbf{H}^r\mathbf{G}(\mathbf{z})^i)\end{bmatrix},$$

which is a $2n\times(m+s)$ matrix. When $2n\geq m+s$, we have the sufficient condition in Theorem 6.3 is $\phi 2(z)$ is persistently exciting. When $2n<m+s$, we get the necessary condition is $\phi 2(z)$ is persistently exciting.

If there exists one k to ensure $\{F_{kl}(Hz)^r, -(h_{2k-1}G_{kq}(z)^r-h_{2k}G_{kq}(z)^i), l=1,2,\ldots,m, q=1,2,\ldots,s\}$ linearly independent or $\{F_{kl}(Hz)^i, -(h_{2k-1}G_{kq}(z)^i+h_{2k}G_{kq}(z)^r), l=1,2,\ldots,m, q=1,2,\ldots,s\}$ linearly independent, then corresponding $a_l-\hat{a}_l\to 0$ and $b_q-\hat{b}_q\to 0$.

If both **A** and **B** are complex vectors, we separate their real and imaginary parts to obtain

$$\begin{aligned}\mathbf{F}(\mathbf{x})^r(\mathbf{A}^r-\hat{\mathbf{A}}^r)+\mathbf{R}(\mathbf{x})^r(\mathbf{A}^i-\hat{\mathbf{A}}^i)-(\mathbf{H}^r\mathbf{G}(\mathbf{z})^r-\mathbf{H}^i\mathbf{G}(\mathbf{z})^i)(\mathbf{B}^r-\hat{\mathbf{B}}^r)\\ -(\mathbf{H}^r\mathbf{Q}(\mathbf{z})^r-\mathbf{H}^i\mathbf{Q}(\mathbf{z})^i)(\mathbf{B}^i-\hat{\mathbf{B}}^i)\to 0,\\ \mathbf{F}(\mathbf{x})^i(\mathbf{A}^r-\hat{\mathbf{A}}^r)+\mathbf{R}(\mathbf{x})^i(\mathbf{A}^i-\hat{\mathbf{A}}^i)-(\mathbf{H}^r\mathbf{G}(\mathbf{z})^i+\mathbf{H}^r\mathbf{G}(\mathbf{z})^i)(\mathbf{B}^r-\hat{\mathbf{B}}^r)\\ -(\mathbf{H}^r\mathbf{Q}(\mathbf{z})^i+\mathbf{H}^r\mathbf{Q}(\mathbf{z})^i)(\mathbf{B}^i-\hat{\mathbf{B}}^i)\to 0.\end{aligned}$$

As $\mathbf{x}\to\mathbf{Hz}$,

$$\phi 2(z)=\begin{bmatrix}\mathbf{F}(\mathbf{Hz})^r & \mathbf{R}(\mathbf{Hz})^r & -(\mathbf{H}^r\mathbf{G}(\mathbf{z})^r-\mathbf{H}^i\mathbf{G}(\mathbf{z})^i) & -(\mathbf{H}^r\mathbf{Q}(\mathbf{z})^r-\mathbf{H}^i\mathbf{Q}(\mathbf{z})^i)\\ \mathbf{F}(\mathbf{Hz})^i & \mathbf{R}(\mathbf{Hz})^i & -(\mathbf{H}^r\mathbf{G}(\mathbf{z})^i+\mathbf{H}^r\mathbf{G}(\mathbf{z})^i) & -(\mathbf{H}^r\mathbf{Q}(\mathbf{z})^i+\mathbf{H}^r\mathbf{Q}(\mathbf{z})^i)\end{bmatrix},$$

which is a $2n\times 2(m+s)$ matrix. If $n\geq m+s$, we have the sufficient condition in Theorem 6.4 is $\phi 2(z)$ is persistently exciting. If $n<m+s$, we get the necessary condition is $\phi 2(z)$ is persistently exciting.

If there exists one k to ensure one of $\{F_{kl}(Hz)^r, R_{kl}(Hz)^r, -(h_{2k-1}G_{kq}(z)^r-h_{2k}G_{kq}(z)^i), -(h_{2k-1}Q_{kq}(z)^r-h_{2k}Q_{kq}(z)^i), l=1,2,\ldots,m, q=1,2,\ldots,s\}$ and $\{F_{kl}(Hz)^i, R_{kl}(Hz)^i, -(h_{2k-1}G_{kq}(z)^i+h_{2k}G_{kq}(z)^r), -(h_{2k-1}Q_{kq}(z)^i+h_{2k}Q_{kq}(z)^r), l=1,2,\ldots,m, q=1,2,\ldots,s\}$ linearly independent, then corresponding $a_l^r-\hat{a}_l^r\to 0$, $a_l^i-\hat{a}_l^i\to 0$, $b_q^r-\hat{b}_q^r\to 0$, and $b_q^i-\hat{b}_q^i\to 0$.

In the case where **A** is real and **B** is complex, we derive

$$\begin{aligned}&\mathbf{F(x)}^r(\mathbf{A}-\hat{\mathbf{A}})-(\mathbf{H}^r\mathbf{G(z)}^r-\mathbf{H}^i\mathbf{G(z)}^i)(\mathbf{B}^r-\hat{\mathbf{B}}^r)-(\mathbf{H}^r\mathbf{Q(z)}^r-\mathbf{H}^i\mathbf{Q(z)}^i)(\mathbf{B}^i-\hat{\mathbf{B}}^i)\to 0,\\&\mathbf{F(x)}^i(\mathbf{A}-\hat{\mathbf{A}})-(\mathbf{H}^r\mathbf{G(z)}^i+\mathbf{H}^r\mathbf{G(z)}^i)(\mathbf{B}^r-\hat{\mathbf{B}}^r)-(\mathbf{H}^r\mathbf{Q(z)}^i+\mathbf{H}^r\mathbf{Q(z)}^i)(\mathbf{B}^i-\hat{\mathbf{B}}^i)\to 0.\end{aligned}$$

Therefore, we choose

$$\phi 2(z)=\begin{bmatrix}\mathbf{F(Hz)}^r & -(\mathbf{H}^r\mathbf{G(z)}^r-\mathbf{H}^i\mathbf{G(z)}^i) & -(\mathbf{H}^r\mathbf{Q(z)}^r-\mathbf{H}^i\mathbf{Q(z)}^i)\\ \mathbf{F(Hz)}^i & -(\mathbf{H}^r\mathbf{G(z)}^i+\mathbf{H}^r\mathbf{G(z)}^i) & -(\mathbf{H}^r\mathbf{Q(z)}^i+\mathbf{H}^r\mathbf{Q(z)}^i)\end{bmatrix},$$

which is a $2n\times(m+2s)$ matrix. If $2n\geq m+2s$, we get the sufficient condition in Theorem 6.5 is $\phi 2(z)$ is persistently exciting. If $2n<m+2s$, we obtain the necessary condition is $\phi 2(z)$ is persistently exciting.

If there exists one k to ensure $\{F_{kl}(Hz)^r, -(h_{2k-1}G_{kq}(z)^r-h_{2k}G_{kq}(z)^i), -(h_{2k-1}Q_{kq}(z)^r-h_{2k}Q_{kq}(z)^i), l=1,2,\ldots,m, q=1,2,\ldots,s\}$ linearly independent or $\{F_{kl}(Hz)^i, -(h_{2k-1}G_{kq}(z)^i+h_{2k}G_{kq}(z)^r), -(h_{2k-1}Q_{kq}(z)^i+h_{2k}Q_{kq}(z)^r), l=1,2,\ldots,m, q=1,2,\ldots,s\}$ linearly independent, then corresponding $a_l-\hat{a}_l\to 0$, $b_q^r-\hat{b}_q^r\to 0$, and $b_q^i-\hat{b}_q^i\to 0$.

If **A** is a complex vector and **B** is a real vector, then we get

$$\begin{aligned}&\mathbf{F(x)}^r(\mathbf{A}^r-\hat{\mathbf{A}}^r)+\mathbf{H(x)}^r(\mathbf{A}^i-\hat{\mathbf{A}}^i)-(\mathbf{H}^r\mathbf{G(z)}^r-\mathbf{H}^i\mathbf{G(z)}^i)(\mathbf{B}^r-\hat{\mathbf{B}}^r)\to 0,\\&\mathbf{F(x)}^i(\mathbf{A}^r-\hat{\mathbf{A}}^r)+\mathbf{H(x)}^i(\mathbf{A}^i-\hat{\mathbf{A}}^i)-(\mathbf{H}^r\mathbf{G(z)}^i+\mathbf{H}^r\mathbf{G(z)}^i)(\mathbf{B}^r-\hat{\mathbf{B}}^r)\to 0.\end{aligned}$$

Therefore, set

$$\phi 2(z)=\begin{bmatrix}\mathbf{F(z)}^r & \mathbf{H(z)}^r & -(\mathbf{H}^r\mathbf{G(z)}^r-\mathbf{H}^i\mathbf{G(z)}^i)\\ \mathbf{F(z)}^i & \mathbf{H(z)}^i & -(\mathbf{H}^r\mathbf{G(z)}^i+\mathbf{H}^r\mathbf{G(z)}^i)\end{bmatrix},$$

which is a $2n\times(2m+s)$ matrix. When $2n\geq 2m+s$, the sufficient condition in Theorem 6.6 is that $\phi_2(z)$ is persistently exciting; when $2n<2m+s$, the corresponding necessary condition is the persistent excitation of $\phi_2(z)$.

It means that if there exists one k to ensure $\{F_{kl}(Hz)^r,\ R_{kl}(Hz)^r, -(h_{2k-1}G_{kq}(z)^r - h_{2k}G_{kq}(z)^i), l=1,2,\ldots,m, q=1,2,\ldots,s\}$ linearly independent or $\{F_{kl}(Hz)^i,\ R_{kl}(Hz)^i, l=1,2,\ldots,m, -(h_{2k-1}G_{kq}(z)^i+h_{2k}G_{kq}(z)^r), q=1,2,\ldots,s\}$ linearly independent, then corresponding $a_l^r-\hat{a}_l^r\to 0$, $a_l^i-\hat{a}_l^i\to 0$ and $b_q-\hat{b}_q\to 0$.

According to the above results, the design of the controller to the case of complex parameter vectors **A** and **B** is similar to the real ones. Hence, we validate the proposed method in the following section using complex systems with real parameter vectors for simplicity.

6.1.4 Numerical Simulations

In the following, we study the uncertain complex Lü system drives an uncertain complex Chen system. The drive system is defined as

$$\begin{cases} \dot{z}_1 = b_1(z_2 - z_1) + d_{21}, \\ \dot{z}_2 = b_2 z_2 - z_1 z_3 + d_{22}, \\ \dot{z}_3 = -b_3 z_3 + (1/2)(\bar{z}_1 z_2 + z_1 \bar{z}_2) + d_{23}, \end{cases} \tag{6.1.12}$$

where z_1, z_2 denote complex state variables, z_3 is real state variable, and $\mathbf{B} = (b_1, b_2, b_3)^{\mathrm{T}}$ is the unknown parameter vector.

We describe the response Chen system as follows

$$\begin{cases} \dot{x}_1 = a_1(x_2 - x_1) + d_{11} + v_1 + jv_2, \\ \dot{x}_2 = (a_2 - a_1)x_1 + a_2 x_2 - x_1 x_3 + d_{12} + v_3 + jv_4, \\ \dot{x}_3 = -a_3 x_3 + (1/2)(\bar{x}_1 x_2 + x_1 \bar{x}_2) + d_{13} + v_5, \end{cases} \tag{6.1.13}$$

where x_1, x_2 denote complex state variables, x_3 is real state variable, and $\mathbf{A} = (a_1, a_{21}, a_2, a_3)^{\mathrm{T}}$ is the unknown parameter vector where $a_{21} = a_2 - a_1$. The designed control inputs are v_1, v_2, v_3, v_4 and v_5.

Following Theorem 6.3, we design the controller as follows,

$$\mathbf{v} = \begin{pmatrix} \hat{a}_1(u_1 - u_3) + \hat{b}_1(h_1(u_3' - u_1') - h_2(u_4' - u_2')) + k_1 e_1 \\ -\hat{a}_{21}u_1 - \hat{a}_2 u_3 + u_1 u_5 + \hat{b}_2(h_3 u_3' - h_4 u_4') - (h_3 u_1' u_5' - h_4 u_2' u_5') + k_2 e_3 \\ \hat{a}_3 u_5 - (u_1 u_3 + u_2 u_4) - h_5 \hat{b}_3 u_5' + h_5(u_1' u_3' + u_2' u_4') + k_3 e_5 \end{pmatrix} + j \begin{pmatrix} \hat{a}_1(u_2 - u_4) + \hat{b}_1(h_2(u_3' - u_1') + h_1(u_4' - u_2')) + k_1 e_2 \\ -\hat{a}_{21}u_2 - \hat{a}_2 u_4 + u_2 u_5 + \hat{b}_2(h_3 u_4' + h_4 u_3') - (h_4 u_1' u_5' + h_3 u_2' u_5') + k_2 e_4 \\ 0 \end{pmatrix},$$

where $h_1 + jh_2, h_3 + jh_4, h_5$ are scaling factors. As $z_3 \in R,\ x_3 \in R$, we set $h_6 = 0$ to prevent the introduction of any new imaginary component in the response system.

We define the parameter adaptation laws as

$$\dot{\hat{\mathbf{A}}} = \begin{pmatrix} \dot{\hat{a}}_1 \\ \dot{\hat{a}}_{21} \\ \dot{\hat{a}}_2 \\ \dot{\hat{a}}_3 \end{pmatrix} = \begin{pmatrix} \eta_1((u_3 - u_1)e_1 + (u_4 - u_2)e_2) \\ \eta_{21}(u_1 e_3 + u_2 e_4) \\ \eta_2(u_3 e_3 + u_4 e_4) \\ -\eta_3 u_5 e_5 \end{pmatrix}, \quad \dot{\mathbf{k}} = \begin{pmatrix} \dot{k}_1 \\ \dot{k}_2 \\ \dot{k}_3 \end{pmatrix} = \begin{pmatrix} -\gamma_1(e_1^2 + e_2^2) \\ -\gamma_2(e_3^2 + e_4^2) \\ -\gamma_3 e_5^2 \end{pmatrix},$$

and

$$\dot{\hat{\mathbf{B}}} = \begin{pmatrix} \dot{\hat{b}}_1 \\ \dot{\hat{b}}_2 \\ \dot{\hat{b}}_3 \end{pmatrix} = \begin{pmatrix} -\tau_1[(h_1(u_3' - u_1') - h_2(u_4' - u_2'))e_1 + (h_2(u_3' - u_1') + h_1(u_4' - u_2'))e_2] \\ -\tau_2[(h_3 u_3' - h_4 u_4')e_3 + (h_3 u_4' + h_4 u_3')e_4] \\ \tau_3 h_5 u_5' e_5 \end{pmatrix}.$$

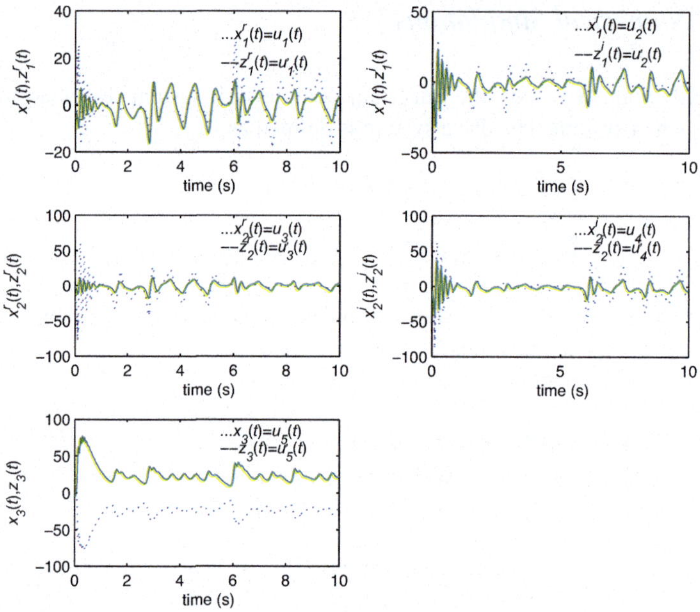

Fig. 6.1 The CMPS between (6.1.12) and (6.1.13) with $\eta_1 = \eta_{21} = \eta_2 = \eta_3 = 10$, $\tau_1 = \tau_2 = \tau_3 = 10$, $\gamma_1 = \gamma_2 = \gamma_3 = 10$ and complex scaling factors $1 + j, 2 - j, -1$

We assign the true values of the unknown parameters as $\mathbf{B} = (29, 21, 2)^T$ and $\mathbf{A} = (27, -4, 23, 1)^T$. The initial condition of (6.1.12) and (6.1.13) are $\mathbf{z(0)} = (-3 - 2j, -2 - 7j, -5)^T$ and $\mathbf{x(0)} = (4 + 15j, -1 - 5j, 30)^T$. The initial values of estimated parameters and control strength are $\hat{\mathbf{A}}(0) = (15,\ 15,\ 15,\ 15)^T$, $\hat{\mathbf{B}}(0) = (10,\ 10,\ 10)^T$ and $\mathbf{k}(0) = (0, 0, 0)^T$. The external disturbances are chosen as $d_{21} = sin(0.25\pi t) + jcos(\pi t), d_{22} = 0.3sin(\pi t) + j0.5cos(\pi t),\ d_{23} = 0.2sin(0.5\pi t)$ and $d_{11} = 0.2cos(1.5\pi t) + j0.5sin(2\pi t),\ d_{12} = 0.5sin(0.5\pi t) + j0.3sin(\pi t), d_{13} = 0.8sin(1.5\pi t)$, then we get $\rho_2 = 1$ and $\rho_1 = 0.8$. Additionally, random unknown interference with amplitude [-5,5] is considered in the simulations.

The differential equations are solved using the fourth-order Runge-Kutta method with a time step of $\Delta t = 10^{-3}$. When the scaling factors are chosen as $1 + j, 2 - j, -1$, the CMPS process between (6.1.12) and (6.1.13) is illustrated in Fig. 6.1. It can be seen that the relationship $x_1 = (1 + j)z_1$ or $x_2 = (2 - j)z_2$ is more complicated than $x_3 = -z_3$. As shown in Fig. 6.2, the FSHPS errors converge asymptotically to zero. The error e_1 depends on h_1 and h_2, while e_2 is associated with h_3 and h_4. Obviously, the probability of an interceptor successfully deciphering two uncorrelated real numbers is lower than that of cracking a single real scaling factor.

Figures 6.3 and 6.4 show the identification processes of $\hat{\mathbf{A}}$ and $\hat{\mathbf{B}}$, whose estimates converge to $(26.9249, -3.9002, 99.9699, -29.8668)^T$ and $(29.5945, 98.2992, -28.9049)^T$, respectively. By neglecting the disturbances, we

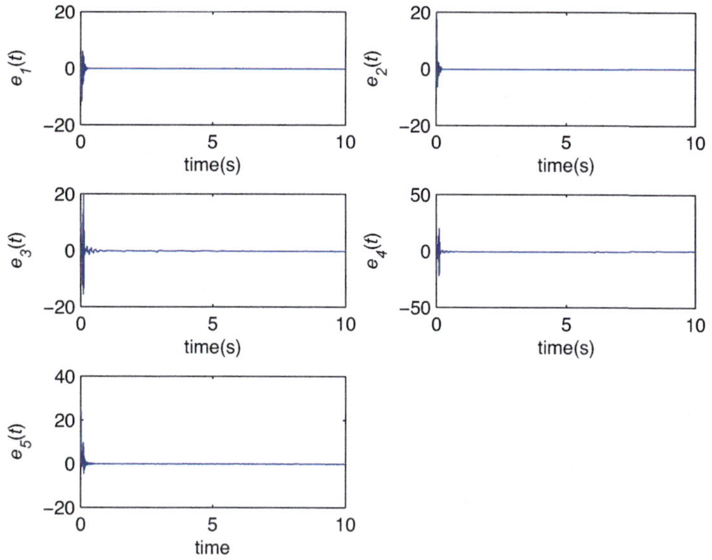

Fig. 6.2 The error dynamic of CMPS between (6.1.12) and (6.1.13) with $\eta_1 = \eta_{21} = \eta_2 = \eta_3 = 10, \tau_1 = \tau_2 = \tau_3 = 10$, $\gamma_1 = \gamma_2 = \gamma_3 = 10$ and complex scaling factors $1 + j, 2 - j, -1$. Here $e_1 = x_1^r - (h_1 z_1^r - h_2 z_1^i) = x_1^r - (z_1^r - z_1^i)$, $e_2 = x_1^i - (h_2 z_1^r + h_1 z_1^i) = x_1^r - (z_1^r + z_1^i)$, $e_3 = x_2^r - (2z_2^r + z_2^i)$, $e_4 = x_2^i - (-z_2^r + 2z_2^i)$, $e_5 = x_3 + z_3$

observe that only a_1, b_1, and a_{21} converge to their true values. The following analysis addresses why this selective convergence occurs.

As $x_1 \to (h_1 + jh_2)z_1$, $x_2 \to (h_3 + jh_4)z_2$ and $x_3 \to h_5 z_3$, the following equations are derived from (6.1.11):

$$z_2[(h_3 + jh_4)\tilde{a}_1 - (h_1 + jh_2)\tilde{b}_1] + z_1[-(h_1 + jh_2)\tilde{a}_1 + (h_1 + jh_2)\tilde{b}_1] \to 0,$$
$$(h_1 + jh_2)z_1\tilde{a}_{21} + (h_3 + jh_4)z_2(\tilde{a}_2 - \tilde{b}_2) \to 0, \quad h_5 z_3(-\tilde{a}_3 + \tilde{b}_3) \to 0.$$

Based on the property of linear independence and given $\mathbf{H} = (1 + j, 2 - j, -1)^{\mathrm{T}}$, it follows that

$$(2 + j)\tilde{a}_1 - (1 + j)\tilde{b}_1 \to 0, (-\tilde{a}_1 + \tilde{b}_1) \to 0, \tilde{a}_{21} \to 0, \tilde{a}_2 - \tilde{b}_2 \to 0, -\tilde{a}_3 + \tilde{b}_3 \to 0.$$

Therefore, the estimated values $\hat{a}_1, \hat{b}_1$ and $\hat{a}_{21}$ converge to the true values, while other estimated values converge to certain constants which satisfy $a_2 - \hat{a}_2 = b_2 - \hat{b}_2$ and $a_3 - \hat{a}_3 = b_3 - \hat{b}_3$. They are in accordance with experiment results as $23 - 99.9699 \approx 21 - 98.2992$ and $1 - (-29.8668) \approx 2 - (-28.9049)$. If $h_1 + jh_2 = h_3 + jh_4$, $\hat{a}_1, \hat{b}_1$ will not converge to their true values. Hence, the convergence of certain parameters to their true values is determined by scaling factors. In practice, we can adjust these scaling factors to derive the true values of unknown parameters.

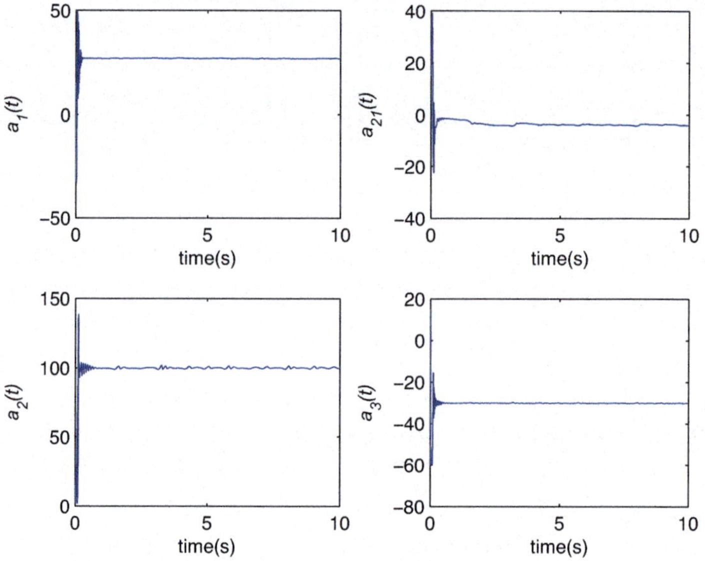

Fig. 6.3 The identification process of unknown parameter vector **B** with $\eta_1 = \eta_{21} = \eta_2 = \eta_3 = 10$, $\tau_1 = \tau_2 = \tau_3 = 10$, $\gamma_1 = \gamma_2 = \gamma_3 = 10$ and complex scaling factors $1 + j, 2 - j, -1$

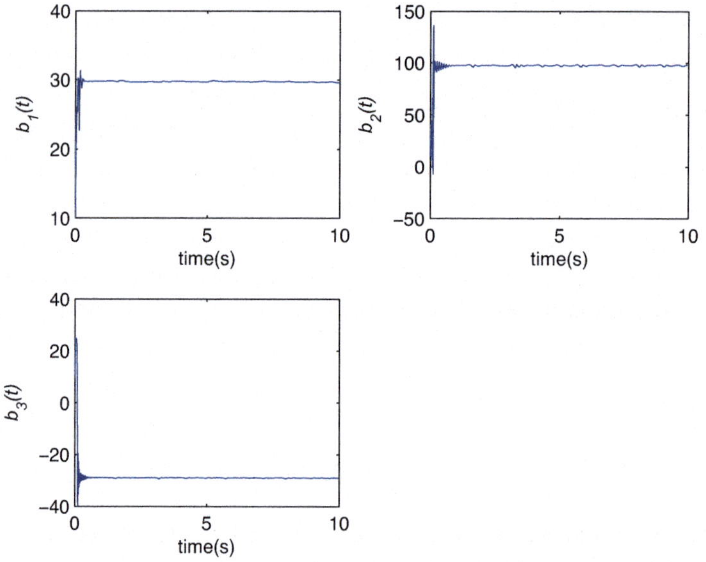

Fig. 6.4 The identification process of unknown parameter vector **B** with $\eta_1 = \eta_{21} = \eta_2 = \eta_3 = 10$, $\tau_1 = \tau_2 = \tau_3 = 10$, $\gamma_1 = \gamma_2 = \gamma_3 = 10$ and complex scaling factors $1 + j, 2 - j, -1$

It also can be explained that

$\phi\mathbf{2}(\mathbf{z}) =$

$$\begin{pmatrix} \begin{array}{c}(h_3u'_3 - h_4u'_4)\\ -(h_1u'_1 - h_2u'_2)\end{array} & 0 & 0 & 0 & \begin{array}{c}h_2(u'_4 - u'_2)\\ -h_1(u'_3 - u'_1)\end{array} & 0 & 0 \\ 0 & h_1u'_1 - h_2u'_2 & h_3u'_3 - h_4u'_4 & 0 & 0 & -(h_3u'_3 - h_4u'_4) & 0 \\ 0 & 0 & 0 & -h_5u'_5 & 0 & 0 & h_5u'_5 \\ \begin{array}{c}(h_3u'_4 + h_4u'_3)\\ -(h_1u'_2 + h_2u'_1)\end{array} & 0 & 0 & 0 & \begin{array}{c}-h_2(u'_3 - u'_1)\\ -h_1(u'_4 - u'_2)\end{array} & 0 & 0 \\ 0 & h_1u'_2 + h_2u'_1 & h_3u'_4 + h_4u'_3 & 0 & 0 & -(h_3u'_4 + h_4u'_3) & 0 \\ 0 & 0 & 0 & 0 & 0 & 0 & 0 \end{pmatrix}$$

is not persistently exciting. It shows that the experiment results are in accordance with our theoretical analysis.

6.2 CMPS of Uncertain Real Chaos and Complex Chaos

6.2.1 Speed-Gradient Method

We consider an n-dimensional chaotic system in the real domain, described by

$$\dot{\mathbf{x}} = \phi(\mathbf{x}, \theta, t), \tag{6.2.1}$$

where $\mathbf{x} = (x_1, x_2, \ldots, x_n)^{\mathrm{T}}$ is a real state vector, θ is a real matrix of unknown parameters. To output the desired signal $\mathbf{x}^*(t)$, we consider the error criterion function $\Phi(t) = \Phi(\mathbf{x}, t)$ (If $\mathbf{x}(t) \to \mathbf{x}^*(t)$, $\Phi(\mathbf{x}, t) \to 0$) which is a scalar smooth nonnegative function. Its time derivative is

$$\omega(\theta, t) = \frac{\partial \Phi(\mathbf{x}, t)}{\partial t} + \nabla_{\mathbf{x}} \Phi(\mathbf{x}, t) \phi(\mathbf{x}, \theta, t). \tag{6.2.2}$$

According to the speed-gradient method, θ evolves along the opposite direction of the gradient of $\omega(\theta, t)$ with respect to θ. The most general form of the resulting algorithm, termed the combined speed-gradient (SG) form, is given by:

$$\frac{d}{dt}(\theta + \varphi(t)) = -\Gamma \nabla_{\theta} \omega(\theta, t), \tag{6.2.3}$$

where Γ is a corresponding positive definite gain matrix, and $\varphi(t)$ is a certain vector function satisfying the pseudogradient condition

$$\varphi^{\mathrm{T}}(t) \nabla_{\theta} \omega(\theta, t) \geq 0. \tag{6.2.4}$$

Equation (6.2.3) can be expressed in the finite-integral form as

$$\theta = -\varphi(t) - \Gamma \int_0^t \nabla_{\theta} \omega(\theta, \alpha) d\alpha. \tag{6.2.5}$$

We now formulate the general stability theorem for the combined SG law (6.2.3).

Lemma 6.1 ([126]) *Consider systems* (6.2.1), (6.2.3) *under the following assumptions*
(Al) $\phi(\mathbf{x}, \theta, t)$ *and* $\nabla_\theta \omega(\theta, t)$ *are bounded together with their partial derivatives on any bounded set of states* $(\mathbf{x}, \theta)$ *of systems* (6.2.1), (6.2.3) *uniformly in* $t \geq 0$ *(regularity condition);*
(A2) $\omega(\theta, t)$ *is convex in* θ *(convexity condition);*
(A3) There exists a real constant matrix θ^* *and scalar uniformly continuous in each bounded region function* $\rho(\mathbf{x}) \geq 0$, $\rho(0) = 0$ *such that inequality* $\omega(\theta^*, t) \leq -\rho(\mathbf{x})$ *holds for all* $\mathbf{x} \in R^n$ *(achievability condition);*
(A4) If $\Phi(\mathbf{x}, t)$ *is bounded then* $\mathbf{x}(t)$ *is bounded as well (boundedness condition);*
Then each trajectory $(\mathbf{x}(t), \theta(t))$ *of systems* (6.2.1), (6.2.3), (6.2.5) *is bounded and* $\lim_{t\to\infty} \rho(\mathbf{x}(t)) = 0$.

6.2.2 CMPS of Real Drive Systems and Complex Response Systems

(I) Mathematical model and problem descriptions.

Let us consider an n-dimensional real drive chaotic system with unknown parameters,

$$\dot{\mathbf{z}} = \mathbf{G}(\mathbf{z})\mathbf{B} + \mathbf{g}(\mathbf{z}), \tag{6.2.6}$$

where $\mathbf{z} = (z_1, z_2, \ldots, z_n)^{\mathrm{T}}$ is a real state vector, $\mathbf{B} = (b_1, b_2, \ldots, b_s)^{\mathrm{T}}$ is an $s \times 1$ real vector of unknown parameters, and $\mathbf{G}(\mathbf{z})$ is an $n \times s$ real matrix, and $\mathbf{g} = (g_1, g_2, \ldots, g_n)^{\mathrm{T}}$ is a vector of real nonlinear function.

We represent the complex response chaotic system as follows

$$\dot{\mathbf{w}} = \mathbf{F}(\mathbf{w})\mathbf{A} + \mathbf{f}(\mathbf{w}) + \mathbf{v}, \tag{6.2.7}$$

where $\mathbf{w} = (w_1, w_2, \ldots, w_n)^{\mathrm{T}} = \mathbf{w}^r + j\mathbf{w}^i$ is a complex state vector. Set $w_1 = u_1 + ju_2$, $w_2 = u_3 + ju_4$, $\ldots$, $w_n = u_{2n-1} + ju_{2n}$, then $\mathbf{w}^r = (u_1, u_3, \ldots, u_{2n-1})^{\mathrm{T}}$, $\mathbf{w}^i = (u_2, u_4, \ldots, u_{2n})^{\mathrm{T}}$. $\mathbf{f} = (f_1, f_2, \ldots, f_n)^{\mathrm{T}}$ is a vector of complex nonlinear function, $\mathbf{F}(\mathbf{w})$ is an $n \times m$ complex matrix and its elements are functions of complex state variables, and $\mathbf{A} = (a_1, a_2, \ldots, a_m)^{\mathrm{T}}$ is an $m \times 1$ real (or complex) vector of unknown parameters. The designed controller is $\mathbf{v} = \mathbf{v}^r + j\mathbf{v}^i$, where $\mathbf{v}^r = (v_1, v_3, \ldots, v_{2n-1})^{\mathrm{T}}$, $\mathbf{v}^i = (v_2, v_4, \ldots, v_{2n})^{\mathrm{T}}$.

Based on the definitions of MPS and FSHPS with real scaling factors for two chaotic complex systems, we define CMPS for real drive and complex response chaotic systems.

Definition 6.4 For the drive system (6.2.6) and response system (6.2.7), if there exists a complex constant matrix $\mathbf{H} = diag\{h_1 + jh_2,\ h_3 + jh_4,\ \ldots,\ h_{2n-1} + jh_{2n}\}$ such that

$$\begin{aligned}\lim_{t\to\infty}\|\mathbf{e}(t)\|^2 &= \lim_{t\to\infty}\|\mathbf{w(t)} - \mathbf{Hz(t)}\|^2\\ &= \lim_{t\to\infty}(\|\mathbf{w}(t)^r - \mathbf{H}^r\mathbf{z}(t)\|^2 + \|\mathbf{w}(t)^i - \mathbf{H}^i\mathbf{z}(t)\|^2)\\ &= 0,\end{aligned}$$

where $\mathbf{e}(t)$ is the error vector, $\mathbf{e}^r(t) = (e_1, e_3, \ldots, e_{2n-1})^{\mathrm{T}}, \mathbf{e}^i(t) = (e_2, e_4, \ldots, e_{2n})^{\mathrm{T}}$, $\mathbf{H}^r = diag\{h_1, h_3, \ldots, h_{2n-1}\}$, $\mathbf{H}^i = diag\{h_2, h_4, \ldots, h_{2n}\}$, then the real drive system (6.2.6) and complex response system (6.2.7) are CMPS, and $h_1 + jh_2,\ h_3 + jh_4,\ \ldots,\ h_{2n-1} + jh_{2n}$ are complex scaling factors. If there exists $w_l \in R\ (l = 1,\ 2, \ldots,\ n)$, we select $h_{2l} = 0$ to avoid increasing a new imaginary part in the response system.

The control objective is to design an adaptive controller $\mathbf{v}$ such that the response complex system (6.2.7) can synchronize the drive real system (6.2.6) asymptotically in the sense of CMPS.

(II) Adaptive CMPS schemes.

If $\mathbf{A}$ is a real vector, based on the SG method, we have

Theorem 6.7 *If the adaptive controller is designed as*

$$\begin{aligned}\mathbf{v} &= -\mathbf{F(w)\hat{A}} - \mathbf{f(w)} + \mathbf{HG(z)\hat{B}} + \mathbf{Hg(z)} + \mathbf{Ke}\\ &= -\mathbf{F(w)}^r\mathbf{\hat{A}} - \mathbf{f(w)}^r + \mathbf{H}^r\mathbf{G(z)\hat{B}} + \mathbf{H}^r\mathbf{g(z)} + \mathbf{Ke}^r\\ &\quad + j[-\mathbf{F(w)}^i\mathbf{\hat{A}} - \mathbf{f(w)}^i + \mathbf{H}^i\mathbf{G(z)\hat{B}} + \mathbf{H}^i\mathbf{g(z)} + \mathbf{Ke}^i],\end{aligned} \tag{6.2.8}$$

where $\mathbf{K} = diag(k_1,\ k_2,\ \ldots,\ k_n)$ *denotes the real control strength matrix and* $\hat{\mathbf{A}}$, $\hat{\mathbf{B}}$ *are the estimated values of* $\mathbf{A}$ *and* $\mathbf{B}$, *respectively; and the adaptive laws are chosen as*

$$\begin{cases}\hat{\mathbf{A}} = -\varphi_{\hat{\mathbf{A}}(t)} - \eta\int_0^t \nabla_{\hat{\mathbf{A}}}\omega(\hat{\mathbf{A}}, \hat{\mathbf{B}}, \alpha)d\alpha,\\ \hat{\mathbf{B}} = -\varphi_{\hat{\mathbf{B}}(t)} - \tau\int_0^t \nabla_{\hat{\mathbf{B}}}\omega(\hat{\mathbf{A}}, \hat{\mathbf{B}}, \alpha)d\alpha,\\ \dot{k}_l = -\gamma_l[e_l^r(t)^2 + e_l^i(t)^2], l = 1, 2, \ldots, n,\end{cases} \tag{6.2.9}$$

where

$$\nabla_{\hat{\mathbf{A}}}\omega(\hat{\mathbf{A}}, \hat{\mathbf{B}}, t) = -(\mathbf{F}^{\mathrm{T}}(\mathbf{w})^r, \mathbf{F}^{\mathrm{T}}(\mathbf{w})^i)\begin{bmatrix}\mathbf{e}^r\\ \mathbf{e}^i\end{bmatrix}, \tag{6.2.10}$$

$$\nabla_{\hat{\mathbf{B}}}\omega(\hat{\mathbf{A}}, \hat{\mathbf{B}}, t) = ((\mathbf{H}^r\mathbf{G(z)})^{\mathrm{T}}, (\mathbf{H}^i\mathbf{G(z)})^{\mathrm{T}})\begin{bmatrix}\mathbf{e}^r\\ \mathbf{e}^i\end{bmatrix}, \tag{6.2.11}$$

and $\varphi_{\hat{\mathbf{A}}(t)} = \lambda \nabla_{\hat{\mathbf{A}}} \omega(\hat{\mathbf{A}}, \hat{\mathbf{B}}, t)$, $\varphi_{\hat{\mathbf{B}}(t)} = \varsigma \nabla_{\hat{\mathbf{B}}} \omega(\hat{\mathbf{A}}, \hat{\mathbf{B}}, t)$ *which satisfy pseudogradient condition,* $\lambda = diag(\lambda_1, \lambda_2, \ldots, \lambda_n)$, $\eta = diag(\eta_1, \eta_2, \ldots, \eta_n)$, $\varsigma = diag(\varsigma_1, \varsigma_2, \ldots, \varsigma_n)$, $\tau = diag(\tau_1, \tau_2, \ldots, \tau_n)$ *and* $\gamma = diag(\gamma_1, \gamma_2, \ldots, \gamma_n)$ *are corresponding convergence factor matrixes and their elements are positive real constants, then the complex response system* (6.2.7) *will synchronize the real drive system* (6.2.6) *in the sense of CMPS asymptotically, and* $\hat{\mathbf{A}}$, $\hat{\mathbf{B}}$ *converge to constant vectors.*

Proof Based on (6.2.6), (6.2.7) and (6.2.8), we conclude

$$
\begin{aligned}
\dot{\mathbf{e}} &= \dot{\mathbf{w}} - \dot{\mathbf{z}} \\
&= \mathbf{F(w)A} + \mathbf{f(w)} - \mathbf{F(w)}\hat{\mathbf{A}} - \mathbf{f(w)} + \mathbf{HG(z)}\hat{\mathbf{B}} + \mathbf{Hg(z)} + \mathbf{Ke} - \mathbf{HG(z)B} - \mathbf{Hg(z)} \\
&= \mathbf{F(w)}(\mathbf{A} - \hat{\mathbf{A}}) - \mathbf{HG(z)}(\mathbf{B} - \hat{\mathbf{B}}) + \mathbf{Ke} \\
&= \mathbf{F(w)}\tilde{\mathbf{A}} - \mathbf{HG(z)}\tilde{\mathbf{B}} + \mathbf{Ke},
\end{aligned} \tag{6.2.12}
$$

where $\tilde{\mathbf{A}} = \mathbf{A} - \hat{\mathbf{A}} = (\tilde{a}_1, \tilde{a}_2, \ldots, \tilde{a}_m)^{\mathrm{T}}$ and $\tilde{\mathbf{B}} = \mathbf{B} - \hat{\mathbf{B}} = (\tilde{b}_1, \tilde{b}_2, \ldots, \tilde{b}_s)^{\mathrm{T}}$ denote the errors between the true values and estimated values of unknown parameters. Setting $\Psi(\mathbf{z}) = \mathbf{HG(z)}$, we get $\Psi(\mathbf{z})^r = \mathbf{H}^r\mathbf{G(z)}$, and $\Psi(\mathbf{z})^i = \mathbf{H}^i\mathbf{G(z)}$. Thus, we obtain $\dot{\mathbf{e}}^r = \mathbf{F(w)}^r\tilde{\mathbf{A}} - \Psi(\mathbf{z})^r\tilde{\mathbf{B}} + \mathbf{Ke}^r$, $\dot{\mathbf{e}}^i = \mathbf{F(w)}^i\tilde{\mathbf{A}} - \Psi(\mathbf{z})^i\tilde{\mathbf{B}} + \mathbf{Ke}^i$.

Consider the following nonnegative Lyapunov function:

$$
\begin{aligned}
V(e,t) = {} & \frac{1}{2}[(\mathbf{e^r})^{\mathrm{T}}\mathbf{e}^r + (\mathbf{e^i})^{\mathrm{T}}\mathbf{e}^i] + \frac{1}{2}\eta^{-1}(\mathbf{A} - \hat{\mathbf{A}} - \varphi_{\hat{A}}(t))^2 \\
& + \frac{1}{2}\tau^{-1}(\mathbf{B} - \hat{\mathbf{B}} - \varphi_{\hat{B}}(t))^2 + \frac{1}{2}\sum_{l=1}^{n}\frac{1}{\gamma_l}(k_l + L)^2,
\end{aligned} \tag{6.2.13}
$$

where L is an arbitrary large positive constant. Using (6.2.9), (6.2.10) and (6.2.11), it follows that $\frac{d}{dt}(\tilde{\mathbf{A}} - \varphi_{\hat{A}}(t)) = -\frac{d}{dt}(\hat{\mathbf{A}} + \varphi_{\hat{A}}(t)) = \eta \nabla_{\hat{\mathbf{A}}} \omega(\hat{\mathbf{A}}, \hat{\mathbf{B}}, t)$ and $\frac{d}{dt}(\tilde{\mathbf{B}} - \varphi_{\hat{B}}(t)) = -\frac{d}{dt}(\hat{\mathbf{B}} + \varphi_{\hat{B}}(t)) = \tau \nabla_{\hat{\mathbf{B}}} \omega(\hat{\mathbf{A}}, \hat{\mathbf{B}}, t)$, then

$$
\begin{aligned}
\dot{V} &= (\dot{\mathbf{e}}^r)^{\mathrm{T}}\mathbf{e}^r + (\dot{\mathbf{e}}^i)^{\mathrm{T}}\mathbf{e}^i - \tilde{\mathbf{A}}[(F^{\mathrm{T}})^r\mathbf{e}^r + (F^{\mathrm{T}})^i\mathbf{e}^i] - \lambda \nabla_{\hat{\mathbf{A}}} \omega(\hat{\mathbf{A}}, \hat{\mathbf{B}}, t)^2 + \tilde{\mathbf{B}}[(\Psi^{\mathrm{T}})^r e^r + (\Psi^{\mathrm{T}})^i \mathbf{e}^i] \\
&\quad - \varsigma \nabla_{\hat{\mathbf{B}}} \omega(\hat{\mathbf{A}}, \hat{\mathbf{B}}, t)^2 - \sum_{l=1}^{n}(k_l + L)[e_l^r(t)^2 + e_l^i(t)^2] \\
&< (\mathbf{F}^r\tilde{\mathbf{A}} - \Psi^r\tilde{\mathbf{B}} + \mathbf{Ke}^r)^{\mathrm{T}}\mathbf{e}^r + (\mathbf{F}^i\tilde{\mathbf{A}} - \Psi^i\tilde{\mathbf{B}} + \mathbf{Ke}^i)^{\mathrm{T}}\mathbf{e}^i - \tilde{\mathbf{A}}[(F^{\mathrm{T}})^r\mathbf{e}^r + (F^{\mathrm{T}})^i\mathbf{e}^i] \\
&\quad + \tilde{\mathbf{B}}[(\Psi^{\mathrm{T}})^r e^r + (\Psi^{\mathrm{T}})^i \mathbf{e}^i] - \sum_{l=1}^{n} k_l[e_l^r(t)^2 + e_l^i(t)^2] - L\|\mathbf{e}\|^2 \\
&= -L\|\mathbf{e}\|^2.
\end{aligned}
$$

Conditions (Al), (A2) are valid since the right-hand sides of (6.2.8), (6.2.9) and (6.2.12) are smooth and linear in $\hat{\mathbf{A}}$ and $\hat{\mathbf{B}}$. Condition (A3) is valid since for the constant matrixs $\mathbf{A}$ and $\mathbf{B}$ we have $\dot{V} < -L\|\mathbf{e}\|^2$, where $L\|\mathbf{e}\|^2 \geq 0$. Condition (A4) is valid because the objective function (6.2.13) is radially unbounded and the trajectories of the drive model (6.2.6) remain bounded. Since all conditions of Lemma 6.1 are satisfied we conclude that $\mathbf{w}(t)$, $\hat{\mathbf{A}}$ and $\hat{\mathbf{B}}$ are bounded, and $\lim_{t\to\infty}\|\mathbf{e(t)}\|^2 = 0$. The

CMPS of system (6.2.6) and system (6.2.7) is realized. From (6.2.9), (6.2.10) and (6.2.11), it follows that when $e(t) \to 0$, $\nabla_{\hat{\mathbf{A}}}\omega(\hat{\mathbf{A}}, \hat{\mathbf{B}}, t) \to 0$ and $\nabla_{\hat{\mathbf{B}}}\omega(\hat{\mathbf{A}}, \hat{\mathbf{B}}, t) \to 0$. Therefore, $\hat{\mathbf{A}}$ and $\hat{\mathbf{B}}$ converge to real constant vectors. The proof is completed. □

If $\mathbf{A}$ is a complex vector, then $\mathbf{A}$ can be expressed in the form of $\mathbf{A} = \mathbf{A}^r + j\mathbf{A}^i$. Thus, (6.2.7) reduces to

$$\begin{aligned}\dot{\mathbf{w}} &= \mathbf{F}(\mathbf{w})\mathbf{A}^r + \mathbf{jF}(\mathbf{w})\mathbf{A}^i + \mathbf{f}(\mathbf{w}) + \mathbf{v} \\ &= \mathbf{F}(\mathbf{w})\mathbf{A}^r + \mathbf{R}(\mathbf{w})\mathbf{A}^i + \mathbf{f}(\mathbf{w}) + \mathbf{v},\end{aligned} \tag{6.2.14}$$

where $\mathbf{R}(\mathbf{w}) = \mathbf{jF}(\mathbf{w})$ is a new $n \times m$ complex matrix. Consequently, we obtain

Theorem 6.8 *If the adaptive controller is designed as*

$$\begin{aligned}\mathbf{v} &= -\mathbf{F}(\mathbf{w})\hat{\mathbf{A}}^r - \mathbf{R}(\mathbf{w})\hat{\mathbf{A}}^i - \mathbf{f}(\mathbf{w}) + \mathbf{HG}(\mathbf{z})\hat{\mathbf{B}} + \mathbf{Hg}(\mathbf{z}) + \mathbf{Ke} \\ &= -\mathbf{F}(\mathbf{w})^r\hat{\mathbf{A}}^r - \mathbf{R}(\mathbf{w})^r\hat{\mathbf{A}}^i - \mathbf{f}(\mathbf{w})^r + \mathbf{H}^r\mathbf{G}(\mathbf{z})\hat{\mathbf{B}} + \mathbf{H}^r\mathbf{g}(\mathbf{z}) + \mathbf{Ke}^r \\ &\quad + j[-\mathbf{F}(\mathbf{w})^i\hat{\mathbf{A}}^r - \mathbf{R}(\mathbf{w})^i\hat{\mathbf{A}}^i - \mathbf{f}(\mathbf{w})^i + \mathbf{H}^i\mathbf{G}(\mathbf{z})\hat{\mathbf{B}} + \mathbf{H}^i\mathbf{g}(\mathbf{z}) + \mathbf{Ke}^i],\end{aligned}$$

and the adaptive laws are selected as

$$\begin{cases}\dot{\hat{\mathbf{A}}}^r = -\lambda\nabla_{\hat{\mathbf{A}}^r}\omega(\hat{\mathbf{A}}^r, \hat{\mathbf{A}}^i, \hat{\mathbf{B}}, t) - \eta\int_0^t \nabla_{\hat{\mathbf{A}}^r}\omega(\hat{\mathbf{A}}^r, \hat{\mathbf{A}}^i, \hat{\mathbf{B}}, \alpha)d\alpha, \\ \dot{\hat{\mathbf{A}}}^i = -\lambda'\nabla_{\hat{\mathbf{A}}^i}\omega(\hat{\mathbf{A}}^r, \hat{\mathbf{A}}^i, \hat{\mathbf{B}}, t) - \eta'\int_0^t \nabla_{\hat{\mathbf{A}}^i}\omega(\hat{\mathbf{A}}^r, \hat{\mathbf{A}}^i, \hat{\mathbf{B}}, \alpha)d\alpha, \\ \dot{\hat{\mathbf{B}}} = -\varsigma\nabla_{\hat{\mathbf{B}}}\omega(\hat{\mathbf{A}}^r, \hat{\mathbf{A}}^i, \hat{\mathbf{B}}, t) - \tau\int_0^t \nabla_{\hat{\mathbf{B}}}\omega(\hat{\mathbf{A}}^r, \hat{\mathbf{A}}^i, \hat{\mathbf{B}}, \alpha)d\alpha, \\ \dot{k}_l = -\gamma_l[e_l^r(t)^2 + e_l^i(t)^2], l = 1, 2, \ldots, n,\end{cases}$$

where

$$\nabla_{\hat{\mathbf{A}}^r}\omega(\hat{\mathbf{A}}^r, \hat{\mathbf{A}}^i, \hat{\mathbf{B}}, t) = -(\mathbf{F}^{\mathrm{T}}(\mathbf{w})^r, \mathbf{F}^{\mathrm{T}}(\mathbf{w})^i)\begin{bmatrix}\mathbf{e}^r \\ \mathbf{e}^i\end{bmatrix},$$

$$\nabla_{\hat{\mathbf{A}}^i}\omega(\hat{\mathbf{A}}^r, \hat{\mathbf{A}}^i, \hat{\mathbf{B}}, t) = -(\mathbf{R}^{\mathrm{T}}(\mathbf{w})^r, \mathbf{R}^{\mathrm{T}}(\mathbf{w})^i)\begin{bmatrix}\mathbf{e}^r \\ \mathbf{e}^i\end{bmatrix},$$

$$\nabla_{\hat{\mathbf{B}}}\omega(\hat{\mathbf{A}}^r, \hat{\mathbf{A}}^i, \hat{\mathbf{B}}, t) = ((\mathbf{H}^r\mathbf{G}(\mathbf{z}))^{\mathrm{T}}, (\mathbf{H}^i\mathbf{G}(\mathbf{z}))^{\mathrm{T}})\begin{bmatrix}\mathbf{e}^r \\ \mathbf{e}^i\end{bmatrix},$$

where $\lambda' = diag(\lambda_1', \lambda_2', \ldots, \lambda_n')$ and $\eta' = diag(\eta_1', \eta_2', \ldots, \eta_n')$ denote corresponding convergence factor matrixes and their elements are positive real constants, then the response system (6.2.14) *synchronizes the real drive chaotic system* (6.2.6) *in the sense of CMPS asymptotically, and $\hat{\mathbf{A}}^r$, $\hat{\mathbf{A}}^i$, $\hat{\mathbf{B}}$ converge to real constant vectors.*

Proof It is similar to the proof in Theorem 6.7 and thus is omitted. □

6.2.3 CMPS of Complex Drive Chaotic System and Real Response Chaotic System

(I) Mathematical model and problem descriptions

The following n-dimensional complex drive chaotic system with unknown parameters can be described as,

$$\dot{\mathbf{y}} = \mathbf{Q}(\mathbf{y})\mathbf{D} + \mathbf{q}(\mathbf{y}), \tag{6.2.15}$$

where $\mathbf{y} = (y_1, y_2, \ldots, y_n)^{\mathrm{T}}$ is a complex state vector. Set $y_1 = u_1' + ju_2'$, $y_2 = u_3' + ju_4'$, $\ldots$, $y_n = u_{2n-1}' + ju_{2n}'$, and $\mathbf{y}^r = (u_1', u_3', \ldots, u_{2n-1}')^{\mathrm{T}}$, $\mathbf{y}^i = (u_2', u_4', \ldots, u_{2n}')^{\mathrm{T}}$. $\mathbf{Q}(\mathbf{y})$ is an $n \times s$ complex matrix and its elements are functions of complex state variables, $\mathbf{D} = (d_1, d_2, \ldots, d_s)^{\mathrm{T}}$ is an $s \times 1$ real (complex) vector of unknown parameters, let $\mathbf{q} = (q_1, q_2, \ldots, q_n)^{\mathrm{T}}$ be a vector whose components are complex nonlinear functions.

We have the real response chaotic system,

$$\dot{\mathbf{x}} = \mathbf{P}(\mathbf{x})\mathbf{C} + \mathbf{p}(\mathbf{x}) + \mathbf{L}. \tag{6.2.16}$$

Let $\mathbf{x} = (x_1, x_2, \ldots, x_n)^{\mathrm{T}}$ be a real state vector. $\mathbf{P}(\mathbf{x})$ denotes an $n \times m$ real matrix, and $\mathbf{C} = (c_1, c_2, \ldots, c_m)^{\mathrm{T}}$ represents an $m \times 1$ real vector of unknown parameters, and $\mathbf{p} = (p_1, p_2, \ldots, p_n)^{\mathrm{T}}$ is a vector of real nonlinear function. $\mathbf{L}$ is the designed controller.

The CMPS between complex drive and real response chaotic systems is specifically defined below.

Definition 6.5 For the drive system (6.2.15) and response system (6.2.16), if there exists a complex constant matrix $\mathbf{H} = diag\{h_1 + jh_2, h_3 + jh_4, \ldots, h_{2n-1} + jh_{2n}\}$ such that

$$\lim_{t\to\infty} \|\mathbf{e}(t)\|^2 = \lim_{t\to\infty} \|\mathbf{x(t)} - \mathbf{H^r y^r(t)} + \mathbf{H^i y^i(t)}\|^2 = 0,$$

where $\mathbf{e}(t) = \mathbf{x}(t) - \mathbf{H^r y^r}(t) + \mathbf{H^i y^i}(t)$ is the real error vector, then the complex drive system (6.2.15) and real response system (6.2.16) are CMPS of real parts. As $\mathbf{x}(t)$ is real, we choose real $\mathbf{L}$ to ensure CMPS of real parts and avoid increasing the imaginary parts of response system. The control objective is to design an adaptive real controller $\mathbf{L}$ such that the response real system (6.2.16) can synchronize the drive complex system (6.2.15) asymptotically in the sense of the CMPS of real parts.

(II) Adaptive CMPS Schemes of real parts

If $\mathbf{D}$ is a real vector, we obtain

Theorem 6.9 *If the adaptive controller is designed as*

$$\mathbf{L} = -\mathbf{P}(\mathbf{x})\hat{\mathbf{C}} - \mathbf{p}(\mathbf{x}) + (\mathbf{H}^r\mathbf{Q}(\mathbf{y})^r - \mathbf{H}^i\mathbf{Q}(\mathbf{y})^i)\hat{\mathbf{D}} + \mathbf{H}^r\mathbf{q}(\mathbf{y})^r - \mathbf{H}^i\mathbf{q}(\mathbf{y})^i + \mathbf{K}\mathbf{e}, \tag{6.2.17}$$

where $\hat{\mathbf{C}}$, $\hat{\mathbf{D}}$ *denote the estimated values of* $\mathbf{C}$ *and* $\mathbf{D}$, *respectively, and the adaptive laws are chosen as*

$$\begin{cases} \dot{\hat{\mathbf{C}}} = -\lambda \nabla_{\hat{\mathbf{C}}} \omega(\hat{\mathbf{C}}, \hat{\mathbf{D}}, t) - \eta \int_0^t \nabla_{\hat{\mathbf{C}}} \omega(\hat{\mathbf{C}}, \hat{\mathbf{D}}, \alpha) d\alpha, \\ \dot{\hat{\mathbf{D}}} = -\varsigma \nabla_{\hat{\mathbf{D}}} \omega(\hat{\mathbf{C}}, \hat{\mathbf{D}}, t) - \tau \int_0^t \nabla_{\hat{\mathbf{D}}} \omega(\hat{\mathbf{C}}, \hat{\mathbf{D}}, \alpha) d\alpha, \\ \dot{k}_l = -\gamma_l [e_l^r(t)^2 + e_l^i(t)^2], l = 1, 2, \ldots, n, \end{cases} \tag{6.2.18}$$

where

$$\nabla_{\hat{\mathbf{C}}} \omega(\hat{\mathbf{C}}, \hat{\mathbf{D}}, t) = -\mathbf{P}^{\mathrm{T}}(\mathbf{x})\mathbf{e}, \tag{6.2.19}$$

$$\nabla_{\hat{\mathbf{D}}} \omega(\hat{\mathbf{C}}, \hat{\mathbf{D}}, t) = (\mathbf{H}^r \mathbf{Q}(\mathbf{y})^r - \mathbf{H}^i \mathbf{Q}(\mathbf{y})^i)^{\mathrm{T}} \mathbf{e}, \tag{6.2.20}$$

then the response system (6.2.16) *and the drive chaotic system* (6.2.15) *will be CMPS for real parts and* $\hat{\mathbf{C}}$, $\hat{\mathbf{D}}$ *converge to real constant vectors.*

Proof It is similar to the proof in Theorem 6.7 and thus is omitted. □

If $\mathbf{D} = \mathbf{D}^r + j\mathbf{D}^i$ is a complex vector, then (6.2.15) can be written as

$$\dot{\mathbf{y}} = \mathbf{Q}(\mathbf{y})\mathbf{D}^r + \mathbf{j}\mathbf{Q}(\mathbf{y})\mathbf{D}^i + \mathbf{q}(\mathbf{y}) = \mathbf{Q}(\mathbf{y})\mathbf{D}^r + \mathbf{O}(\mathbf{y})\mathbf{D}^i + \mathbf{q}(\mathbf{y}), \tag{6.2.21}$$

where $\mathbf{O}(\mathbf{y}) = \mathbf{j}\mathbf{Q}(\mathbf{y})$ is a new $n \times m$ complex matrix. It follows that

Theorem 6.10 *If the adaptive controller is designed as*

$$\begin{aligned} \mathbf{L} = & -\mathbf{P}(\mathbf{x})\hat{\mathbf{C}} - \mathbf{p}(\mathbf{x}) + (\mathbf{H}^r \mathbf{Q}(\mathbf{y})^r - \mathbf{H}^i \mathbf{Q}(\mathbf{y})^i)\hat{\mathbf{D}}^r \\ & + (\mathbf{H}^r \mathbf{O}(\mathbf{y})^r - \mathbf{H}^i \mathbf{O}(\mathbf{y})^i)\hat{\mathbf{D}}^i + \mathbf{H}^r \mathbf{q}(\mathbf{y})^r - \mathbf{H}^i \mathbf{q}(\mathbf{y})^i + \mathbf{K}\mathbf{e} \end{aligned}$$

and the adaptive laws are selected as

$$\begin{cases} \dot{\hat{\mathbf{C}}} &= -\lambda \nabla_{\hat{\mathbf{C}}} \omega(\hat{\mathbf{C}}, \hat{\mathbf{D}}^r, \hat{\mathbf{D}}^i, t) - \eta \int_0^t \nabla_{\hat{\mathbf{C}}} \omega(\hat{\mathbf{C}}, \hat{\mathbf{D}}^r, \hat{\mathbf{D}}^i, \alpha) d\alpha, \\ \dot{\hat{\mathbf{D}}}^r &= -\varsigma \nabla_{\hat{\mathbf{D}}^r} \omega(\hat{\mathbf{C}}, \hat{\mathbf{D}}^r, \hat{\mathbf{D}}^i, t) - \tau \int_0^t \nabla_{\hat{\mathbf{D}}^r} \omega(\hat{\mathbf{C}}, \hat{\mathbf{D}}^r, \hat{\mathbf{D}}^i, \alpha) d\alpha, \\ \dot{\hat{\mathbf{D}}}^i &= -\varsigma' \nabla_{\hat{\mathbf{D}}^i} \omega(\hat{\mathbf{C}}, \hat{\mathbf{D}}^r, \hat{\mathbf{D}}^i, t) - \tau' \int_0^t \nabla_{\hat{\mathbf{D}}^i} \omega(\hat{\mathbf{C}}, \hat{\mathbf{D}}^r, \hat{\mathbf{D}}^i, \alpha) d\alpha, \\ \dot{k}_l &= -\gamma_l [e_l^r(t)^2 + e_l^i(t)^2], l = 1, 2, \ldots, n, \end{cases}$$

where

$$\nabla_{\hat{\mathbf{C}}} \omega(\hat{\mathbf{C}}, \hat{\mathbf{D}}^r, \hat{\mathbf{D}}^i, t) = -\mathbf{P}^{\mathrm{T}}(\mathbf{x})\mathbf{e},$$

$$\nabla_{\hat{\mathbf{D}}^r} \omega(\hat{\mathbf{C}}, \hat{\mathbf{D}}^r, \hat{\mathbf{D}}^i, t) = (\mathbf{H}^r \mathbf{Q}(\mathbf{y})^r - \mathbf{H}^i \mathbf{Q}(\mathbf{y})^i)^{\mathrm{T}} \mathbf{e},$$

$$\nabla_{\hat{\mathbf{D}}^i} \omega(\hat{\mathbf{C}}, \hat{\mathbf{D}}^r, \hat{\mathbf{D}}^i, t) = (\mathbf{H}^r \mathbf{O}(\mathbf{y})^r - \mathbf{H}^i \mathbf{O}(\mathbf{y})^i)^{\mathrm{T}} \mathbf{e},$$

where $\varsigma' = diag(\varsigma_1', \varsigma_2', \ldots, \varsigma_n')$ *and* $\tau' = diag(\tau_1', \tau_2', \ldots, \tau_n')$ *are convergence factor matrixs and its elements are positive real constants, then the response system* (6.2.16) *and the drive chaotic system* (6.2.19) *will be CMPS for real parts, and* $\hat{\mathbf{C}}$, $\hat{\mathbf{D}}^r$, $\hat{\mathbf{D}}^i$ *converge to real constant vectors.*

Proof It is similar to the proof in Theorem 6.7 and thus is omitted. □

It is evident that the CMPS of the imaginary parts are similar to those of the real parts, and thus is omitted.

6.2.4 Simulations

(I) CMPS of real Rössler system and complex Lorenz system

To investigate CMPS of real drive system and complex response system, we consider that real hyperchaotic Rössler system [128] drives complex hyperchaotic Lorenz system [129]. The drive hyperchaotic Rössler system is designed as follows

$$\begin{cases} \dot{z}_1 = -z_2 - z_3, \\ \dot{z}_2 = z_1 + b_1 z_2 + z_4, \\ \dot{z}_3 = b_2 + z_1 z_3, \\ \dot{z}_4 = -b_3 z_3 + b_4 z_4, \end{cases} \tag{6.2.22}$$

where

$$\mathbf{G}(\mathbf{z}) = \begin{pmatrix} 0 & 0 & 0 & 0 \\ z_2 & 0 & 0 & 0 \\ 0 & 1 & 0 & 0 \\ 0 & 0 & -z_3 & z_4 \end{pmatrix}, \quad \mathbf{g}(\mathbf{z}) = \begin{pmatrix} -z_2 - z_3 \\ z_1 + z_4 \\ z_1 z_3 \\ 0 \end{pmatrix},$$

z_1, z_2 and z_3 denote real state variables, $\mathbf{B} = (b_1, b_2, b_3, b_4)^{\mathrm{T}}$ represents the unknown parameter vector.

Consider the complex response hyperchaotic Lorenz system

$$\begin{cases} \dot{w}_1 = a_1(w_2 - w_1), \\ \dot{w}_2 = a_2 w_1 - w_2 - w_1 w_3 + w_4, \\ \dot{w}_3 = -a_3 w_3 + (1/2)(\bar{w}_1 w_2 + w_1 \bar{w}_2), \\ \dot{w}_4 = a_4 w_1 + a_5 w_2, \end{cases} \tag{6.2.23}$$

where

$$\mathbf{F}(\mathbf{w}) = \begin{pmatrix} w_2 - w_1 & 0 & 0 & 0 & 0 \\ 0 & w_1 & 0 & 0 & 0 \\ 0 & 0 & -w_3 & 0 & 0 \\ 0 & 0 & 0 & w_1 & w_2 \end{pmatrix}, \quad \mathbf{f}(\mathbf{w}) = \begin{pmatrix} 0 \\ -w_2 - w_1 w_3 + w_4 \\ (1/2)(\bar{w}_1 w_2 + w_1 \bar{w}_2) \\ 0 \end{pmatrix},$$

$w_1 = u_1 + ju_2, w_2 = u_3 + ju_4, w_4 = u_6 + ju_7$ represent complex state variables, and $w_3 = u_5$ is a real state variable, a_1, a_2, a_3, a_4, a_5 denote real unknown parameters (Fig. 6.7).

By applying Theorem 6.7, we construct the controller

$$\mathbf{v} = -\mathbf{F}(\mathbf{w})\hat{\mathbf{A}} - \mathbf{f}(\mathbf{w}) + \mathbf{H}\mathbf{G}(\mathbf{z})\hat{\mathbf{B}} + \mathbf{H}\mathbf{g}(\mathbf{z}) + \mathbf{K}\mathbf{e}$$

$$= \begin{pmatrix} -\hat{a}_1(w_2 - w_1) + (h_1 + jh_2)(-z_2 - z_3) + k_1(e_1 + je_2) \\ -\hat{a}_2 w_1 + w_2 + w_1 w_3 - w_4 + (h_3 + jh_4)(z_1 + \hat{b}_1 z_2 + z_4) + k_2(e_3 + je_4) \\ \hat{a}_3 w_3 - (1/2)(\bar{w}_1 w_2 + w_1 \bar{w}_2) + h_5(\hat{b}_2 + z_1 z_3) + k_3 e_5 \\ -\hat{a}_4 w_1 - \hat{a}_5 w_2 + (h_6 + jh_7)(-\hat{b}_3 z_3 + \hat{b}_4 z_4) + k_4(e_6 + je_7) \end{pmatrix}$$

$$= \begin{pmatrix} -\hat{a}_1(u_3 - u_1) + h_1(-z_2 - z_3) + k_1 e_1 \\ -\hat{a}_2 u_1 + u_3 + u_1 u_5 - u_6 + h_3(z_1 + \hat{b}_1 z_2 + z_4) + k_2 e_3 \\ \hat{a}_3 u_5 - (u_1 u_3 + u_2 u_4) + h_5(\hat{b}_2 + z_1 z_3) + k_3 e_5 \\ -\hat{a}_4 u_1 - \hat{a}_5 u_3 + h_6(-\hat{b}_3 z_3 + \hat{b}_4 z_4) + k_4 e_6 \end{pmatrix}$$

$$+ j \begin{pmatrix} -\hat{a}_1(u_4 - u_2) + h_2(-z_2 - z_3) + k_1 e_2 \\ -\hat{a}_2 u_2 + u_4 + u_2 u_5 - u_7 + h_4(z_1 + \hat{b}_1 z_2 + z_4) + k_2 e_4 \\ 0 \\ -\hat{a}_4 u_2 - \hat{a}_5 u_4 + h_7(-\hat{b}_3 z_3 + \hat{b}_4 z_4) + k_4 e_7 \end{pmatrix},$$

where $h_1 + jh_2, h_3 + jh_4, h_5, h_6 + jh_7$ are scaling factors, $e_1 = u_1 - h_1 z_1$, $e_2 = u_2 - h_2 z_1$, $e_3 = u_3 - h_3 z_2$, $e_4 = u_4 - h_4 z_2$, $e_5 = u_5 - h_5 z_3$, $e_6 = u_6 - h_6 z_4$, and $e_7 = u_7 - h_7 z_4$.

The adaptive laws are given by (6.2.9), where

$$\nabla_{\hat{\mathbf{A}}}\omega(\hat{\mathbf{A}}, \hat{\mathbf{B}}, t) = \begin{pmatrix} -[(u_3 - u_1)e_1 + (u_4 - u_2)e_2] \\ -[u_1 e_3 + u_2 e_4] \\ u_5 e_5 \\ -(u_1 e_6 + u_2 e_7) \\ -(u_3 e_6 + u_4 e_7) \end{pmatrix}, \quad \nabla_{\hat{\mathbf{B}}}\omega(\hat{\mathbf{A}}, \hat{\mathbf{B}}, t) = \begin{pmatrix} h_3 z_2 e_3 + h_4 z_2 e_4 \\ h_5 e_5 \\ -h_6 z_3 e_6 - h_7 z_3 e_7 \\ h_6 z_4 e_6 + h_7 z_4 e_7 \end{pmatrix},$$

and

$$\dot{\mathbf{k}} = \begin{pmatrix} \dot{k}_1 \\ \dot{k}_2 \\ \dot{k}_3 \\ \dot{k}_4 \end{pmatrix} = \begin{pmatrix} -\gamma_1(e_1^2 + e_2^2) \\ -\gamma_2(e_3^2 + e_4^2) \\ -\gamma_3 e_5^2 \\ -\gamma_4(e_6^2 + e_7^2) \end{pmatrix}.$$

Let $\mathbf{A} = (14, 35, 3, -5, -4)^{\mathrm{T}}$ and $\mathbf{B} = (0.25, 3, 0.5, 0.05)^{\mathrm{T}}$. The initial condition of drive system (6.2.22) and response system (6.2.23) are set as $\mathbf{z(0)} = (-20, 0, 0, 15)^{\mathrm{T}}$ and $\mathbf{w(0)} = (-1 - 2j, -3 - 4j, -5, -6 - 7j)^{\mathrm{T}}$. The parameter estimates and control strength are initialized as $\hat{\mathbf{A}}(0) = (10,\ 10,\ 10,\ 10,\ 10)^{\mathrm{T}}$, $\hat{\mathbf{B}}(0) = (1,\ 1,\ 1,\ 1)^{\mathrm{T}}$ and $\mathbf{k}(0) = (0,\ 0,\ 0,\ 0)^{\mathrm{T}}$. We employ a fourth-order Runge-Kutta solver with a step size $\Delta t = 10^{-3}s$ to numerically integrate the differential equations. The CMPS process of systems (6.2.22) and (6.2.23) is displayed in Fig. 6.5, where the solid line shows the states of the drive system and the dotted line presents the states of the response system. Their chaotic behaviors are depicted in Figs. 6.6 and 6.7, where the red line shows the trajectory of the drive system and the blue line presents the trajectory of the response system. The real parts (w_1^r, w_2^r, w_3) of

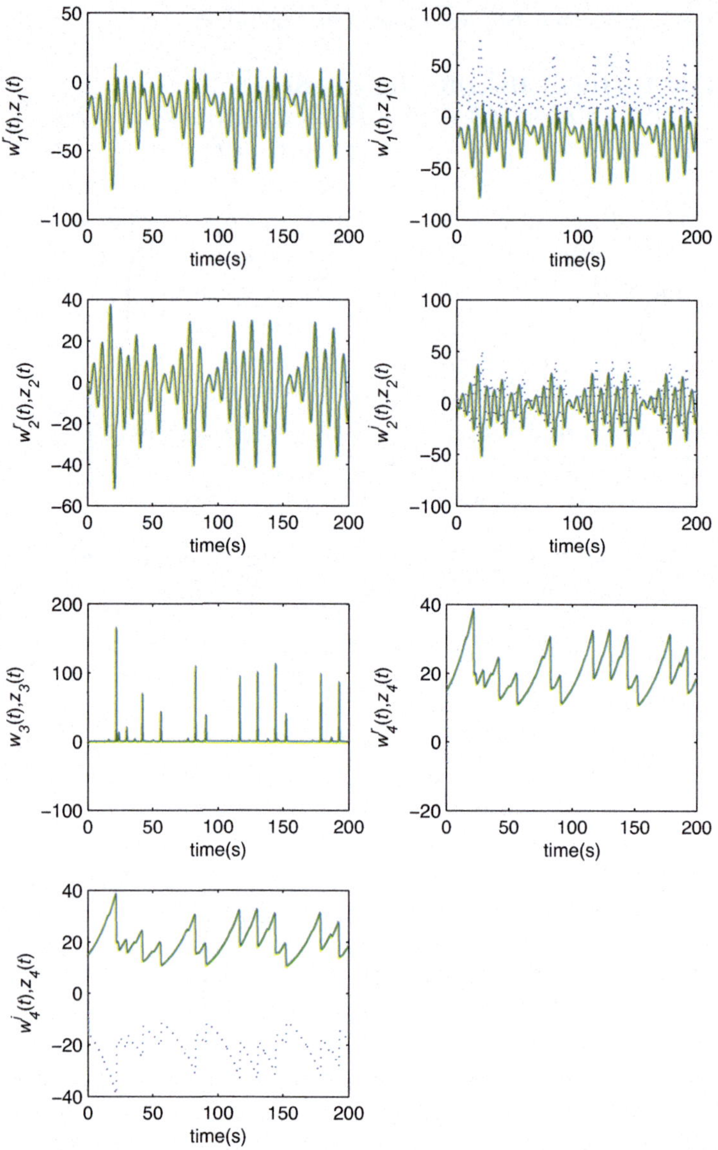

Fig. 6.5 The CMPS of system (6.2.22) and (6.2.23) with $\eta_1 = 0.05$, $\eta_2 = \eta_3 = 01$, $\eta_4 = \eta_5 = 0.1$, $\lambda_1 = \lambda_2 = \lambda_3 = \lambda_4 = \lambda_5 = 0.01$, $\varsigma_1 = \varsigma_2 = \varsigma_3 = \varsigma_4 = 0.01$, $\tau_1 = 0.05$, $\tau_2 = \tau_3 = 1$, $\tau_4 = 0.1$, $\gamma_1 = \gamma_2 = \gamma_3 = \gamma_4 = 10$ and complex scaling factors $1 - j$, $1 - j$, 1, $1 - j$

the response system (6.2.23) completely synchronize (z_1, z_2, z_3) of the drive system (6.2.22) in Fig. 6.6 as $h_1 = h_3 = h_5 = 1$, while the imaginary parts (w_1^i, w_2^i, w_4^i) of

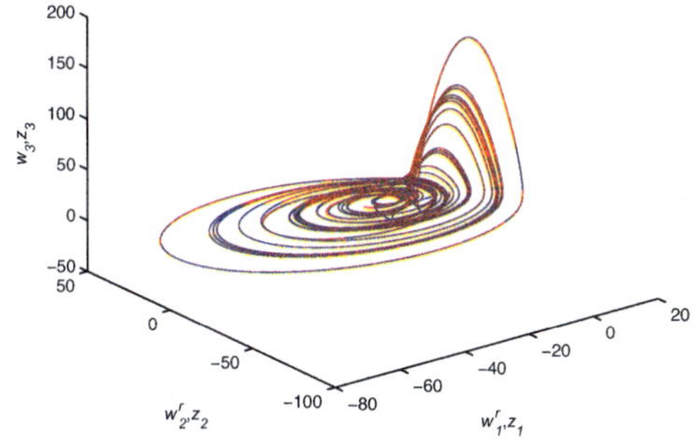

Fig. 6.6 The real parts (w_1^r, w_2^r, w_3) of system (6.2.23) synchronize (z_1, z_2, z_3) of system (6.2.22) as $h_1 = h_3 = h_5 = 1$

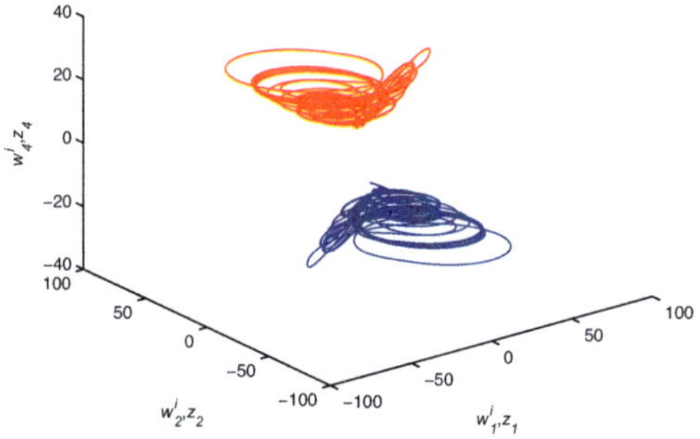

Fig. 6.7 The imaginary parts (w_1^i, w_2^i, w_4^i) of system (6.2.23) anti-synchronize (z_1, z_2, z_4) of system (6.2.22) as $h_2 = h_4 = h_7 = -1$

the response system (6.2.23) anti-synchronize (z_1, z_2, z_4) of the drive system (6.2.22) in Fig. 6.7 as $h_2 = h_4 = h_7 = -1$. Figure 6.8 shows that the errors of CMPS converge asymptotically to zero.

Figures 6.9 and 6.10 display that the processes of parameters identification of $\hat{\mathbf{A}}$ and $\hat{\mathbf{B}}$. The estimated vector $\hat{\mathbf{A}}$ converges to $(14.0383, 34.9475, 2.9451, -5.0000, -4.0000)^{\mathrm{T}}$, while $\hat{\mathbf{B}}$ converges to $(0.2071, 3.0332, 0.5231, 0.0523)^{\mathrm{T}}$, which are near the true values.

(II) CMPS of the complex Lorenz drive system and real Lorenz response system

To observe CMPS behaviors of the complex drive system and real response system, we analyze that the uncertain complex Lorenz system drives the uncertain real Lorenz system. Hence, the complex drive system is described as

$$\begin{cases} \dot{y}_1 = d_1(y_2 - y_1), \\ \dot{y}_2 = d_2 y_1 - y_1 y_3 - y_2, \\ \dot{y}_3 = -d_3 y_3 + (1/2)(\bar{y}_1 y_2 + y_1 \bar{y}_2), \end{cases} \tag{6.2.24}$$

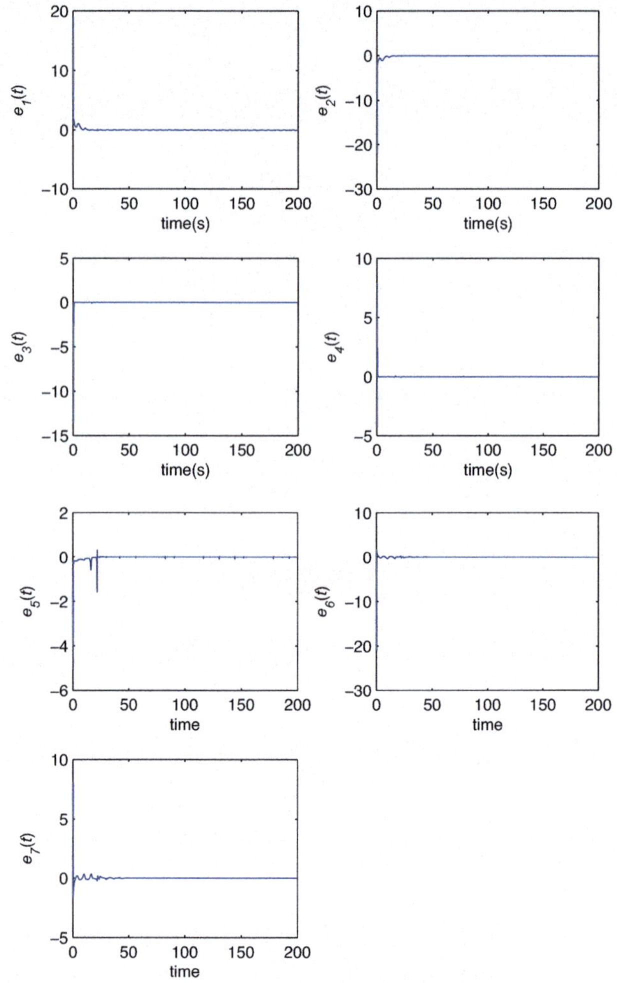

Fig. 6.8 The error dynamic of CMPS of systems (6.2.22) and (6.2.23)

where

$$\mathbf{Q}(\mathbf{y}) = \begin{pmatrix} y_2 - y_1 & 0 & 0 \\ 0 & y_1 & 0 \\ 0 & 0 & -y_3 \end{pmatrix}, \quad \mathbf{q}(\mathbf{y}) = \begin{pmatrix} 0 \\ -y_1 y_3 - y_2 \\ (1/2)(\bar{y}_1 y_2 + y_1 \bar{y}_2) \end{pmatrix},$$

$y_1 = u_1' + ju_2'$, $y_2 = u_3' + ju_4'$ denote complex state variables, $y_3 = u_5'$ is a real state variable, and $\mathbf{D} = (d_1, d_2, d_3)^{\mathrm{T}}$ is the unknown parameter vector.

The real response Lorenz system is defined as

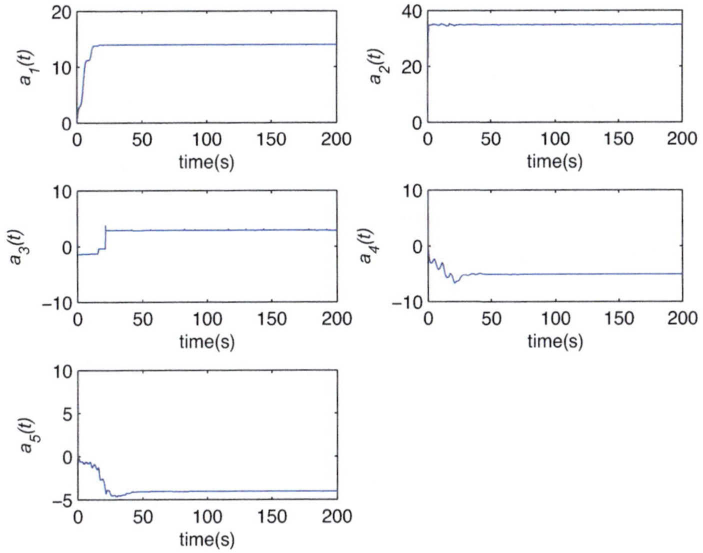

Fig. 6.9 The identification process of unknown parameter vector **A**

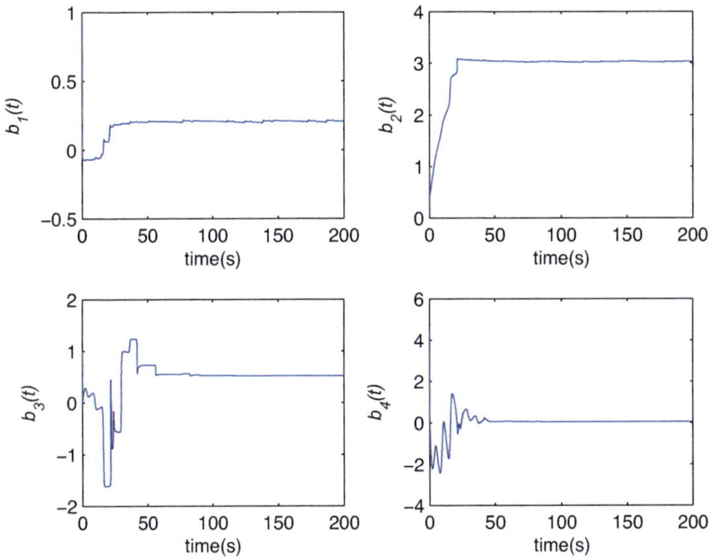

Fig. 6.10 The identification process of unknown parameter vector **B**

$$\begin{cases} \dot{x}_1 = c_1(x_2 - x_1), \\ \dot{x}_2 = c_2 x_1 - x_2 - x_1 x_3, \\ \dot{x}_3 = -c_3 x_3 + x_1 x_2, \end{cases} \tag{6.2.25}$$

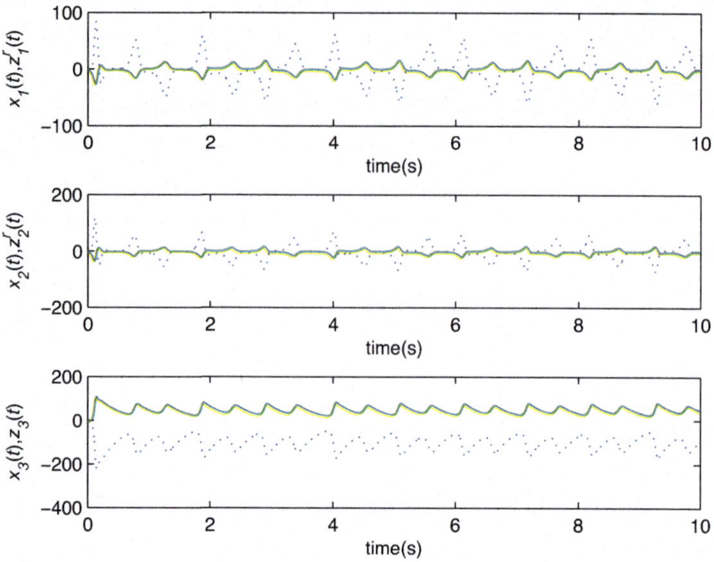

Fig. 6.11 The CMPS of systems (6.2.24) and (6.2.25) with $\eta_1 = \eta_2 = \eta_3 = 1$, $\lambda_1 = \lambda_2 = \lambda_3 = 0.01$, $\varsigma_1 = \varsigma_2 = \varsigma_3 = 0.01$, $\tau_1 = \tau_2 = \tau_3 = 1$, $\gamma_1 = \gamma_2 = \gamma_3 = 1$ and complex scaling factors $-2+j, -2+j, -2$

here

$$\mathbf{P}(\mathbf{x}) = \begin{pmatrix} x_2 - x_1 & 0 & 0 \\ 0 & x_1 & 0 \\ 0 & 0 & -x_3 \end{pmatrix}, \quad \mathbf{p}(\mathbf{x}) = \begin{pmatrix} 0 \\ -x_2 - x_1 x_3 \\ x_1 x_2 \end{pmatrix},$$

x_1, x_2, x_3 denote real state variables, and c_1, c_2, c_3 represent real unknown parameters.

Based on Theorem 6.9, the controller is designed as

$$\begin{aligned} \mathbf{v} &= -\mathbf{P}(\mathbf{x})\hat{\mathbf{C}} - \mathbf{p}(\mathbf{x}) + (\mathbf{H}^r\mathbf{Q}(\mathbf{y})^r - \mathbf{H}^i\mathbf{Q}(\mathbf{y})^i)\hat{\mathbf{D}} + \mathbf{H}^r\mathbf{q}(\mathbf{y})^r - \mathbf{H}\mathbf{q}(\mathbf{y})^i + \mathbf{K}\mathbf{e} \\ &= \begin{pmatrix} -\hat{c}_1(x_2 - x_1) + [h_1(u_3' - u_1') - h_2(u_4' - u_2')]\hat{d}_1 + k_1 e_1 \\ -\hat{c}_2 x_1 + x_2 + x_1 x_3 + (h_3 u_1' - h_4 u_2')\hat{d}_2 - h_3(u_1' u_5' + u_3') + h_4(u_2' u_5' + u_4') + k_2 e_2 \\ \hat{c}_3 x_3 - x_1 x_2 + h_5(-\hat{d}_3 u_5' + u_1' u_3' + u_2' u_4') + k_3 e_3 \end{pmatrix}, \end{aligned}$$

where $h_1 + jh_2, h_3 + jh_4, h_5$ are scaling factors, $e_1 = x_1 - (h_1 u_1' - h_2 u_2')$, $e_2 = x_2 - (h_3 u_3' - h_4 u_4')$, and $e_3 = x_3 - h_5 u_5'$.

The adaptive laws are chosen as (6.2.18) where

$$\nabla_{\hat{\mathbf{C}}}\omega(\hat{\mathbf{C}}, \hat{\mathbf{D}}, t) = \begin{pmatrix} -(x_2 - x_1)e_1 \\ -x_1 e_2 \\ x_3 e_3 \end{pmatrix}, \quad \nabla_{\hat{\mathbf{D}}}\omega(\hat{\mathbf{C}}, \hat{\mathbf{D}}, t) = \begin{pmatrix} e_1[h_1(u_3' - u_1') - h_2(u_4' - u_2')] \\ e_2(h_3 u_1' - h_4 u_2') \\ -u_5' e_5 \end{pmatrix},$$

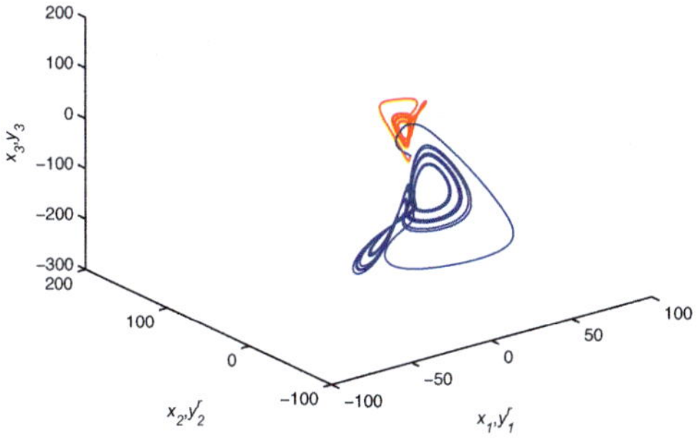

Fig. 6.12 The response system (6.2.25) synchronizes the real parts $(-2y_1^r - y_1^i, -2y_2^r - y_2^i, -2y_3)$ of the product that the drive system (6.2.24) multiplied by complex scaling factors $-2 + j, -2 + j, -2$

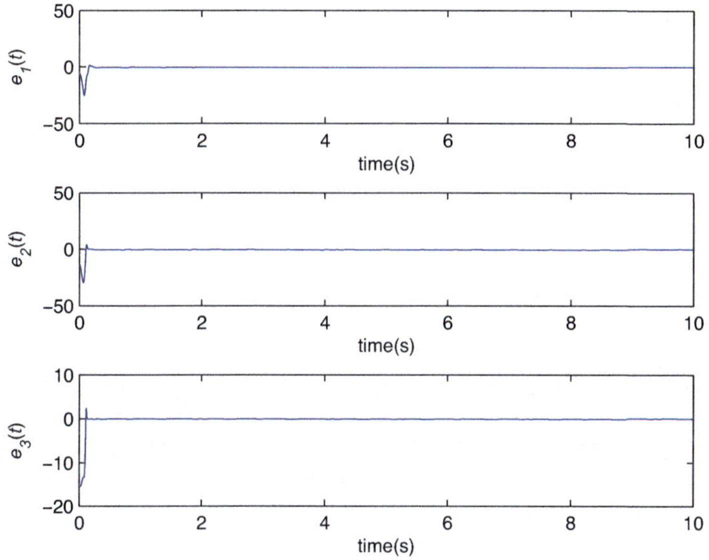

Fig. 6.13 The error dynamic of CMPS with $-2 + j, -2 + j, -2$ of systems (6.2.24) and (6.2.25). Here $e_1 = x_1 - (-2y_1^r - y_1^i)$, $e_2 = x_2 - (-2y_2^r - y_2^i)$ and $e_3 = x_3 - (-2y_3)$

and

$$\dot{\mathbf{k}} = \begin{pmatrix} \dot{k}_1 \\ \dot{k}_2 \\ \dot{k}_3 \end{pmatrix} = \begin{pmatrix} -\gamma_1 e_1^2 \\ -\gamma_2 e_2^2 \\ -\gamma_3 e_3^2 \end{pmatrix}.$$

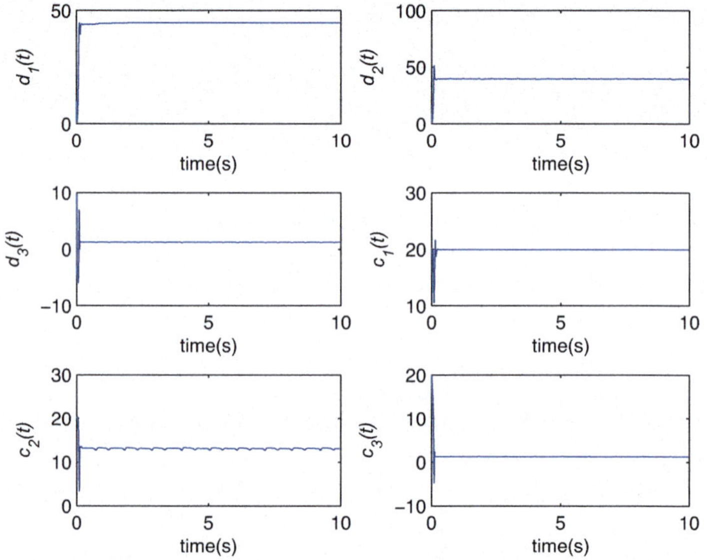

Fig. 6.14 The identification processes of unknown parameter vector **C** and **D**

Let $\mathbf{C} = (10,\ 28,\ 8/3)^{\mathrm{T}}$ and $\mathbf{D} = (35,\ 55,\ 8/3)^{\mathrm{T}}$. The initial conditions are chosen as $\mathbf{y(0)} = (-1-2j, -3-4j, -5)^{\mathrm{T}}$ and $\mathbf{x(0)} = (-1, -3, -5)^{\mathrm{T}}$. The initial values of estimated parameters and control strength are set as $\hat{\mathbf{C}}(0) = (20,\ 20,\ 20)^{\mathrm{T}}$, $\hat{\mathbf{D}}(0) = (10,\ 10,\ 10)^{\mathrm{T}}$ and $\mathbf{k}(0) = (0,\ 0,\ 0)^{\mathrm{T}}$. We solve the differential equations with a fourth-order Runge-Kutta method by using a time step of with $\Delta t = 10^{-3}s$. The CMPS process of systems (6.2.24) and (6.2.25) is depicted in Fig. 6.11, where the solid line shows the states of the drive system and the dotted line presents the states of the response system. Their chaotic behaviors are displayed in Fig. 6.12, where the red line shows the trajectory of the drive system and the blue line presents the trajectory of the response system. From Fig. 6.12, the response system (6.2.25) synchronizes the real parts of the product of **H** and the complex drive system (6.2.24). Figure 6.13 illustrates that the errors of CMPS converge asymptotically to zero. As displayed in Fig. 6.14, the processes of parameters identification of $\hat{\mathbf{C}}$ and $\hat{\mathbf{D}}$. Evidently, the estimated values of $\hat{\mathbf{C}}$ and $\hat{\mathbf{D}}$ converge to certain constants.

Chapter 7
Complex Function Projective Synchronization of Complex Chaotic System

Function projective synchronization (FPS) is a type of chaotic synchronization where the state variables of two coupled dynamical systems synchronize up to a scaling function [130–134]. Complex function projective synchronization (CFPS) is a unique type of synchronization to complex chaotic systems. The chapter presents CFPS of complex chaotic systems. This chapter includes two sections. Section 7.1 is the CFPS scheme of coupled chaotic complex systems. Section 7.2 gives some numerical simulations.

7.1 The CFPS Scheme of Coupled Chaotic Complex Systems

A general form of controlled coupled complex chaotic systems can be described by n-dimensional differential equations

$$\begin{cases} \dot{\mathbf{y}} = \mathbf{f}(\mathbf{y}, \mathbf{z}), \\ \dot{\mathbf{z}} = \mathbf{g}(\mathbf{y}, \mathbf{z}), \\ \dot{\mathbf{x}} = \mathbf{p}(\mathbf{x}, \mathbf{z}) + \mathbf{v}, \end{cases} \tag{7.1.1}$$

where $\mathbf{y} = (y_1, y_2, \ldots, y_q)^{\mathrm{T}}$ and $\mathbf{z} = (z_1, z_2, \ldots, z_{n-q})^{\mathrm{T}}$ denote complex state vectors of the drive system. The state vector $\mathbf{z}$ is treated as the coupling vector. $\mathbf{x} = (x_1, x_2, \ldots, x_q)^{\mathrm{T}}$ denotes a complex state vector of the response system. The designed controller is $\mathbf{v} = \mathbf{v}^r + j\mathbf{v}^i$, where $\mathbf{v}^r = (v_1, v_3, \ldots, v_{2q-1})^{\mathrm{T}}$, $\mathbf{v}^i = (v_2, v_4, \ldots, v_{2q})^{\mathrm{T}}$.

Based on the definitions of modified function projective synchronization (MFPS) [135–137] and generalized function projective synchronization (GFPS) [138, 139], the definition of CFPS for complex chaotic systems is given as follows.

S. Liu et al., *Chaos in Complex Dynamical System: Theory and Application*,
https://doi.org/10.1007/978-981-95-7762-0_7

Definition 7.1 For the complex chaotic state variables, there is said to be complex function projective synchronization (CFPS) between $\mathbf{x(t)}$ and $\mathbf{y(t)}$ if there exists a complex vector function $\mathbf{v}$ such that

$$\lim_{t\to\infty}\|\mathbf{x(t)}-\mathbf{H(y,z,t)y(t)}\|^2 = \|\mathbf{x}(t)^r-\mathbf{H(y,z,t)}^r\mathbf{y}(t)^r+\mathbf{H(y,z,t)}^i\mathbf{y}(t)^i\|^2$$
$$+\|\mathbf{x}(t)^i-\mathbf{H(y,z,t)}^r\mathbf{y}(t)^i-\mathbf{H(y,z,t)}^i\mathbf{y}(t)^r\|^2=0,$$

where $\mathbf{H(y,z,t)} = diag\{h_1(y,z,t), h_2(y,z,t), \ldots, h_q(y,z,t)\}, h_l(y,z,t) : \mathbb{C}^n \to \mathbb{C}$ $(l = 1, 2, \ldots, q)$ is a bounded complex function.

The error vector of CFPS between $\mathbf{x(t)}$ and $\mathbf{y(t)}$ is $\mathbf{e(t)} = \mathbf{x(t)} - \mathbf{H(y,z,t)y(t)}$ where $\mathbf{e} = \mathbf{e}^r + j\mathbf{e}^i$, and

$$\mathbf{e}^r = (e_1, e_3, \ldots, e_{2q-1})^{\mathrm{T}} = \mathbf{x}(t)^r-\mathbf{H(y,z,t)}^r\mathbf{y}(t)^r+\mathbf{H(y,z,t)}^i\mathbf{y}(t)^i,$$
$$\mathbf{e}^i = (e_2, e_4, \ldots, e_{2q})^{\mathrm{T}} = \mathbf{x}(t)^i-\mathbf{H(y,z,t)}^r\mathbf{y}(t)^i-\mathbf{H(y,z,t)}^i\mathbf{y}(t)^r.$$

Remark 7.1 The matrix $\mathbf{H(y,z,t)}$ is called a complex scaling function matrix, and $h_1(y,z,t)$, $h_2(y,z,t)$, ..., $h_q(y,z,t)$ are called complex scaling function factors. If there is no time variable in $\mathbf{H(y,z,t)}$, i.e. $h_1(y,z,t) = \psi_1(y,z)$, $h_2(y,z,t) = \psi_2(y,z)$, ..., $h_q(y,z,t) = \psi_q(y,z)$, the CFPS becomes GFPS. Especially, $\psi_1(y,z) = \psi_2(y,z) = \cdots = \psi_q(y,z)$, the GFPS is simplified to the function projective synchronization (FPS) [140, 141]. If there is no $\mathbf{y}$ and $\mathbf{z}$ in $\mathbf{H(y,z,t)}$, i.e. $h_1(y,z,t) = d_1(t)$, $h_2(y,z,t) = d_2(t)$, ..., $h_q(y,z,t) = d_q(t)$, it is called MFPS. In fact, MFPS is the same as GFPS.

Remark 7.2 If $h_1(y,z,t) = \delta_1$, $h_2(y,z,t) = \delta_2$, ..., $h_q(y,z,t) = \delta_q$ (δ_l is a real constant or complex constant, $l = 1, 2, \ldots, q$), the problem will become modified projective synchronization (MPS) with real scaling factors (if $\forall\delta_l \in R$) or complex scaling factors (if $\exists\delta_l \notin R$) for complex dynamical systems. If $\delta_1 = \delta_2 = \cdots = \delta_q = \delta$, the problem will be reduced to PS. In particular, if $\delta = 1$ and $\delta = -1$, the problem is further simplified to CS and AS, respectively.

Hence, the CFPS can contain most existing works and extend previous works. The control objective is to design a controller $\mathbf{v}$ that ensures the synchronization of $\mathbf{x(t)}$ with $\mathbf{y(t)}$ in the CFPS sense.

From the definition of CFPS, it follows that $e_l(t) = x_l(t) - h_l(\mathbf{y},\mathbf{z},t)z_l(t)$ $(l = 1, 2, \ldots, q)$, then

$$\begin{aligned}\dot{e}_l(t) &= \dot{x}_l(t) - \frac{d(h_l(\mathbf{y},\mathbf{z},t)y_l(t))}{dt}\\ &= \dot{x}_l(t) - ((\frac{\partial h_l(y,z,t)}{\partial t} + \nabla_{\mathbf{y}}h_l(\mathbf{y},\mathbf{z},t)\dot{\mathbf{y}} + \nabla_{\mathbf{z}}h_l(\mathbf{y},\mathbf{z},t)\dot{\mathbf{z}})y_l(t) + h_l(y,z,t)\dot{y}_l(t)).\end{aligned} \tag{7.1.2}$$

Set $J_l(t) = \frac{d(h_l(\mathbf{y},\mathbf{z},t)y_l(t))}{dt}$, then $\mathbf{J} = \frac{d(\mathbf{H(y,z,t)y(t)})}{dt} = (J_1, J_2, \ldots, J_q)^{\mathrm{T}}$ is a complex vector.

Lemma 7.1 ([117]) *If* $\varepsilon : [0, \infty] \to R$ *satisfies* $\lim_{t\to\infty} \int_0^t \varepsilon^2(\tau)d\tau < \infty$, $\dot{\varepsilon}(t)$ *exists and it is bounded, then* $\lim_{t\to\infty} \varepsilon(t) = 0$.

Theorem 7.1 *For given complex function scaling matrix* $\mathbf{H}$ *and any initial conditions* $\mathbf{y}(0), \mathbf{z}(0), \mathbf{x}(0)$, *the CFPS of coupled chaotic complex system* (7.1.1) *will occur by the following control law*

$$\mathbf{v} = -\mathbf{p}(\mathbf{x}, \mathbf{z}) + \mathbf{J} - \mathbf{K}\mathbf{e}, \tag{7.1.3}$$

where $\mathbf{K} = diag(k_1, k_2, \ldots, k_q)$ *is the control strength matrix, and all its elements are positive real numbers.*

Proof From (7.1.1), (7.1.2) and (7.1.3), it follows that

$$\dot{\mathbf{e}} = \dot{\mathbf{x}} - \frac{d(\mathbf{H}(\mathbf{y}, \mathbf{z}, \mathbf{t})\mathbf{y}(\mathbf{t}))}{dt} = \dot{\mathbf{x}} - \mathbf{J} = \mathbf{p}(\mathbf{x}, \mathbf{z}) - \mathbf{p}(\mathbf{x}, \mathbf{z}) + \mathbf{J} - \mathbf{K}\mathbf{e} - \mathbf{J} = -\mathbf{K}\mathbf{e}.$$

Then we have $\dot{\mathbf{e}}^r = -\mathbf{K}\mathbf{e}^r$ and $\dot{\mathbf{e}}^i = -\mathbf{K}\mathbf{e}^i$. The matrix $\mathbf{K}$ regulates the convergence rate of the synchronization process.

Consider the following nonnegative Lyapunov function:

$$V(e, t) = \frac{1}{2}[(\mathbf{e}^r)^{\mathrm{T}}\mathbf{e}^r + (\mathbf{e}^i)^{\mathrm{T}}\mathbf{e}^i],$$

then we get that

$$\dot{V} = (\dot{\mathbf{e}}^r)^{\mathrm{T}}\mathbf{e}^r + (\dot{\mathbf{e}}^i)^{\mathrm{T}}\mathbf{e}^i = -\sum_{l=1}^{q} k_l e_l^r(t)^2 - \sum_{l=1}^{q} k_l e_l^i(t)^2 = -\sum_{l=1}^{q} k_l[e_l^r(t)^2 + e_l^i(t)^2] \le -k_{min}\|\mathbf{e}\|^2,$$

where $k_{min} = min(k_1, k_2, \ldots, k_q)$.

By integrating both sides, we obtain

$$\int_0^t \|\mathbf{e}\|^2 dt = \int_0^t \sum_{l=1}^{q} [e_l^r(t)^2 + e_l^i(t)^2]dt \le [V(0) - V(t)]/k_{min} \le V(0)/k_{min} < \infty.$$

Since $\dot{\mathbf{e}}$ exists and it is bounded, $\dot{\mathbf{e}}^r$ and $\dot{\mathbf{e}}^i$ exist and they are bounded. Then

$$\frac{d(\|\mathbf{e}(t)\|)}{dt} = 0.5\|\mathbf{e}(t)\|^{-0.5}[2(\dot{\mathbf{e}}^r)^{\mathrm{T}}\mathbf{e}^r + 2(\dot{\mathbf{e}}^i)^{\mathrm{T}}\mathbf{e}^i] = \frac{(\dot{\mathbf{e}}^r)^{\mathrm{T}}\mathbf{e}^r + (\dot{\mathbf{e}}^i)^{\mathrm{T}}\mathbf{e}^i}{\|\mathbf{e}(t)\|}.$$

Hence, we get the derivative of $\|\mathbf{e}(t)\|$ exists and it is bounded. It follows that $\lim_{t\to+\infty} \|\mathbf{e}(t)\| = 0$ from Lemma 7.1. Therefore, CFPS with $\mathbf{H}$ is realized for coupled chaotic system (7.1.1). The proof is completed. □

Remark 7.3 Note that the complex function scaling matrix **H** has no effect on the time derivative $\dot{V}$, thus one can adjust **H** arbitrarily without worrying about the control robustness.

Hence, the complex function scaling matrix **H** can be used as information signals for the following communication scheme. In practice, $q = 1$ is typically set to reduce controller complexity. These aspects will be illustrated in the following example.

7.2 A Novel Communication Scheme

We study the application of CFPS to secure communication in theory. The block diagram of the proposed communication scheme is presented in Fig. 7.1, with L1 and L2 acting as the transmitter and receiver, respectively.

In Fig. 7.1, the transmitted signal is

$$\begin{aligned} s(t) &= J(t) + D\varepsilon(t) + jD\varepsilon(t) \\ &= \frac{dh(t)y(t)}{dt} + D\varepsilon(t) + jD\varepsilon(t) \\ &= \dot{h}y + h\dot{y} + D\varepsilon + jD\varepsilon \\ &= (\dot{h}^r y^r - \dot{h}^i y^i + h^r \dot{y}^r - h^i \dot{y}^i + D\varepsilon) + j(\dot{h}^r y^i + \dot{h}^i y^r + h^r \dot{y}^i + h^i \dot{y}^r + D\varepsilon), \end{aligned}$$

where $D\varepsilon(t)$ simulates the noise both appeared in the communication channel and produced by the noise source. The scaling function factor $h(t)$ serves as the information signal. Another state vector **z** is sent to the receiver. At the receiver end, the controller **v** is designed as $\mathbf{v} = -p(x, z) + s(t) - Ke$. As CFPS occurs, $x(t)$ will converge to $h(t)y(t)$. It is evident that the recovered signal $h_g(t) = x(t)/y(t)$. To enhance the security of communications, we may also utilize $m(t)$ as the information signal, making $h(t)$ a function of the information signal. For digital signals, we introduce an encoder and a decoder, indicated by a dotted line.

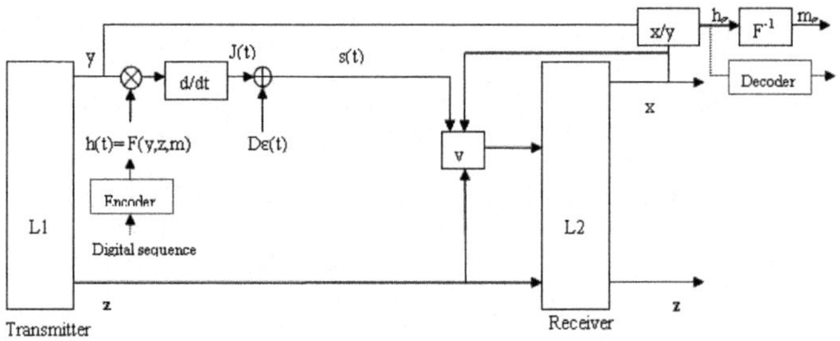

Fig. 7.1 The block diagram of our communication scheme

Remark 7.4 The real part and imaginary part of $s(t)$ double the number of real variables, which increase the contents of the transmitted information in secure communication.

Remark 7.5 The real parts or the imaginary parts of $s(t)$ are related to all parts of drive system and information signal. It is no longer to transmit each information signal using separate channel. The transmitted signals are arbitrary combinations of the real parts and the imaginary parts of state variables of drive system, which increases the security of secure communication.

The proposed scheme exhibits a number of favorable characteristics. Firstly, given that the transmitted signal $s(t)$ involves complex numerical operations, any unauthorized third party lacking substantial information about the transmitter generator's characteristics and the noise profile cannot decode the information signal. This inherent complexity enhances the communication security. Secondly, compared with CS that usually used for chaos communication, CFPS can be observed in the nonidentical and even different dynamical systems (including the systems with the different dimension of the phase space). At the same time, CS requires an identity of the transmitting and receiving generators being located in the different sides of the communication channel. It is a very big conceptual problem, especially for a long time of operation of the devices [142]. Thirdly, as for digital symbols, we convert M binary bits into one complex scaling factor by corresponding computation, which increases the rate of the information transmission. Finally, the noise is used to provide the additional masking of the information signal and there is a simple filter in the receiver end to assure the accuracy of recovered information.

In summary, CFPS is highly effective for secure communication, offering high security, rapid transmission, low BER, and strong noise robustness.

7.3 Numerical Simulations

The following coupled complex Lorenz system with a controller v is described as:

$$L1: \begin{cases} \dot{y} = a_1(z_1 - y), \\ \dot{z}_1 = a_2 y - y z_2 - z_1, \\ \dot{z}_2 = -a_3 z_2 + (1/2)(\bar{y} z_1 + y \bar{z}_1), \end{cases} \tag{7.3.1}$$

and

$$L2: \dot{x} = a_1(z_1 - x) + v, \tag{7.3.2}$$

where $y = u_1' + ju_2'$, $z_1 = u_3' + ju_4'$ are complex state variables of drive system $L1$, $x = u_1 + ju_2$ denotes the complex state variable of response system $L2$, the coupling variable is z_1 and the real variable is $z_2 = u_5'$, and $(a_1, a_2, a_3)^{\mathrm{T}}$ represents the real parameter vector.

Following Theorem 7.1, we design the controller as

$$\begin{cases} v^r = -a_1(z_1^r - x^r) + J^r - ke^r + D\varepsilon, \\ v^i = -a_1(z_1^i - x^i) + J^i - ke^i + D\varepsilon, \end{cases} \tag{7.3.3}$$

where $J(t) = \frac{d(h(t)y(t))}{dt}$, $e^r = x^r - (h^r y^r - h^i y^i)$, and $e^i = x^i - (h_i y^r + h_r y^i)$. The initial conditions are chosen as $y(0) = 1 + 2j$, $z_1(0) = 3 + 4j$, $z_2(0) = 5$, and $x(0) = -1 - 2j$. The parameters are set to $a_1 = 35, a_2 = 55, a_3 = 8/3$.

7.3.1 Analog Signals

To begin, an analogous signal is transmitted. For simplicity, the function $h(t)$ is chosen as $2sin(0.5\pi t) + jcos(\pi t)$, yielding

$$\begin{cases} J^r = u_1'\pi cos(0.5\pi t) + u_2'\pi sin(\pi t) + 2a_1 sin(0.5\pi t)(u_3' - u_1') - a_1 cos(\pi t)(u_4' - u_2'), \\ J^i = u_2'\pi cos(0.5\pi t) - u_1'\pi sin(\pi t) + a_1 cos(\pi t)(u_3' - u_1') + 2a_1 sin(0.5\pi t)(u_4' - u_2'). \end{cases} \tag{7.3.4}$$

The noise $\varepsilon(t)$ is modeled as a stochastic Gaussian process described by the following probability distribution

$$q(\varepsilon) = \frac{1}{\sqrt{2\pi}\sigma} exp(-\frac{(\varepsilon - \varepsilon_0)^2}{2\sigma^2}), \tag{7.3.5}$$

where $\varepsilon_0 = 0$ and $\sigma = 1$ are the mean value and variance [142].

Consequently, it is impossible to recover the information signal $h(t)$ from the transmitted signal without access to the dynamics of y, z_1 and $D\varepsilon(t)$. Setting $D = 5$, we find that the transmitted signal completely masks the information signal, as illustrated in Fig. 7.2. Figure 7.3 describes the time evolution of $\mathbf{x}$ and $\mathbf{y}$ for a control strength $k = 1500$, with $y(t)$ and $x(t)$ traced by the blue and red lines, respectively. Figure 7.4 shows the controller output $\mathbf{v}$. The error vector of CFPS is presented in Fig. 7.5, where the error vector converges to zero quickly as time increases. These results demonstrate the successful achievement of CFPS under the desired function scaling matrix. As shown in Fig. 7.6, the information signal $h(t)$ is accurately recovered by $h_g(t)$, confirming the effectiveness of the proposed scheme.

7.3.2 Digital Signals

A binary digital sequence, generated using a random function in Matlab, is transmitted in this section. As illustrated in Fig. 7.7, the M binary bits are first converted into corresponding scaling factors through a 2^M-ary scheme before being fed into the

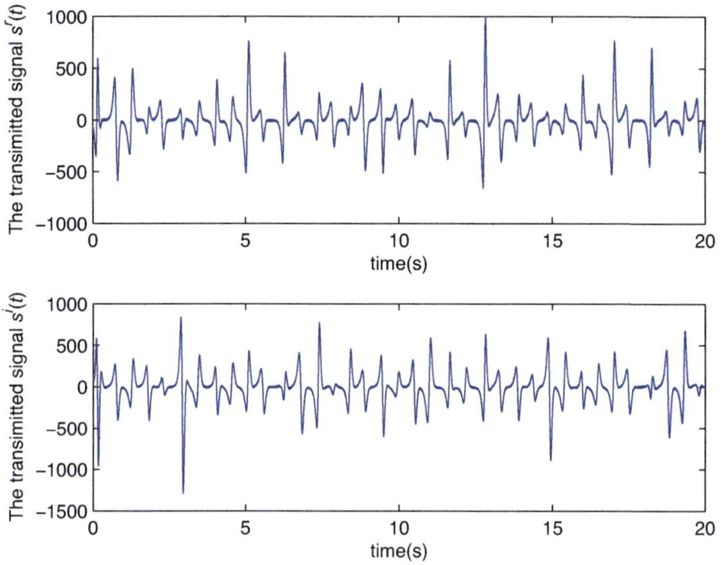

Fig. 7.2 The transmitted signal $s(t)$

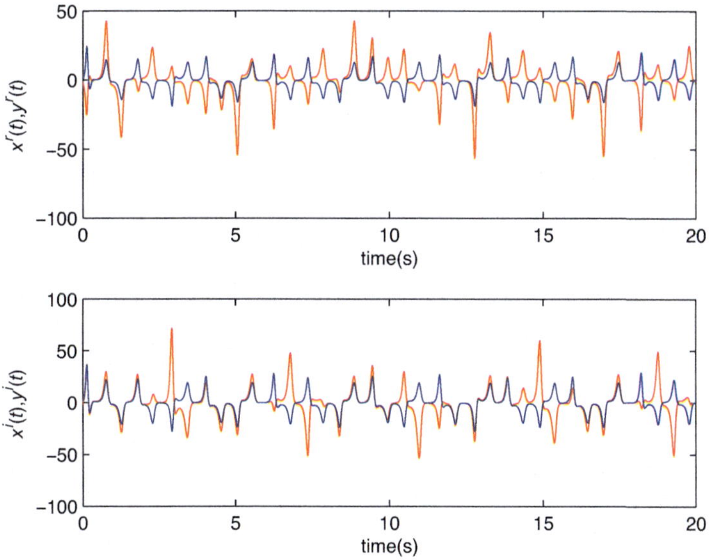

Fig. 7.3 The time evolution of $\mathbf{x}$ and $\mathbf{y}$ with $k = 1500$

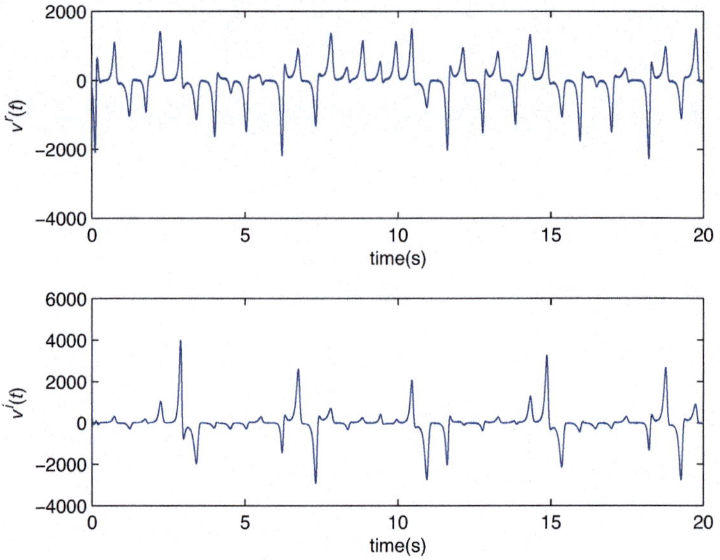

Fig. 7.4 The output of the controller **v**

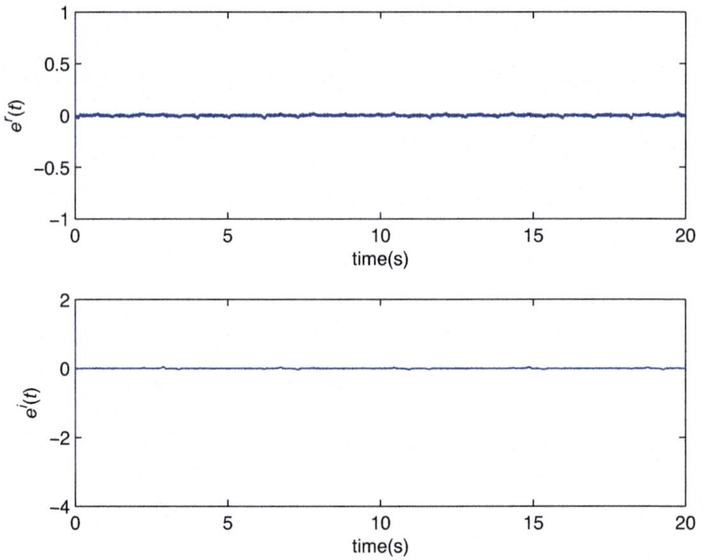

Fig. 7.5 The error vector of CFPS of coupled complex Lorenz systems (7.3.1) and (7.3.2)

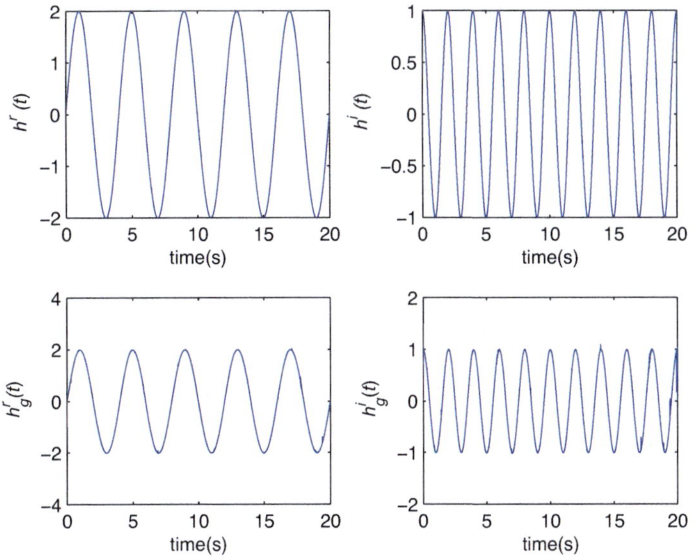

Fig. 7.6 The information signal $h(t)$ and the recovered signal $h_g(t)$

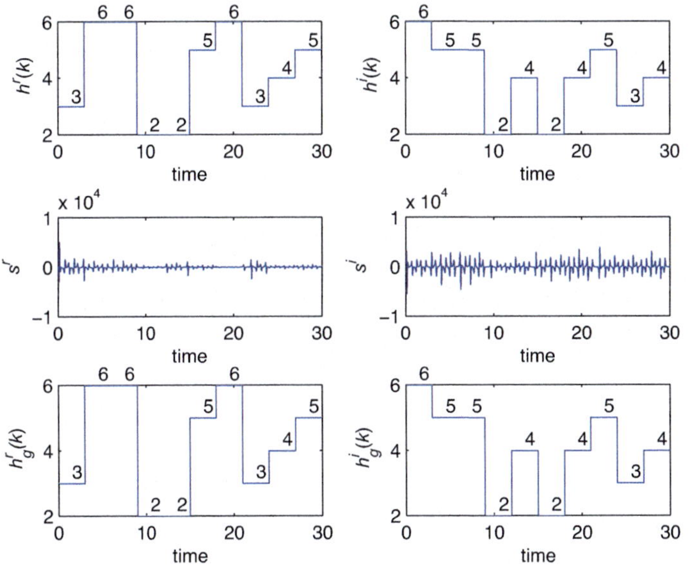

Fig. 7.7 The digital signals transmitted without noise

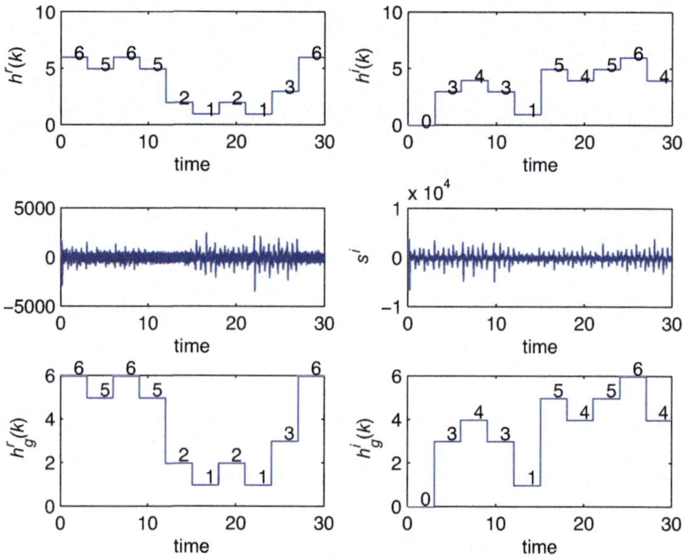

Fig. 7.8 The digital signals transmitted with stochastic Gaussian noise (7.3.5) $D = 200$

controller (7.3.3). Here the assignment of scaling factors to digital symbols is carried out by 8-ary, i.e., $M = 3$. The symbol duration used here is equal to 3 time-units. The computed scaling factors transmitted to the receiver and the corresponding transmitted signal $s(t)$ are shown in Fig. 7.7 with $D = 0$, where $s(t)$ is an analogous signal and completely covers up the binary digital sequence. This indicates that the relationship between the drive and response states converges to a constant complex value. Consequently, digital symbols can theoretically be retrieved with perfect accuracy, yielding a bit error rate of zero.

Figure 7.8 illustrates the transmission process of the computed scaling factors under stochastic Gaussian noise. The corresponding transmitted signal $s(t)$ is expressed as

$$\begin{cases} s^r(t) = J^r + D\varepsilon(t) = a_1[h^r(z_1^r - y^r) - h^i(z_1^i - y^i)] + D\varepsilon(t), \\ s^r(t) = J^i + D\varepsilon(t) = a_1[h^i(z_1^r - y^r) + h^r(z_1^i - y^i)] + D\varepsilon(t). \end{cases} \tag{7.3.6}$$

As evident, $s(t)$ constitutes a composite of the real and imaginary components from both the state variables of the drive system and the information signal, thereby employing dual channels for simultaneous transmission of two signal sequences. Therefore, it is more complicated than real signal, and it increases the contents and the security of secure communication. Moreover, the information signal $h(t)$ is masked by chaotic signals and noises. As the complexity of ideal white noise is 1, the complexity of $s(t)$ tends to 1 with $D \to +\infty$. As shown in Fig. 7.8 with $D = 200$, the transmitted signal $s(t)$ closely resembles noise, making it difficult to extract the information signal.

To retrieve the transmitted symbols from the varying scaling factor and reduce the effect of noise, the basic idea of the filter is to compute the average value of the scaling factors for each symbol duration before rounding off to the nearest integer that is bounded between 0 and $2^M - 1$. Consequently, the proposed scheme ensures low-BER recovery of digital symbols. As illustrated in Fig. 7.8, both digital sequences are accurately recovered even at a noise amplitude of 200. This indicates that the BER tends toward zero, with the signal to noise ratio (SNR) being approximately 10. Compared with the communication scheme using PS with real signal [143–145] (where an error bit occurred when SNR equals to 10), this communication scheme based on CFPS has lower BER for transmitting digital symbols.

Chapter 8
The Characteristics and Synchronization of Time-Delay Complex Chaotic System

Time-delay is frequently encountered in a variety of practical chaotic systems, such as chaos communication [146–150]. The behaviors of time-delay chaotic system are greatly different from those of the original system [151–155]. Self-time-delay synchronization (STDS) implies that the synchronization between the time-delay system and the original system while maintaining the structure and parameters of systems unchanged, thus these various problems produced by time-delay in practice are avoided. This chapter explores the characteristics and synchronization of time-delay complex chaotic system, and includes three sections. Section 8.1 presents the characteristics of time-delay complex Lorenz system. Section 8.2 studies STDS of time-delay coupled complex chaotic system and its applications to communication.

8.1 The Characteristics of Time-Delay Complex Lorenz System

The complex Lorenz system is depicted as

$$\begin{cases} \dot{y}_1 = a_1(y_2 - y_1), \\ \dot{y}_2 = a_2 y_1 - y_1 y_3 - y_2, \\ \dot{y}_3 = -a_3 y_3 + (1/2)(\bar{y}_1 y_2 + y_1 \bar{y}_2), \end{cases} \tag{8.1.1}$$

where $y_1 = u_1 + ju_2$, $y_2 = u_3 + ju_4$ are complex-variable states, $y_3 = u_5$ is a real-variable state, $(a_1, a_2, a_3)^{\mathrm{T}}$ is the real parameter vector.

Without changing the structure and parameters of the complex Lorenz system, we give the following one-time-delay system:

S. Liu et al., *Chaos in Complex Dynamical System: Theory and Application*,
https://doi.org/10.1007/978-981-95-7762-0_8

$$\begin{cases} \dot{x}_1 = a_1(x_2 - x_1), \\ \dot{x}_2 = a_2x_1 - x_1x_3 - x_2, \\ \dot{x}_3 = -a_3x_3(t-\tau) + (1/2)(\bar{x}_1x_2 + x_1\bar{x}_2), \end{cases} \tag{8.1.2}$$

where $\tau = \tau(t)(0 \leq \tau(t) \leq \tau_m)$ is a bounded time-varying lag, $x_1 = u'_1 + ju'_2$ and $x_2 = u'_3 + ju'_4$ are complex-variable states, and $x_3 = u'_5$ is a real-variable state. Separating the real and imaginary parts of system (8.1.2), we obtain the following real versions:

$$\begin{cases} \dot{u'}_1 = a_1(u'_3 - u'_1), \\ \dot{u'}_2 = a_1(u'_4 - u'_2), \\ \dot{u'}_3 = a_2u'_1 - u'_1u'_5 - u'_3, \\ \dot{u'}_4 = a_2u'_2 - u'_2u'_5 - u'_4, \\ \dot{u'}_5 = -a_3u'_5(t-\tau) + (u'_1u'_3 + u'_2u'_4). \end{cases} \tag{8.1.3}$$

In the following, we investigate the basic dynamical properties of system (8.1.3).

I. Chaotic behavior

Their different chaotic attractor projections and states are shown in Fig. 8.1 (τ = 1s) and Fig. 8.2 (τ = 2cos(t)). It is clear that time-delay complex Lorenz system is chaotic. Here, we adopt $a_1 = 14$, $a_2 = 35$, $a_3 = 3.7$.

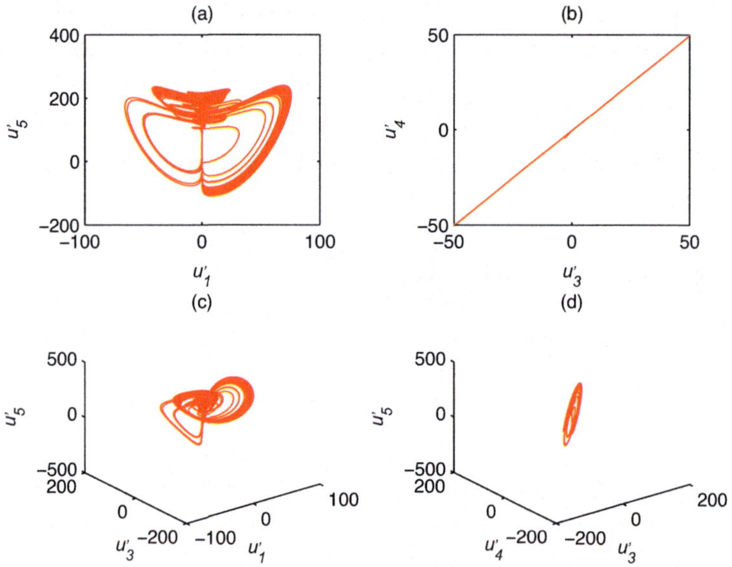

Fig. 8.1 The different chaotic attractor projections of time-delay complex Lorenz system (8.1.3) with τ = 1s

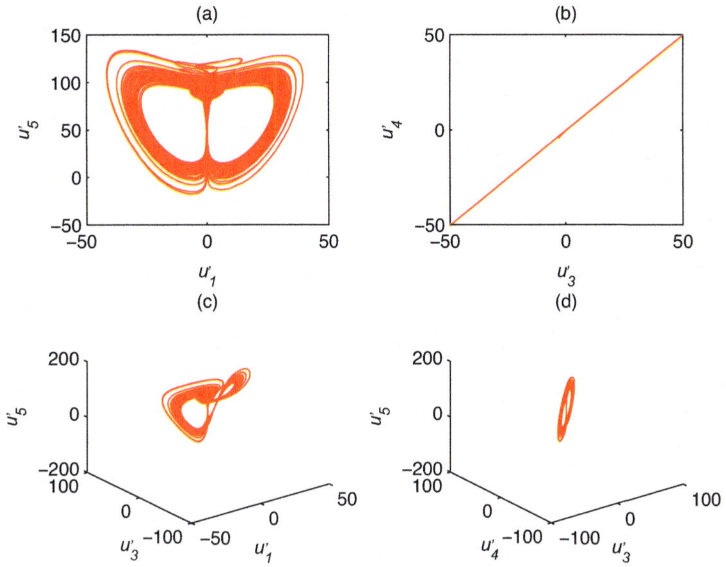

Fig. 8.2 The different chaotic attractor projections of time-delay complex Lorenz system (8.1.3) with $\tau = 2\cos(t)$s

II. The sensitivity of initial condition

When $\tau = 1$s, we select two close initial conditions such as $(-1, -2, -3, -4, 1)^T$ and $(-1.001, -2.001, -3, -4, 1)^T$, and obtain the evolution of states shown as Fig. 8.3. Time-delay complex Lorenz system displays sensitive dependence on initial conditions.

III. Lyapunov exponent

According to the definition of Lyapunov exponent, we obtain five Lyapunov exponents of system (8.1.3) with τ varying from 0 to 1.5, which are plotted in Fig. 8.4. Obviously, it is clearly that the characteristics of time-delay complex Lorenz system and number of positive Lyapunov exponents are related to the time-delay τ. It also indicates that the simple structure of time-delay chaotic system can generate highly stochastic and unpredictable time series, which will improve encryption performance of chaos-communication.

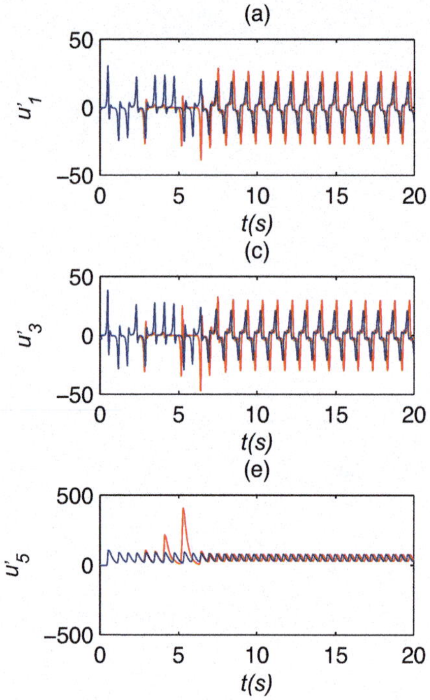

Fig. 8.3 The sensitivity of initial condition for time-delay complex Lorenz system (8.1.3) with $\tau = 1\mathrm{s}$

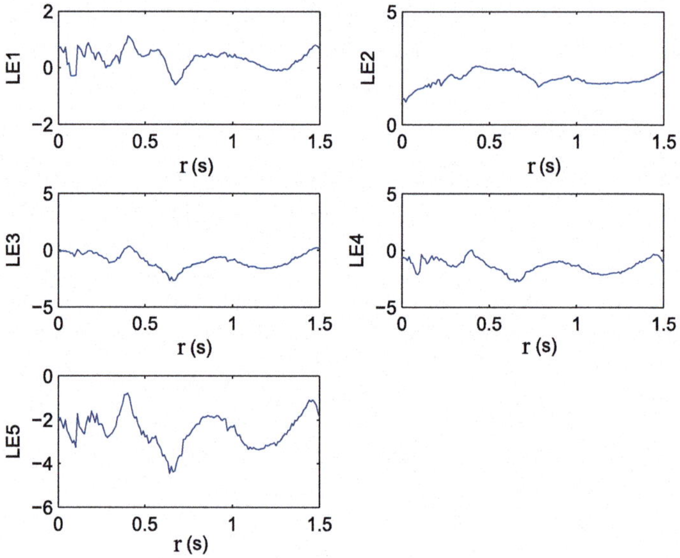

Fig. 8.4 Lyapunov exponents of system (8.1.3) with τ varying

8.2 Self-Time-Delay Synchronization of Time-Delay Complex Lorenz System

8.2.1 The Definition of Self-Time-Delay Synchronization

Definition 8.1 For two n-dimensional complex chaotic systems (8.2.1) and (8.2.2) as follows,

$$\dot{\mathbf{y}}(t) = f(\mathbf{y}(t)) \leftarrow \textit{original system}, \tag{8.2.1}$$

$$\dot{\mathbf{x}}(t) = f(\mathbf{x}(t-\tau)) + \mathbf{v}(\mathbf{x}(t-\tau), \mathbf{x}(t), \mathbf{y}(t)) \leftarrow \textit{time} - \textit{delay system}, \tag{8.2.2}$$

where $\mathbf{x(t)} = \{x_1(t), x_2(t), \ldots, x_n(t)\}^{\mathrm{T}}$ and $\mathbf{y(t)} = \{y_1(t), y_2(t), \ldots, y_n(t)\}^{\mathrm{T}}$ are complex state vectors, $\tau = \{\tau_1, \tau_2, \ldots, \tau_n\}^{\mathrm{T}}$ $(\tau_i \geq 0, i = 1, 2, \ldots, n)$ is the time lag vector. There is said to be self-time-delay synchronization (STDS) between $\mathbf{x(t)}$ and $\mathbf{y(t)}$, if there exists the controller v such that

$$\lim_{t\to\infty} \|\mathbf{x(t)} - \mathbf{y(t)}\|^2 = \lim_{t\to\infty} (\|\mathbf{x(t)}^r - \mathbf{y(t)}^r\|^2 + \|\mathbf{x(t)}^i - \mathbf{y(t)}^i\|^2) = 0.$$

Remark 8.1 If $\tau = 0$, STDS becomes CS. STDS contains CS and extends CS, thus the research on STDS further expands the scope of synchronization.

Remark 8.2 It is different from LS [156–160]. LS implies that the state variables of the two chaotic systems become synchronized but with a time lag with respect to each other [25]. Neither of these two systems is time-delay system. STDS implies the state variables of the time-delay system and the original system become synchronized without time lag.

8.2.2 The STDS Controller of Time-Delay Coupled Complex Chaotic Systems

The time-delay coupled complex chaotic system is expressed by an n-dimensional differential equation

$$\begin{cases} \dot{\mathbf{y}}(t) = \mathbf{A}\mathbf{y}(t) + \mathbf{f}(\mathbf{y(t)}, \mathbf{z(t)}), \\ \dot{\mathbf{z}}(t) = \mathbf{g}(\mathbf{y(t)}, \mathbf{z(t)}), \\ \dot{\mathbf{x}}(t) = \mathbf{A}\mathbf{x}(t) + \mathbf{f}(\mathbf{x(t)}, \mathbf{z(t-\tau)}) + \mathbf{v}, \end{cases} \tag{8.2.3}$$

where the state of original system is broken into two parts $(\mathbf{y}, \mathbf{z})$, and $\mathbf{y} = (y_1, y_2, \ldots, y_q)^{\mathrm{T}}$ and $\mathbf{z} = (z_1, z_2, \ldots, z_{n-q})^{\mathrm{T}}$ are complex state vectors. The state vector $\mathbf{z}$ is regarded as the coupling vector. Here, $\mathbf{x} = (x_1, x_2, \ldots, x_q)^{\mathrm{T}}$ is a complex state vector of time-delay system, $\mathbf{A}$ is a real $q \times q$ matrix, $\mathbf{f}(\mathbf{y}, \mathbf{z})$ and $\mathbf{g}(\mathbf{y}, \mathbf{z})$ are $q \times 1$ and $(n-q) \times 1$ complex function vector, respectively. The designed controller is $\mathbf{v} = \mathbf{v}^r + j\mathbf{v}^i$, where $\mathbf{v}^r = (v_1, v_3, \ldots, v_{2q-1})^{\mathrm{T}}$, $\mathbf{v}^i = (v_2, v_4, \ldots, v_{2q})^{\mathrm{T}}$. The goal is to design the controller $\mathbf{v}$ so that $\mathbf{x(t)}$ and $\mathbf{y(t)}$ are asymptotically synchronized in the sense of STDS.

Let $\mathbf{e}(t) = \mathbf{x}(t) - \mathbf{y}(t)$. For simplicity, we always adopt $\mathbf{e}, \mathbf{z}_\tau, \mathbf{x}, \mathbf{y}$ instead of $\mathbf{e}(t), \mathbf{z}(t-\tau), \mathbf{x}(t), \mathbf{y}(t)$ respectively. Following the definition of STDS, we obtain the complex error dynamical system as

$$\dot{\mathbf{e}} = \dot{\mathbf{x}} - \dot{\mathbf{y}} = \mathbf{A}\mathbf{e} + \mathbf{v} + \mathbf{f}(\mathbf{x}, \mathbf{z}_\tau) - \mathbf{f}(\mathbf{y}, \mathbf{z}). \tag{8.2.4}$$

According to the idea of active control method, we have

Theorem 8.1 *For any initial conditions* $\mathbf{y}(0), \mathbf{z}(0), \mathbf{x}(0)$, *the STDS of coupled chaotic complex system* (8.2.3) *will occur by the control law* (8.2.5) *as below*

$$\mathbf{v} = -\mathbf{f}(\mathbf{x}, \mathbf{z}_\tau) + \mathbf{f}(\mathbf{y}, \mathbf{z}) - \mathbf{K}\mathbf{e}, \tag{8.2.5}$$

where $\mathbf{K} = diag(k_1, k_2, \ldots, k_q)$ *is the real positive constant control strength matrix.*

Proof Substituting (8.2.5) into (8.2.4), we have

$$\dot{\mathbf{e}} = (\mathbf{A} - \mathbf{K})\mathbf{e} = (\mathbf{A} - \mathbf{K})\mathbf{e}^r + j(\mathbf{A} - \mathbf{K})\mathbf{e}^i.$$

Thus, $\dot{\mathbf{e}}^r = (\mathbf{A} - \mathbf{K})\mathbf{e}^r$, $\dot{\mathbf{e}}^i = (\mathbf{A} - \mathbf{K})\mathbf{e}^i$. The control strength matrix $\mathbf{K}$ is used to adjust the convergence rate of synchronization.

The control strength matrix $\mathbf{K}$ is selected appropriately to ensure that the real parts of all the eigenvalues of $(\mathbf{A} - \mathbf{K})$ are negative. Based on the Lyapunov stable theory, we obtain that $\lim_{t\to\infty} \mathbf{e}^r = 0$ and $\lim_{t\to\infty} \mathbf{e}^i = 0$. Therefore, $\mathbf{x(t)}$ can synchronize $\mathbf{y(t)}$ asymptotically in the sense of STDS. The proof is completed. □

Remark 8.3 Most of the typical chaotic complex systems can be written as the system (8.2.3), such as complex Lorenz system, complex Chen system, complex Lü system, complex Van der Pol oscillator, complex Duffing system and so on.

Remark 8.4 The convergence rate of error system can be flexibly regulated by choosing the eigenvalues of matrix $\mathbf{A} - \mathbf{K}$. The larger the absolute values of negative real parts of eigenvalues are, the faster convergence speed of errors is, i.e. the speed of realizing synchronization of system (8.2.3) increases.

8.2.3 A Novel Communication Scheme Based on STDS

This section presents a theoretical investigation into the application of STDS for secure communication. We adopt $L1$ as the transmitter and $L2$ as the receiver by the following time-delay coupled complex chaotic system

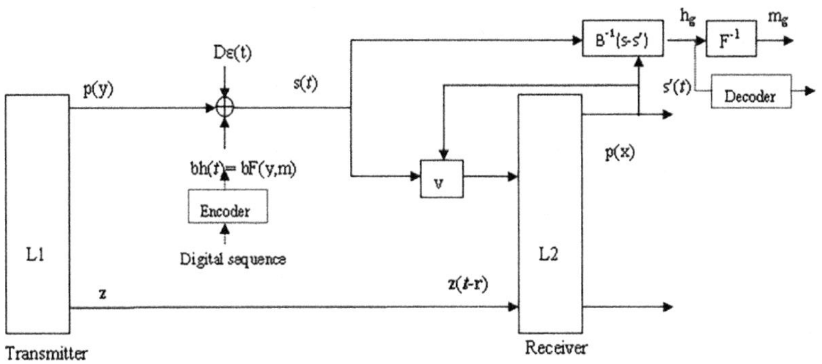

Fig. 8.5 The block diagram of our communication scheme

$$L1:\begin{cases}\dot{\mathbf{y}} = \mathbf{Ay} + \mathbf{f}(\mathbf{y}, \mathbf{z}),\\ \dot{\mathbf{z}} = \mathbf{g}(\mathbf{y}, \mathbf{z}),\\ \mathbf{s} = \mathbf{p}(\mathbf{y}) + \mathbf{Bh},\end{cases} \tag{8.2.6}$$

$$\begin{aligned} L2:\dot{\mathbf{x}} &= \mathbf{Ax} + \mathbf{f}(\mathbf{x}, \mathbf{z}_\tau) + \mathbf{v},\\ \mathbf{s}' &= \mathbf{p}(\mathbf{x}),\end{aligned} \tag{8.2.7}$$

where $\mathbf{h} = (h_1, h_2, \ldots, h_q)^{\mathrm{T}}$ denotes the information signal vector, and $\mathbf{B} = diag(b_1, b_2, \ldots, b_q)$ is its parameter matrix. The transmitted signal (i.e. the output of the transmitter) is $\mathbf{s}$ and the output of the receiver is $\mathbf{s}'$. Here, $\mathbf{p}$ is a $q \times 1$ complex function vector.

The block diagram of our communication scheme is shown in Fig. 8.5. To minimize controller complexity in practice, the parameter q is usually fixed as $q = 1$. Thus, the information signal is denoted as $h(t)$ and its parameter is b. These will be illustrated in Fig. 8.5 and numerical simulation. Considering $D\varepsilon(t)$, which simulates the noise both from the communication channel and the noise source, the transmitted signal is given by $s(t) = p(y) + bh + D\varepsilon(t)$. It is transmitted by a costly line with ignorable time-delay. The other state vector $\mathbf{z}$ is transmitted to the receiver by ordinary cheap lines with time-delay τ. At the receiver end, the controller v takes the form of (8.2.5).

With the occurrence of STDS, $x(t)$ will approach $y(t)$, and the recovered signal $h_g \approx b^{-1}(s - s') = b^{-1}[p(y) - p(x) + bh + D\varepsilon(t)] \to h$. The error $e_m(t) = h(t) - h_g(t) = b^{-1}[p(y) - p(x) + D\varepsilon(t)]$. Obviously, if there exists no noise, the signal is retrieved accurately with no errors in theory. If there exists noise, $e_m(t)$ deceases as b increases.

To enhance the security of communication, the information signal can be set as $m(t)$. Consequently, $h(t)$ is defined as a function of this signal, namely $h(t) = F(y, m)$. As for digital signals, we add encoder and decoder presented by dotted line.

Additionally, we can employ simple filters at the receive end to enhance the accuracy of the information recovery process.

8.2.4 Numerical Simulation

$L1$ and $L2$ are described by the following coupled time-delay complex Lorenz system

$$L1: \begin{cases} \dot{y} &= -a_1 y + a_1 z_1, \\ \dot{z}_1 &= a_2 y - y z_2 - z_1, \\ \dot{z}_2 &= -a_3 z_2 + (1/2)(\bar{y} z_1 + y \bar{z}_1), \\ s &= cy + bh + D\varepsilon(t), \end{cases} \quad (8.2.8)$$

$$L2: \begin{cases} \dot{x} &= -a_1 x + a_1 z_1(t - \tau_1) + v, \\ s' &= cx, \end{cases} \quad (8.2.9)$$

where $y = u_1' + ju_2', z_1 = u_3' + ju_4'$ denote complex state variables of original system $L1, x = u_1 + ju_2$ is the complex state variable of time-delay system $L2$. The coupling variable is z_1, and the real variable is $z_2 = u_5'$. Here, τ_1 is the constant time-delay and $(a_1, a_2, a_3)^{\mathrm{T}}$ is the real parameter vector. The function $p(y)$ of (8.2.6) is $cy(t)$ where c is an adjusted parameter.

Based on (8.2.5), (8.2.8) and (8.2.9), the controller is given by

$$v = a_1[z_1 - z_1(t - \tau_1)] - ke, \quad (8.2.10)$$

where $e = x - y$.

The noise $\varepsilon(t)$ is the stochastic Gaussian process which is designed by the following probability distribution

$$\rho(\varepsilon) = \frac{1}{\sqrt{2\pi}\sigma} exp(-\frac{(\varepsilon - \varepsilon_0)^2}{2\sigma^2}), \quad (8.2.11)$$

where $\varepsilon_0 = 0$ and $\sigma = 1$ are the mean value and variance.

Set the initial condition as $y(0) = 2 + 1j, z_1(0) = 5 + 3j, z_2(0) = 4, x(0) = -1 - 2j$. Specify parameters $a_1 = 14$, $a_2 = 35$, $a_3 = 3.7$, and time-delay $\tau_1 = 1s$.

I. Analog signals

The state vector and error vector of STDS are displayed in Fig. 8.6 with the control strength $k = 1000$, where $x(t)$ approaches $y(t)$ and the error vector converges to zero quickly as time increases. It is clear that the state variable $x(t)$ of time-delay system $L2$ synchronizes the state variable $y(t)$ of original system $L1$, i.e. STDS takes place. Figure 8.7 describes the controller $\mathbf{v}$. We transmit the famous melody "Song of Joy" written by Beethoven with $b = 100$ and $c = 10$. It contains two components which

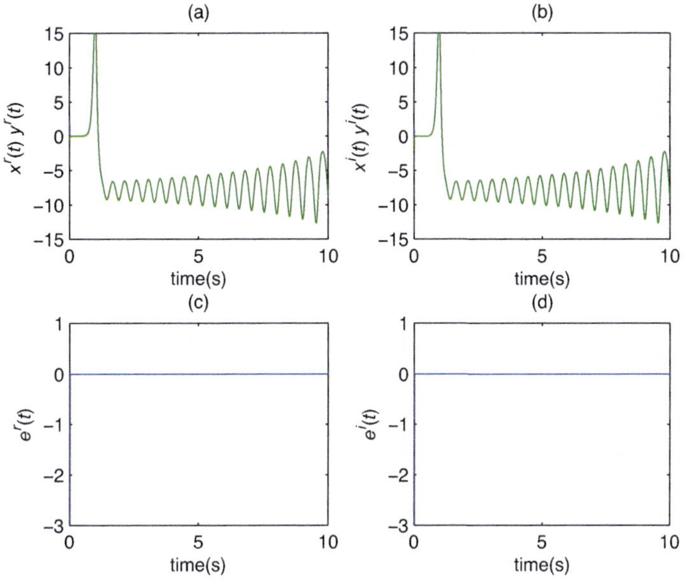

Fig. 8.6 The state and error vector of STDS for analog signals

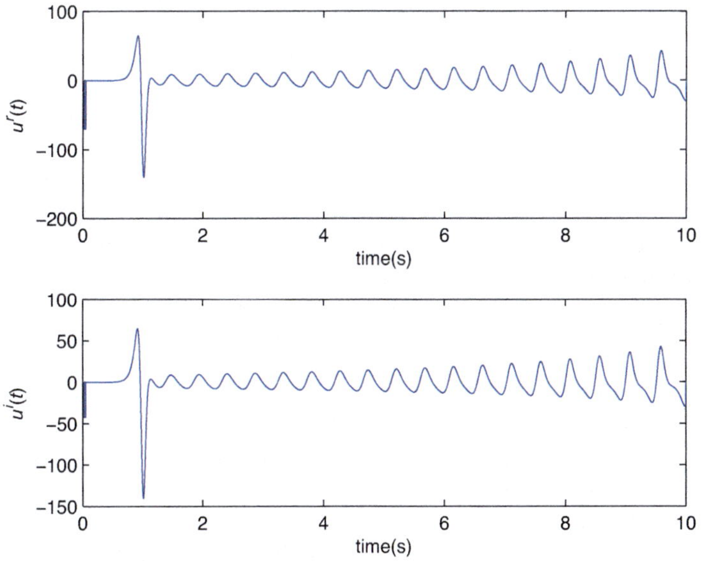

Fig. 8.7 The STDS controller for analog signals

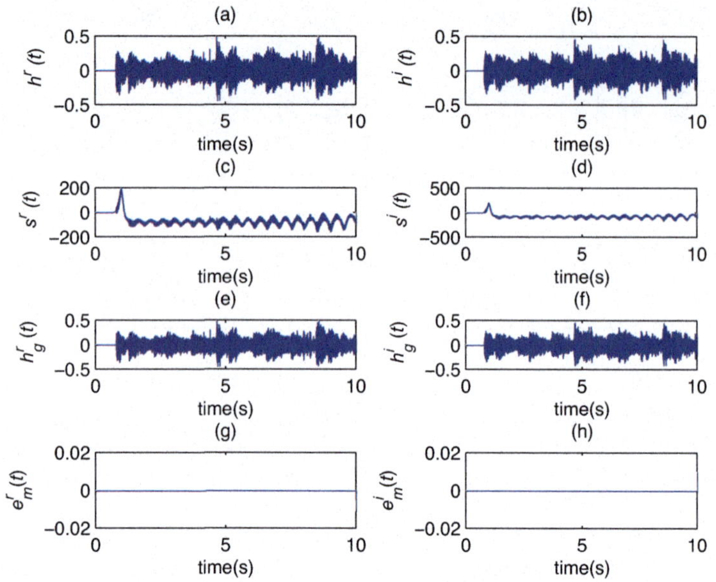

Fig. 8.8 The transmission process of the famous melody "Song of Joy" with $D = 0$

are read by Matlab software shown in Figs. 8.8(a)–(b) and included by the real part and imaginary part of $s(t)$ respectively.

In the absence of noise, the transmitted signal $s(t) = cy(t) + bh(t)$ is such that it completely covers up the information signal, as depicted in Figs. 8.8(c)–(d). The recovered signal $h_g(t)$ is displayed in Figs. 8.8(e)–(f). Figures 8.8(g)–(h) show the error $e_m(t)$. In theory, the information signal $h(t)$ is perfectly reconstructed without any error.

If we add the noise with $D = 5$, $b = 1000$ and $c = 100$, the error $e_m(t)$ is shown in Fig. 8.9, where the amplitude of $e_m(t)$ is about 0.02. Hence, the achieved recovery accuracy of 96% for the information signal $h(t)$ indicates a certain level of robustness against noise in this communication scheme.

II. Digital signals

In this section, binary sequences produced by random functions in MATLAB are transmitted. As shown in Fig. 8.5, M binary bits must first be converted into corresponding analog signals by 2^M-ary and zero-order holder, and then input into the controller (8.2.10). Here the assignment of binary bits to analog signals is carried out by 8-ary, i.e., $M = 3$. It means the symbol duration is equal to 3 time-units. The state vector and error vector of STDS are depicted in Fig. 8.10 with the control strength $k = 0.1$, which indicates STDS takes place. The controller $\mathbf{v}$ is shown in Fig. 8.11. The computed analog signals transmitted to the receiver and the corresponding transmitted signal $s(t)$ are displayed in Fig. 8.12 with $D = 0$, $b = 2$ and $c = 1$, where $s(t)$ is an analogous signal and completely covers up the binary digital sequence. The

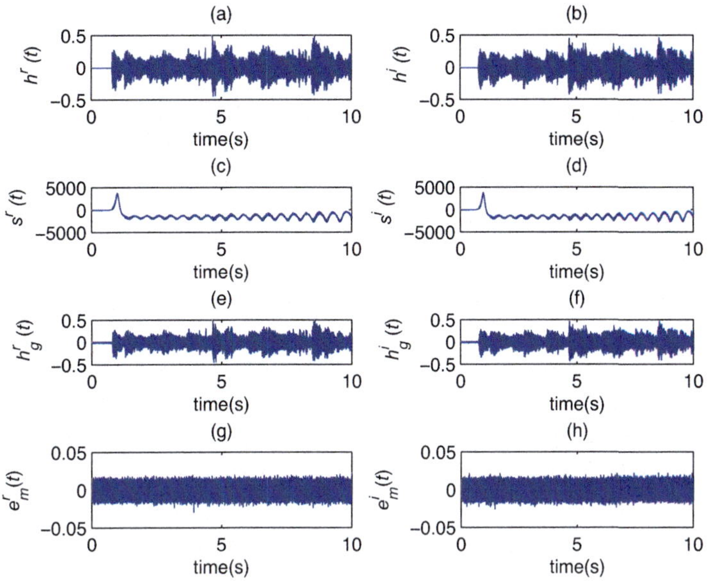

Fig. 8.9 The transmission process of the famous melody "Song of Joy" with $D = 5$

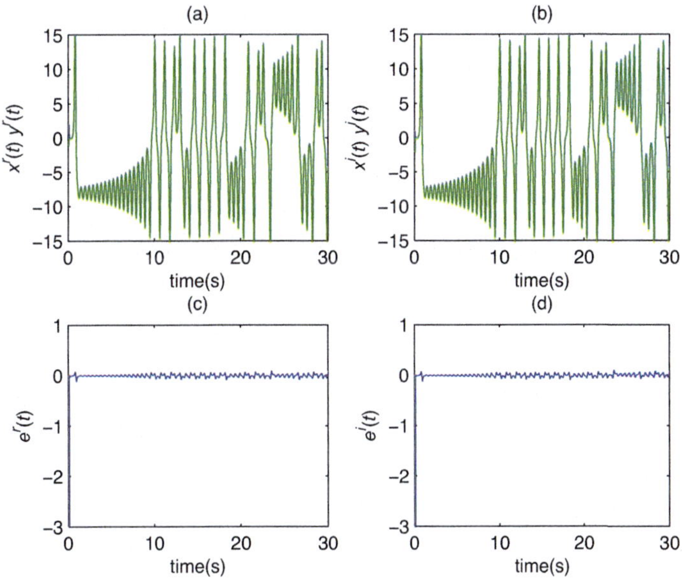

Fig. 8.10 The state and error vector of STDS for digital signals

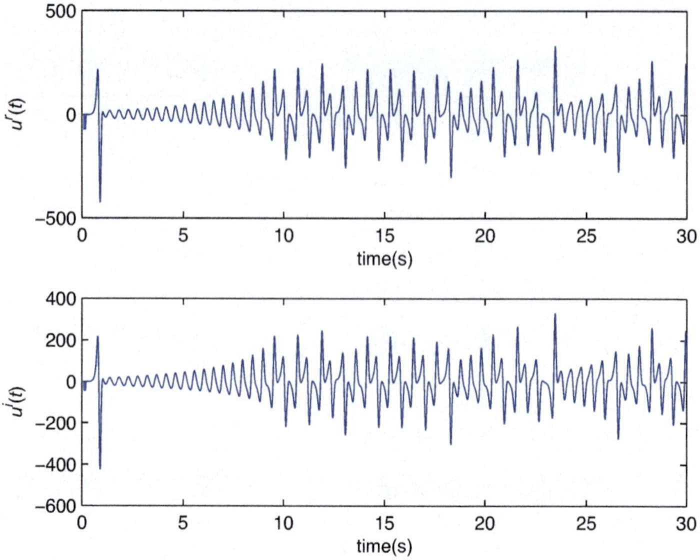

Fig. 8.11 The STDS controller for digital signals

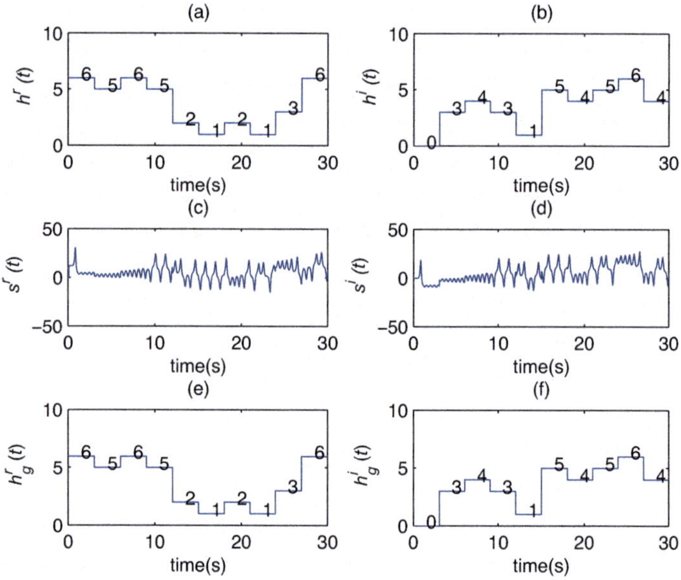

Fig. 8.12 The digital signals transmitted without noise

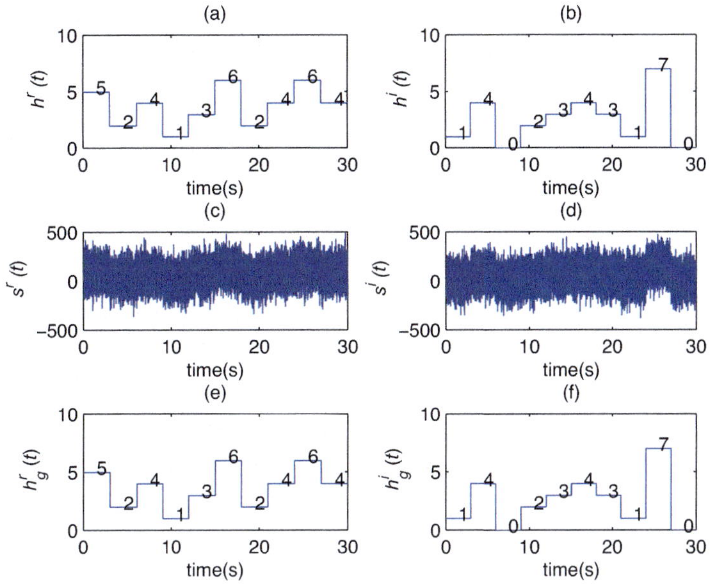

Fig. 8.13 The digital signals transmitted with stochastic Gaussian noise (8.2.11) $D = 200$

digital symbols are retrieved accurately with no errors in theory, i.e. bit error rate ($BER = \frac{the\ number\ of\ error\ bits}{totalbit}$) is zero in theory.

When the stochastic Gaussian noise (8.2.11) with $D = 100$ is introduced into the channel, the corresponding transmitted signal shown in Fig. 8.13 (Here, $b = 20$ and $c = 1$) is completely like a noise. It is difficult to extract the information signal. To retrieve the information signal from transmitted symbols and reduce the effect of noise, we add a simple filter. The basic idea of the filter is to compute the average value of the recovered analogous signal for each symbol duration before rounding off to the nearest integer that is bounded between 0 and $2^M - 1$. Thus, the scheme enables both accurate retrieval of digital symbols with a low BER and demonstrates strong robustness against noise.

Chapter 9
Adaptive Complex Modified Projective Synchronization and Complex Modified Hybrid Projective Synchronization of Complex Chaotic Systems

This chapter focuses on the exploration of adaptive complex modified projective synchronization (CMPS) and complex modified hybrid projective synchronization (CMHPS) for complex chaotic systems. Section 9.1 introduces the adaptive CMPS of complex chaotic (and hyperchaotic) systems with uncertain complex parameters. Section 9.2 presents a novel four-wing hyperchaotic complex system, along with its CMHPS in different dimensions.

9.1 Adaptive CMPS of Complex Chaotic (Hyperchaotic) Systems with Uncertain Complex Parameters

9.1.1 The Definition of CMPS of Complex Nonlinear Systems with Complex Parameters and Problem Descriptions

First, we focus on a class of n-dimensional complex chaotic (hyperchaotic) master-slave systems with uncertain or deterministic complex parameters

$$\dot{\mathbf{z}}(t) = \mathbf{G}(\mathbf{z},t), \tag{9.1.1}$$

$$\dot{\mathbf{w}}(t) = \mathbf{F}(\mathbf{w},t), \tag{9.1.2}$$

where $\mathbf{z} = (z_1, z_2, \ldots, z_n)^T \in \mathbb{C}^n$ and $\mathbf{w} = (w_1, w_2, \ldots, w_n)^{\mathrm{T}} \in \mathbb{C}^n$ are complex state vectors.

S. Liu et al., *Chaos in Complex Dynamical System: Theory and Application*,
https://doi.org/10.1007/978-981-95-7762-0_9

Table 9.1 Types of synchronization

Settings the matrix M	Type of synchronization
$M = diag\{\delta_1, \delta_2, \ldots, \delta_n\} \in \mathbb{C}^{n\times n}$	CMPS
$M = diag\{\delta, \delta, \ldots, \delta\} \in \mathbb{C}^{n\times n}$	CPS
$M = diag\{\delta_1, \delta_2, \ldots, \delta_n\} \in \mathbb{R}^{n\times n}$	MPS

Subsequently, we propose the definition of CMPS for complex systems with complex parameters with reference to the CMPS definition applicable to complex systems with real parameters [163, 164].

Definition 9.1 For given master system (9.1.1) and slave system (9.1.2) with any initial condition $\mathbf{z}(0)$ and $\mathbf{w}(0)$, it is said to achieve CMPS with $\mathbf{M}$ between $\mathbf{w}(t)$ and $\mathbf{z}(t)$, if there exists a n-dimensional complex matrix $\mathbf{M}$ such that

$$\lim_{t\to+\infty} \|\mathbf{w}(t) - \mathbf{M}\mathbf{z}(t)\| = 0, \tag{9.1.3}$$

where $\mathbf{M} = diag\{\delta_1, \delta_2, \ldots, \delta_n\}$, $\delta_l = \delta_l^r + j\delta_l^i (l = 1, 2, \ldots n)$ are complex constants.

Remark 9.1 The matrix $\mathbf{M}$ is called the complex scaling matrix. Several kinds of synchronization are special cases of our definition as in Table 9.1.

Hence, this new category of complex synchronization is a generalization of the CPS, MPS, and CMPS schemes proposed for complex chaotic systems with real parameters in existing recent studies.

We consider a general n-dimensional complex master chaotic or hyperchaotic system with uncertain complex parameters, formulated as

$$\dot{\mathbf{z}}(t) = \mathbf{G}(\mathbf{z},t) = \Phi(\mathbf{z})\mathbf{A} + \phi(\mathbf{z}), \tag{9.1.4}$$

where $\mathbf{z} = \mathbf{z}^r + j\mathbf{z}^i \in \mathbb{C}^n$, $\mathbf{z}^r = (z_1^r, z_2^r, \ldots, z_n^r)^{\mathrm{T}}$, $\mathbf{z}^i = (z_1^i, z_2^i, \ldots, z_n^i)^{\mathrm{T}}$. $\mathbf{A} = (a_1, a_2, \ldots, a_s)^{\mathrm{T}} \in \mathbb{C}^s$ is a s-dimensional complex vector of unknown parameters, $\Phi(\mathbf{z})$ is a $n \times s$ complex matrix and its elements are functions of complex state variables, $\phi = (\phi_1, \phi_2, \ldots, \phi_n)^{\mathrm{T}}$ is a n-dimensional vector of the complex nonlinear function. On the other hand, the n-dimensional complex slave chaotic system with the controller is depicted as

$$\dot{\mathbf{w}}(t) = \mathbf{F}(\mathbf{w},t) + +\mathbf{v} = \Psi(\mathbf{w})\mathbf{B} + \psi(\mathbf{w}) + \mathbf{v}, \tag{9.1.5}$$

where $\mathbf{w} = \mathbf{w}^r + j\mathbf{w}^i \in \mathbb{C}^n$, $\mathbf{w}^r = (w_1^r, w_2^r, \ldots, w_n^r)^{\mathrm{T}}$, $\mathbf{w}^i = (w_1^i, w_2^i, \ldots, w_n^i)^{\mathrm{T}}$. $\mathbf{B} = (b_1, b_2, \ldots, b_q)^{\mathrm{T}} \in \mathbb{C}^q$ is a q-dimensional complex vector of unknown parameters, $\Psi(\mathbf{w})$ is a $n \times q$ complex matrix and its elements are functions of complex state variables, $\psi = (\psi_1, \psi_2, \ldots, \psi_n)^T$ is a n-dimensional vector of the complex nonlinear function. The controller $\mathbf{v} = \mathbf{v}^r + j\mathbf{v}^i \in \mathbb{C}^n$ needs to be designed.

If CMPS error system is defined as

$$\mathbf{e}(t) = \mathbf{w}(t) - \mathbf{Mz}(t), \tag{9.1.6}$$

where $\mathbf{e} = \mathbf{e}^r + j\mathbf{e}^i \in \mathbb{C}^n$ and $\mathbf{e}^r = (e_1^r, e_2^r, \ldots, e_n^r)^{\mathrm{T}}$, $\mathbf{e}^i = (e_1^i, e_2^i, \ldots, e_n^i)^{\mathrm{T}}$, the essential goal of this section is to design complex update laws of unknown parameter vectors and complex adaptive controller $\mathbf{v}$ such that the synchronization error tends to zero, i.e.

$$\lim_{t\to+\infty} \|\mathbf{e}(t)\| = \lim_{t\to+\infty} \|\mathbf{w}(t) - \mathbf{Mz}(t)\| = 0.$$

Remark 9.2 Most of the well-studied complex chaotic and hyperchaotic systems can be written as the form of system (9.1.4), such as complex Lorenz system, complex Chen system, complex Lü system, complex Van der Pol oscillator, complex Duffing system, and complex hyperchaotic Lorenz system, complex hyperchaotic Lü system, etc.

9.1.2 CMPS of Complex Chaotic (Hyperchaotic) Systems with Uncertain Complex Parameters

Theorem 9.1 *For given complex scaling matrix* $\mathbf{M}$ *and any initial conditions* $\mathbf{w}(0), \mathbf{z}(0)$, *if the complex adaptive controller is designed as*

$$\mathbf{v} = -\Psi(\mathbf{w})\hat{\mathbf{B}} - \psi(\mathbf{w}) + \mathbf{M}(\Phi(\mathbf{z}(t))\hat{\mathbf{A}} + \phi(\mathbf{z}(t))) - \mathbf{Ke}, \tag{9.1.7}$$

and the complex update laws of complex parameter vectors are selected as

$$\begin{cases} \dot{\hat{\mathbf{A}}} &= -(\mathbf{M}\Phi(\mathbf{z}(t)))^{\mathrm{H}}\,\mathbf{e} - \eta_a\,\tilde{\mathbf{A}}, \\ \dot{\hat{\mathbf{B}}} &= \Psi^{\mathrm{H}}(\mathbf{w})\,\mathbf{e} - \eta_b\,\tilde{\mathbf{B}}, \end{cases} \tag{9.1.8}$$

where $\mathbf{K} = diag\{k_1, k_2, \ldots, k_n\}$, $\eta_a = diag\{\eta_{a_1}, \eta_{a_2}, \ldots, \eta_{a_s}\}$, $\eta_b = diag\{\eta_{b_1}, \eta_{b_2}, \ldots, \eta_{b_q}\}$ *are real positive definite matrices,* $k_i, (i = 1, 2, \ldots, n)$ *are coupling strengths,* $\hat{\mathbf{A}}$, $\hat{\mathbf{B}}$ *are the estimations of unknown parameter vectors* $\mathbf{A}$ *and* $\mathbf{B}$, $\tilde{\mathbf{A}} = \hat{\mathbf{A}} - \mathbf{A}$ *and* $\tilde{\mathbf{B}} = \hat{\mathbf{B}} - \mathbf{B}$ *are the parameter error vectors, respectively, then ACMPS between the slave system* (9.1.5) *and master system* (9.1.4) *is achieved asymptotically, and* $\hat{\mathbf{A}}$, $\hat{\mathbf{B}}$ *converge to complex constant vectors* $\mathbf{A}$ *and* $\mathbf{B}$, *respectively,* (9.1.8) *is said to be complex update laws of unknown complex parameter vectors* $\mathbf{A}$ *and* $\mathbf{B}$.

Proof Insertion of (9.1.4), (9.1.5) and (9.1.7) into (9.1.6) gives

$$\begin{aligned}\dot{\mathbf{e}}(t) &= \dot{\mathbf{w}}(t) - \mathbf{M}\dot{\mathbf{z}}(t)\\ &= \Psi(\mathbf{w})\mathbf{B} + \psi(\mathbf{w}) - \Psi(\mathbf{w})\hat{\mathbf{B}} - \psi(\mathbf{w}) + \mathbf{M}(\Phi(\mathbf{z}(t))\hat{\mathbf{A}} + \phi(\mathbf{z}(t)))\\ &\quad -\mathbf{M}(\Phi(\mathbf{z}(t))\mathbf{A} + \phi(\mathbf{z}(t))) - \mathbf{K}\mathbf{e}\\ &= \mathbf{M}\Phi(\mathbf{z}(t))(\hat{\mathbf{A}} - \mathbf{A}) - \Psi(\mathbf{w})(\hat{\mathbf{B}} - \mathbf{B}) - \mathbf{K}\mathbf{e}\\ &= \mathbf{M}\Phi(\mathbf{z}(t))\tilde{\mathbf{A}} - \Psi(\mathbf{w})\tilde{\mathbf{B}} - \mathbf{K}\mathbf{e}.\end{aligned} \tag{9.1.9}$$

Consider the following candidate Lyapunov function defined as

$$V(\mathbf{e}, \tilde{\mathbf{A}}, \tilde{\mathbf{B}}, t) = \frac{1}{2}(\mathbf{e}^{\mathrm{H}}\mathbf{e} + \tilde{\mathbf{A}}^{\mathrm{H}}\tilde{\mathbf{A}} + \tilde{\mathbf{B}}^{\mathrm{H}}\tilde{\mathbf{B}}), \tag{9.1.10}$$

utilizing the complex update laws (9.1.8) and the relations $\dot{\tilde{\mathbf{A}}} = \dot{\hat{\mathbf{A}}}$, $\dot{\tilde{\mathbf{B}}} = \dot{\hat{\mathbf{B}}}$, the time derivative of V along the trajectories of the error system (9.1.9) reads

$$\begin{aligned}\dot{V}(t) &= \frac{1}{2}[(\dot{\mathbf{e}})^{\mathrm{H}}\mathbf{e} + \mathbf{e}^{\mathrm{H}}(\dot{\mathbf{e}}) + \dot{\tilde{\mathbf{A}}}^{\mathrm{H}}\tilde{\mathbf{A}} + \tilde{\mathbf{A}}^{\mathrm{H}}\dot{\tilde{\mathbf{A}}} + \dot{\tilde{\mathbf{B}}}^{\mathrm{H}}\tilde{\mathbf{B}} + \tilde{\mathbf{B}}^{\mathrm{H}}\dot{\tilde{\mathbf{B}}}]\\ &= \frac{1}{2}[(\mathbf{M}\Phi(\mathbf{z}(t))\tilde{\mathbf{A}} - \Psi(\mathbf{w})\tilde{\mathbf{B}} - \mathbf{K}\mathbf{e})^{\mathrm{H}}\mathbf{e} + \mathbf{e}^{\mathrm{H}}(\mathbf{M}\Phi(\mathbf{z}(t))\tilde{\mathbf{A}} - \Psi(\mathbf{w})\tilde{\mathbf{B}} - \mathbf{K}\mathbf{e})\\ &\quad +((-\mathbf{M}\Phi(\mathbf{z}(t)))^{\mathrm{H}}\,\mathbf{e} - \eta_a\,\tilde{\mathbf{A}})^{\mathrm{H}}\,\tilde{\mathbf{A}} + \tilde{\mathbf{A}}^{\mathrm{H}}\,((-\mathbf{M}\Phi(\mathbf{z}(t)))^{\mathrm{H}}\,\mathbf{e} - \eta_a\,\tilde{\mathbf{A}})\\ &\quad +(\Psi^{\mathrm{H}}(\mathbf{w})\,\mathbf{e} - \eta_b\,\tilde{\mathbf{B}})^{\mathrm{H}}\tilde{\mathbf{B}} + \tilde{\mathbf{B}}^{\mathrm{H}}(\Psi^{\mathrm{H}}(\mathbf{w})\mathbf{e} - \eta_b\,\tilde{\mathbf{B}})]\\ &= -\mathbf{e}^{\mathrm{H}}\mathbf{K}\mathbf{e} - \tilde{\mathbf{A}}^{\mathrm{H}}\eta_a\tilde{\mathbf{A}} - \tilde{\mathbf{B}}^{\mathrm{H}}\eta_b\tilde{\mathbf{B}} < 0.\end{aligned} \tag{9.1.11}$$

Since Lyapunov function $V(t)$ is positive definite, and $\dot{V}(t)$ is negative definite, based on the Lyapunov stability theory, the error vector $\mathbf{e}(t) \to 0$ as $t \to \infty$. So, one arrives at ACMPS with desired complex scaling matrix $\mathbf{M}$ between two complex systems (9.1.4) and (9.1.5) by using the controller (9.1.7) and complex update laws (9.1.8). According to (9.1.8), at $t \to 0$, the parameter errors $\tilde{\mathbf{A}}$ and $\tilde{\mathbf{B}}$ also converge to zero, which shows the estimations $\hat{\mathbf{A}}$ and $\hat{\mathbf{B}}$ of unknown complex parameter vectors in both the master and slave systems are also identified. The proof is completed. □

Theorem 9.2 *Assume the structure of systems* (9.1.4) *and* (9.1.5) *is identical, i.e.* $\mathbf{G}(\cdot) = \mathbf{F}(\cdot)$, $\mathbf{g}(\cdot) = \mathbf{f}(\cdot)$, *and* $\mathbf{A} = \mathbf{B}$. *Then, for given complex scaling matrix* $\mathbf{M}$ *and any initial conditions* $\mathbf{w}(0)$, $\mathbf{z}(0)$, *ACMPS between the slave system* (9.1.5) *and master system* (9.1.4) *is achieved asymptotically, and estimation vector* $\hat{\mathbf{A}}$ *converges to complex constant vector* $\mathbf{A}$ *under the following complex adaptive controller and complex parameter update law as*

$$\mathbf{v} = (\mathbf{M}(\Phi(\mathbf{z}(t)) - \Phi(\mathbf{w}))\hat{\mathbf{A}} - \phi(\mathbf{w}) + \mathbf{M}\phi(\mathbf{z}(t)) - \mathbf{K}\mathbf{e}, \tag{9.1.12}$$

$$\dot{\hat{\mathbf{A}}} = (\Phi(\mathbf{w}) - \mathbf{M}\Phi(\mathbf{z}(t)))^{\mathrm{H}}\,\mathbf{e} - \eta_a\,\tilde{\mathbf{A}}, \tag{9.1.13}$$

where $\eta_a = diag\{\eta_{a_1}, \eta_{a_2}, \ldots, \eta_{a_s}\}$ *and* $\mathbf{K} = diag\{k_1, k_2, \ldots, k_n\}$ *are real positive definite matrices,* k_i, $(i = 1, 2, \ldots, n)$ *are coupling strengths,* $\tilde{\mathbf{A}} = \hat{\mathbf{A}} - \mathbf{A}$ *is the parameter error vector.*

Proof Define the following candidate Lyapunov function as

$$V(\mathbf{e}, \tilde{\mathbf{A}}, t) = \frac{1}{2}(\mathbf{e}^{\mathrm{H}}\mathbf{e} + \tilde{\mathbf{A}}^{\mathrm{H}}\tilde{\mathbf{A}}). \tag{9.1.14}$$

Subsequently, the proof follows similarly to that presented in Theorem 9.1 and is therefore omitted. □

Corollary 9.1 *Assume the parameter vector* **A** *of the master system* (9.1.4) *is known. Then, for given complex scaling matrix* **M** *and any initial conditions* $\mathbf{w}(0), \mathbf{z}(0)$, *CMPS between the slave system* (9.1.5) *and master system* (9.1.4) *is achieved asymptotically, and estimation vector* $\hat{\mathbf{B}}$ *converges to complex constant vector* **B** *under the following complex adaptive controller and complex parameter update law as*

$$\mathbf{v} = -\Psi(\mathbf{w})\hat{\mathbf{B}} - \psi(\mathbf{w}) + \mathbf{M}(\Phi(\mathbf{z}(t))\mathbf{A} + \phi(\mathbf{z}(t))) - \mathbf{K}\mathbf{e}, \tag{9.1.15}$$

$$\dot{\hat{\mathbf{B}}} = \Psi^{\mathrm{H}}(\mathbf{w})\, \mathbf{e} - \eta_b\, \tilde{\mathbf{B}}, \tag{9.1.16}$$

where $\eta_b = diag\{\eta_{b_1}, \eta_{b_2}, \ldots, \eta_{b_q}\}$ *and* $\mathbf{K} = diag\{k_1, k_2, \ldots, k_n\}$ *are real positive definite matrices,* k_i, $(i = 1, 2, \ldots, n)$ *are coupling strengths and* $\tilde{\mathbf{B}} = \hat{\mathbf{B}} - \mathbf{B}$ *is the parameter error vector, respectively.*

Corollary 9.2 *Assume the parameter vector* **B** *of the slave system* (9.1.5) *is known. Then, for given complex scaling matrix* **M** *and any initial conditions* $\mathbf{w}(0), \mathbf{z}(0)$, *ACMPS between the slave system* (9.1.5) *and master system* (9.1.4) *is achieved asymptotically, and estimation* $\hat{\mathbf{A}}$ *converges to the complex constant vector* **A** *under the following complex adaptive controller and complex parameter update law as*

$$\mathbf{v} = -\Psi(\mathbf{w})\mathbf{B} - \psi(\mathbf{w}) + \mathbf{M}(\Phi(\mathbf{z}(t))\hat{\mathbf{A}} + \phi(\mathbf{z}(t))) - \mathbf{K}\mathbf{e}, \tag{9.1.17}$$

$$\dot{\hat{\mathbf{A}}} = -(\mathbf{D}(t)\mathbf{F}(\mathbf{z}(t)))^{\mathrm{H}}\, \mathbf{e} - \eta_a\, \tilde{\mathbf{A}}, \tag{9.1.18}$$

where $\eta_a = diag\{\eta_{a_1}, \eta_{a_2}, \ldots, \eta_{a_s}\}$ *and* $\mathbf{K} = diag\{k_1, k_2, \ldots, k_n\}$ *are real positive definite matrices,* k_i, $(i = 1, 2, \ldots, n)$ *are coupling strengths and* $\tilde{\mathbf{A}} = \hat{\mathbf{A}} - \mathbf{A}$ *is the parameter error vector, respectively.*

Corollary 9.3 *Assume both parameter matrices* **A** *and* **B** *are known. Then, for given complex scaling matrix* **M** *and any initial conditions* $\mathbf{w}(0), \mathbf{z}(0)$, *CMPS between the slave system* (9.1.5) *and master system* (9.1.4) *is achieved asymptotically under the following controller as*

$$\mathbf{v} = -\Psi(\mathbf{w})\mathbf{B} - \psi(\mathbf{w}) + \mathbf{M}(\Phi(\mathbf{z}(t))\mathbf{A} + \phi(\mathbf{z}(t))) - \mathbf{K}\mathbf{e}, \tag{9.1.19}$$

where $\mathbf{K} = diag\{k_1, k_2, \ldots, k_n\}$ *is real positive definite matrix,* k_i, $(i = 1, 2, \ldots, n)$ *are coupling strengths.*

Remark 9.3 In contrast to the methods adopted in [165–169], we propose an ACMPS scheme for complex chaotic systems with uncertain complex parameters. Notably, the complex update laws for uncertain parameters and the complex adaptive controller are designed without separating the real and imaginary parts of complex state variables or complex parameters, rendering the conclusions more concise and practically applicable.

Remark 9.4 Note that the complex scaling matrix **M** has no effect on the time derivative $\dot{V}(t)$, thus one can adjust complex scaling matrix arbitrarily without altering the control robustness. Hence, the complex scaling matrix can be used as information signal for the communication scheme. In particular, when **A** and **B** are real, Theorem 9.1 and Corollarys 9.1–9.3 are also applied to achieve chaotic synchronization of complex systems with real parameters.

9.1.3 Numerical Examples

In this section, for the purpose of validating the effectiveness and feasibility of the synchronization scheme introduced in Sect. 9.1.2, examples are provided for two cases: namely, CMPS between two identical chaotic complex systems with uncertain complex parameters and CMPS between two nonidentical chaotic complex systems with uncertain complex parameters.

I. CMPS of Two Nonidentical Complex Nonlinear Systems with Uncertain Complex Parameters

To exemplify ACMPS for two nonidentical systems, we suppose that a complex Lorenz system with uncertain complex parameters acts as the drive system to a complex chaotic Chen system. Hence, the master system is formulated in the form:

$$\begin{cases} \dot{z}_1 &= a_1(z_2 - z_1), \\ \dot{z}_2 &= a_2 z_1 - a_3 z_2 - z_1 z_3, \\ \dot{z}_3 &= \frac{1}{2}(\bar{z}_1 z_2 + z_1 \bar{z}_2) - a_4 z_3, \end{cases} \tag{9.1.20}$$

where

$$\Phi(\mathbf{z}) = \begin{pmatrix} z_2 - z_1 & 0 & 0 & 0 \\ 0 & z_1 & -z_2 & 0 \\ 0 & 0 & 0 & -z_3 \end{pmatrix}, \quad \phi(\mathbf{z}) = \begin{pmatrix} 0 \\ -z_1 z_3 \\ (1/2)(\bar{z}_1 z_2 + z_1 \bar{z}_2) \end{pmatrix},$$

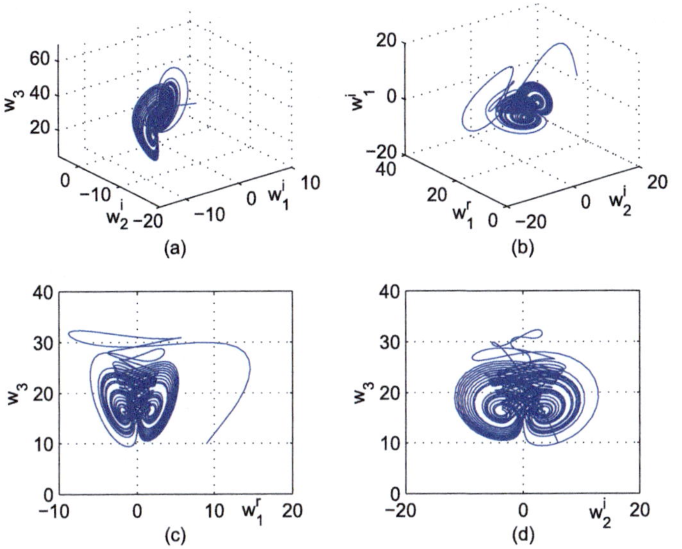

Fig. 9.1 Phase portraits of the attractors for the complex Chen system (9.1.21) with parameters $b_1 = 28,\ b_2 = 22,\ b_3 = 1$ and initial value $w(0) = (9 + 12j, 12 + 6j, 10)^T$: **a** on (w_1^r, w_2^i, w_3) space; **b** on $(w_2^i, w_1^r, w_1^i))$ space; **c** on (w_1^r, w_3) plane; **d** on (w_2^i, w_3) plane

and $z_1 = z_1^r + jz_1^i, z_2 = z_2^r + jz_2^i$ are complex state variables and z_3 is a real state variable, $\mathbf{A} = (a_1,\ a_2,\ a_3,\ a_4)^{\mathrm{T}}$ is an unknown complex parameter vector. System (9.1.20) is chaotic when $a_1 = 3,\ a_2 = 60 + 0.2j,\ a_3 = 1 - 0.6j,\ a_4 = 0.8$.

The complex slave system with the controller is written in the form as

$$\begin{cases} \dot{w}_1 &= b_1(w_2 - w_1) + v_1, \\ \dot{w}_2 &= (b_2 - b_1)w_1 + b_2 w_2 - w_1 w_3 + v_2, \\ \dot{w}_3 &= -b_3 w_3 + (1/2)(\bar{w}_1 w_2 + w_1 \bar{w}_2) + v_3, \end{cases} \tag{9.1.21}$$

where

$$\Psi(\mathbf{w}) = \begin{pmatrix} w_2 - w_1 & 0 & 0 \\ -w_1 & w_1 + w_2 & 0 \\ 0 & 0 & -w_3 \end{pmatrix}, \quad \psi(\mathbf{w}) = \begin{pmatrix} 0 \\ -w_1 w_3 \\ (1/2)(\bar{w}_1 w_2 + w_1 \bar{w}_2) \end{pmatrix}.$$

System (9.1.21) is chaotic when $b_1 = 28, b_2 = 22, b_3 = 1$, as shown in Fig. 9.1.

Based on Theorem 9.1, the controller (9.1.22) is formulated as follows:

$$\begin{aligned}\mathbf{v} &= -\Psi(\mathbf{w})\hat{\mathbf{B}} - \psi(\mathbf{w}) + \mathbf{M}(\Phi(\mathbf{z}(t))\hat{\mathbf{A}} + \phi(\mathbf{z}(t))) - \mathbf{K}\mathbf{e} \\ &= \begin{pmatrix} -\hat{b}_1(w_2(t)-w_1(t))+\delta_1\hat{a}_1(z_2(t)-z_1(t))-k_1e_1 \\ (\hat{b}_1-\hat{b}_2)w_1(t)-\hat{b}_2w_2(t)+w_1(t)w_3(t)+\delta_2(\hat{a}_2z_1(t)-\hat{a}_3z_2(t)-z_1(t)z_3(t))-k_2e_2 \\ \hat{b}_3w_3(t)-\frac{1}{2}(\bar{w}_1(t)w_2(t)+w_1(t)\bar{w}_2(t))+\delta_3(-\hat{a}_4z_3(t)+\frac{1}{2}(\bar{z}_1(t)z_2(t)+z_1(t)\bar{z}_2(t)))-k_3e_3 \end{pmatrix},\end{aligned} \tag{9.1.22}$$

the complex update laws of complex parameter vectors are taken as (9.1.23), where

$$\dot{\hat{\mathbf{A}}} = -(\mathbf{M}\Phi(\mathbf{z}(t))^{\mathrm{H}}\,\mathbf{e} - \eta_a\,\tilde{\mathbf{A}} = \begin{pmatrix} -\bar{\delta}_1(\bar{z}_2(t)-\bar{z}_1(t))e_1 - \eta_{a_1}(\hat{a}_1-a_1) \\ -\bar{\delta}_2\bar{z}_1(t)e_2 - \eta_{a_2}(\hat{a}_2-a_2) \\ \bar{\delta}_2\bar{z}_2(t)e_2 - \eta_{a_3}(\hat{a}_3-a_3) \\ \bar{\delta}_3\bar{z}_3(t)e_3 - \eta_{a_4}(\hat{a}_4-a_4) \end{pmatrix}, \tag{9.1.23}$$

and

$$\dot{\hat{\mathbf{B}}} = \Psi^{\mathrm{H}}(\mathbf{w})\,\mathbf{e} - \gamma_b\,\tilde{\mathbf{B}} = \begin{pmatrix} (\bar{w}_2(t)-\bar{w}_1(t))e_1 - \bar{w}_1(t)e_2 - \eta_{b_1}(\hat{b}_1-b_1) \\ (\bar{w}_1(t)+\bar{w}_2(t))e_2 - \eta_{b_2}(\hat{b}_2-b_2) \\ -\bar{w}_3(t)e_3 - \eta_{b_3}(\hat{b}_3-b_3) \end{pmatrix}. \tag{9.1.24}$$

We choose the values of the unknown parameters as $\mathbf{A} = (a_1, a_2, a_3, a_4)^{\mathrm{T}} = (3, 60+0.2j, 1-0.6j, 0.8)^{\mathrm{T}}$ and $\mathbf{B} = (b_1, b_2, b_3)^{\mathrm{T}} = (28, 22, 1)^{\mathrm{T}}$, respectively. The initial conditions for the master system (9.1.20) and slave system (9.1.21) are randomly selected as $\mathbf{z(0)} = (1+3j, 10+j, 25)^{\mathrm{T}}$, $\mathbf{w(0)} = (9+12j, 12+6j, 10)^{\mathrm{T}}$. For the estimated parameters and control strengths, their initial values are taken as $\hat{\mathbf{A}}(0) = (\hat{a}_1(0), \hat{a}_2(0), \hat{a}_3(0), \hat{a}_4(0))^{\mathrm{T}} = (1-j, 30, 1+5j, 1+j)^{\mathrm{T}}$, $\hat{\mathbf{B}}(0) = (\hat{b}_1(0), \hat{b}_2(0), \hat{b}_3(0))^{\mathrm{T}} = (15+2j, 18+3j, 19-j)^{\mathrm{T}}$. The complex scaling matrix is chosen as

$$\mathbf{M} = \begin{pmatrix} 1+j & 0 & 0 \\ 0 & 2-j & 0 \\ 0 & 0 & 1 \end{pmatrix}.$$

The constants are chosen as $\eta_a = diag\{10, 8, 15, 16\}$, $\eta_b = diag\{16, 16, 15\}$ and $\mathbf{K} = diag\{15, 20, 12\}$, and complex scaling factors are $1+j$, $2-j$, 1. Here $e_1 = w_1 - (1+j)z_1$, $e_2 = w_2 - (2-j)z_2$, $e_3 = w_3 - z_3$.

The CMPS error of the complex Lorenz master system (9.1.20) and Chen slave system (9.1.21) converges asymptotically to zero as in Fig. 9.2. The identification processes of parameters $\mathbf{A}$ and $\mathbf{B}$ are shown in Figs. 9.3 and 9.4, respectively. The parameter estimations $\hat{\mathbf{A}}$ and $\hat{\mathbf{B}}$ gradually converge to the selected values $\mathbf{A} =$

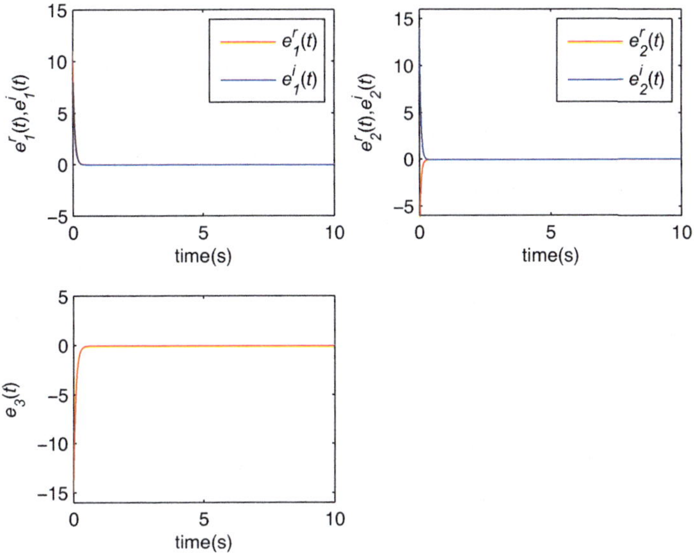

Fig. 9.2 The dynamics of CMPS error between two nonidentical complex Lorenz master system (9.1.20) and Chen slave system (9.1.21) with the controller (9.1.22) and complex parameter update laws (9.1.23) and (9.1.24)

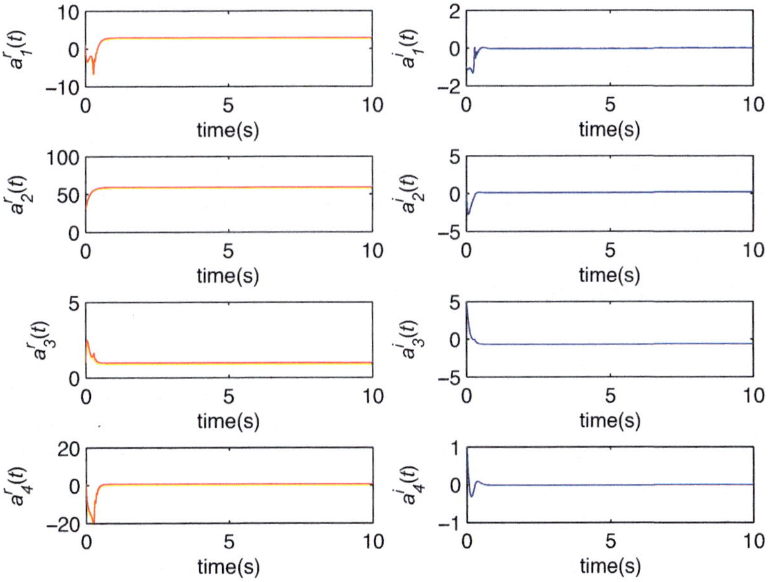

Fig. 9.3 The identification process of unknown complex parameter vector $\mathbf{A} = (a_1, a_2, a_3, a_4)^T$ for the system (9.1.20)

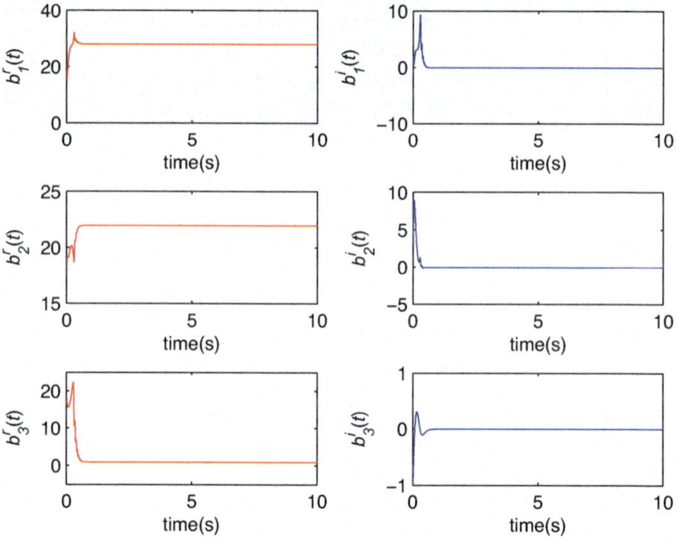

Fig. 9.4 The identification process of unknown parameter vector $\mathbf{B} = (b_1, b_2, b_3)^{\mathrm{T}}$ for the system (9.1.21)

$(a_1, a_2, a_3, a_4)^{\mathrm{T}} = (3, 60 + 0.2j, 1 - 0.6j, 0.8)^{\mathrm{T}}$ and $\mathbf{B} = (b_1, b_2, b_3)^{\mathrm{T}} = (28, 22, 1)^{\mathrm{T}}$ as $t \to \infty$, respectively.

As expected, the aforementioned findings verify that CMPS has been accomplished between the complex Lorenz master system (9.1.20) (featuring uncertain complex parameters) and the complex Chen slave system (9.1.21), with all unknown parameters in both systems accurately identified by means of the designed controller (9.1.22) and the complex parameter update laws (9.1.23) and (9.1.24).

II. CMPS of Two Identical Complex Nonlinear Systems with Uncertain Complex Parameters

To illustrate complex modified projective synchronization (CMPS) for two identical systems, we adopt the complex chaotic Lorenz system with uncertain complex parameters as both the master and slave systems. Thus, the complex Lorenz system (9.1.20) serves as the master system, and the slave system integrated with the controller is expressed in the form:

$$\begin{cases} \dot{w}_1 &= a_1(w_2 - w_1) + v_1, \\ \dot{w}_2 &= a_2 w_1 - a_3 w_2 - w_1 w_3 + v_2, \\ \dot{w}_3 &= -a_4 w_3 + \frac{1}{2}(\bar{w}_1 w_2 + w_1 \bar{w}_2) + v_3, \end{cases} \tag{9.1.25}$$

where

$$\Phi(\mathbf{w}) = \begin{pmatrix} w_2 - w_1 & 0 & 0 & 0 \\ 0 & w_1 & -w_2 & 0 \\ 0 & 0 & 0 & -w_3 \end{pmatrix}, \quad \phi(\mathbf{w}) = \begin{pmatrix} 0 \\ -w_1 w_3 \\ (1/2)(\bar{w}_1 w_2 + w_1 \bar{w}_2) \end{pmatrix},$$

and $w_1 = w_1^r + jw_1^i$, $w_2 = w_2^r + jw_2^i$ are complex state variables, and w_3 is a real state variable, $\mathbf{A} = (a_1, a_2, a_3, a_4)^{\mathrm{T}}$ is unknown complex parameter vector, $v = (v_1, v_2, v_3)^{\mathrm{T}}$ is the controller to be designed.

Design the controller (9.1.26) according to Theorem 9.2 as follows

$$\begin{aligned} \mathbf{v} &= (\mathbf{M}(\Phi(\mathbf{z}(t)) - \Phi(\mathbf{w}))\hat{\mathbf{A}} - \phi(\mathbf{w}) + \mathbf{M}\phi(\mathbf{z}(t) - \mathbf{K}\mathbf{e} \\ &= \begin{pmatrix} [\delta_1(z_2(t)-z_1(t))-(w_2(t)-w_1(t))]\hat{a}_1-k_1e_1 \\ [\delta_2 z_1(t)-w_1(t)]\hat{a}_2+[-\delta_2 z_2(t)+w_2(t)]\hat{a}_3-\delta_2 z_1(t)z_3(t)+w_1(t)w_3(t)-k_2e_2 \\ [-\delta_3 z_3(t)+w_3(t)]\hat{a}_4+\frac{1}{2}[\delta_3(\bar{z}_1(t)z_2(t)+z_1(t)\bar{z}_2(t))-(\bar{w}_1(t)w_2(t)+w_1(t)\bar{w}_2(t))]-k_3e_3 \end{pmatrix}, \end{aligned} \tag{9.1.26}$$

the complex parameter update law is given as (9.1.27)

$$\begin{aligned} \dot{\hat{\mathbf{A}}} &= (\Phi(\mathbf{w}) - \mathbf{M}\Phi(\mathbf{z}(t)))^{\mathrm{H}}\,\mathbf{e} - \eta_a\,\tilde{\mathbf{A}} \\ &= \begin{pmatrix} [(\bar{w}_2(t)-\bar{w}_1(t))-\bar{\delta}_1(\bar{z}_2(t)-\bar{z}_1(t))]e_1-\eta_{a_1}(\hat{a}_1-a_1) \\ (\bar{w}_1(t) - \bar{\delta}_2\bar{z}_1(t))e_2 - \eta_{a_2}(\hat{a}_2 - a_2) \\ (-\bar{w}_2(t) + \bar{\delta}_2\bar{z}_2(t))e_2 - \eta_{a_3}(\hat{a}_3 - a_3) \\ (-\bar{w}_3(t) + \bar{\delta}_3\bar{z}_3(t))e_3 - \eta_{a_4}(\hat{a}_4 - a_4) \end{pmatrix}. \end{aligned} \tag{9.1.27}$$

The values of unknown parameters are chosen as $\mathbf{A} = (a_1, a_2, a_3, a_4)^{\mathrm{T}} = (3, 60 + 0.2j, 1 - 0.6j, 0.8)^{\mathrm{T}}$. The initial condition of master system (9.1.20) and slave system (9.1.26) are randomly chosen as $\mathbf{z}(\mathbf{0}) = (2 + 0.02j, 1 + 0.2j, -1)^{\mathrm{T}}$, $\mathbf{w}(\mathbf{0}) = (1 + 3j, 10 + j, 25)^{\mathrm{T}}$. The initial value of estimated parameter is $\hat{\mathbf{A}}(0) = (\hat{a}_1(0), \hat{a}_2(0), \hat{a}_3(0), \hat{a}_4(0))^{\mathrm{T}} = (1 - j,\ 30 - j,\ 2 + j,\ 1 + j)^{\mathrm{T}}$. The complex scaling matrix is selected as

$$\mathbf{M} = \begin{pmatrix} 0.8\exp(j\pi/4) & 0 & 0 \\ 0 & \exp(j\pi/10) & 0 \\ 0 & 0 & -1 \end{pmatrix}.$$

The constants are chosen as $\eta_a = diag\{10, 8, 15, 16\}$ and $\mathbf{K} = diag\{5, 20, 12\}$, and complex scaling factors are $0.8\exp(j\pi/4)$, $\exp(j\pi/10)$, -1. Here $e_1 = w_1 - (0.8\exp(j\pi/4))z_1$, $e_2 = w_2 - (\exp(j\pi/10))z_2$, $e_3 = w_3 + z_3$. As shown in Fig. 9.5, the CMPS error between systems (9.1.26) and (9.1.20) converges asymptotically to zero. The parameter identification processes of $\mathbf{A}$ are illustrated in Fig. 9.6, where $\hat{\mathbf{A}}$ (the parameter estimation) gradually approaches the selected value $\mathbf{A} = (a_1, a_2, a_3, a_4)^{\mathrm{T}} = (3, 60 + 0.2j, 1 - 0.6j, 0.8)^{\mathrm{T}}$.

As anticipated, the aforementioned results demonstrate that complex modified projective synchronization (CMPS) has been successfully achieved between the two identical complex Lorenz systems (9.1.25) and (9.1.20), and all unknown parameters

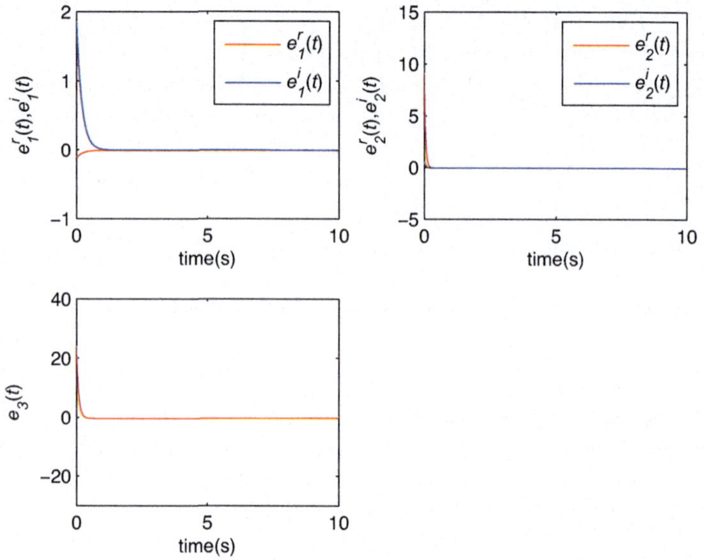

Fig. 9.5 The dynamics of CMPS error between two identical complex Lorenz systems (9.1.20) and (9.1.25) with the controller (9.1.26) and complex parameter update law (9.1.27)

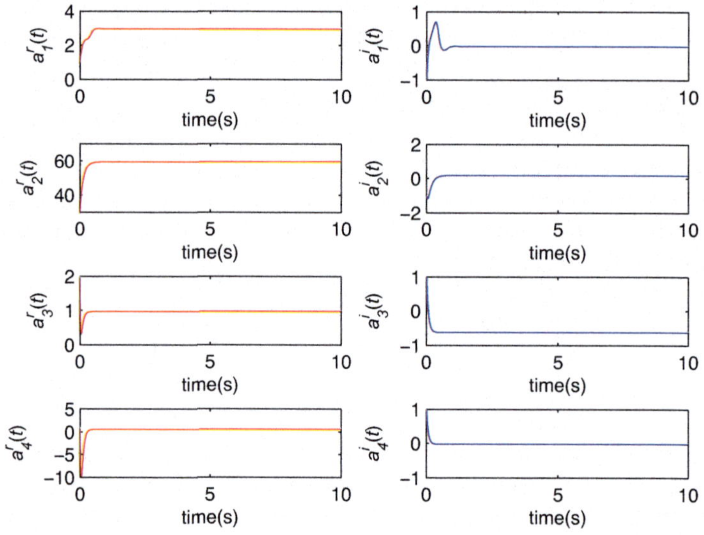

Fig. 9.6 The identification process of unknown complex parameter vector $\mathbf{A} = (a_1, a_2, a_3, a_4)^{\mathrm{T}}$ for systems (9.1.25) and (9.1.20)

have been accurately identified via the designed controller (9.1.26) and the complex parameter update laws (9.1.27).

9.2 A Novel Four-Wing Hyperchaotic Complex System and Its Complex Modified Hybrid Projective Synchronization with Different Dimensions

9.2.1 A New Hyperchaotic Complex Dynamic System

I. The Proposed Complex Dynamic System

In 2012, Dadras et al. introduced a novel 4D smooth autonomous hyperchaotic nonlinear system with real variables, commonly referred to as the Dadras system for short:

$$\begin{cases} \dot{x} = \alpha x - yz + w, \\ \dot{y} = xz - \beta y, \\ \dot{z} = xy + xw - \gamma z, \\ \dot{w} = -y, \end{cases} \tag{9.2.1}$$

where $(x, y, z, w)^{\mathrm{T}}$ is the real state vector, and α, β and γ are positive constant parameters. System (9.2.1) can generate a four-wing hyperchaotic attractor when $\alpha = 8$, $\beta = 40$ and $\gamma = 14.9$ in Fig. 9.7 and a four-wing chaotic attractor when $\alpha = 8, \beta = 40, \gamma = 49$. In this paper we discuss the complex extension of the Dadras system (9.2.1), which is expressed by

$$\begin{cases} \dot{z}_1 &= az_1 - z_2 z_3 + z_4, \\ \dot{z}_2 &= z_1 z_3 - bz_2, \\ \dot{z}_3 &= \frac{1}{2}[\bar{z}_1(z_2 + z_4) + z_1(\bar{z}_2 + \bar{z}_4)] - cz_3, \\ \dot{z}_4 &= -\frac{1}{2}(z_2 + \bar{z}_2), \end{cases} \tag{9.2.2}$$

where a, b and c are real constant parameters, $z_1 = u_1 + ju_2$ and $z_2 = u_3 + ju_4$ are complex state variables, $z_3 = u_5$ and $z_4 = u_6$ are real state variables. The complex Dadras system can be rewritten as a real first-order ordinary differential equation (ODE) of the form

$$\begin{cases} \dot{u}_1 &= au_1 - u_3 u_5 + u_6, \\ \dot{u}_2 &= au_2 - u_4 u_5, \\ \dot{u}_3 &= u_1 u_5 - bu_3, \\ \dot{u}_4 &= u_2 u_5 - bu_4, \\ \dot{u}_5 &= u_1(u_3 + u_6) + u_2 u_4 - cu_5, \\ \dot{u}_6 &= -u_3. \end{cases} \tag{9.2.3}$$

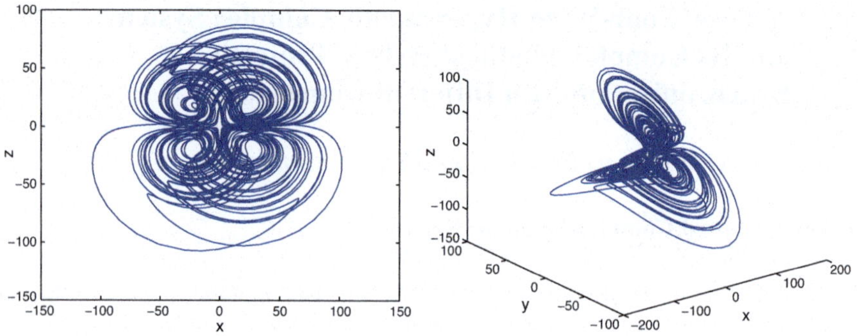

Fig. 9.7 The four-wing hyperchaotic attractors of real Dadras system (9.2.1) for $\alpha = 8$, $\beta = 40$ and $\gamma = 14.9$

II. Dynamic Properties of the New System

System (9.2.3) has the following dynamic properties.

(1) Symmetry and invariance

System (9.2.3) is symmetrical about the u_5-axis, due to its invariance under the coordinates transform $(u_1,u_2,u_3,u_4,u_5,u_6) \rightarrow (-u_1,-u_2,-u_3,-u_4,u_5,-u_6)$. In particular, the symmetry about the u_5-axis for any choice of parameters a, b and c is accurate.

(2) Dissipation and the existence of attractor

For system (9.2.3), one has

$$\nabla V = \sum_{k=1}^{6} \frac{\partial \dot{u}_k}{\partial u_k} = 2a - 2b - c.$$

Hence, in order to ensure system (9.2.3) be dissipative, it is required that $2a - 2b - c < 0$. Under this condition, system (9.2.3) converges exponentially

$$\frac{dV}{dt} = (2a - 2b - c)V \Rightarrow V = V_0 e^{(2a-2b-c)t},$$

where V_0 denotes the initial volume element. This indicates that every volume enclosing the system trajectory shrinks exponentially to zero as $t \to \infty$ at a rate of $2a - 2b - c$. In fact, numerical simulations have demonstrated that the system orbits are ultimately confined to a specific zero-volume limit set, with the asymptotic motion of the system settling onto an attractor.

(3) Equilibria and stability

The equilibria of system (9.2.3) can be found by solving the following algebraic equations simultaneously: $au_1 - u_3u_5 + u_6 = 0$, $au_2 - u_4u_5 = 0$, $u_1u_5 - bu_3 = 0$,

$u_2u_5 - bu_4 = 0$, $u_1(u_3 + u_6) + u_2u_4 - cu_5 = 0$, $-u_3 = 0$. Hence, one has five isolated fixed equilibria: $E_1 = (0, 0, 0, 0, 0, 0)$, $E_{2,3} = (\pm\sqrt{ac}, \pm\sqrt{ab}, 0, 0, \sqrt{bc}, 0)$, and $E_{4,5} = (\pm\sqrt{ac}, \mp\sqrt{ab}, 0, 0, -\sqrt{bc}, 0)$. Here, the stability of the zero equilibrium E_1 is discussed. The Jacobian matrix of system (9.2.3) is

$$J = \begin{bmatrix} a & 0 & -u_5 & 0 & -u_3 & 1 \\ 0 & a & 0 & -u_5 & -u_4 & 0 \\ u_5 & 0 & -b & 0 & u_1 & 0 \\ 0 & u_5 & 0 & -b & u_2 & 0 \\ u_3 + u_6 & u_4 & u_1 & u_2 & -c & u_1 \\ 0 & 0 & -1 & 0 & 0 & 0 \end{bmatrix} \xrightarrow{E_1=(0,0,0,0,0,0)} J_{E_1} = \begin{bmatrix} a & 0 & 0 & 0 & 0 & 1 \\ 0 & a & 0 & 0 & 0 & 0 \\ 0 & 0 & -b & 0 & 0 & 0 \\ 0 & 0 & 0 & -b & 0 & 0 \\ 0 & 0 & 0 & 0 & -c & 0 \\ 0 & 0 & -1 & 0 & 0 & 0 \end{bmatrix}.$$

So, the eigenvalues of the Jacobian matrix J_{E_1} are obtained as follows

$$|\mu I - J_{E_1}| = 0 \Rightarrow \mu_1 = \mu_2 = a, \mu_3 = 0, \mu_4 = \mu_5 = -b, \mu_6 = -c.$$

Since a, b and c are all positive real numbers, one can easily find $\mu_1 = \mu_2 > 0$, $\mu_3 = 0$, $\mu_4 = \mu_5 < 0$, $\mu_6 < 0$, implying that the equilibrium E_1 is unstable.

(4) Lyapunov exponents and fractal dimensions

Numerical simulations are performed using MATLAB software, with the fourth-order Runge–Kutta integration algorithm employed to solve the differential equations. The initial condition is set as $x_0 = (10 + j, 10 + j, 10, 1)^{\mathrm{T}}$ and system (9.2.3) is capable of displaying both four-wing hyperchaotic and four-wing chaotic attractors.

(i) Four-wing hyperchaotic attractor

By setting the parameters to $a = 10$, $b = 40$ and $c = 14.9$, system (9.2.3) exhibits hyperchaotic behavior and can generate a four-wing hyperchaotic attractor. Consequently, the complex Dadras system (9.2.2) is capable of displaying a four-wing hyperchaotic attractor, as illustrated in Fig. 9.8.

The Lyapunov exponents have been calculated as

$$LE_1 = 2.80065, \quad LE_2 = 0.703862, \quad LE_3 = 0,$$

$$LE_4 = -0.184465, \quad LE_5 = -29.7874, \quad LE_6 = -48.4024.$$

And the fractal dimension is

$$D_L = j + \frac{1}{|LE_{j+1}|} \sum_{i=1}^{j} LE_i = 4 + \frac{LE_1 + LE_2 + LE_3 + LE_4}{|LE_5|} = 4.11146.$$

Both Lyapunov exponents and fractal dimension indicate that the system (9.2.2) is hyperchaotic with the aforementioned set of parameters.

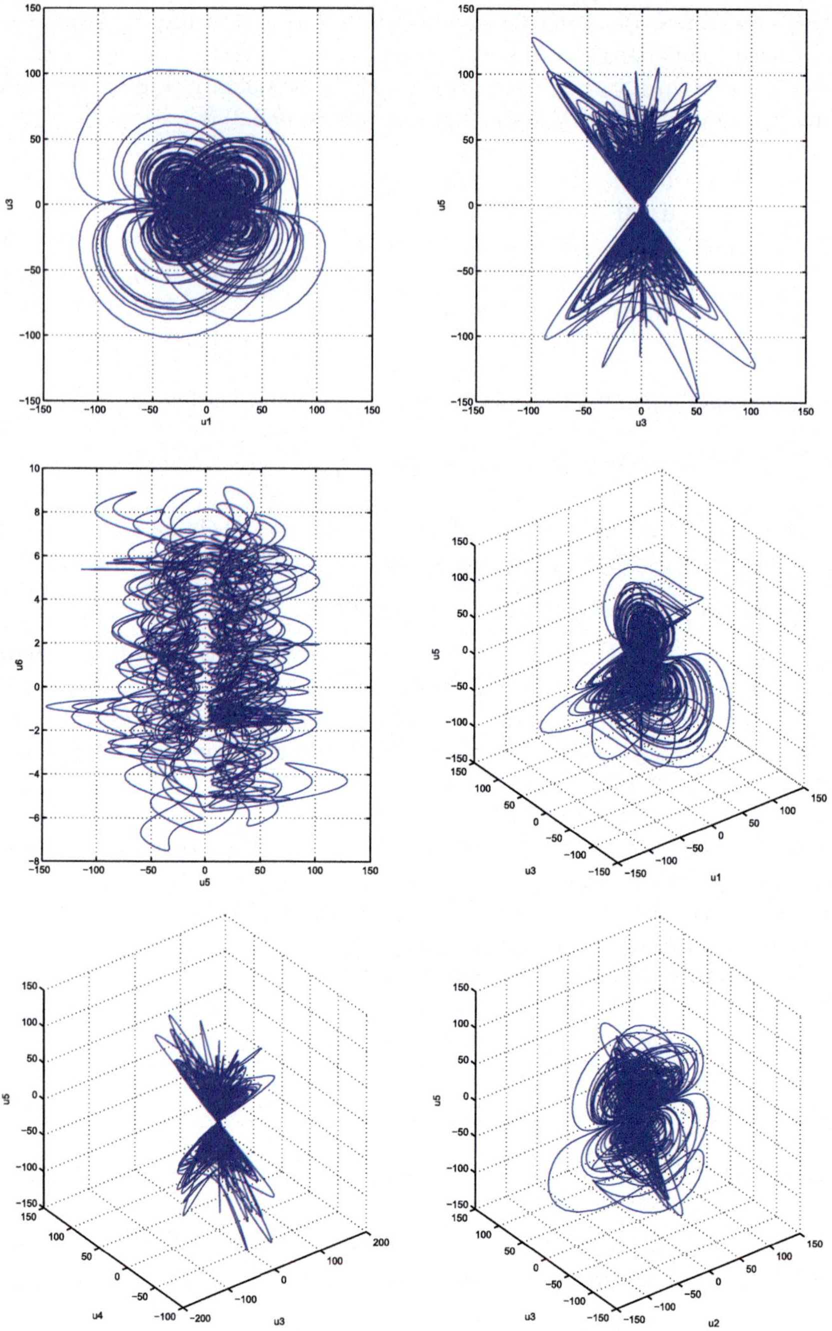

Fig. 9.8 The hyperchaotic attractors of complex Dadras system (9.2.2) with $a = 10$, $b = 40$ and $c = 14.9$

(ii) Four-wing chaotic attractor
Setting the parameters as $a = 10,\ b = 40$ and $c = 49$, system (9.2.3) has generated a four-wing chaotic attractor. So complex Dadras system (9.2.2) can display four-wing chaotic attractor in Fig. 9.9. The Lyapunov exponents of this system in this case are

$$LE_1 = 0.915193, \quad LE_2 = 0, \quad LE_3 = -0.0910834,$$

$$LE_4 = -2.8148, \quad LE_5 = -30.0099, \quad LE_6 = -77.0274.$$

And its fractal dimension is

$$D_L = j + \frac{1}{|LE_{j+1}|} \sum_{i=1}^{j} LE_i = 3 + \frac{LE_1 + LE_2 + LE_3}{|LE_4|} = 3.292777.$$

Both Lyapunov exponents and fractal dimension indicate that the system (9.2.2) is chaotic with the aforementioned set of parameters.

(5) Poincaré map

As an important analysis technique, the Poincaré map can reflect bifurcation and folding properties of hyperchaos. We have taken

$$\Xi_1 = \{(u_1, u_2, u_3, u_4, u_5, u_6)^T \in R^6 : u_1 = 0\},$$

$$\Xi_2 = \{(u_1, u_2, u_3, u_4, u_5, u_6)^T \in R^6 : u_2 = 0\},$$

$$\Xi_3 = \{(u_1, u_2, u_3, u_4, u_5, u_6)^T \in R^6 : u_3 = 0\},$$

$$\Xi_4 = \{(u_1, u_2, u_3, u_4, u_5, u_6)^T \in R^6 : u_4 + u_5 + u_6 = 0\},$$

as cross sections and the system parameters are set as $a = 10, b = 40, c = 14.9$. Figure 9.10 shows projections of the Poincaré map on (u_1, u_2, u_3) space and (u_1, u_2), (u_1, u_3), (u_2, u_3) planes. From Fig. 9.10, one can see that the Poincaré map here consists of several limbs with various bifurcations in different directions, which indicates that the system has extremely rich dynamics. Also, the Poincaré maps show that the branches are jointed and united as a single attractor. This proves the existence of the four-wing hyperchaotic attractor of complex Dadras system (9.2.2) in Fig. 9.8.

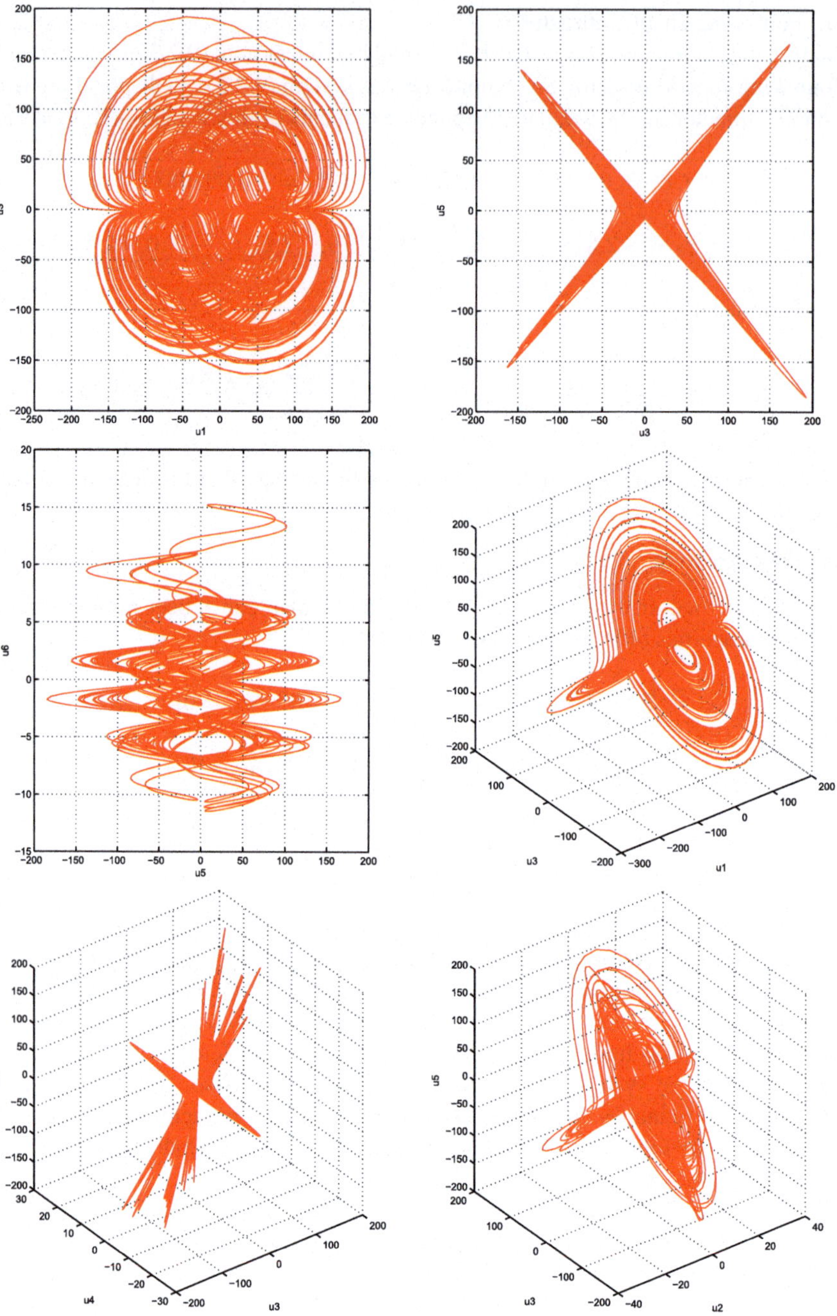

Fig. 9.9 The chaotic attractors of complex Dadras system (9.2.2) with $a = 10$, $b = 40$ and $c = 49$

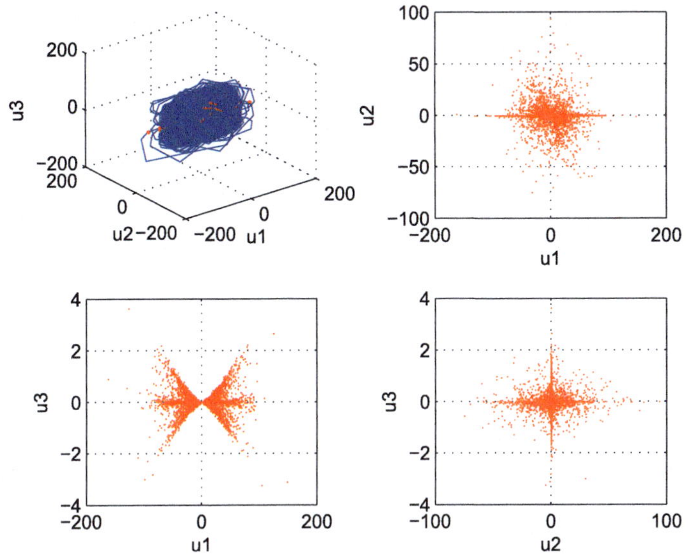

Fig. 9.10 Projections of Poincaré maps of the four-wing hyperchaotic attractors of complex Dadras system (9.2.2) with parameters $a = 10$, $b = 40$ and $c = 14.9$ on different crossing sections

9.2.2 CMHPS Between Hyperchaotic and Chaotic Complex Systems with Complex Parameters and with Different Dimensions

I. Scheme of CMHPS

First, we consider a class of m-dimensional complex chaotic (hyperchaotic) drive system as

$$\dot{x}(t) = Ax + F(x), \tag{9.2.4}$$

and n-dimensional response system with the controller as

$$\dot{y}(t) = By + G(y) + P(x, y), \tag{9.2.5}$$

where $x = x^r + jx^i \in \mathbb{C}^m$ and $y = y^r + jy^i \in \mathbb{C}^n$ are complex states vectors, $A \in \mathbb{C}^{m\times m}$ and $B \in \mathbb{C}^{n\times n}$ are the coefficient matrices of x and y, while F and G are the nonlinear parts respectively, $P = (p_1, p_2, \ldots, p_n)^{\mathrm{T}}$ is the controller to be designed.

Next we consider the definition of CMHPS with a complex transformation matrix of complex systems with complex parameters based on that of CMPS of complex systems with real parameters.

Definition 9.2 For the drive system (9.2.4) and response system (9.2.5), it is said to be CMHPS with Λ between $x(t)$ and $y(t)$, if there exists a norm bounded matrix $\Lambda \in \mathbb{C}^{n\times m}$, such that

$$\lim_{t\to+\infty} \|y(t) - \Lambda x(t)\| = 0,$$

while the matrix Λ is defined as the complex transformation matrix of the drive system (9.2.4).

If the error of CMHPS is defined as

$$e(t) = y(t) - \Lambda x(t), \tag{9.2.6}$$

the objective of this section is to design a controller P to ensure that the synchronization error tends to zero, i.e.

$$\lim_{t\to+\infty} \|e(t)\| = \lim_{t\to+\infty} \|y(t) - \Lambda x(t)\| = 0. \tag{9.2.7}$$

Remark 9.5 Most of the classical chaotic (hyperchaotic) complex system can be formed as system (9.2.4), such as the complex Lorenz system, complex Chen system, complex Lü system, complex hyperchaotic Lorenz system, complex Duffing system.

Lemma 9.1 *For a matrix $Q \in \mathbb{C}^{n\times n}$, all of the real parts of its eigenvalues are negative, denoted $Re(\lambda_i(Q)) < 0, i = 1, 2, \ldots, n$, then*

$$\lim_{t\to+\infty} exp(Qt) = 0. \tag{9.2.8}$$

Theorem 9.3 *For given complex transformation matrix Λ and initial conditions $x(0)$, $y(0)$, the complex nonlinear controller is designed as*

$$P = -G(y) + \Lambda(F(x) + Ax) - B\Lambda x - Ke, \tag{9.2.9}$$

then, CMHPS between the response system (9.2.5) and drive system (9.2.4) will occur with desired complex transformation matrix Λ asymptotically if $Re(\lambda_i(B - K)) < 0 (i = 1, 2, \ldots, n)$, where $K \in \mathbb{C}^{n\times n}$ is the complex control gain matrix.

Proof Insertion of (9.2.4) and (9.2.5) into (9.2.6) gives

$$\dot{e}(t) = Be + B\Lambda x + G(y) - \Lambda Ax - \Lambda F(x) + P. \tag{9.2.10}$$

Substituting the controller (9.2.9) into system (9.2.10), we can obtain

$$\dot{e}(t) = (B - K)e(t),$$

or

$$\dot{e}(t) - (B - K)e(t) = 0. \tag{9.2.11}$$

By multiplying $exp(-(B - K)t)$ on both sides of (9.2.11), we have that the left-hand side and the right-hand side of (9.2.11) take the form, respectively

$$\frac{d(exp(-(B-K)t)e(t))}{dt} = 0. \tag{9.2.12}$$

Integrating two sides of (9.2.12) on time interval $[0, t]$, and with the initial condition $e(0) = \xi$, we can get

$$e(t) = exp((B-K)t)\,\xi. \tag{9.2.13}$$

Then, there exists a suitable $K \in \mathbb{C}^{n\times n}$ such that $Re(\lambda_i(B - K)) < 0(i = 1, 2, \ldots, n)$, to each B, and we get as follows by Lemma 9.1,

$$\lim_{t\to+\infty} e(t) = \lim_{t\to+\infty} exp((B-K)t)\,\xi = 0. \tag{9.2.14}$$

Thus, we arrive at CMHPS with desired complex transformation matrix Λ between the complex systems (9.2.5) and (9.2.4) by using the controller (9.2.9) asymptotically. The proof is completed. □

Remark 9.6 Note that one can adjust complex transformation matrix Λ arbitrarily without worrying about the control robustness, and $K \in \mathbb{C}^{n\times n}$ is the complex control gain matrix. In particular, if A and B are real, Theorem 9.3 is also applied to achieve CMHPS of complex chaotic systems with real parameters.

Remark 9.7 In particular, if the dimension of the drive system equals to that of the response system, i.e. $m = n$, and the transformation matrix Λ is diagonal, the CMHPS can be simplified to CFSHPS [170], CMPS, CPS, FSHPS, MPS and PS with the same dimension.

Remark 9.8 When the scaling matrix $\Lambda = 0$, the synchronization problem reduces to the control of a complex chaotic (or hyperchaotic) system.

II. Realization of Synchronization

Throughout the present section, we will demonstrate the effectiveness of the proposed method. Two examples are used respectively to analyze two cases: increased-order synchronization ($m < n$) and reduced order synchronization ($m > n$).

(1) CMHPS of complex hyperchaotic Dadras drive system and complex chaotic Lorenz response system with complex parameters

For the purpose of observing reduced-order CMHPS behaviors, it is assumed that the new 4D hyperchaotic complex Dadras system (9.2.2) drives the 3D chaotic complex

Lorenz system with complex parameters. Therefore, the drive system reads in the form

$$\begin{cases} \dot{x}_1 = a_1x_1 - x_2x_3 + x_4, \\ \dot{x}_2 = -a_2x_2 + x_1x_3, \\ \dot{x}_3 = \frac{1}{2}[\bar{x}_1(x_2 + x_4) + x_1(\bar{x}_2 + \bar{x}_4)] - a_3x_3, \\ \dot{x}_4 = -\frac{1}{2}(x_2 + \bar{x}_2), \end{cases} \tag{9.2.15}$$

where

$$A = \begin{bmatrix} a_1 & 0 & 0 & 1 \\ 0 & -a_2 & 0 & 0 \\ 0 & 0 & -a_3 & 0 \\ 0 & -\frac{1}{2} & 0 & 0 \end{bmatrix}, F(x) = \begin{bmatrix} -x_2x_3 \\ x_1x_3 \\ \frac{1}{2}[\bar{x}_1(x_2 + x_4) + x_1(\bar{x}_2 + \bar{x}_4)] \\ -\frac{1}{2}\bar{x}_2 \end{bmatrix},$$

$x_1 = x_1^r + jx_1^i, x_2 = x_2^r + jx_2^i$ are complex state variables and x_3, x_4 are real state variables.

The complex response with the controller writes in the form as

$$\begin{cases} \dot{y}_1 = b_1(y_2 - y_1) + p_1, \\ \dot{y}_2 = b_2y_1 - b_3y_2 - y_1y_3 + p_2, \\ \dot{y}_3 = \frac{1}{2}(\bar{y}_1y_2 + y_1\bar{y}_2) - b_4y_3 + p_3, \end{cases} \tag{9.2.16}$$

where

$$B = \begin{bmatrix} -b_1 & b_1 & 0 \\ b_2 & -b_3 & 0 \\ 0 & 0 & -b_4 \end{bmatrix}, G(y) = \begin{bmatrix} 0 \\ -y_1y_3 \\ \frac{1}{2}(\bar{y}_1y_2 + y_1\bar{y}_2) \end{bmatrix},$$

$y_1 = y_1^r + jy_1^i, y_2 = y_2^r + jy_2^i$ are complex state variables and y_3 is a real state variable, $P = (p_1, p_2, p_3)^{\mathrm{T}}$ is the complex controller to be decided.

Complex matrix Λ can be chosen as

$$\Lambda = \begin{bmatrix} -j & 0 & 0 & 0 \\ 0 & j & 0 & 0 \\ 0 & 0 & 0 & -1 \end{bmatrix},$$

the error system $e(t) = y(t) - \Lambda x(t)$ is obtained as

$$\begin{cases} e_1 = y_1 + jx_1, \\ e_2 = y_2 - jx_2, \\ e_3 = y_3 + x_4. \end{cases}$$

In order to satisfy $Re(\lambda_i(B - K)) < 0 (i = 1, 2, 3)$, we can choose the complex control gain matrix as

$$K = \begin{bmatrix} -b_1 - \lambda_1 & b_1 & 0 \\ b_2 & -b_3 - \lambda_2 & 0 \\ 0 & 0 & -b_4 - \lambda_3 \end{bmatrix},$$

where all of the real parts of $\lambda_1, \lambda_2, \lambda_3$ are negative.

Design the complex state feedback controller according to (9.2.10) in Theorem 9.3 as follows

$$\begin{aligned} P &= -G(y) + \Lambda(F(x) + Ax) - B\Lambda x - Ke \\ &= \begin{pmatrix} -j((a_1 + b_1)x_1 + b_1x_2 - x_2x_3 + x_4) + (b_1 + \lambda_1)e_1 - b_1e_2 \\ j(b_2x_1 + (b_3 - a_2)x_2 + x_1x_3) + y_1y_3 - b_2e_1 + (b_3 + \lambda_2)e_2 \\ \frac{1}{2}(x_2 + \bar{x}_2) - b_4x_4 - \frac{1}{2}(\bar{y}_1y_2 + y_1\bar{y}_2) + (b_4 + \lambda_3)e_3 \end{pmatrix}. \end{aligned} \tag{9.2.17}$$

The parameters of drive system (9.2.15) and response system (9.2.16) are $a_1 = 10,\ a_2 = 40,\ a_3 = 14.9$ and $b_1 = 5,\ b_2 = 60 + 0.2j,\ b_3 = 1 - 0.6j,\ b_4 = 1.8$, respectively. The initial values are randomly chosen as $x_0 = (10 + j, 10 + j, 10, 1)^T$ and $y_0 = (2 + 0.2j, 15 + 0.2j, -5)^T$, respectively. All of the eigenvalues of $B - K$ are taken as $\lambda_1 = -6 - 5j, \lambda_2 = -1 - j, \lambda_3 = -10$. The simulation results are demonstrated in Fig. 9.11, where the blue line presents the states of drive system (9.2.15) and the red line presents the states of response system (9.2.16). The errors of CMHPS converge asymptotically to zero as in Fig. 9.12, where the red line shows the real parts of the errors and the blue line presents the imaginary parts of the errors.

Therefore, the above findings confirm that CMHPS has been effectively realized between the 4D hyperchaotic complex Dadras system (9.2.15) and the 3D complex Lorenz system (9.2.16) featuring complex parameters.

(2) CMHPS of complex chaotic Lü drive system and complex hyperchaotic Dadras response system

In order to observe increased order CMHPS behaviors, it is assumed that the 3D complex chaotic Lü system drives the new 4D hyperchaotic complex Dadras system. The drive system for complex Lü system can be described as

$$\begin{cases} \dot{x}_1 &= a_1(x_2 - x_1), \\ \dot{x}_2 &= a_2x_2 - x_1x_3, \\ \dot{x}_3 &= \frac{1}{2}(\bar{x}_1x_2 + x_1\bar{x}_2) - a_3x_3, \end{cases} \tag{9.2.18}$$

where

$$A = \begin{bmatrix} -a_1 & a_1 & 0 \\ 0 & a_2 & 0 \\ 0 & 0 & -a_3 \end{bmatrix}, \quad F(x) = \begin{bmatrix} 0 \\ -x_1x_3 \\ \frac{1}{2}(\bar{x}_1x_2 + x_1\bar{x}_2) \end{bmatrix},$$

$x_1 = x_1^r + jx_1^i, x_2 = x_2^r + jx_2^i$ are complex state variables and x_3 is a real state variable. When $a_1 = 29,\ a_2 = 21,\ a_3 = 2$, complex Lü system (9.2.18) is chaotic.

The complex response system with the controller is described as

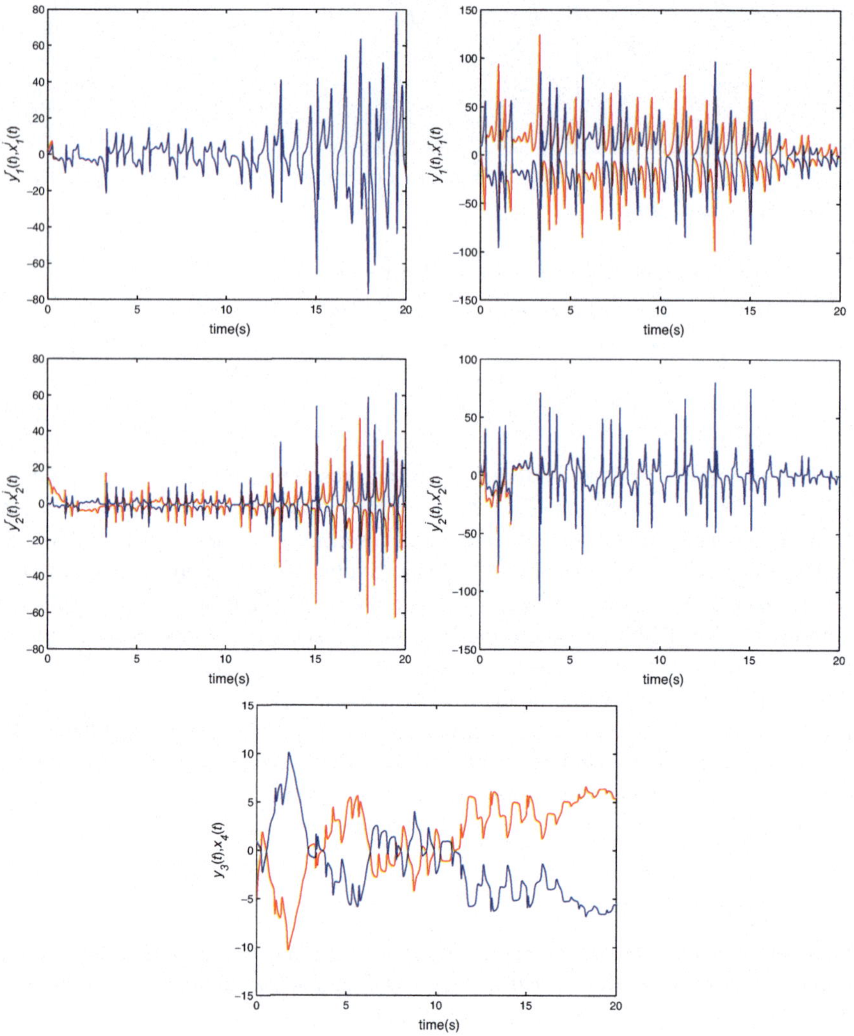

Fig. 9.11 Reduced order synchronization between 4-dimensional complex hyperchaotic Dadras drive system (9.2.15) (blue line) and 3-dimensional complex chaotic Lorenz response system (9.2.17) (red line) with the controller (9.2.9)

$$\begin{cases} \dot{y}_1 &= b_1 y_1 - y_2 y_3 + y_4 + p_1, \\ \dot{y}_2 &= y_1 y_3 - b_2 y_2 + p_2, \\ \dot{y}_3 &= \frac{1}{2}[\bar{y}_1(y_2 + y_4) + y_1(\bar{y}_2 + \bar{y}_4)] - b_3 y_3 + p_3, \\ \dot{y}_4 &= -\frac{1}{2}(y_2 + \bar{y}_2) + p_4, \end{cases} \tag{9.2.19}$$

where

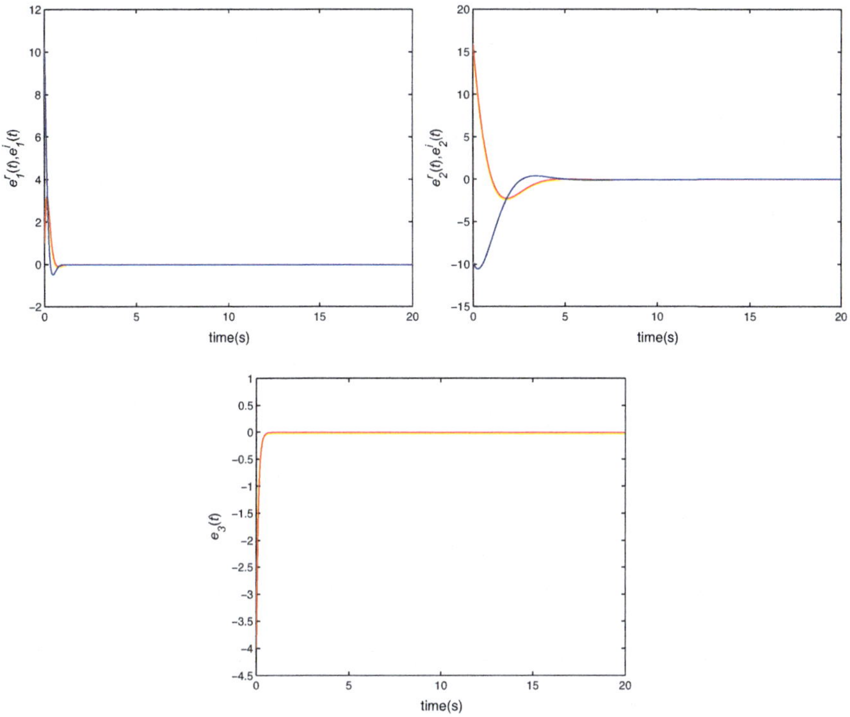

Fig. 9.12 CMHPS error dynamics between 4-dimensional complex hyperchaotic Dadras drive system (9.2.15) and 3-dimensional complex chaotic Lorenz response system (9.2.17)

$$B = \begin{bmatrix} b_1 & 0 & 0 & 1 \\ 0 & -b_2 & 0 & 0 \\ 0 & 0 & -b_3 & 0 \\ 0 & -\frac{1}{2} & 0 & 0 \end{bmatrix}, G(y) = \begin{bmatrix} -y_2 y_3 \\ y_1 y_3 \\ \frac{1}{2}[\bar{y}_1(y_2 + y_4) + y_1(\bar{y}_2 + \bar{y}_4)] \\ -\frac{1}{2}\bar{y}_2 \end{bmatrix},$$

$y_1 = y_1^r + jy_1^i$, $y_2 = y_2^r + jy_2^i$ are complex state variables, y_3, y_4 are real state variables, and $P = (p_1, p_2, p_3, p_4)^{\mathrm{T}}$ is the complex controller to be decided. Complex transformation matrix can be taken as

$$\Lambda = \begin{bmatrix} j & 0 & 0 \\ 0 & -j & 0 \\ 0 & 0 & 1 \\ 0 & 0 & -1 \end{bmatrix},$$

the error system $e(t) = y(t) - \Lambda x(t)$ is obtained as

$$
\begin{cases}
e_1 = y_1 - jx_1, \\
e_2 = y_2 + jx_2, \\
e_3 = y_3 - x_3, \\
e_4 = y_4 + x_3.
\end{cases}
$$

In order to satisfy $Re(\lambda_i(B - K)) < 0 (i = 1, 2, 3, 4)$, we can choose the complex control gain matrix as

$$
K = \begin{bmatrix} b_1 - \lambda_1 & 0 & 0 & 1 \\ 0 & -b_2 - \lambda_2 & 0 & 0 \\ 0 & 0 & -b_3 - \lambda_3 & 0 \\ 0 & -\frac{1}{2} & 0 & -\lambda_4 \end{bmatrix},
$$

where all of the real parts of $\lambda_1, \lambda_2, \lambda_3, \lambda_4$ are negative.

Design the controller according to (9.2.10) in Theorem 9.3 as follows

$$
\begin{aligned}
P &= -G(y) + \Lambda(F(x) + Ax) - B\Lambda x - Ke \\
&= \begin{pmatrix} j(-(a_1 + b_1)x_1 + a_1x_2) + x_3 + y_2y_3 - (b_1 - \lambda_1)e_1 - e_4 \\ -j((a_2 + b_2)x_2 - x_1x_3) - y_1y_3 + (b_2 + \lambda_2)e_2 \\ \frac{1}{2}(\bar{x}_1x_2 + x_1\bar{x}_2 - \bar{y}_1(y_2 + y_4) - y_1(\bar{y}_2 + \bar{y}_4)) + (b_3 - a_3)x_3 + (b_3 + \lambda_3)e_3 \\ \frac{1}{2}(\bar{y}_2 - jx_2 - \bar{x}_1x_2 - x_1\bar{x}_2) + a_3x_3 + \frac{1}{2}e_2 + \lambda_4e_4 \end{pmatrix}.
\end{aligned}
\tag{9.2.20}
$$

The parameters of drive system (9.2.18) and response system (9.2.19) are $a_1 = 29,\ a_2 = 21,\ a_3 = 2$ and $b_1 = 10,\ b_2 = 40,\ b_3 = 14.9$, respectively. The initial values are randomly chosen as $x_0 = (4 - 2j, 3 + j, -10)^T$ and $y_0 = (10 + j, 10 + j, 10, 1)^T$, respectively. All of the eigenvalues of $B - K$ are taken as $\lambda_1 = -1 - j, \lambda_2 = -3 - j, \lambda_3 = -500, \lambda_4 = -2$. The simulation results are demonstrated in Fig. 9.13, where the blue line presents the states of drive system (9.2.18) and the red line presents the states of response system (9.2.19). The errors of CMHPS converge asymptotically to zero as in Fig. 9.14, where the red line shows the real parts of the errors and the blue line presents the imaginary parts of the errors.

Hence, the above results show that CMHPS has been achieved between the complex chaotic Lü drive system (9.2.18) and the complex hyperchaotic Dadras response system (9.2.19).

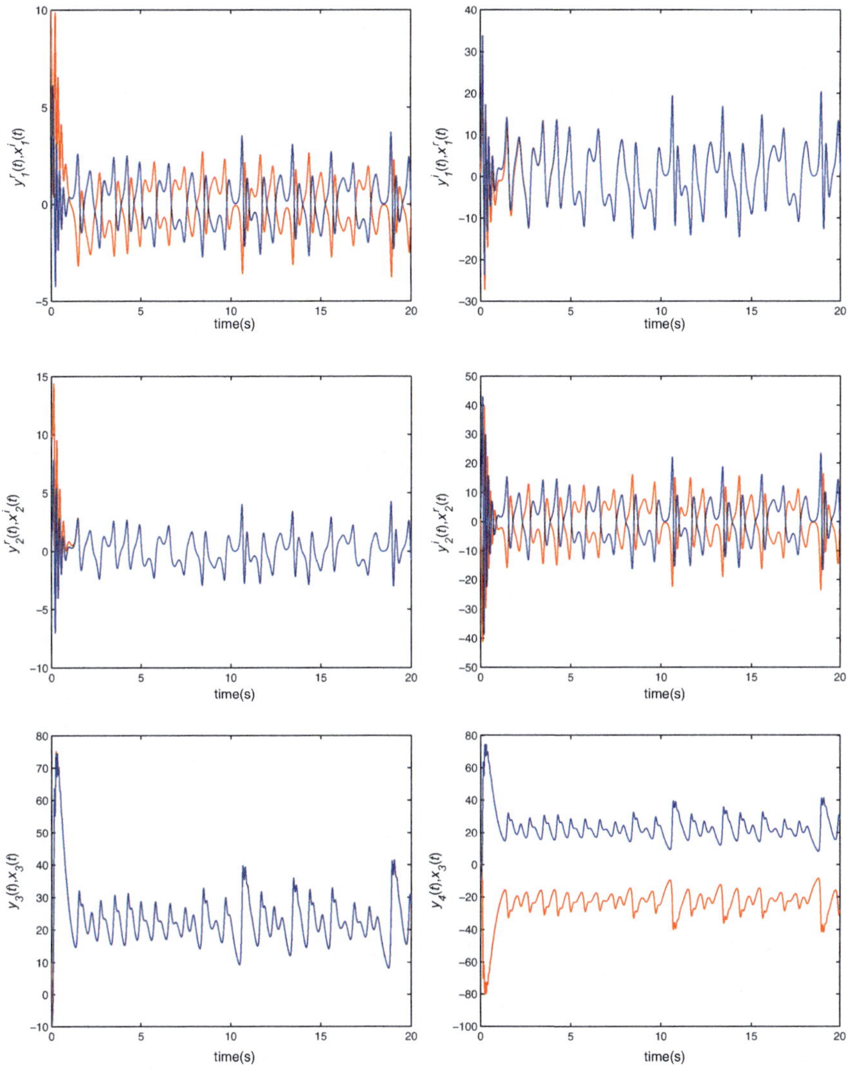

Fig. 9.13 Increased order synchronization between 3-dimensional complex chaotic Lü drive system (9.2.18) (blue line) and 4-dimensional complex hyperchaotic Dadras response system (9.2.19) (red line) with the controller (9.2.20)

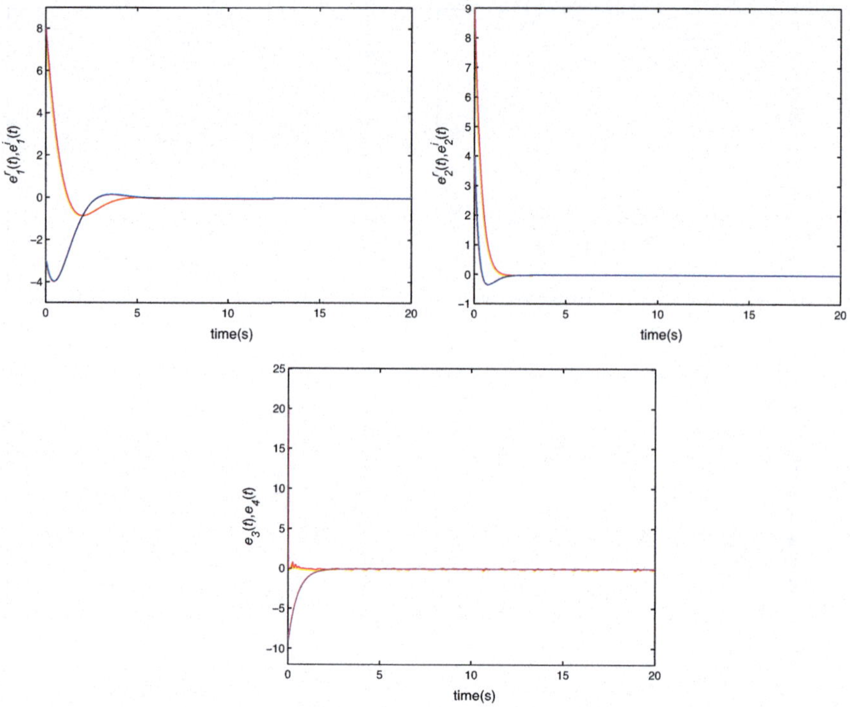

Fig. 9.14 CMHPS error dynamics between 3-dimensional complex chaotic Lü drive system (9.2.18) and 4-dimensional complex hyperchaotic Dadras response system (9.2.20)

Chapter 10
Generalized Combination Complex Synchronization of New Hyperchaotic Complex Systems

Combination complex synchronization is a specialized form of synchronization observed in complex chaotic systems, characterized by a unique coupling mechanism where multiple drive systems interact with a single response system [172, 173]. In this chapter, based on two drive complex systems and one response complex system with different dimensions, we propose generalized combination complex synchronization and design a general controller. Additionally, we investigate generalized combination complex synchronization between real systems and complex systems via two complex scaling matrices. In the following, a new hyperchaotic complex system is presented and its dynamical properties are discussed.

10.1 A New Hyperchaotic Complex Lü-like system

The Lü-like or Pan system was proposed by Zhou et al. [174], which can be expressed as

$$\begin{cases} \dot{x} = a(y - x), \\ \dot{y} = cx - xz, \\ \dot{z} = xy - bz, \end{cases} \tag{10.1.1}$$

where a, b, and c denote real constants, $(x, y, z)^T$ is a real state vector. With the parameters set to $a = 10$, $b = 2$ and $c = 16$, system (10.1.1) exhibits chaotic behavior, as illustrated in Figs. 10.1(a)–(b).

Through the introduction of a state feedback controller, a new modified hyperchaotic Lü-like or Pan system has been constructed in [175], which is governed by the following equations

S. Liu et al., *Chaos in Complex Dynamical System: Theory and Application*,
https://doi.org/10.1007/978-981-95-7762-0_10

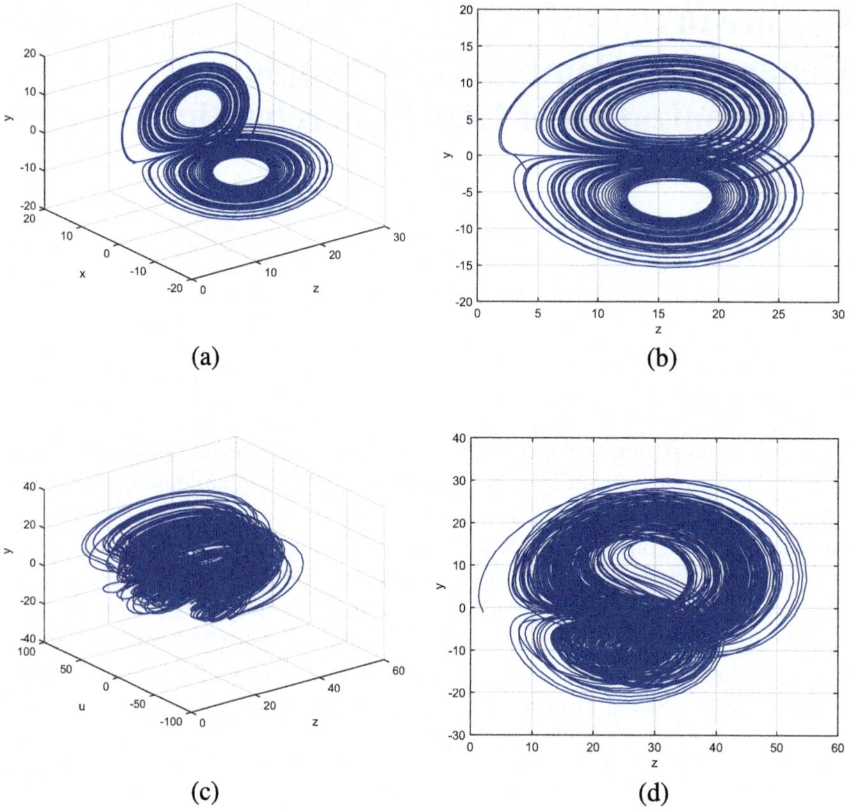

Fig. 10.1 Attractors of chaotic system (10.1.1) and hyperchaotic system (10.1.2)

$$\begin{cases} \dot{x} = a(y - x), \\ \dot{y} = cx - xz + u, \\ \dot{z} = xy - bz, \\ \dot{u} = -dy, \end{cases} \tag{10.1.2}$$

where a, b, c, and d are real constants. The hyperchaotic attractors of system (10.1.2) are shown in Figs. 10.1(c)–(d) with $a = 10$, $b = 8/3$, $c = 28$, and $d = 10$.

This work proposes the complex extension of the modified hyperchaotic Lü-like system, which is formulated by

$$\begin{cases} \dot{x}_1 = a(x_2 - x_1), \\ \dot{x}_2 = cx_1 - x_1x_3 + x_4, \\ \dot{x}_3 = \frac{1}{2}(\overline{x}_1x_2 + x_1\overline{x}_2) - bx_3, \\ \dot{x}_4 = -\frac{d}{2}(\overline{x}_2 + x_2), \end{cases} \tag{10.1.3}$$

where $a, b, c,$ and d are real positive parameters, $x_1 = m_1 + jm_2$ and $x_2 = m_3 + jm_4$ are complex variables, $x_3 = m_5$ and $x_4 = m_6$ are real variables. The hyperchaotic complex Lü-like system can be rewritten as follows

$$\begin{cases} \dot{m}_1 = a(m_3 - m_1), \\ \dot{m}_2 = a(m_4 - m_2), \\ \dot{m}_3 = cm_1 - m_1m_5 + m_6, \\ \dot{m}_4 = cm_2 - m_2m_5, \\ \dot{m}_5 = m_1m_3 + m_2m_4 - bm_5, \\ \dot{m}_6 = -dm_3. \end{cases} \tag{10.1.4}$$

10.2 Dynamical Properties of the New Hyperchaotic Complex Lü-Like System

In the following, we investigate the basic dynamical properties of system (10.1.4).

10.2.1 Symmetry and Invariance

It is clear that system (10.1.4) is symmetric about m_5-axis, which means it is invariant for the coordinate transformation of $(m_1, m_2, m_3, m_4, m_5, m_6) \longrightarrow (-m_1, -m_2, -m_3, -m_4, m_5, -m_6)$.

10.2.2 Dissipation

System (10.1.4) is dissipative, with its divergence defined as

$$\nabla V = \sum_{k=1}^{6} \frac{\partial \dot{m}_k}{\partial m_k} = -2a - b.$$

Thus, the dissipation condition requires $2a + b > 0$.

10.2.3 Equilibria and Stability

The equilibria of system (10.1.4) can be obtained by solving the equations $\dot{m}_1 = 0$, $\dot{m}_2 = 0$, $\dot{m}_3 = 0$, $\dot{m}_4 = 0$, $\dot{m}_5 = 0$, and $\dot{m}_6 = 0$. Then we have three equilibrium points: $E_1 = (0, 0, 0, 0, 0, 0)$ and $E_{2,3} = (0, \pm\sqrt{bc}, 0, \pm\sqrt{bc}, 0, 0)$. In order to investigate the stability of the zero equilibrium point E_1, we obtian the Jacobian of system (10.1.4) at E_1 as follows

$$J_{E_1} = \begin{pmatrix} -a & 0 & a & 0 & 0 & 0 \\ 0 & -a & 0 & a & 0 & 0 \\ c & 0 & 0 & 0 & 0 & 1 \\ 0 & c & 0 & 0 & 0 & 0 \\ 0 & 0 & 0 & 0 & -b & 0 \\ 0 & 0 & -d & 0 & 0 & 0 \end{pmatrix}.$$

A straightforward computation yields the characteristic polynomial of Jacobian matrix J_{E_1} as

$$(\lambda + b)(\lambda^2 + a\lambda - ac)(\lambda^3 + a\lambda^2 - (ac - d)\lambda + ad) = 0.$$

According to the Routh-Hurwitz Theorem, we conclude that E_1 will be stable when $a > 0$, $b > 0$, $c < 0$, and $d > 0$. Otherwise, it is an unstable fixed point. Similarly, we can discuss the stability of the equilibrium points E_2 and E_3.

10.2.4 Lyapunov Exponents and Fractal Dimensions

By means of Runge-Kutta method of order 4 in MATLAB environment, we obtain Lyapunov exponents shown in Fig. 10.2 using $a = 10, b = 8/3, c = 28, d = 10$ and the initial condition $x(0) = (1 + j, 1 + j, 2, 3)$. For system (10.1.4), the six Lyapunov exponents are found be $\lambda_1 = 0.689209$, $\lambda_2 = 0.314361$, $\lambda_3 = -0.009003 \approx$

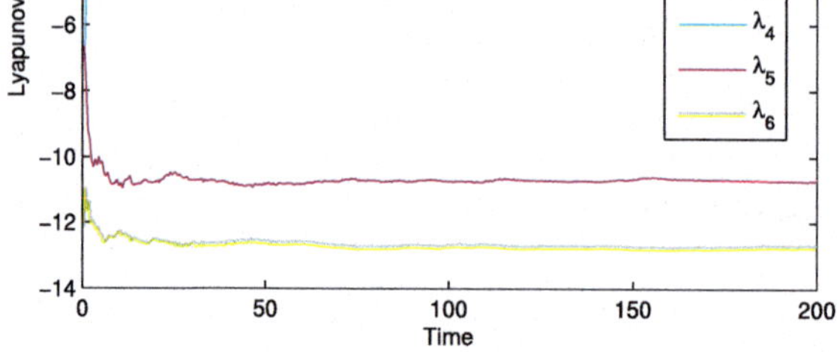

Fig. 10.2 Lyapunov exponents of system (10.1.4) with $(a, b, c, d) = (10, 8/3, 28, 10)$

0, $\lambda_4 = -0.275494$, $\lambda_5 = -10.697494$, and $\lambda_6 = -12.685651$. The presence of two positive exponents (λ_1 and λ_2) confirms the hyperchaotic behavior of system (10.1.4). Thus, we can calculate the fractal dimension [176] as follows

$$D = j + \frac{1}{|\lambda_{j+1}|}\sum_{i=1}^{j}\lambda_i = 4 + \frac{\lambda_1 + \lambda_2 + \lambda_3 + \lambda_4}{|\lambda_5|} = 4.06722.$$

10.2.5 Hyperchaotic Behavior and Attractors

Under the parameter set $(a, b, c, d) = (10, 8/3, 28, 10)$ and the initial condition $x(0) = (1 + j, 1 + j, 2, 3)$, System (10.1.4) is hyperchaotic. The corresponding hyperchaotic attractors are depicted in various phased planes and projections in Fig. 10.3.

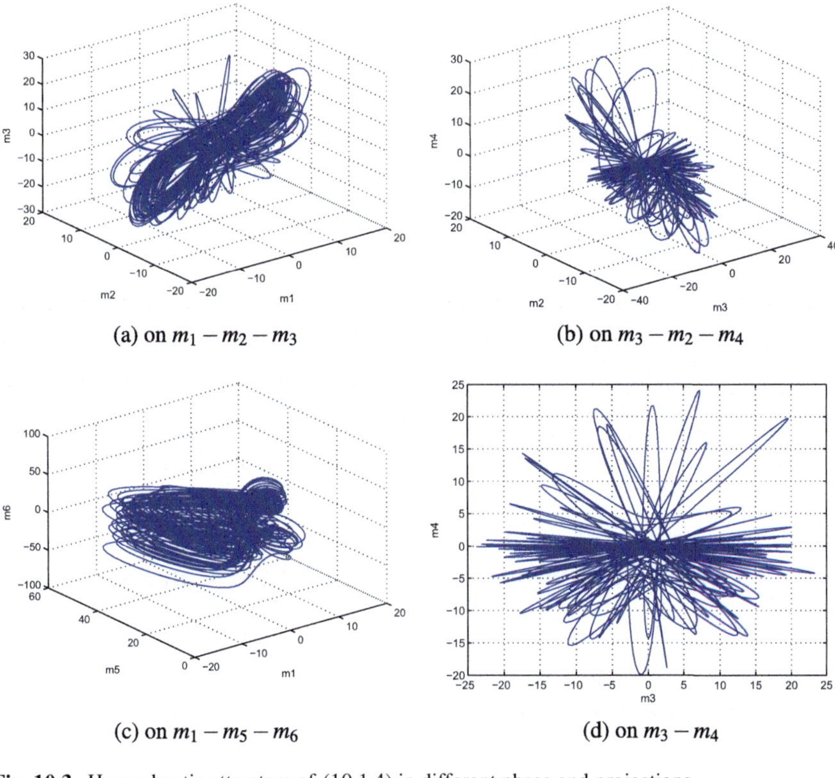

(a) on $m_1 - m_2 - m_3$

(b) on $m_3 - m_2 - m_4$

(c) on $m_1 - m_5 - m_6$

(d) on $m_3 - m_4$

Fig. 10.3 Hyperchaotic attractors of (10.1.4) in different phase and projections

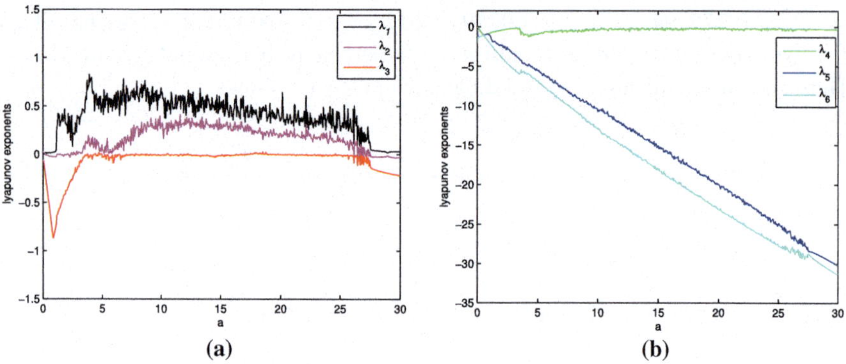

Fig. 10.4 The Lyapunov exponents of system (10.1.4) with a varying

We proceed to numerically identify the parameter ranges of system (10.1.4) for which chaotic attractors exist under the aforementioned conditions, considering the following two cases.

(i) Fix $b = 8/3$, $c = 28$, $d = 10$ and let a vary: To observe the Lyapunov exponents spectrum clearly, we plot the values of λ_1, λ_2, and λ_3 in Fig. 10.4(a), while the values of λ_4, λ_5, and λ_6 are shown in Fig. 10.4(b). Figure 10.4(a) shows that system (10.1.4) has hyperchaotic attractors for $a \in [3.3, 27.4]$, while it has chaotic attractors when $a \in [1.2, 3.2]$. And the above results can be demonstrated by the bifurcation diagram which is plotted in Fig. 10.5(a), while Figs. 10.5(b)–(d) depict attractors of system (10.1.4).

(ii) Fix $a = 10$, $c = 28$, $d = 10$ and let b vary: The values of λ_1, λ_2, and λ_3 are described in Fig. 10.6(a), while the values of λ_4, λ_5, and λ_6 are shown in Fig. 10.6(b). From Fig. 10.6(a), we obtain that system (10.1.4) has hyperchaotic attractors for $b \in [0.51, 4.99]$, chaotic attractors for $b \in [0.18, 0.36]$ and $[0.44, 0.5]$, and solutions of system (10.1.4) that approach fixed points for $b \in [0.37, 0.39]$. Corresponding bifurcation diagram with the step size of 0.1 is plotted in Fig. 10.7(a). Meanwhile, the attractors of system (10.1.4) are depicted in Figs. 10.7(b)–(c).

10.3 Generalized Combination Complex Synchronization

This section aims to propose generalized combination complex synchronization and design a general controller. Then two simulation examples are given to verify the effectiveness of the schemes.

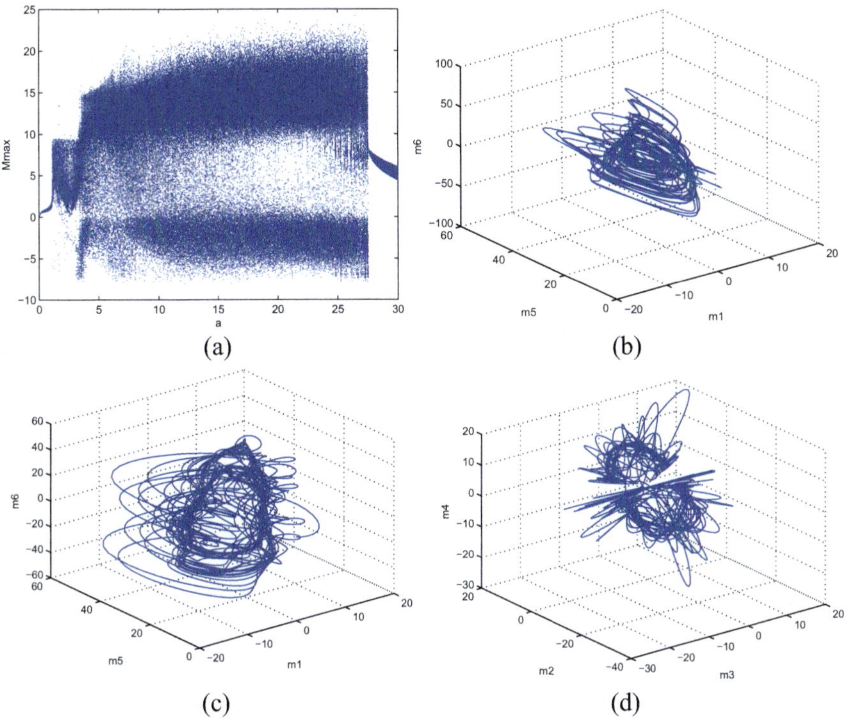

Fig. 10.5 Bifurcation diagram and attractors of system (10.1.4) with a varying

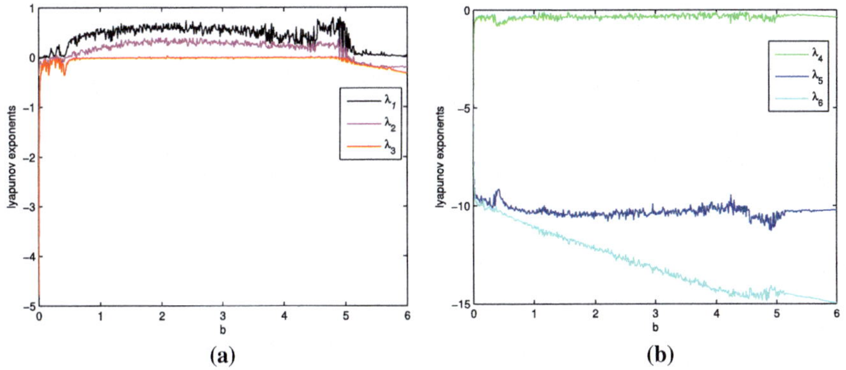

Fig. 10.6 Lyapunov exponents of system (10.1.4) with b varying

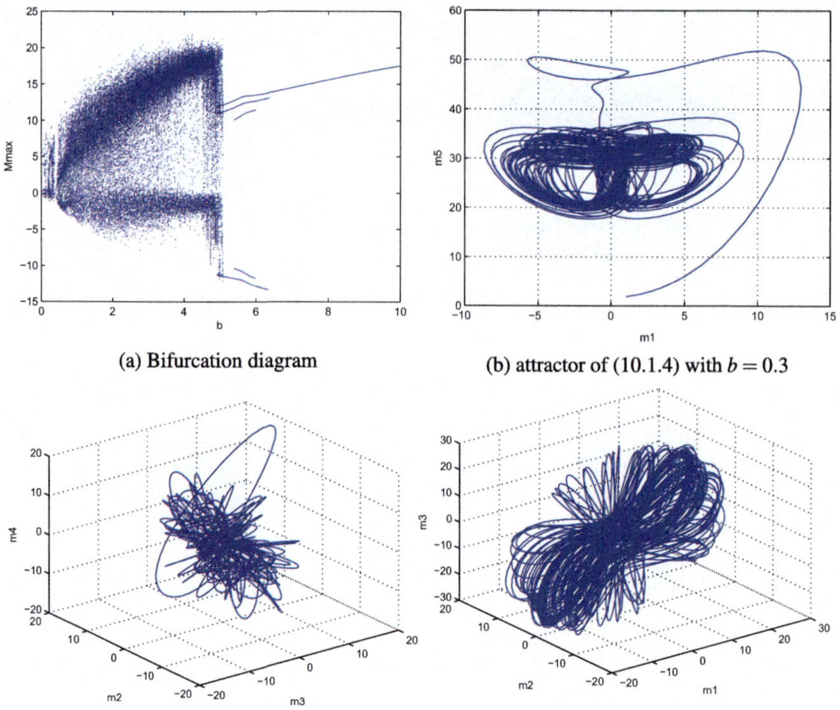

(a) Bifurcation diagram

(b) attractor of (10.1.4) with $b = 0.3$

(c) attractor of (10.1.4) with $b = 0.8$

(d) attractor of (10.1.4) with $b = 0.48$

Fig. 10.7 Bifurcation diagram and attractors of system (10.1.4) with b varying

10.3.1 Scheme of Generalized Combination Complex Synchronization

Take an n_1-dimensional chaotic complex system as the first drive system

$$\dot{x} = Ax + f(x), \tag{10.3.1}$$

the second drive chaotic complex system with n_2-dimensions is described as

$$\dot{y} = By + g(y), \tag{10.3.2}$$

while an n-dimensional response chaotic complex system is defined as

$$\dot{z} = Cz + h(z) + U(x, y, z), \tag{10.3.3}$$

where $x = x^r + jx^i \in \mathbb{C}^{n_1\times 1}$, $y = y^r + jy^i \in \mathbb{C}^{n_2\times 1}$, and $z = z^r + jz^i \in \mathbb{C}^{n\times 1}$ are the state complex vectors, $A \in \mathbb{R}^{n_1\times n_1}$, $B \in \mathbb{R}^{n_2\times n_2}$, and $C \in \mathbb{R}^{n\times n}$ are parameter matrices, while f, g, and h are nonlinear complex functions and U is a controller to be designed.

The generalized combination complex synchronization is defined as follows.

Definition 10.1 For two drive systems (10.3.1), (10.3.2), and one response system (10.3.3), they are said to achieve generalized combination complex synchronization if there exist two complex matrices $M_1 = M_1^r + jM_1^i \in \mathbb{C}^{n\times n_1}$ and $M_2 = M_2^r + jM_2^i \in \mathbb{C}^{n\times n_2}$, such that

$$\lim_{t\to\infty} ||e|| = \lim_{t\to\infty} ||z - M_1x - M_2y|| = 0,$$

where $||.||$ denotes the matrix norm, $e = e^r + je^i$ is the error vector, $e^r = z^r - M_1^r x^r + M_1^i x^i - M_2^r y^r + M_2^i y^i$, $e^i = z^i - M_1^r x^i - M_1^i x^r - M_2^r y^i - M_2^i y^r$, the complex matrices M_1 and M_2 are called the scaling matrices.

Remark 10.1 If the dimensions of two drive systems (10.3.1) and (10.3.2) equal to that of the response system (10.3.3), i.e., $n = n_1 = n_2$, then the proposed synchronization will be combination complex synchronization.

Remark 10.2 If the scaling matrix $M_1 = O_{n\times n_1}$ or $M_2 = O_{n\times n_2}$, then we can achieve complex projective synchronization. If $M_1^i = O_{n\times n_1}$ or $M_2^i = O_{n\times n_2}$, then combination synchronization can be carried out. If $M_1 = O_{n\times n_1}$ and $M_2 = O_{n\times n_2}$, then synchronization problem will be turned into a chaos control problem.

Remark 10.3 Definition 10.1 can be generalized to three or more chaotic complex systems. Additionally, drive systems and response systems can be identical or different.

Theorem 10.1 *If the control law is chosen as follows*

$$U = -C(M_1x + M_2y) - h(z) + M_1(Ax + f(x)) + M_2(By + g(y)) - Ke, \quad (10.3.4)$$

where K is a complex control gain matrix, then generalized combination complex synchronization between two drive systems (10.3.1), (10.3.2), *and the response system* (10.3.3) *can be achieved if and only if all eigenvalues of $C - K$ satisfy $Re(\lambda_i(C - K)) < 0$ $(i = 1, 2, \ldots, n)$.*

Proof Based on Definition 10.1, the error vector between two drive systems (10.3.1), (10.3.2), and one response system (10.3.3) can be derived as follows

$$e(t) = z - M_1x - M_2y. \quad (10.3.5)$$

By calculating the derivative of the error vector (10.3.5) and with the designed controller (10.3.4), it can be concluded that

$$\dot{e}(t) - (C - K)e(t) = 0. \tag{10.3.6}$$

Multiplying $\exp((-C - K)t)$ on both sides of (10.3.6), it follows that

$$\frac{d(\exp(-(C - K)t)e(t))}{dt} = 0. \tag{10.3.7}$$

Integrating equation (10.3.7) from 0 to t, it is straightforward obtained that

$$e(t) = \exp((C - K)t)e(0), \tag{10.3.8}$$

where $e(0)$ is an initial condition. Therefore, taking the limit on both sides of (10.3.8) and applying Lemma 9.1, owing to the condition that $\mathrm{Re}(\lambda_i(C - K)) < 0$, we find that

$$\lim_{t\to\infty} e(t) = \lim_{t\to\infty} \exp((C - K)t)e(0) = 0.$$

Therefore, generalized combination complex synchronization is realized between two drive systems (10.3.1), (10.3.2), and the response system (10.3.3) with the controller (10.3.4). This completes the proof. □

Our analysis now turns to generalized combination complex synchronization between real chaos and complex chaos. We consider two cases which include two real systems driving one complex system and two complex systems driving one real system.

Corollary 10.1 *Suppose that two drive systems* (10.3.1) *and* (10.3.2) *are chaotic real systems, i.e.,* $x \in \mathbb{R}^{n_1\times 1}$, $y \in \mathbb{R}^{n_2\times 1}$, *and the response system is a chaotic complex system* (10.3.3), *i.e.,* $z \in \mathbb{C}^{n\times 1}$. *Then generalized combination complex synchronization between two drive real systems* (10.3.1), (10.3.2), *and one response complex system* (10.3.3) *can occur with the designed controller*

$$U = -C(M_1x + M_2y) - h(z) + M_1(Ax + f(x)) + M_2(By + g(y)) - Ke,$$

where $K \in \mathbb{C}^{n\times n}$, *all eigenvalues of* $C - K$ *satisfy* $Re(\lambda_i(C - K)) < 0$ $(i = 1, 2, \ldots, n)$.

Corollary 10.2 *Assume that two drive systems* (10.3.1) *and* (10.3.2) *are chaotic complex systems, i.e.,* $x \in \mathbb{C}^{n_1\times 1}$, $y \in \mathbb{C}^{n_2\times 1}$, *and the response system* (10.3.3) *is a chaotic real system, where* $z \in \mathbb{R}^{n\times 1}$. *Since* $z(t)$ *is real, we choose a real controller* U *to ensure combination synchronization of real parts and avoid increasing the imaginary parts of the response system. Consequently, the error vector is defined as*

$$e = z - M_1^r x^r + M_1^i x^i - M_2^r y^r + M_2^i y^i.$$

If the real controller is designed as

$$U = -h(z) - C(M_1^r x^r - M_1^i x^i + M_2^r y^r - M_2^i y^i) + M_1^r(Ax^r + f^r(x)) - M_1^i(Ax^i + f^i(x))$$
$$+ M_2^r(By^r + g^r(y)) - M_2^i(By^i + g^i(y)) - Ke,$$

where $K \in \mathbb{R}^{n \times n}$, *all eigenvalues of* $C - K$ *satisfy* $Re(\lambda_i(C - K)) < 0$ $(i = 1, 2, \ldots, n)$, *then two drive complex systems* (10.3.1), (10.3.2), *and the response real system* (10.3.3) *are generalized combination complex synchronization of real parts.*

Theorem 10.1 yields the following corollaries which we present without proof.

Corollary 10.3 *(I) Suppose* $M_1 = O_{n \times n_1}$. *If the controller is defined as follows*

$$U = -CM_2 y - h(z) + M_2(By + g(y)) - Ke,$$

where K *is a control gain matrix, then complex projective synchronization between two different dimensional systems* (10.3.2) *and* (10.3.3) *can be realized if and only if all eigenvalues of* $C - K$ *satisfy* $Re(\lambda_i(C - K)) < 0$ $(i = 1, 2, \ldots, n)$.
(II) Suppose $M_2 = O_{n \times n_2}$. *If the controller is designed as follows*

$$U = -CM_1 x - h(z) + M_1(Ax + f(x)) - Ke,$$

where K *is a control gain matrix, then complex projective synchronization between two different dimensional systems* (10.3.1) *and* (10.3.3) *can occur if and only if all eigenvalues of* $C - K$ *satisfy* $Re(\lambda_i(C - K)) < 0$ $(i = 1, 2, \ldots, n)$.

Corollary 10.4 *Assume two scaling matrices* $M_1 = O_{n \times n_1}$ *and* $M_2 = O_{n \times n_2}$. *If the complex controller is designed as*

$$U = -h(z) - Ke,$$

where K *is a control gain matrix and all eigenvalues of* $C - K$ *satisfy* $Re(\lambda_i(C - K)) < 0$ $(i = 1, 2, \ldots, n)$, *then the equilibrium point of the response system* (10.3.3) *is asymptotically stable.*

10.3.2 Numerical Examples

This subsection presents two numerical examples to demonstrate the feasibility and effectiveness of the proposed results. We first investigate synchronization between two 3D chaotic complex systems and a 4D hyperchaotic complex system.

(I) Synchronization Between Two Drive Chaotic Complex Systems and a Response Hyperchaotic Complex System

Now, we configure the complex Lü system and complex Lorenz system to drive a hyperchaotic complex Lü-like system. Thus, two drive systems are described as:

$$\begin{cases} \dot{x}_1 = a_1(x_2 - x_1), \\ \dot{x}_2 = -x_1x_3 + a_2x_2, \\ \dot{x}_3 = \frac{1}{2}(\bar{x}_1x_2 + x_1\bar{x}_2) - a_3x_3, \end{cases} \tag{10.3.9}$$

$$\begin{cases} \dot{y}_1 = b_1(y_2 - y_1), \\ \dot{y}_2 = b_2y_1 - y_2 - y_1y_3, \\ \dot{y}_3 = \frac{1}{2}(\bar{y}_1y_2 + y_1\bar{y}_2) - b_3y_3, \end{cases} \tag{10.3.10}$$

where $A = \begin{pmatrix} -a_1 & a_1 & 0 \\ 0 & a_2 & 0 \\ 0 & 0 & -a_3 \end{pmatrix}$, $f(x) = \begin{pmatrix} 0 \\ -x_1x_3 \\ \frac{1}{2}(\bar{x}_1x_2 + x_1\bar{x}_2) \end{pmatrix}$, $B = \begin{pmatrix} -b_1 & b_1 & 0 \\ b_2 & -1 & 0 \\ 0 & 0 & -b_3 \end{pmatrix}$, $g(y) = \begin{pmatrix} 0 \\ -y_1y_3 \\ \frac{1}{2}(\bar{y}_1y_2 + y_1\bar{y}_2) \end{pmatrix}$; $x_1 = x_1^r + jx_1^i$, $x_2 = x_2^r + jx_2^i$, $y_1 = y_1^r + jy_1^i$, and $y_2 = y_2^r + jy_2^i$ are complex variables, x_3 and y_3 are real variables. Systems (10.3.9) and (10.3.10) exhibit chaotic behavior under the given parameters $(a_1, a_2, a_3) = (40, 22, 5)$ and $(b_1, b_2, b_3) = (14, 35, 3.7)$, respectively, see Figs. 10.8(a)–(b).

The proposed hyperchaotic complex Lü-like system is designated as the response system

$$\begin{cases} \dot{z}_1 = c_1(z_2 - z_1) + U_1, \\ \dot{z}_2 = c_2z_1 - z_1z_3 + z_4 + U_2, \\ \dot{z}_3 = \frac{1}{2}(\bar{z}_1z_2 + z_1\bar{z}_2) - c_3z_3 + U_3, \\ \dot{z}_4 = -\frac{c_4}{2}(\bar{z}_2 + z_2) + U_4, \end{cases} \tag{10.3.11}$$

where $C = \begin{pmatrix} -c_1 & c_1 & 0 & 0 \\ c_2 & 0 & 0 & 1 \\ 0 & 0 & -c_3 & 0 \\ 0 & -\frac{c_4}{2} & 0 & 0 \end{pmatrix}$, $h(z) = \begin{pmatrix} 0 \\ -z_1z_3 \\ \frac{1}{2}(\bar{z}_1z_2 + z_1\bar{z}_2) \\ -\frac{c_4}{2}\bar{z}_2 \end{pmatrix}$; $z_1 = z_1^r + jz_1^i$ and $z_2 = z_2^r + jz_2^i$ are complex variables, z_3 and z_4 are real variables. $U = (U_1, U_2, U_3, U_4)^T$ is a controller to be determined.

Two complex scaling matrices M_1 and M_2 are chosen as

$$M_1 = \begin{pmatrix} j & 1 & -2 \\ 1 & -j & 1 \\ 0 & 0 & -2 \\ 0 & 0 & 1 \end{pmatrix}, \quad M_2 = \begin{pmatrix} -j & 1 & -1 \\ -4 & j & 1 \\ 0 & 0 & 1 \\ 0 & 0 & 2 \end{pmatrix}.$$

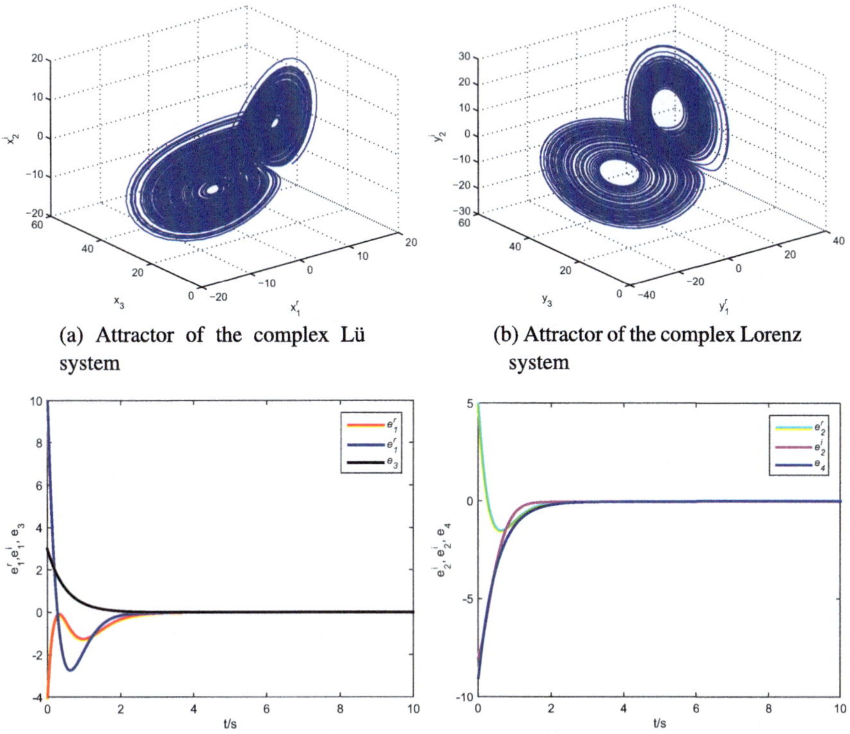

(a) Attractor of the complex Lü system

(b) Attractor of the complex Lorenz system

(c) Synchronization errors

(d) Synchronization errors

Fig. 10.8 Chaotic attractors and synchronization errors

Hence, we obtain the following error system $e = z - M_1x - M_2y$

$$\begin{cases} e_1 = z_1 - x_2 + 2x_3 - y_2 + y_3 + j(y_1 - x_1), \\ e_2 = z_2 - x_1 - x_3 + 4y_1 - y_3 + j(x_2 - y_2), \\ e_3 = z_3 + 2x_3 - y_3, \\ e_4 = z_4 - x_3 - 2y_3. \end{cases}$$

A control matrix is chosen as $K = \begin{pmatrix} -7 & 6 & -1-j & 0 \\ 29 & 3 & 0 & -1-j \\ 0 & 0 & -2/3 & 0 \\ 0 & -5 & -3 & 1 \end{pmatrix}$, and with the choice of $(c_1, c_2, c_3, c_4) = (10, 28, 8/3, 10)$ we have $C - K = \begin{pmatrix} -3 & 4 & 1+j & 0 \\ -1 & -3 & 0 & 2+j \\ 0 & 0 & -2 & 0 \\ 0 & 0 & 3 & -1 \end{pmatrix}$. We compute the eigenvalues of $C - K$ as $\lambda_1 = -3 + 2j$,

$\lambda_2 = -3 - 2j, \lambda_3 = -2$, and $\lambda_4 = -1$. It is evident that all of eigenvalues of $C - K$ satisfy $\mathrm{Re}(\lambda_i(C - K)) < 0\ (i = 1, 2, 3, 4)$.

Following Theorem 10.1, we design a complex controller as follows

$$\begin{cases} U_1 = (a_2 + c_1)x_2 + (c_1 - 1)y_2 + (2a_3 - 3c_1)x_3 + (b_3 - 2c_1)y_3 - c_1x_1 + 7e_1 \\ \quad +(b_2 + 4c_1)y_1 - x_1x_3 - 2x_1^r x_2^r - 2x_1^i x_2^i - y_1y_3 - y_1^r y_2^r - y_1^i y_2^i - 6e_2 \\ \quad +(1 + j)e_3 + j((c_1 - a_1)x_1 + (b_1 - c_1)y_1 - (b_1 + c_1)y_2 + (a_1 + c_1)x_2), \\ U_2 = (a_1 - c_2)x_2 - (c_2 + 4b_1)y_2 + (2c_2 - 1 - a_3)x_3 + (c_2 - 2 - b_3)y_3 - a_1x_1 \\ \quad +4b_1y_1 + z_1z_3 + x_1^r x_2^r + x_1^i x_2^i + y_1^r y_2^r + y_1^i y_2^i - \frac{11}{3}e_1 - 3e_2 + (1 + j)e_4 \\ \quad +j(x_1x_3 - y_1y_3 + (b_2 + c_2)y_1 - c_2x_1 - y_2 - a_2x_2), \\ U_3 = 2(a_3 - c_3)x_3 + (c_3 - b_3)y_3 - z_1^r z_2^r - z_1^i z_2^i - 2x_1^r x_2^r - 2x_1^i x_2^i + y_1^r y_2^r + y_1^i y_2^i + 26e_3, \\ U_4 = 10z_2^r - a_3x_3 - 2b_3y_3 + x_1^r x_2^r + x_1^i x_2^i + 2y_1^r y_2^r + 2y_1^i y_2^i + 3e_3 - e_4. \end{cases}$$

For the numerical simulations, the initial values of the drive and response systems are set to $x(0) = (-2 + j, 3 - 4j, -2)^T$, $y(0) = (4 - j, 3 + 2j, 5)^T$, and $z(0) = (-1 + 2j, -16 - 3j, 12, -1)^T$, respectively. As shown in Figs. 10.3.4(c)–(d), the synchronization errors converge to zero, confirming the realization of synchronization between two drive chaotic complex systems and a response hyperchaotic complex system.

(II) Synchronization Between Two Drive Hyperchaotic Complex Systems and a Response Chaotic Real System

We now turn to the synchronization between two 4-dimensional hyperchaotic complex systems and a 3-dimensional chaotic real system. In this configuration two hyperchaotic complex Lü-like systems drive the real Lorenz system [1]. Thus, two drive systems are modeled as follows

$$\begin{cases} \dot{x}_1 = a_1(x_2 - x_1), \\ \dot{x}_2 = a_2x_1 - x_1x_3 + x_4, \\ \dot{x}_3 = \frac{1}{2}(\overline{x}_1x_2 + x_1\overline{x}_2) - a_3x_3, \\ \dot{x}_4 = -\frac{a_4}{2}(\overline{x}_2 + x_2), \end{cases} \tag{10.3.12}$$

$$\begin{cases} \dot{y}_1 = b_1(y_2 - y_1), \\ \dot{y}_2 = b_2y_1 - y_1y_3 + y_4, \\ \dot{y}_3 = \frac{1}{2}(\overline{y}_1y_2 + y_1\overline{y}_2) - b_3y_3, \\ \dot{y}_4 = -\frac{b_4}{2}(\overline{y}_2 + y_2), \end{cases} \tag{10.3.13}$$

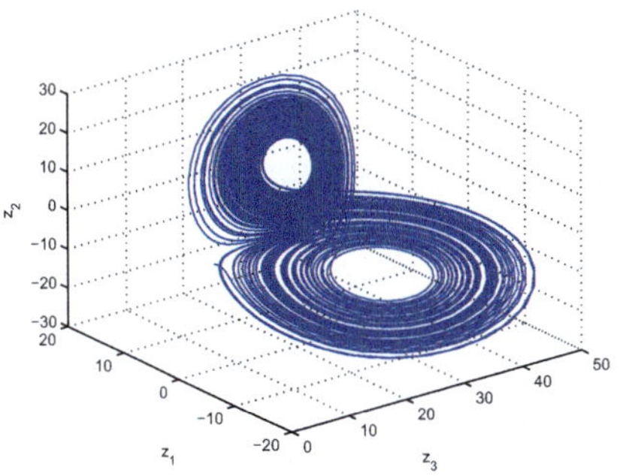

(a) Attractor of system (10.3.14)

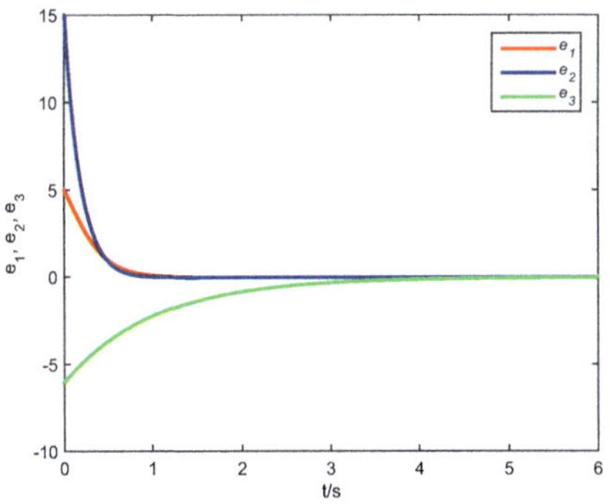

(b) Synchronization errors

Fig. 10.9 Attractors and synchronization errors

where $A = \begin{pmatrix} -a_1 & a_1 & 0 & 0 \\ a_2 & 0 & 0 & 1 \\ 0 & 0 & -a_3 & 0 \\ 0 & -\frac{a_4}{2} & 0 & 0 \end{pmatrix}$, $f(x) = \begin{pmatrix} 0 \\ -x_1x_3 \\ \frac{1}{2}(\bar{x}_1x_2 + x_1\bar{x}_2) \\ -\frac{a_4}{2}\overline{x}_2 \end{pmatrix}$, $g(y) = \begin{pmatrix} 0 \\ -y_1y_3 \\ \frac{1}{2}(\bar{y}_1y_2 + y_1\bar{y}_2) \\ -\frac{b_4}{2}\overline{y}_2 \end{pmatrix}$, $B = \begin{pmatrix} -b_1 & b_1 & 0 & 0 \\ b_2 & 0 & 0 & 1 \\ 0 & 0 & -b_3 & 0 \\ 0 & -\frac{b_4}{2} & 0 & 0 \end{pmatrix}$; $x_1 = x_1^r + jx_1^i$, $x_2 = x_2^r + jx_2^i$, $y_1 = y_1^r + jy_1^i$, and $y_2 = y_2^r + jy_2^i$ are complex variables, x_3, x_4, y_3, and y_4 are real variables.

The real Lorenz system is expressed as

$$\begin{cases} \dot{z}_1 = c_1(z_2 - z_1) + U_1, \\ \dot{z}_2 = c_2z_1 - z_1z_3 - z_2 + U_2, \\ \dot{z}_3 = z_1z_2 - c_3z_3 + U_3, \end{cases} \tag{10.3.14}$$

where $C = \begin{pmatrix} -c_1 & c_1 & 0 \\ c_2 & -1 & 0 \\ 0 & 0 & -c_3 \end{pmatrix}$, $h(z) = \begin{pmatrix} 0 \\ -z_1z_3 \\ z_1z_2 \end{pmatrix}$; z_1, z_2, and z_3 are real variables, $U = (U_1, U_2, U_3)^T$ is a real controller which will be designed later. Under the choice of $(c_1, c_2, c_3) = (10, 28, 8/3)$, system (10.3.14) exhibits chaotic behavior, as shown in Fig. 10.9(a).

Assume two complex scaling matrices M_1 and M_2 as follows

$$M_1 = \begin{pmatrix} j & 1 & -j & 2j \\ 2 & j & 4j & -j \\ 1 & 0 & 1+j & j \end{pmatrix} = \begin{pmatrix} 0 & 1 & 0 & 0 \\ 2 & 0 & 0 & 0 \\ 1 & 0 & 1 & 0 \end{pmatrix} + j\begin{pmatrix} 1 & 0 & -1 & 2 \\ 0 & 1 & 4 & -1 \\ 0 & 0 & 1 & 1 \end{pmatrix},$$

$$M_2 = \begin{pmatrix} 1 & j & 3j & j \\ j & -2 & j & 2j \\ 0 & 1 & j & 1+j \end{pmatrix} = \begin{pmatrix} 1 & 0 & 0 & 0 \\ 0 & -2 & 0 & 0 \\ 0 & 1 & 0 & 1 \end{pmatrix} + j \begin{pmatrix} 0 & 1 & 3 & 1 \\ 1 & 0 & 1 & 2 \\ 0 & 0 & 1 & 1 \end{pmatrix}.$$

Thus the real error system $e = z - M_1^r x^r + M_1^i x^i - M_2^r y^r + M_2^i y^i$ is written in the following form

$$\begin{cases} e_1 = z_1 - x_2^r + x_1^i - y_1^r + y_2^i, \\ e_2 = z_2 - 2x_1^r + x_2^i + 2y_2^r + y_1^i, \\ e_3 = z_3 - x_1^r - x_3 - y_2^r - y_4. \end{cases}$$

A real control matrix is chosen as $K = \begin{pmatrix} -5 & 9 & 0 \\ 29 & 4 & 0 \\ 0 & 0 & -5/3 \end{pmatrix}$ and let $(c_1, c_2, c_3) =$ $(10, 28, 8/3)$, then we have $C - K = \begin{pmatrix} -5 & 1 & 0 \\ -1 & -5 & 0 \\ 0 & 0 & -1 \end{pmatrix}$.

Based on Corollary 10.2, the real controller can be constructed as

$$\begin{cases} U_1 = (a_1 - c_1)(x_1^i - x_2^i) - (b_1 - c_1)y_1^r + (a_2 - 2c_1)x_1^r + (b_1 + 2c_1)y_2^r + 5e_1 \\ \quad +(c_1 - b_2)y_1^i + c_1(x_2^r - y_2^i) - x_1^r x_3 + x_4 + y_1^i y_3 - 9e_2, \\ U_2 = z_1 z_3 + (2a_1 - c_2)x_2^r + (c_2 - a_2)x_1^i - (2b_2 + c_2)y_1^r + (c_2 - b_1)y_2^i - 29e_1 \\ \quad +2(1 - a_1)x_1^r - x_2^i - 2y_2^r + x_1^i x_3 + (b_1 - 1)y_1^i + 2y_1^r y_3 - 2y_4 - 4e_2, \\ U_3 = (c_3 - a_1)x_1^r - z_1 z_2 + (c_3 - a_3)x_3 + (c_3 - b_4)y_2^r + (c_3 + 1)y_4 + a_1 x_2^r \\ \quad +x_1^r x_2^r + x_1^i x_2^i + b_2 y_1^r - y_1^r y_3 + \frac{5}{3}e_3. \end{cases}$$

For the numerical simulations, the initial values of the drive and response systems are chosen as $x(0) = (10 - 8j, -5 + 6j, 4, 3)^T$, $y(0) = (-2 - 4j, 1 - 3j, -5, -2)^T$, and $z(0) = (5, 15, -6)^T$, respectively. All of the eigenvalues of $C - K$ are $\lambda_1 = -5 + j, \lambda_2 = -5 - j$, and $\lambda_3 = -1$, which satisfy $\mathrm{Re}(\lambda_i(C - K)) < 0$ $(i = 1, 2, 3)$. The errors of synchronization converge asymptotically to zero, which is shown in Fig. 10.9(b). As observed in Fig. 10.9(b), the synchronization errors are seen to converge asymptotically to zero. Hence, synchronization between two drive hyperchaotic complex systems and a response chaotic real system is achieved in the real parts.

Chapter 11
Synchronization and Anti-synchronization of N-Coupled Complex Chaotic Systems with Ring Connection

Synchronization of coupled chaotic systems [177–182] with ring connection extends the traditional mode of synchronization from one-to-one into one-to-many. And this synchronization is realized at lower cost and has better flexibility and practicability [183–186]. This chapter discusses synchronization and anti-synchronization of N-coupled complex chaotic systems with ring connection. Based on the direct design method and antisymmetric structure, the appropriate controllers are designed to ensure the occurrence of synchronization and anti-synchronization in an array of N-coupled general complex chaotic systems described by a unified mathematical expression with ring connection. Section 11.1 presents the theoretical analysis. Section 11.2 shows the main results. Section 11.3 explores the synchronization and anti-synchronization of 4-coupled complex PMSM systems.

11.1 Theoretical Analysis

Take the following complex chaotic system as

$$\begin{cases} \dot{X}_1 &= A_1 X_1 + G_1(X_1), \\ \dot{X}_2 &= A_2 X_2 + G_2(X_2), \\ &\vdots \\ \dot{X}_N &= A_N X_N + G_N(X_N), \end{cases} \tag{11.1.1}$$

where $X_1, X_2, \ldots, X_N \in \mathbb{C}^n$ $(N > 2)$ are complex state vectors of chaotic systems and $X_k = X_k^r + jX_k^i$; $A_1, A_2, \ldots, A_N \in \mathbb{R}^{n\times n}$ are constant matrices $(A_k \neq A_l,\ k \neq l)$; $G_k(X_k)$ are the continuous complex nonlinear functions $(G_k \neq G_l,\ k \neq l)$. Now, we consider the case that multiple complex chaotic systems are coupled in a ring structure as shown in Fig. 11.1. In this model, the state variables of the first complex

S. Liu et al., *Chaos in Complex Dynamical System: Theory and Application*,
https://doi.org/10.1007/978-981-95-7762-0_11

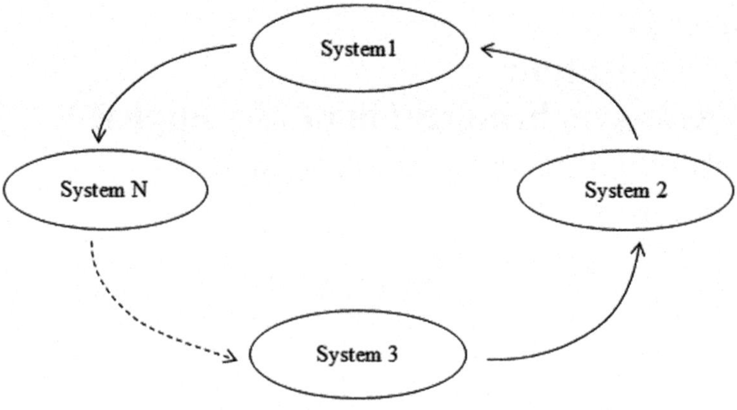

Fig. 11.1 Diagram of coupled complex chaotic systems with ring connection

chaotic system couple the Nth system, and the state variables of the second system couple the first,...analogically, until the state variables of the Nth system couple those of the $(N-1)$th. Thus, N-coupled complex chaotic systems with ring connection can be depicted as

$$\begin{cases} \dot{X}_1 = A_1X_1 + G_1(X_1) + Q_1(X_N - X_1), \\ \dot{X}_2 = A_2X_2 + G_2(X_2) + Q_2(X_1 - X_2), \\ \quad\vdots \\ \dot{X}_N = A_NX_N + G_N(X_N) + Q_N(X_{N-1} - X_N), \end{cases} \tag{11.1.2}$$

where $Q_k = \text{diag}(q_{k1}, q_{k2}, \ldots, q_{kn})$ are n-dimensional real diagonal matrices and $q_{kl} \geq 0$ are the ideal gains which represent the coupled parameters ($k = 1, 2, \ldots, N$; $l = 1, 2, \ldots, n$).

We choose the first system as the aim and add controllers to the remaining systems. Consequently, system (11.1.2) can be rewritten as

$$\begin{cases} \dot{X}_1 = A_1X_1 + G_1(X_1) + Q_1(X_N - X_1), \\ \dot{X}_2 = A_2X_2 + G_2(X_2) + Q_2(X_1 - X_2) + U_1, \\ \dot{X}_3 = A_3X_3 + G_3(X_3) + Q_3(X_2 - X_3) + U_2, \\ \quad\vdots \\ \dot{X}_N = A_NX_N + G_N(X_N) + Q_N(X_{N-1} - X_N) + U_{N-1}, \end{cases} \tag{11.1.3}$$

where $U_1 = (u_{11}, u_{12}, \ldots, u_{1n})^T, U_2 = (u_{21}, u_{22}, \ldots, u_{2n})^T, \ldots, U_{N-1} = (x_{(N-1)1}, x_{(N-1)2}, \ldots, x_{(N-1)n})^T$ $(N > 2)$ are controllers to be designed. We define the synchronization error vector is $e_k = X_{k+1} - \delta X_k$ $(k = 1, 2, \ldots, N-1; \delta = 1, -1)$. Thus, we obtain that

$$\begin{cases} e_{k1} = x_{(k+1),1} - \delta x_{k1}, \\ e_{k2} = x_{(k+1),2} - \delta x_{k2}, \\ \quad \vdots \\ e_{kn} = x_{(k+1),n} - \delta x_{kn}. \end{cases}$$

Evidently, the error system e_k reduces to the complete synchronization case when $\delta = 1$, and to the anti-synchronization case when $\delta = -1$. Considering system (11.1.3), we obtain the following error dynamical system:

$$\begin{cases} \dot{e}_1^r &= (A_2 - \delta^{N-1}Q_1 - Q_2)e_1^r - \delta^{N-2}Q_1 e_2^r - \cdots - \delta^2 Q_1 e_{N-2}^r + U_1^r - \delta Q_1 e_{N-1}^r \\ & \quad +(\delta(A_2 - A_1) + (1-\delta)Q_2 - (\delta^N - \delta)Q_1)X_1^r + G_2^r(X_2) - \delta G_1^r(X_1), \\ \dot{e}_2^r &= Q_2 e_1^r + (A_3 - Q_3)e_2^r + (\delta(A_3 - A_2) + (\delta - 1)(Q_2 - Q_3))X_2^r \\ & \quad +G_3^r(X_3) - \delta G_2^r(X_2) + U_2^r - \delta U_1^r, \\ & \vdots \\ \dot{e}_{N-1}^r &= Q_{N-1}e_{N-2}^r + (A_N - Q_N)e_{N-1}^r + G_N^r(X_N) - \delta G_{N-1}^r(X_{N-1}) \\ & \quad +(\delta(A_N - A_{N-1}) + (\delta - 1)(Q_{N-1} - Q_N))X_{N-1}^r + U_{N-1}^r - \delta U_{N-2}^r, \end{cases} \tag{11.1.4}$$

and

$$\begin{cases} \dot{e}_1^i &= (A_2 - \delta^{N-1}Q_1 - Q_2)e_1^i - \delta^{N-2}Q_1 e_2^i - \cdots - \delta^2 Q_1 e_{N-2}^i - \delta Q_1 e_{N-1}^i \\ & \quad +(\delta(A_2 - A_1) + (1-\delta)Q_2 - (\delta^N - \delta)Q_1)X_1^i + G_2^i(X_2) - \delta G_1^i(X_1) + U_1^i, \\ \dot{e}_2^i &= Q_2 e_1^i + (A_3 - Q_3)e_2^i + (\delta(A_3 - A_2) + (\delta - 1)(Q_2 - Q_3))X_2^i \\ & \quad +G_3^i(X_3) - \delta G_2^i(X_2) + U_2^i - \delta U_1^i, \\ & \vdots \\ \dot{e}_{N-1}^i &= Q_{N-1}e_{N-2}^i + (A_N - Q_N)e_{N-1}^i + G_N^i(X_N) - \delta G_{N-1}^i(X_{N-1}) \\ & \quad +(\delta(A_N - A_{N-1}) + (\delta - 1)(Q_{N-1} - Q_N))X_{N-1}^i + U_{N-1}^i - \delta U_{N-2}^i. \end{cases} \tag{11.1.5}$$

Thus, the essential goal of this paper is to design complex controllers $U_k = U_k^r + jU_k^i$ $(k = 1, 2, \ldots, N-1)$ such that the error dynamical error $e(t)$ tends to zero as $t \to \infty$, namely,

$$\lim_{t\to\infty} ||e^r(t)|| = 0 \text{ and } \lim_{t\to\infty} ||e^i(t)|| = 0,$$

then

$$\lim_{t\to\infty} ||e(t)|| = 0.$$

This demonstrates that both synchronization and anti-synchronization of the N-coupled complex chaotic systems are achieved.

11.2 Main Results

A direct design method proposed in [187–189] has two advantages: (i) it presents an easy procedure for choosing proper controllers in chaos synchronization; (ii) it constructs simple controllers easily to implement. Hence, in this paper, we apply this method to transform the error system into a stable system with antisymmetric structure and further realize synchronization and anti-synchronization of chaotic systems with complex variables.

We design the controllers U_k^r and U_k^i $(k = 1, 2, \ldots, N-1)$ as

$$\begin{cases} U_1^r &= V_1^r - (\delta(A_2 - A_1) + (1-\delta)Q_2 - (\delta^N - \delta)Q_1)X_1^r - G_2^r(X_2) + \delta G_1^r(X_1), \\ U_2^r &= V_2^r - (\delta(A_3 - A_2) + (\delta - 1)(Q_2 - Q_3))X_2^r - G_3^r(X_3) + \delta G_2^r(X_2) + \delta U_1^r, \\ &\vdots \\ U_{N-1}^r &= V_{N-1}^r - (\delta(A_N - A_{N-1}) + (\delta - 1)(Q_{N-1} - Q_N))X_{N-1}^r - G_N^r(X_N) \\ &\quad +\delta G_{N-1}^r(X_{N-1}) + \delta U_{N-2}^r, \end{cases} \tag{11.2.1}$$

and

$$\begin{cases} U_1^i &= V_1^i - (\delta(A_2 - A_1) + (1-\delta)Q_2 - (\delta^N - \delta)Q_1)X_1^i - G_2^i(X_2) + \delta G_1^i(X_1), \\ U_2^i &= V_2^i - (\delta(A_3 - A_2) + (\delta - 1)(Q_2 - Q_3))X_2^i - G_3^i(X_3) + \delta G_2^i(X_2) + \delta U_1^i, \\ &\vdots \\ U_{N-1}^i &= V_{N-1}^i - (\delta(A_N - A_{N-1}) + (\delta - 1)(Q_{N-1} - Q_N))X_{N-1}^i - G_N^i(X_N) \\ &\quad +\delta G_{N-1}^i(X_{N-1}) + \delta U_{N-2}^i, \end{cases} \tag{11.2.2}$$

where

$$\begin{pmatrix} V_1^r \\ V_2^r \\ \vdots \\ V_{N-1}^r \end{pmatrix} = \Theta \begin{pmatrix} e_1^r \\ e_2^r \\ \vdots \\ e_{N-1}^r \end{pmatrix}, \quad \begin{pmatrix} V_1^i \\ V_2^i \\ \vdots \\ V_{N-1}^i \end{pmatrix} = \Theta \begin{pmatrix} e_1^i \\ e_2^i \\ \vdots \\ e_{N-1}^i \end{pmatrix}, \tag{11.2.3}$$

and Θ is a real coefficient matrix. Then, the error dynamical systems (11.1.4) and (11.1.5) can be depicted as

$$\dot{e}^r(t) = \Phi(t)e^r(t) \quad \text{and} \quad \dot{e}^i(t) = \Phi(t)e^i(t), \tag{11.2.4}$$

where $\Phi(t) = \Theta +$

$$\begin{pmatrix} A_2 - \delta^{N-1}Q_1 - Q_2 & -\delta^{N-2}Q_1 & \cdots & -\delta^2 Q_1 & -\delta Q_1 \\ Q_2 & A_3 - Q_3 & \cdots & 0 & 0 \\ \vdots & \vdots & \ddots & \vdots & \vdots \\ 0 & 0 & \cdots & Q_{N-1} & A_N - Q_N \end{pmatrix}.$$

It is evident that the matrix Θ which ensures the asymptotic stability of the error dynamical system (11.2.4) at the origin is not unique. Here, Θ is generally treated as a state-dependent real coefficient matrix. Consequently, the sufficient stability conditions for system (11.2.4) can be derived by transforming it into a stable system with an antisymmetric structure. We first introduce the following lemma before proceeding to the main result.

Theorem 11.1 *Synchronization and anti-synchronization of N-coupled complex chaotic systems* (11.1.2) *will occur by using the control laws* (11.2.1) *and* (11.2.2)*, if the state dependent real coefficient matrix $\Phi(t) = \Phi_1(t) + \Phi_2$ satisfies the following assumptions:*

$$\begin{aligned} &\Phi_1^T(t) = -\Phi_1(t), \\ &\Phi_2 = diag(-\varphi_1, -\varphi_2, \ldots, -\varphi_{N-1}), \end{aligned} \tag{11.2.5}$$

where $\varphi_k = diag(\varphi_{k1}, \varphi_{k2}, \ldots, \varphi_{kn})$ $(k = 1, 2, \ldots, N-1)$ and $\varphi_{kl} > 0$ $(l = 1, 2, \ldots, n)$.

Proof A positive definite Lyapunov candidate function is chosen as

$$V(t) = \frac{1}{2}\left(e^r(t)\right)^T e^r(t) + \frac{1}{2}(e^i(t))^T e^i(t).$$

Then, the time derivative of $V(t)$ along the trajectory of the error system (11.2.4) is as follows:

$$\dot{V}(t) = \frac{1}{2}\left((e^r(t))^T(\Phi^T(t) + \Phi(t))e^r(t)\right) + \frac{1}{2}\left((e^i(t))^T(\Phi^T(t) + \Phi(t))e^i(t)\right). \tag{11.2.6}$$

Substitution of (11.2.5) into (11.2.6) yields

$$\begin{aligned} \dot{V}(t) &= \frac{1}{2}\left[(e^r(t))^T(\Phi_2^T + \Phi_2)e^r(t)\right] + \frac{1}{2}\left[(e^i(t))^T(\Phi_2^T + \Phi_2)e^i(t)\right] \\ &= (e^r(t))^T\Phi_2 e^r(t) + (e^i(t))^T\Phi_2 e^i(t) \\ &= \left((e_1^r(t))^T, \ldots, (e_{N-1}^r(t))^T\right)\begin{pmatrix} -\varphi_1 & 0 & \cdots & 0 \\ 0 & -\varphi_2 & \cdots & 0 \\ \vdots & \vdots & \ddots & \vdots \\ 0 & 0 & \cdots & -\varphi_{N-1} \end{pmatrix}\begin{pmatrix} e_1^r(t) \\ e_2^r(t) \\ \vdots \\ e_{N-1}^r(t) \end{pmatrix} \\ &\quad + \left((e_1^i(t))^T, \ldots, (e_{N-1}^i(t))^T\right)\begin{pmatrix} -\varphi_1 & 0 & \cdots & 0 \\ 0 & -\varphi_2 & \cdots & 0 \\ \vdots & \vdots & \ddots & \vdots \\ 0 & 0 & \cdots & -\varphi_{N-1} \end{pmatrix}\begin{pmatrix} e_1^i(t) \\ e_2^i(t) \\ \vdots \\ e_{N-1}^i(t) \end{pmatrix} \end{aligned}$$

$$\begin{aligned}
&= -(e_1^r(t))^T\varphi_1 e_1^r(t) - (e_2^r(t))^T\varphi_2 e_2^r(t) - \cdots - (e_{N-1}^r(t))^T\varphi_{N-1}e_{N-1}^r(t) \\
&\quad -(e_1^i(t))^T\varphi_1 e_1^i(t) - (e_2^i(t))^T\varphi_2 e_2^i(t) - \cdots - (e_{N-1}^i(t))^T\varphi_{N-1}e_{N-1}^i(t) \\
&\le -\varphi_{min}((e_1^r(t))^T e_1^r(t) + (e_2^r(t))^T e_2^r(t) + \cdots + (e_{N-1}^r(t))^T e_{N-1}^r(t) \\
&\quad +(e_1^i(t))^T e_1^i(t) + \cdots + (e_{N-1}^i(t))^T e_{N-1}^i(t)) \\
&= -\varphi_{min}||e(t)||^2,
\end{aligned}$$

where $\varphi_{min} = \min_{1\le k\le N-1, 1\le l\le n}\{\varphi_{kl}\}$. Thus, taking the integral of both sides, we have

$$\int_0^t ||e(t)||^2 dt \le [V(0) - V(t)]/\varphi_{min} \le V(0)/\varphi_{min} < \infty.$$

As $\dot{e}^r(t)$ and $\dot{e}^i(t)$ exist and they are bounded. Then,

$$\frac{d(||e(t)||)}{dt} = \frac{1}{2}||e(t)||^{-1}\left[2(\dot{e}^r(t))^T e^r(t) + 2(\dot{e}^i(t))^T e^i(t)\right] = \frac{(\dot{e}^r(t))^T e^r(t) + (\dot{e}^i(t))^T e^i(t)}{||e(t)||}.$$

Thus, the derivative of $||e(t)||$ is confirmed to exist and be bounded. Application of Lemma 7.1 gives $\lim_{t\to\infty}||e(t)|| = 0$. That implies that the closed-loop system (11.2.4) is globally asymptotically stable. Consequently, synchronization and anti-synchronization are achieved for the N-coupled complex chaotic systems (11.1.2). This completes the proof. □

As a consequence of Theorem 11.1, we can obtain the following corollary.

Corollary 11.1 *If the structures of general complex chaotic systems are identical, i.e., $A_k = A_l = A$ and $G_k(\cdot) = G_l(\cdot) = G(\cdot)$, then complex controllers $U_k = U_k^r + jU_k^i$ can be constructed as*

$$\begin{cases}
U_1 &= V_1 - ((1-\delta)Q_2 - (\delta^N - \delta)Q_1)X_1 - G(X_2) + \delta G(X_1), \\
U_2 &= V_2 - (\delta - 1)(Q_2 - Q_3)X_2 - G(X_3) + \delta G(X_2) + \delta U_1, \\
&\vdots \\
U_{N-1} &= V_{N-1} - (\delta - 1)(Q_{N-1} - Q_N)X_{N-1} - G(X_N) + \delta G(X_{N-1}) + \delta U_{N-2},
\end{cases}$$

where $V_k = V_k^r + jV_k^i$ $(k = 1, 2, \ldots, N-1)$ are defined as in (11.2.3) *and Θ is a real coefficient matrix. Similar to the aforementioned discussion, we can achieve synchronization and anti-synchronization of N-coupled identical complex chaotic systems with ring connection based on antisymmetric structure.*

As demonstrated in the above analysis, the error dynamical systems (11.1.4) and (11.1.5) can be transformed into system (11.2.4) using the complex control laws U_k. Furthermore, we can convert system (11.2.4) into a stable system with an antisymmetric structure by selecting an appropriate real coefficient matrix Θ. It is known that the antisymmetric structure in Theorem 11.1 is the generalization of the tridiagonal structure [188, 189]. The error system constructed with the antisymmetric structure is more convenient than the one with tridiagonal structure when the original system has some zero elements at the tridiagonal position and nonzero elements at other positions. Additionally, we observe that the selection of the coefficient matrix plays a crucial role in achieving both synchronization and anti-synchronization of N-coupled complex chaotic systems with ring connection.

11.3 Numerical Simulations

The complex permanent magnet synchronous motor (PMSM) system presented in [190–192] is described as follows

$$\begin{cases} \dot{x}_1 = a(x_2 - x_1), \\ \dot{x}_2 = bx_1 - x_2 - x_1x_3, \\ \dot{x}_3 = \frac{1}{2}(\bar{x}_1x_2 + x_1\bar{x}_2) - x_3, \end{cases} \tag{11.3.1}$$

where x_1 and x_2 are complex variables. When $a = 11$, $b = 20$, system (11.3.1) is chaotic and the phase portraits are shown in Fig. 11.2.

We now focus on the synchronization and anti-synchronization of N-coupled complex PMSM systems with ring connection. Here, 4-coupled complex PMSM systems with the designed controllers can be depicted as

$$\begin{cases} \dot{x}_{11} = a(x_{12} - x_{11}) + q_{11}(x_{41} - x_{11}), \\ \dot{x}_{12} = bx_{11} - x_{12} - x_{11}x_{13} + q_{12}(x_{42} - x_{12}), \\ \dot{x}_{13} = \frac{1}{2}(\bar{x}_{11}x_{12} + x_{11}\bar{x}_{12}) - x_{13} + q_{13}(x_{43} - x_{13}), \end{cases}$$

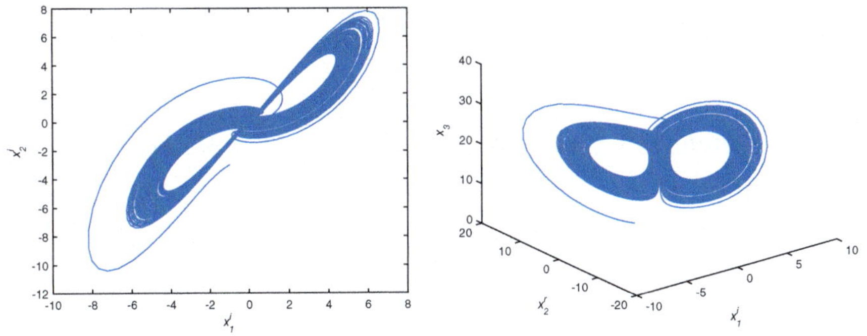

Fig. 11.2 Chaotic behavior of the complex PMSM system (11.3.1)

$$\begin{cases} \dot{x}_{21} = a(x_{22} - x_{21}) + q_{21}(x_{11} - x_{21}) + u_{11}, \\ \dot{x}_{22} = bx_{21} - x_{22} - x_{21}x_{23} + q_{22}(x_{12} - x_{22}) + u_{12}, \\ \dot{x}_{23} = \frac{1}{2}(\bar{x}_{21}x_{22} + x_{21}\bar{x}_{22}) - x_{23} + q_{23}(x_{13} - x_{23}) + u_{13}, \end{cases}$$

$$\begin{cases} \dot{x}_{31} = a(x_{32} - x_{31}) + q_{31}(x_{21} - x_{31}) + u_{21}, \\ \dot{x}_{32} = bx_{31} - x_{32} - x_{31}x_{33} + q_{32}(x_{22} - x_{32}) + u_{22}, \\ \dot{x}_{33} = \frac{1}{2}(\bar{x}_{31}x_{32} + x_{31}\bar{x}_{32}) - x_{33} + q_{33}(x_{23} - x_{33}) + u_{23}, \end{cases}$$

and

$$\begin{cases} \dot{x}_{41} = a(x_{42} - x_{41}) + q_{41}(x_{31} - x_{41}) + u_{31}, \\ \dot{x}_{42} = bx_{41} - x_{42} - x_{41}x_{43} + q_{42}(x_{32} - x_{42}) + u_{32}, \\ \dot{x}_{43} = \frac{1}{2}(\bar{x}_{41}x_{42} + x_{41}\bar{x}_{42}) - x_{43} + q_{43}(x_{33} - x_{43}) + u_{33}, \end{cases}$$

where

$$A_k = A = \begin{pmatrix} -a & a & 0 \\ b & -1 & 0 \\ 0 & 0 & -1 \end{pmatrix}, \quad G_k(X_k) = G(X_k) = \begin{pmatrix} 0 \\ -x_{k1}x_{k3} \\ \frac{1}{2}(\bar{x}_{k1}x_{k2} + x_{k1}\bar{x}_{k2}) \end{pmatrix},$$

$Q_k = \mathrm{diag}(q_{k1}, q_{k2}, q_{k3})$ are the coupled matrices ($k = 1, 2, 3, 4$), and $U_k = (u_{k1}, u_{k2}, u_{k3})^T$ ($k = 1, 2, 3$) are the complex control inputs.

The synchronization error is defined as $e_k = X_{k+1} - \delta X_k$ ($k = 1, 2, 3$), leading to the following error dynamical system

$$\begin{cases} \dot{e}_1 = (A - \delta^3 Q_1 - Q_2)e_1 - \delta^2 Q_1 e_2 - \delta Q_1 e_3 + G(X_2) - \delta G(X_1) + U_1 \\ \quad +((1-\delta)Q_2 - (\delta^4 - \delta)Q_1)X_1, \\ \dot{e}_2 = Q_2 e_1 + (A - Q_3)e_2 + (\delta - 1)(Q_2 - Q_3)X_2 + G(X_3) - \delta G(X_2) \\ \quad +U_2 - \delta U_1, \\ \dot{e}_3 = Q_3 e_2 + (A - Q_4)e_3 + (\delta - 1)(Q_3 - Q_4)X_3 + G(X_4) - \delta G(X_3) \\ \quad +U_3 - \delta U_2. \end{cases}$$

11.3.1 Synchronization of 4-Coupled Complex PMSM Systems

When $\delta = 1$, the error system reduces to the following complete synchronization error system.

$$\dot{e}(t) = \begin{pmatrix} A - Q_1 - Q_2 & -Q_1 & -Q_1 \\ Q_2 & A - Q_3 & 0 \\ 0 & Q_3 & A - Q_4 \end{pmatrix} \begin{pmatrix} e_1(t) \\ e_2(t) \\ e_3(t) \end{pmatrix} + \begin{pmatrix} G_2(X_2) - G_1(X_1) + U_1 \\ G_3(X_3) - G_2(X_2) + U_2 - U_1 \\ G_4(X_4) - G_3(X_3) + U_3 - U_2 \end{pmatrix}.$$

Choose the real coefficient matrix $\Theta = \begin{pmatrix} \Theta_1 \\ \Theta_2 \\ \Theta_3 \end{pmatrix}$ as follows:

$$
\Theta_1 = \begin{pmatrix} 0 & 0 & 0 & 0 & x_{11}^r & 0 & 0 & 0 & 0 \\ -a-b & 0 & 0 & 0 & 0 & 0 & 0 & 0 & 0 \\ 0 & 0 & 0 & 0 & 0 & 0 & 0 & 0 & 0 \end{pmatrix},
$$

$$
\Theta_2 = \begin{pmatrix} q_{11}-q_{21} & 0 & 0 & 0 & 0 & 0 & -q_{31} & 0 & 0 \\ -x_{11}^r & q_{12}-q_{22} & 0 & -a-b & 0 & 0 & 0 & -q_{32} & 0 \\ 0 & 0 & q_{13}-q_{23} & 0 & 0 & 0 & 0 & 0 & -q_{33} \end{pmatrix},
$$

$$
\Theta_3 = \begin{pmatrix} q_{11} & 0 & 0 & 0 & 0 & 0 & 0 & 0 & 0 \\ 0 & q_{12} & 0 & 0 & 0 & 0 & -a-b & 0 & 0 \\ 0 & 0 & q_{13} & 0 & 0 & 0 & 0 & 0 & 0 \end{pmatrix}.
$$

The complex controllers $U_k = (u_{k1}, u_{k2}, u_{k3})^T$ $(k = 1, 2, 3)$ are thus given by

$$
\begin{cases} u_{11} = x_{11}^r e_{22}, \\ u_{12} = -(a+b)e_{11} + x_{21}x_{23} - x_{11}x_{13}, \\ u_{13} = -x_{21}^r x_{22}^r - x_{21}^i x_{22}^i + x_{11}^r x_{12}^r + x_{11}^i x_{12}^i, \end{cases}
$$

$$
\begin{cases} u_{21} = (q_{11}-q_{21})e_{11} - q_{31}e_{31} + x_{11}^r e_{22}, \\ u_{22} = (q_{12}-q_{22})e_{12} - (a+b)(e_{21}+e_{11}) - q_{32}e_{32} - x_{11}^r e_{11} + x_{31}x_{33} - x_{11}x_{13}, \\ u_{23} = (q_{13}-q_{23})e_{13} - q_{33}e_{33} - x_{31}^r x_{32}^r - x_{31}^i x_{32}^i + x_{11}^r x_{12}^r + x_{11}^i x_{12}^i, \end{cases}
$$

and

$$
\begin{cases} u_{31} = (2q_{11}-q_{21})e_{11} - q_{31}e_{31} + x_{11}^r e_{22}, \\ u_{32} = (2q_{12}-q_{22})e_{12} - (a+b)(e_{11}+e_{21}+e_{31}) - q_{32}e_{32} - x_{11}^r e_{11} + x_{41}x_{43} \\ \qquad -x_{11}x_{13}, \\ u_{33} = (2q_{13}-q_{23})e_{13} - q_{33}e_{33} - x_{41}^r x_{42}^r - x_{41}^i x_{42}^i + x_{11}^r x_{12}^r + x_{11}^i x_{12}^i. \end{cases}
$$

Therefore, we have $\dot{e}(t) = (\Phi_1(t) + \Phi_2)e(t)$, where $\Phi_2 = \mathrm{diag}(-a - q_{11} - q_{21}, -1 - q_{12} - q_{22}, -1 - q_{13} - q_{23}, -a - q_{31}, -1 - q_{32}, -1 - q_{33}, -a - q_{41}, -1 - q_{42}, -1 - q_{43})$ and

$$
\Phi_1(t) = \begin{pmatrix} 0 & a & 0 & -q_{11} & x_{11}^r & 0 & -q_{11} & 0 & 0 \\ -a & 0 & 0 & 0 & -q_{12} & 0 & 0 & -q_{12} & 0 \\ 0 & 0 & 0 & 0 & 0 & -q_{13} & 0 & 0 & -q_{13} \\ q_{11} & 0 & 0 & 0 & a & 0 & -q_{31} & 0 & 0 \\ -x_{11}^r & q_{12} & 0 & -a & 0 & 0 & 0 & -q_{32} & 0 \\ 0 & 0 & q_{13} & 0 & 0 & 0 & 0 & 0 & -q_{33} \\ q_{11} & 0 & 0 & q_{31} & 0 & 0 & 0 & a & 0 \\ 0 & q_{12} & 0 & 0 & q_{32} & 0 & -a & 0 & 0 \\ 0 & 0 & q_{13} & 0 & 0 & q_{33} & 0 & 0 & 0 \end{pmatrix}.
$$

If the conditions

$$-a - q_{11} - q_{21} < 0, \quad -1 - q_{12} - q_{22} < 0, \quad -1 - q_{13} - q_{23} < 0, \quad -a - q_{31} < 0, \quad -1 - q_{32} < 0,$$
$$-1 - q_{33} < 0, \quad -a - q_{41} < 0, \quad -1 - q_{42} < 0, \quad -1 - q_{43} < 0$$

hold, then the error system is asymptotically stable due to Theorem 11.1. Consequently, complete synchronization of 4-coupled complex PMSM systems is achieved. The simulation results are performed under the initial conditions $X_1(0) = (2 - j, 4 - 3j, 1)^T$, $X_2(0) = (4 - 2j, 1 + 2j, 8)^T$, $X_3(0) = (-3 + 4j, -2 + 5j, -1)^T$, and $X_4(0) = (5 - 5j, 4 - 2j, 3)^T$. With $q_{k1} = q_{k2} = q_{k3} = 1$ $(k = 1, 2, 3, 4)$, the simulation results are depicted in Figs. 11.3 and 11.4. Figure 11.3 shows the state variables for complete synchronization of the 4-coupled complex PMSM systems. As observed in Fig. 11.4, the errors of synchronization converge asymptotically to zero in a quite short time. Therefore, we achieve complete synchronization of 4-coupled complex PMSM systems with ring connection.

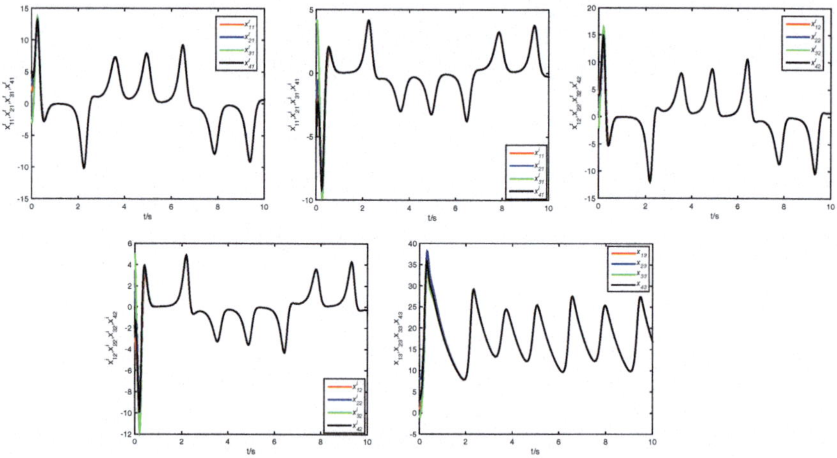

Fig. 11.3 The state trajectories x_{k1}, x_{k2}, and x_{k3} $(k = 1, 2, 3, 4)$ for synchronization of 4-coupled complex PMSM systems

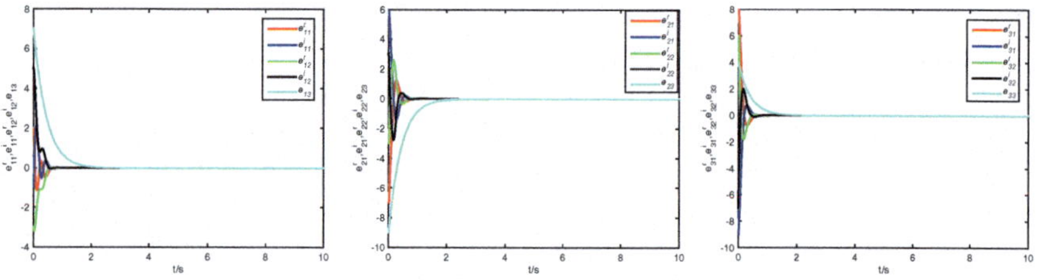

Fig. 11.4 The error dynamics of synchronization among 4-coupled complex PMSM systems

11.3.2 Anti-synchronization of 4-Coupled Complex PMSM Systems

When $\delta = -1$, the error system reduces to the following anti-synchronization error system.

$$\begin{cases} \dot{e}_1 = (A + Q_1 - Q_2)e_1 - Q_1(e_2 - e_3) + 2(Q_2 - Q_1)X_1 + G(X_2) + G(X_1) + U_1, \\ \dot{e}_2 = Q_2 e_1 + (A - Q_3)e_2 + 2(Q_3 - Q_2)X_2 + G(X_3) + G(X_2) + U_2 + U_1, \\ \dot{e}_3 = Q_3 e_2 + (A - Q_4)e_3 + 2(Q_4 - Q_3)X_3 + G(X_4) + G(X_3) + U_3 + U_2. \end{cases}$$

Choose the real coefficient matrix $\Theta = \begin{pmatrix} \Theta_1 \\ \Theta_2 \\ \Theta_3 \end{pmatrix}$ as follows:

$$\Theta_1 = \begin{pmatrix} 0 & -a-b & 0 & q_{11}-q_{21} & 0 & 0 & -q_{11} & 0 & 0 \\ 0 & 0 & 0 & 0 & q_{12}-q_{22} & 0 & 0 & -q_{12} & 0 \\ 0 & 0 & 0 & 0 & 0 & q_{13}-q_{23} & 0 & 0 & -q_{13} \end{pmatrix},$$

$$\Theta_2 = \begin{pmatrix} 0 & 0 & 0 & 0 & -a-b & 0 & -q_{31} & 0 & 0 \\ 0 & 0 & 0 & 0 & 0 & 0 & 0 & -q_{32} & 0 \\ 0 & 0 & 0 & 0 & 0 & 0 & 0 & 0 & -q_{33} \end{pmatrix},$$

$$\Theta_3 = \begin{pmatrix} 0 & 0 & 0 & 0 & 0 & 0 & 0 & -a-b-x_{32}^r & 0 \\ 0 & 0 & 0 & 0 & 0 & 0 & x_{32}^r & 0 & 0 \\ 0 & 0 & 0 & 0 & 0 & 0 & 0 & 0 & 0 \end{pmatrix},$$

we design the complex controllers $U_k = (u_{k1}, u_{k2}, u_{k3})^T$ $(k = 1, 2, 3)$ as

$$\begin{cases} u_{11} = -(a+b)e_{12} + (q_{11} - q_{21})e_{21} - q_{11}e_{31} - 2(q_{21} - q_{11})x_{11}, \\ u_{12} = (q_{12} - q_{22})e_{22} - q_{12}e_{32} - 2(q_{22} - q_{12})x_{12} + x_{21}x_{23} + x_{11}x_{13}, \\ u_{13} = (q_{13} - q_{23})e_{23} - q_{13}e_{33} - 2(q_{23} - q_{13})x_{13} - x_{21}^r x_{22}^r - x_{21}^i x_{22}^i - x_{11}^r x_{12}^r - x_{11}^i x_{12}^i, \end{cases}$$

$$\begin{cases} u_{21} = (a+b)(e_{12} - e_{22}) + (q_{11} - q_{31})e_{31} - (q_{11} - q_{21})(e_{21} + 2x_{11}) + 2(q_{21} - q_{31})x_{21}, \\ u_{22} = (q_{12} - q_{32})e_{32} - (q_{12} - q_{22})(e_{22} + 2x_{12}) + 2(q_{22} - q_{32})x_{22} + x_{31}x_{33} - x_{11}x_{13}, \\ u_{23} = (q_{13} - q_{33})e_{33} - (q_{13} - q_{23})(e_{23} + 2x_{13}) + 2(q_{23} - q_{33})x_{23} - x_{31}^r x_{32}^r - x_{31}^i x_{32}^i \\ \qquad + x_{11}^r x_{12}^r + x_{11}^i x_{12}^i, \end{cases}$$

and

$$\begin{cases} u_{31} = (a+b)(e_{22} - e_{32} - e_{12}) + (q_{31} - q_{11})e_{31} + (q_{11} - q_{21})(e_{21} + 2x_{11}) \\ \qquad - x_{32}^r e_{32} + 2(q_{31} - q_{41})x_{31} - 2(q_{21} - q_{31})x_{21}, \\ u_{32} = (q_{32} - q_{12})e_{32} + (q_{12} - q_{22})(e_{22} + 2x_{12}) + x_{32}^r e_{31} + 2(q_{32} - q_{42})x_{32} \\ \qquad - 2(q_{22} - q_{32})x_{22} + x_{41}x_{43} + x_{11}x_{13}, \\ u_{33} = (q_{33} - q_{13})e_{33} + (q_{13} - q_{23})(e_{23} + 2x_{13}) + 2(q_{33} - q_{43})x_{33} \\ \qquad - 2(q_{23} - q_{33})x_{23} - x_{41}^r x_{42}^r - x_{41}^i x_{42}^i - x_{11}^r x_{12}^r - x_{11}^i x_{12}^i. \end{cases}$$

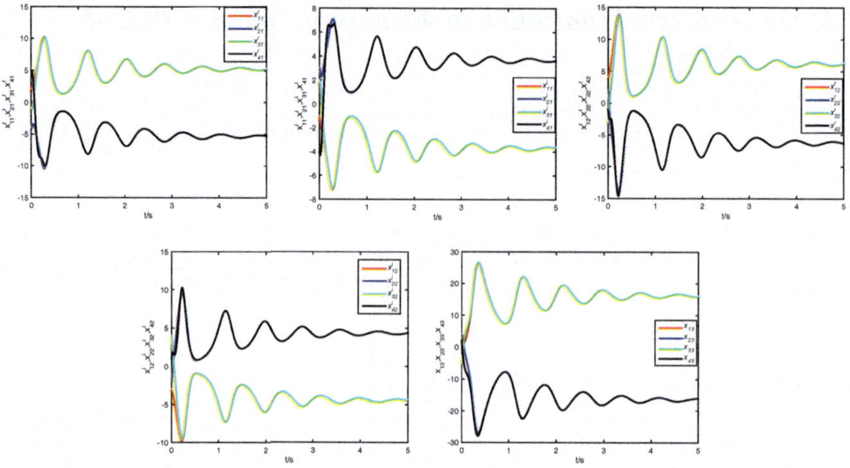

Fig. 11.5 The state trajectories x_{k1}, x_{k2}, and x_{k3} $(k = 1, 2, 3, 4)$ for anti-synchronization of 4-coupled complex PMSM systems

We thus derive the error system $\dot{e}(t) = \Phi(t)e(t) = (\Phi_1(t) + \Phi_2)e(t)$, where $\Phi_2 = \mathrm{diag}(-a + q_{11} - q_{21}, -1 + q_{12} - q_{22}, -1 + q_{13} - q_{23}, -a - q_{31}, -1 - q_{32}, -1 - q_{33}, -a - q_{41}, -1 - q_{42}, -1 - q_{43})$ and

$$\Phi_1(t) = \begin{pmatrix} 0 & -b & 0 & -q_{21} & 0 & 0 & 0 & 0 & 0 \\ b & 0 & 0 & 0 & -q_{22} & 0 & 0 & 0 & 0 \\ 0 & 0 & 0 & 0 & 0 & -q_{23} & 0 & 0 & 0 \\ q_{21} & 0 & 0 & 0 & -b & 0 & -q_{31} & 0 & 0 \\ 0 & q_{22} & 0 & b & 0 & 0 & 0 & -q_{32} & 0 \\ 0 & 0 & q_{23} & 0 & 0 & 0 & 0 & 0 & -q_{33} \\ 0 & 0 & 0 & q_{31} & 0 & 0 & 0 & -b - x_{32}^r & 0 \\ 0 & 0 & 0 & 0 & q_{32} & 0 & b + x_{32}^r & 0 & 0 \\ 0 & 0 & 0 & 0 & 0 & q_{33} & 0 & 0 & 0 \end{pmatrix}.$$

If the conditions

$$-a + q_{11} - q_{21} < 0, \quad -1 + q_{12} - q_{22} < 0, \quad -1 + q_{13} - q_{23} < 0, \quad -a - q_{31} < 0,$$
$$-1 - q_{32} < 0, \quad -1 - q_{33} < 0, \quad -a - q_{41} < 0, \quad -1 - q_{42} < 0, \quad -1 - q_{43} < 0$$

hold, then the error system is asymptotically stable under complex controllers U_k $(k = 1, 2, 3)$ according to Theorem 11.1. Hence, anti-synchronization of 4-coupled complex PMSM systems is realized.

The simulation results are obtained with the initial condition $X_1(0) = (2 - j, 4 - 3j, 1)^T, X_2(0) = (-4 + 3j, 2 - j, 4)^T, X_3(0) = (-1 + 2j, -3 + 4j, -5)^T$, and $X_4(0) = (5 - 4j, -1 + j, 2)^T$. Furthermore, selecting $q_{11} = q_{12} = q_{13} = q_{31} =$

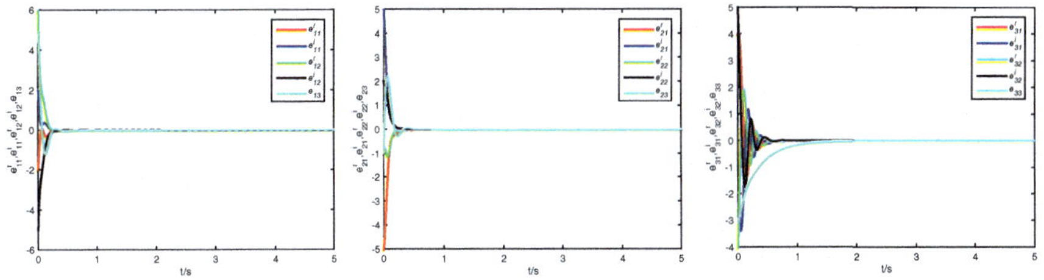

Fig. 11.6 The error dynamics of anti-synchronization among 4-coupled complex PMSM systems

$q_{32} = q_{33} = 1$, $q_{21} = q_{22} = q_{23} = 21$, and $q_{41} = q_{42} = q_{43} = 2$, the corresponding simulation results are shown in Figs. 11.5 and 11.6. Figure 11.5 displays the state variables for anti-synchronization of the 4-coupled complex PMSM systems. As observed in Fig. 11.6, the errors of synchronization converge asymptotically to zero in a quite short period. Thus, anti-synchronization of the 4-coupled complex PMSM systems is realized.

Chapter 12
Robust Finite-Time Stabilization of Unified Chaotic Complex System with Certain and Uncertain Parameters

This chapter addresses the robust finite-time stabilization [193, 194] of a unified chaotic complex system, considering both certain and uncertain parameters. In practical engineering applications, it is often desirable to stabilize a system as quickly as possible, making the study of finite-time stabilization for chaotic systems particularly important. Achieving finite-time stability in controlled systems relies on finite-time control techniques [195], which have been shown to offer enhanced robustness. The chapter is structured as follows: Sect. 12.1 introduces the definition and relevant lemmas for finite-time stabilization of chaotic complex systems. Section 12.2 provides a description of the unified chaotic complex system. Finally, Sect. 12.3 presents the design of a finite-time stability controller for such systems with unknown parameters.

12.1 Definition and Lemmas of Finite-Time Stabilization for Chaotic Complex System

The precise definition of finite-time stability for chaotic complex systems is given, along with some necessary lemmas.

Definition 12.1 We focus on the dynamics described by the following complex system:

$$\dot{z}(t) = f(z), \tag{12.1.1}$$

where $z(t) \in \mathbb{R}^n$ represents the system state, and there exists a constant $T > 0$ (which may depend on the initial state $z(0)$) for which

$$\lim_{t \to T} \|z(t)\| = \lim_{t \to T} \sqrt{\|z^r(t)\|^2 + \|z^i(t)\|^2} = 0,$$

S. Liu et al., *Chaos in Complex Dynamical System: Theory and Application*,
https://doi.org/10.1007/978-981-95-7762-0_12

and $\|z(t)\| = \sqrt{\|z^r(t)\|^2 + \|z^i(t)\|^2} \equiv 0$, if $t \geq T$, then the complex system (12.1.1) is finite-time stable.

Lemma 12.1 *Suppose a continuous and positive-definite function $V(t)$ satisfies the following differential inequality [196]:*

$$\dot{V}(t) \leq -cV^{\eta} \quad \forall \geq t_0, \quad V(t_0) \geq 0, \tag{12.1.2}$$

where $c > 0$ and $0 < \eta < 1$ are constants, it follows that for any given t_0, $V(t)$ satisfies the inequality:

$$V^{1-\eta}(t) \leq V^{1-\eta}(t_0) - c(1-\eta)(t - t_0), \quad t_0 \leq t \leq t_1, \tag{12.1.3}$$

and

$$V(t) \equiv 0, \quad \forall t \geq t_1 \tag{12.1.4}$$

with

$$t_1 = t_0 + \frac{V^{1-\eta}(t_0)}{c(1-\eta)}. \tag{12.1.5}$$

Proof The following differential equation is considered:

$$\dot{X}(t) = -cX^{\eta}(t), \quad X(t_0) = V(t_0). \tag{12.1.6}$$

Despite the fact that this differential equation does not satisfy the global Lipschitz condition, a unique solution for Eq. (12.1.6) remains attainable and is given by:

$$X^{1-\eta}(t) = X^{1-\eta}(t_0) - c(1-\eta)(t - t_0). \tag{12.1.7}$$

Consequently, it follows from Comparison Lemma 13 that

$$V^{1-\eta}(t) = V^{1-\eta}(t_0) - c(1-\eta)(t - t_0), t_0 \leq t \leq t_1, \tag{12.1.8}$$

and

$$V(t) = 0, \quad \forall t \geq t_1$$

with t_1 given in (12.1.5). □

Lemma 12.2 *When a, b and $c < 1$ are all positive numbers, the following inequality holds:*

$$(a+b)^c \leq a^c + b^c. \tag{12.1.9}$$

Proof It is straightforward to prove that the function $f(\lambda) = \lambda^c + (1-\lambda)^c$ arrives its minimal value 1 at $\lambda = \frac{1}{2}$ when $0 < \lambda < 1$ and $0 < c < 1$. Thus, $f(\lambda) \geq 1$. Since a and b are positive numbers, namely, $a + b \neq 0$, set $\lambda = \frac{a}{a+b}$. Then $f(a, b) = (\frac{a}{a+b})^c + (1 - \frac{a}{a+b})^c = \frac{a^c}{(a+b)^c} + \frac{b^c}{(a+b)^c} \geq 1$. Hence, $(a+b)^c \leq a^c + b^c$. □

12.2 Description of the Unified Chaotic Complex System

Building on the unified chaotic system [197–199], we consider the following general chaotic complex system:

$$\begin{cases} \dot{z}_1 = (25\alpha + 10)(z_2 - z_1), \\ \dot{z}_2 = (28 - 35\alpha)z_1 + (29\alpha - 1)z_2 - z_1 z_3, \\ \dot{z}_3 = \frac{1}{2}(\bar{z}_1 z_2 + z_1 \bar{z}_2) - (8 + \alpha)z_3/3, \end{cases} \tag{12.2.1}$$

here, z_1, z_2, and z_3 represent the state variables, and $\alpha \in [0, 1]$ is the system parameter. System (12.2.1) is referred to as the unified chaotic complex system since it exhibits chaotic behavior for all $\alpha \in [0, 1]$. Depending on the value of α, the system is classified as follows: For $0 \leq \alpha < 0.8$, it is termed the generalized complex Lorenz system. The specific case of the complex Lorenz system ($\alpha = 0$) has been studied in Ref. [200]. When $\alpha = 0.8$, the system corresponds to the complex Lü system [201]. For $0.8 < \alpha \leq 1$, it is referred to as the generalized complex Chen system. The complex Chen system at $\alpha = 1$ has also been analyzed in Ref. [200]. In the subsequent sections, we investigate the finite-time stability of the unified chaotic complex system.

12.3 Controller Design of Finite-Time Stability for the Determined Unified Chaotic Complex System

Consider the controlled chaotic complex system:

$$\begin{cases} \dot{z}_1 = (25\alpha + 10)(z_2 - z_1) + u_1, \\ \dot{z}_2 = (28 - 35\alpha)z_1 + (29\alpha - 1)z_2 - z_1 z_3 + u_2, \\ \dot{z}_3 = \frac{1}{2}(\bar{z}_1 z_2 + z_1 \bar{z}_2) - (8 + \alpha)z_3/3 + u_3. \end{cases} \tag{12.3.1}$$

In the following, we design a complex controller vector to achieve finite-time stabilization of the uncertain unified chaotic complex system (12.3.1). The design procedure comprises two main steps:

Step 1 Let the complex controller $u_1 = -(25\alpha + 10)z_2 - [(z_1^r)^\beta + j(z_1^i)^\beta]$, where $\beta = \frac{q}{p}$ and $p > q$ are positive odd integers. Introducing the controller u_1, the first equation of (12.3.1) is

$$\dot{z}_1 = -(25\alpha + 10)z_1 - [(z_1^r)^\beta + j(z_1^i)^\beta]. \tag{12.3.2}$$

Consider the candidate Lyapunov function $V_1 = \frac{1}{2}z_1\tilde{z}_1$, where $\tilde{z}$ denotes the conjugate of variable z, so $V_1 = \frac{1}{2}[(z_1^r)^2 + (z_1^i)^2]$. The time derivative of V_1 along the trajectory of V_1 along the trajectory of (12.3.2) is

$$\begin{aligned}
\dot{V}_1 &= z_1^r[-(25\alpha + 10)z_1^r - (z_1^r)^\beta] + z_1^i[-(25\alpha + 10)z_1^i - (z_1^i)^\beta] \\
&= -(25\alpha + 10)(z_1^r)^2 - (z_1^r)^{\beta+1} - (25\alpha + 10)(z_1^i)^2 - (z_1^i)^{\beta+1} \\
&\le -(z_1^r)^{\beta+1} - (z_1^i)^{\beta+1} \\
&= -(\frac{1}{2})^{-\frac{1+\beta}{2}}[\frac{1}{2}(z_1^r)^2]^{\frac{1+\beta}{2}} - (\frac{1}{2})^{-\frac{1+\beta}{2}}[\frac{1}{2}(z_1^i)^2]^{\frac{1+\beta}{2}} \\
&= -(\frac{1}{2})^{-\frac{1+\beta}{2}}\{[\frac{1}{2}(z_1^r)^2]^{\frac{1+\beta}{2}} + [\frac{1}{2}(z_1^i)^2]^{\frac{1+\beta}{2}}\}.
\end{aligned} \tag{12.3.3}$$

According to Lemma 12.1, we have

$$[\frac{1}{2}(z_1^r)^2]^{\frac{1+\beta}{2}} + [\frac{1}{2}(z_1^i)^2]^{\frac{1+\beta}{2}} \ge \{\frac{1}{2}[(z_1^r)^2 + (z_1^i)^2]\}^{\frac{1+\beta}{2}},$$

then

$$-[\frac{1}{2}(z_1^r)^2]^{\frac{1+\beta}{2}} - [\frac{1}{2}(z_1^i)^2]^{\frac{1+\beta}{2}} \le -\{\frac{1}{2}[(z_1^r)^2 + (z_1^i)^2]\}^{\frac{1+\beta}{2}}.$$

Hence, we get

$$\begin{aligned}
\dot{V}_1 &= -(\frac{1}{2})^{-\frac{1+\beta}{2}}\{[\frac{1}{2}(z_1^r)^2]^{\frac{1+\beta}{2}} + [\frac{1}{2}(z_1^i)^2]^{\frac{1+\beta}{2}}\} \\
&\le -\{\frac{1}{2}[(z_1^r)^2 + (z_1^i)^2]\}^{-\frac{1+\beta}{2}} \cdot (\frac{1}{2})^{-\frac{1+\beta}{2}} \\
&= (\frac{1}{2})^{-\frac{1+\beta}{2}} V_1^{\frac{1+\beta}{2}}.
\end{aligned} \tag{12.3.4}$$

Since $0 < \beta < 1$, then $0 < \frac{1+\beta}{2} < 1$. From Lemma 12.1, the state z_1 will reach $z_1 = 0$ at a finite time $T_1 = \frac{z_1^r(0)+z_1^i(0)}{1-\beta}$.

Step 2 Let $u_2 = -L_1 z_2 - [(z_2^r)^\beta + j(z_2^i)^\beta]$, where $L_1 \ge 29\alpha - 1$ and $u_3 = -L_2 z_3 - z_3^\beta$, where $L_2 \ge -\frac{8+\alpha}{3}$. Then when $t > T_1$, the last two equations of system (12.3.1) become

$$\begin{cases}
\dot{z}_2 = (29\alpha - 1)z_2 - L_1 z_2 - [(z_2^r)^\beta + j(z_2^i)^\beta], \\
\dot{z}_3 = -(\frac{8+\alpha}{3})z_3 - L_2 z_3 - z_3^\beta.
\end{cases} \tag{12.3.5}$$

Choose Lyapunov function for (12.3.5) as follows $V_2 = \frac{1}{2}(z_2\tilde{z}_2 + z_3^2) = \frac{1}{2}[(z_1^r)^2 + (z_1^i)^2 + z_3^2]$. The derivative of V_2 along the trajectories of (12.3.5) is

$$
\begin{aligned}
\dot{V}_2 &= \dot{z}_2^r z_2^r + \dot{z}_2^i z_2^i + \dot{z}_3 z_3 \\
&= [(29\alpha - 1)z_2^r - L_1 z_2^r - (z_2^r)^\beta]z_2^r + [(29\alpha - 1)z_2^i - L_1 z_2^i - (z_2^i)^\beta]z_2^i \\
&\quad +[-(\frac{8+\alpha}{3})z_3 - L_2 z_3 - z_3^\beta]z_3 \\
&= -[L_1 - (29\alpha - 1)][(z_2^r)^2 + (z_2^i)^2] - [L_2 + \frac{8+\alpha}{3}]z_3^2 - (z_2^r)^{\beta+1} \\
&\quad -(z_2^i)^{\beta+1} - z_3^{\beta+1}.
\end{aligned}
\tag{12.3.6}
$$

Also $L_1 \geq 29\alpha - 1$ and $L_2 \geq -\frac{8+\alpha}{3}$, hence

$$
\begin{aligned}
\dot{V}_2 \leq\ & -(z_2^r)^{\beta+1} - (z_2^i)^{\beta+1} - z_3^{\beta+1} \\
=\ & -(\frac{1}{2})^{-\frac{1+\beta}{2}}[\frac{1}{2}(z_2^r)^2]^{\frac{1+\beta}{2}} - (\frac{1}{2})^{-\frac{1+\beta}{2}}[\frac{1}{2}(z_2^i)^2]^{\frac{1+\beta}{2}} - (\frac{1}{2})^{-\frac{1+\beta}{2}}(\frac{1}{2}z_3^2)^{\frac{1+\beta}{2}}.
\end{aligned}
\tag{12.3.7}
$$

Similarly, according to Lemma 12.1 we have

$$
\begin{aligned}
\dot{V}_2 \leq\ & -\{\frac{1}{2}[(z_2^r)^2 + (z_2^i)^2 + z_3^2]\}^{\frac{1+\beta}{2}} \cdot (\frac{1}{2})^{-\frac{1+\beta}{2}} \\
=\ & -\{\frac{1}{2}[z_2\tilde{z}_2 + z_3^2]\}^{\frac{1+\beta}{2}} \cdot (\frac{1}{2})^{-\frac{1+\beta}{2}} \\
=\ & (\frac{1}{2})^{-\frac{1+\beta}{2}} V_2^{\frac{1+\beta}{2}}.
\end{aligned}
\tag{12.3.8}
$$

Since $0 < \beta < 1$, it follows that $0 < \frac{\beta+1}{2} < 1$. According to Lemma 12.2, the complex state variables z_2 and z_3 will converge to $z_2 = 0$ and $z_3 = 0$ within a finite time T_2. Thereafter (i.e., for $t \geq T_2$), the state variables of system (12.3.2) will remain at $z_2 = 0$, $z_3 = 0$. This implies that the controlled unified chaotic system with certain parameters is finite-time stable under the controllers $u_2 = -L_1 z_2 - [(z_2^r)^\beta + j(z_2^i)^\beta]$ and $u_3 = -L_2 z_3 - z_3^\beta$.

To validate the effectiveness of the proposed approach, we provide numerical simulations demonstrating finite-time stability of the unified chaotic systems with certain parameters. All simulations employ the fourth-order Runge-Kutta method with a fixed time step of 0.001. The initial condition of the chaotic system is set as $(z_1(0), z_2(0), z_3(0)) = (-5 + 3j, 5 - 4j, 10)$. The controller parameters are selected as $L_1 = 29\alpha + 1$, $L_2 = \frac{8+\alpha}{3} + 1$, $\beta = \frac{1}{3}$. Simulation results for the specific complex Lorenz, Lü, and Chen systems are shown in Figs. 12.1, 12.2 and 12.3, respectively. These results clearly indicate that the controlled chaotic complex systems achieve finite-time stability.

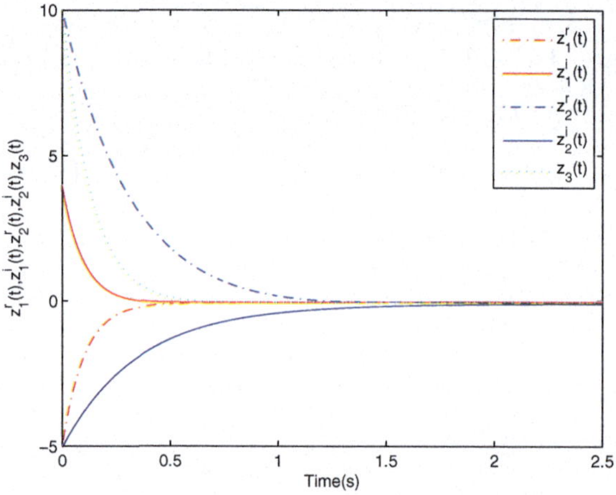

Fig. 12.1 The real and imaginary part responses of complex state variables of controlled complex Lorenz system ($\alpha = 0$)

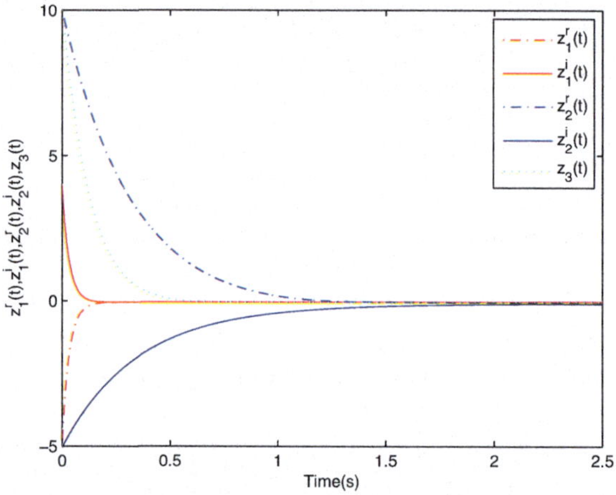

Fig. 12.2 The real and imaginary part responses of complex state variables of controlled complex Lü system ($\alpha = 0.8$)

12.4 Controller Design of Finite-Time Stability for the Unified Chaotic Complex Systems with Unknown Parameters

In practical applications, the parameters of chaotic complex systems are often subject to variations caused by external disturbances. Such variations can significantly degrade system performance [202], making it essential to account for their effects in the control design process. This section aims to develop a control law that achieves

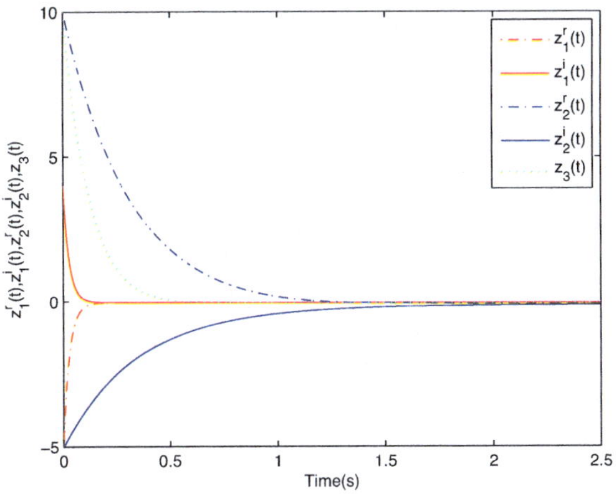

Fig. 12.3 The real and imaginary part responses of complex state variables of controlled complex Chen system ($\alpha = 1$)

finite-time stability for the unified chaotic complex systems in the presence of unknown parameters.

We now consider the following unified chaotic complex system with uncertain parameters:

$$\begin{cases} \dot{z}_1 = (25\alpha + 10)(z_2 - z_1) + u_1', \\ \dot{z}_2 = (28 - 35\alpha + \Delta_1)z_1 + (29\alpha - 1 + \Delta_2)z_2 - z_1 z_3 + u_2', \\ \dot{z}_3 = \frac{1}{2}(\tilde{z}_1 z_2 + z_1 \tilde{z}_2) - (\frac{8+\alpha}{3} + \Delta_3)z_3 + u_3', \end{cases} \quad (12.4.1)$$

where Δ_i $(i = 1, 2, 3)$ represent bounded uncertain parameters satisfying $|\Delta_i| \leq \delta_i$ for $i = 1, 2, 3$. In what follows, we design a robust complex controller vector $\mathbf{U} = (u_1', u_2', u_3')^{\mathrm{T}}$ to achieve finite-time stabilization of the uncertain unified chaotic complex system (12.4.1). The design procedure comprises two steps:

Step 1 Let the complex controller $u_1' = -(25\alpha + 10)z_2 - [(z_1^r)^\beta + j(z_1^i)^\beta]$, where $\beta = \frac{q}{p}$ with $p > q$ being positive odd integers. Substituting controller u_1' into the first equation of (12.4.1) yields

$$\dot{z}_1 = -(25\alpha + 10)z_1 - [(z_1^r)^\beta + j(z_1^i)^\beta]. \quad (12.4.2)$$

Building upon the analysis of system (12.3.2), we conclude that system (12.4.2) is likewise finite-time stable. This implies that $z_1 \equiv 0$ after a finite time T_3.

Step 2 Let $u_2' = -L_2 z_2 - [(z_2^r)^\beta + j(z_2^i)^\beta]$, where $L_2 \geq 29\alpha - 1 + \delta_2$ and $u_3' = -L_3 z_3 - z_3^\beta$, where $L_3 \geq -\frac{8+\alpha}{3} + \delta_3$. Then for $t > T_1$, the last two equations of system (12.4.1) become

$$\begin{cases} \dot{z}_2 = (29\alpha - 1 + \Delta_2)z_2 - L_2 z_2 - [(z_2^r)^\beta + j(z_2^i)^\beta], \\ \dot{z}_3 = -(\frac{8+\alpha}{3} + \Delta_3)z_3 - L_3 z_3 - z_3^\beta. \end{cases} \tag{12.4.3}$$

Choose Lyapunov function for (12.4.3) as $V_2 = \frac{1}{2}(z_2\tilde{z}_2 + z_3^2) = \frac{1}{2}[(z_1^r)^2 + (z_1^i)^2 + z_3^2]$. The derivative of V_2 along the trajectories of (12.4.3) is

$$\begin{aligned} \dot{V}_2 &= \dot{z}_2^r z_2^r + \dot{z}_2^i z_2^i + \dot{z}_3 z_3 \\ &= [(29\alpha - 1 + \Delta_2)z_2^r - L_2 z_2^r - (z_2^r)^\beta]z_2^r + [(29\alpha - 1 + \Delta_2)z_2^i - L_2 z_2^i - (z_2^i)^\beta]z_2^i \\ &\quad + [-(\frac{8+\alpha}{3} + \Delta_3)z_3 - L_3 z_3 - z_3^\beta]z_3 \\ &= (29\alpha - 1 + \Delta_2)(z_2^r)^2 - L_2(z_2^r)^2 - (z_2^r)^{\beta+1} + (29\alpha - 1 + \Delta_2)(z_2^i)^2 - L_2(z_2^i)^2 \\ &\quad -(z_2^i)^{\beta+1} - (\frac{8+\alpha}{3} + \Delta_3)z_3^2 - L_3 z_3^2 - z_3^{\beta+1} \\ &= -[L_2 - (29\alpha - 1 + \Delta_2)][(z_2^r)^2 + (z_2^i)^2] - [L_3 + \frac{8+\alpha}{3} + \Delta_3]z_3^2 - (z_2^r)^{\beta+1} \\ &\quad -(z_2^i)^{\beta+1} - z_3^{\beta+1}. \end{aligned} \tag{12.4.4}$$

Also $L_2 \geq 29\alpha - 1 + \delta_2$ and $L_3 \geq -\frac{8+\alpha}{3} + \delta_3$, hence

$$\begin{aligned} \dot{V}_2 &\leq -(z_2^r)^{\beta+1} - (z_2^i)^{\beta+1} - z_3^{\beta+1} \\ &= -(\frac{1}{2})^{-\frac{1+\beta}{2}}[\frac{1}{2}(z_2^r)^2]^{\frac{1+\beta}{2}} - (\frac{1}{2})^{-\frac{1+\beta}{2}}[\frac{1}{2}(z_2^i)^2]^{\frac{1+\beta}{2}} - (\frac{1}{2})^{-\frac{1+\beta}{2}}(\frac{1}{2}z_3^2)^{\frac{1+\beta}{2}}. \end{aligned} \tag{12.4.5}$$

Similarly, according to Lemma 12.1, we have

$$\begin{aligned} \dot{V}_2 &\leq -\{\frac{1}{2}[(z_2^r)^2 + (z_2^i)^2 + z_3^2]\}^{\frac{1+\beta}{2}} \cdot (\frac{1}{2})^{-\frac{1+\beta}{2}} \\ &= -\{\frac{1}{2}[z_2\tilde{z}_2 + z_3^2]\}^{\frac{1+\beta}{2}} \cdot (\frac{1}{2})^{-\frac{1+\beta}{2}} \\ &= (\frac{1}{2})^{-\frac{1+\beta}{2}} V_2^{\frac{1+\beta}{2}}. \end{aligned} \tag{12.4.6}$$

Given that $0 < \beta < 1$, it follows that $0 < \frac{\beta+1}{2} < 1$. According to Lemma 12.1, the complex state variables z_2 and z_3 will converge to $z_2 = 0$ and $z_3 = 0$ within a finite time T_4. For $t \geq T_4$, the state variables of system (12.3.4) will remain at $z_2 = 0$, $z_3 = 0$. This demonstrates that the controlled unified chaotic complex system with uncertain parameters achieves finite-time stability under the action of the controllers $u_2' = -L_2 z_2 - [(z_2^r)^\beta + j(z_2^i)^\beta]$ and $u_3' = -L_3 z_3 - z_3^\beta$. To verify the effectiveness of the proposed control scheme, we provide numerical simulations illustrating the finite-time stabilization of the unified chaotic complex systems with uncertain parameters. In all cases, the initial condition of the chaotic complex system is set as $(z_1(0), z_2(0), z_3(0)) = (-5 + 3j, 5 - 4j, 10)$. The uncertainties are specified as $\Delta_1 = -0.5\sin(z_1^r) + 0.5j\cos(z_1^i)$, $\Delta_2 = 0.8\cos(z_2^r) - 0.2j\sin(z_2^i)$, $\Delta_3 = \cos(z_3)$. The corresponding bounds of the uncertainties are given by $\delta_1 = \delta_2 =$

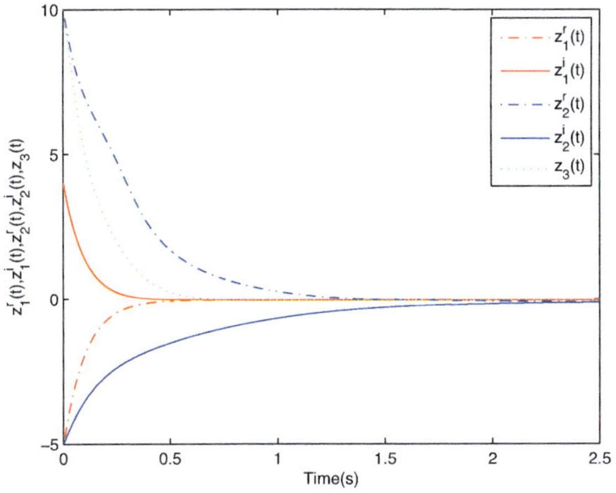

Fig. 12.4 The real and imaginary part responses of complex state variables of controlled uncertain complex Lorenz system ($\alpha = 0$)

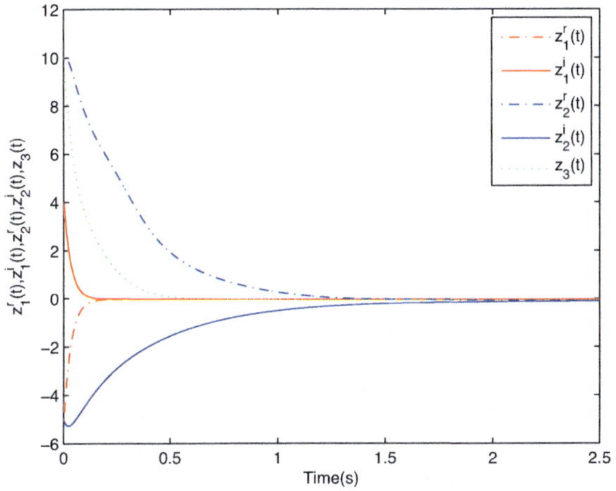

Fig. 12.5 The real and imaginary part responses of complex state variables of controlled uncertain complex Lü system ($\alpha = 0.8$)

$\delta_3 = 1$. The controller parameters are selected as $L_2 = 29\alpha + 1$, $L_3 = \frac{8+\alpha}{3} + 1$, and $\beta = \frac{1}{3}$. Finite-time stabilization of the uncertain complex Lorenz, Lü and Chen systems, respectively, and Chen systems is demonstrated in Figs. 12.4, 12.5, and 12.6, respectively. These results collectively demonstrate the strong robustness of the proposed controller vector against system uncertainties, confirming that the controlled uncertain chaotic complex systems achieve finite-time stability.

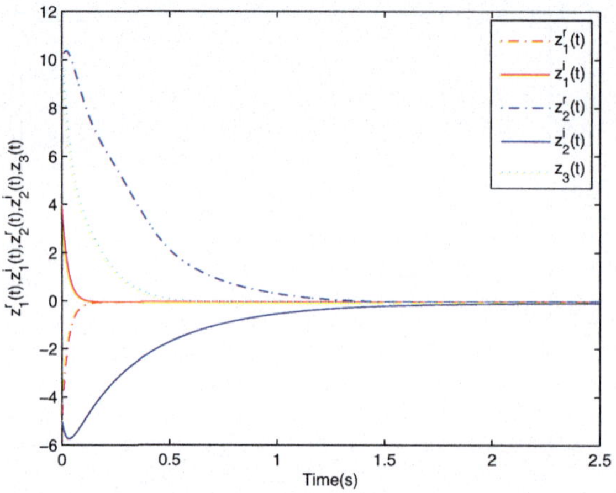

Fig. 12.6 The real and imaginary part responses of complex state variables of controlled uncertain complex Chen system ($\alpha = 1$)

Chapter 13
Anti-synchronization of Fractional-Order Chaotic Complex Systems

Chaotic behavior and synchronization of fractional-order dynamical systems have been extensively studied over the last decade [203–211]. In this chapter, we propose a new fractional-order chaotic complex system and study its dynamical properties including symmetry, equilibria and their stability, and chaotic attractors. Secondly, anti-synchronization of different fractional-order chaotic complex systems is considered based on the stability theory of fractional-order systems. Additionally, we investigate adaptive control for anti-synchronization of a class of uncertain fractional-order chaotic complex systems described by a unified mathematical expression.

13.1 A New Fractional-Order Complex Lü System

Mahmound et al. [79] proposed the complex Lü system, which can be expressed as

$$\begin{cases} \dot{x_1} = a_1(x_2 - x_1), \\ \dot{x_2} = -x_1x_3 + a_2x_2, \\ \dot{x_3} = \frac{1}{2}(\bar{x}_1x_2 + x_1\bar{x}_2) - a_3x_3, \end{cases} \tag{13.1.1}$$

where $x = (x_1, x_2, x_3)^T$ is the vector of state variables, $x_1 = m_1 + jm_2$, $x_2 = m_3 + jm_4$ are complex variables and $x_3 = m_5$ is real variable, $a_i > 0$ $(i = 1, 2, 3)$ are the system real parameters. With the parameters set to $a_1 = 40$, $a_2 = 22$, $a_3 = 5$, system (13.1.1) exhibits chaos, as depicted in Fig. 13.1.

In this chapter, the derivative operator in system (13.1.1) is redefined in terms of the fractional order α $(0 < \alpha \leq 1)$. Thus, the fractional-order complex Lü system can be described by

S. Liu et al., *Chaos in Complex Dynamical System: Theory and Application*,
https://doi.org/10.1007/978-981-95-7762-0_13

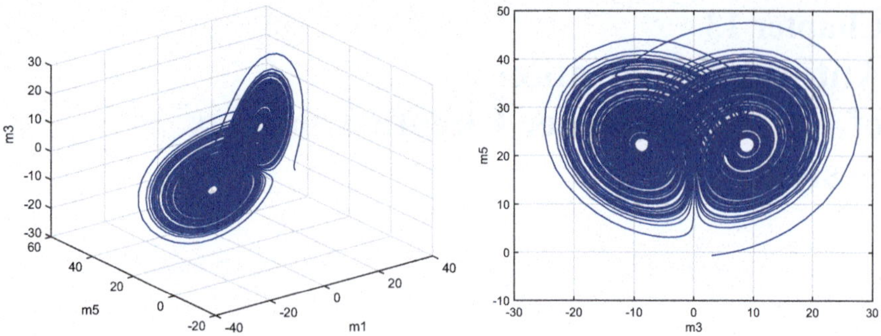

Fig. 13.1 Chaotic attractor of complex Lü system with $(a_1, a_2, a_3) = (40, 22, 5)$

$$\begin{cases} D_*^{\alpha_1} x_1 = a_1(x_2 - x_1), \\ D_*^{\alpha_2} x_2 = -x_1 x_3 + a_2 x_2, \\ D_*^{\alpha_3} x_3 = \frac{1}{2}(\bar{x}_1 x_2 + x_1 \bar{x}_2) - a_3 x_3, \end{cases} \tag{13.1.2}$$

when the derivative order $\alpha_i = 1$ $(i = 1, 2, 3)$, system (13.1.1) will be the common integer-order complex Lü system. The Caputo differential operator is a linear operator [212], i.e. $D_*^{\alpha}(ay(t) + bz(t)) = aD_*^{\alpha}(y(t)) + bD_*^{\alpha}(z(t))$, for arbitrary constants a and b. Therefore, $D_*^{\alpha_i}(x_i) = D_*^{\alpha_i}(m_{2i-1} + jm_{2i}) = D_*^{\alpha_i} m_{2i-1} + jD_*^{\alpha_i} m_{2i}$, $(i = 1, 2, 3)$.

13.2 Dynamics of the Fractional-Order Complex Lü System

We separate the real and imaginary parts of the new system to arrive at the following system:

$$\begin{cases} D_*^{\alpha_1} m_1 = a_1(m_3 - m_1), \\ D_*^{\alpha_1} m_2 = a_1(m_4 - m_2), \\ D_*^{\alpha_2} m_3 = -m_1 m_5 + a_2 m_3, \\ D_*^{\alpha_2} m_4 = -m_2 m_5 + a_2 m_4, \\ D_*^{\alpha_3} m_5 = m_1 m_3 + m_2 m_4 - a_3 m_5. \end{cases} \tag{13.2.1}$$

13.2.1 Symmetry and Invariance

It is obvious that the new system is symmetric about m_5-axis, which means it is invariant for the coordinate transformation of $(m_1, m_2, m_3, m_4, m_5) \longrightarrow (-m_1, -m_2, -m_3, -m_4, m_5)$.

13.2.2 Equilibria and Stability

The equilibria can be calculated by solving the equations $D_*^{\alpha_j} m_i = 0$ ($j = 1, 2, 3; i = 1, 2, \ldots, 5$). Then we obtain that this system has an isolated equilibria $E_0 = (0, 0, 0, 0, 0)$ and nontrivial equilibria $E_\theta = (rcos\theta, rsin\theta, rcos\theta, rsin\theta, a_2)$, where $r = \sqrt{a_2 a_3}$, $\theta \in [0, 2\pi]$.

The equilibrium E_0 is stable when $a_2 < 0$ and unstable when $a_2 > 0$. For E_θ, the Jacobian matrix is given by:

$$J_{E_\theta} = \begin{pmatrix} -a_1 & 0 & a_1 & 0 & 0 \\ 0 & -a_1 & 0 & a_1 & 0 \\ -a_2 & 0 & a_2 & 0 & -r\cos\theta \\ 0 & -a_2 & 0 & a_2 & -r\sin\theta \\ r\cos\theta & r\sin\theta & r\cos\theta & r\sin\theta & -a_3 \end{pmatrix}.$$

Thus, the characteristic polynomial of Jacobian matrix is $\lambda(\lambda + a_1 - a_2)(\lambda^3 + (a_1 + a_3 - a_2)\lambda^2 + a_1 a_3 \lambda + 2a_1 a_2 a_3) = 0$ when $a_2 > 0$. Based on the fractional-order Routh-Hurwitz conditions, E_θ will be stable when the inequality $(a_1 + a_3 - a_2) \cdot (a_1 a_3) > 2a_1 a_2 a_3$ is satisfied.

13.2.3 Chaotic Behavior and Attractors

Using the Adams-Bashforth-Moulton predictor-corrector scheme, we find that chaotic behaviors exist in the fractional-order complex Lü system. In the numerical simulations, the system parameters are set to $(a_1, a_2, a_3) = (42, 22, 5)$, and an initial value is $(x_1, x_2, x_3)^T = (1 + 2j, 3 + 4j, 5)^T$. The new system displays diverse motions when the fractional orders α_i $(i = 1, 2, 3)$ are varied. There are many effective algorithms for the calculation of the Lyapunov exponents (LEs) [213, 214]. In this study, we calculate the Largest Lyapunov exponents (LLEs) using the Wolf algorithm.

(1) Commensurate order $\alpha_1 = \alpha_2 = \alpha_3 = \alpha$

The bifurcation diagram is numerically computed over the range $\alpha \in [0.9, 1]$ with a step size of 0.0002. As shown in Fig. 13.2, it is found that chaotic range is $\alpha \in [0.928, 1]$. To identify the route to chaos, the time history of m_5 is described in Figs. 13.3(a)–(d). As clearly observed in Fig. 13.3(a), the state variables are stable at the fixed point at $\alpha = 0.922$. With increasing α, the system exhibits intermittent dynamical behavior, illustrated in Figs. 13.3(b)–(c). As α is further increased, the motion becomes chaotic as shown for $\alpha = 0.928$, where the LLE is 0.0154. Phase portraits at $\alpha = 0.927$ and 0.928 are shown in Figs. 13.4(a)–(b), respectively. Numerical results demonstrate that the minimum order for chaos generation is 4.64, which corresponds to $\alpha = 0.928$.

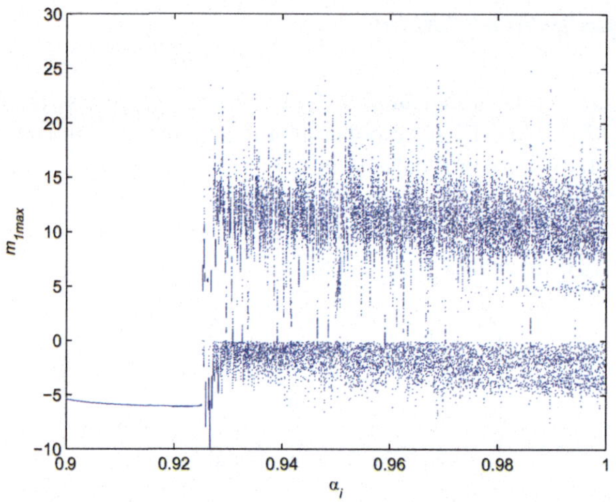

Fig. 13.2 Bifurcation diagram of the new system with $\alpha_1 = \alpha_2 = \alpha_3 = \alpha \in [0.9, 1]$

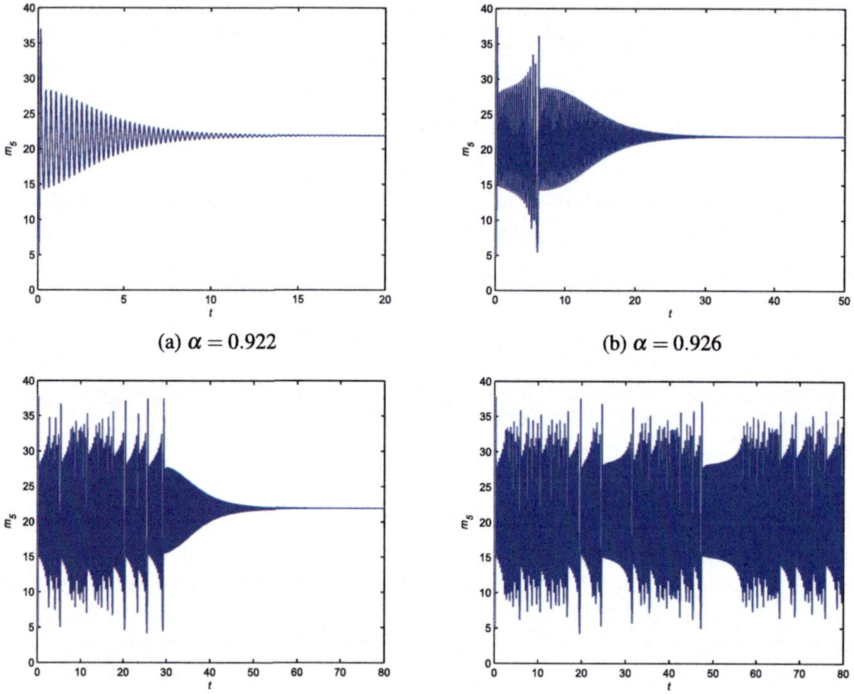

(a) $\alpha = 0.922$

(b) $\alpha = 0.926$

(c) $\alpha = 0.927$

(d) $\alpha = 0.928$

Fig. 13.3 Time histories of the new system at $\alpha_1 = \alpha_2 = \alpha_3 = \alpha$

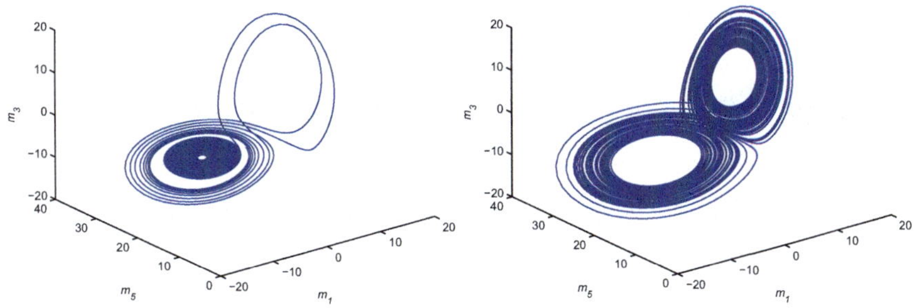

Fig. 13.4 Phase portraits of the new system at $\alpha_1 = \alpha_2 = \alpha_3 = \alpha$

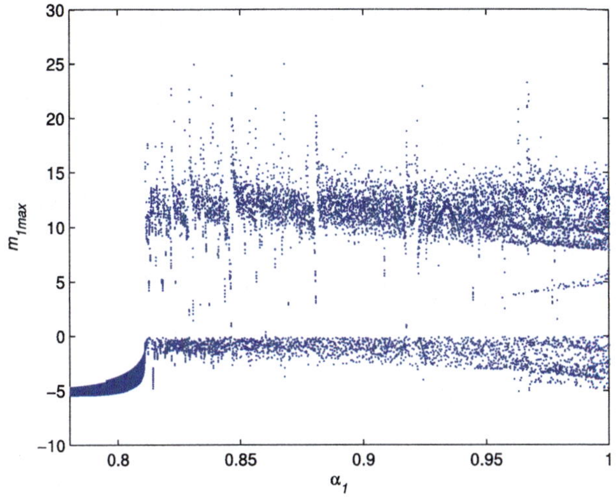

Fig. 13.5 Bifurcation diagram of the new system with $\alpha_1 \in [0.78, 1]$

(2) $\alpha_2 = \alpha_3 = 1$, and let α_1 vary

The bifurcation diagram is numerically computed over $\alpha_1 \in [0.78, 1]$ with a step size of 0.0002. As shown in Fig. 13.5, chaotic motions occur within the range $\alpha_1 \in [0.812, 1]$. Figure 13.6 displays the time history of m_5. When $\alpha_1 = 0.807$, the state variables are stable at the fixed point as depicted in Fig. 13.6(a). With increasing α_1, the system exhibits intermittent dynamical behavior, illustrated in Figs. 13.6(b)–(c). As α_1 is further increased, the motion becomes chaotic as shown for $\alpha_1 = 0.812$, where the LLE is $\lambda = 0.0899$. In this case, the chaotic threshold for this system occurs at an order of 4.624, with the parameter $\alpha_1 = 0.812$.

(3) $\alpha_1 = \alpha_3 = 1$, and let α_2 vary

The bifurcation diagram is numerically computed over the range of $\alpha_2 \in [0.75, 1]$ with a step size of 0.0005. Figure 13.7(a) shows that the chaotic zone covers most of the range of $\alpha_2 \in [0.805, 1]$. To examine the dynamical behavior of the system,

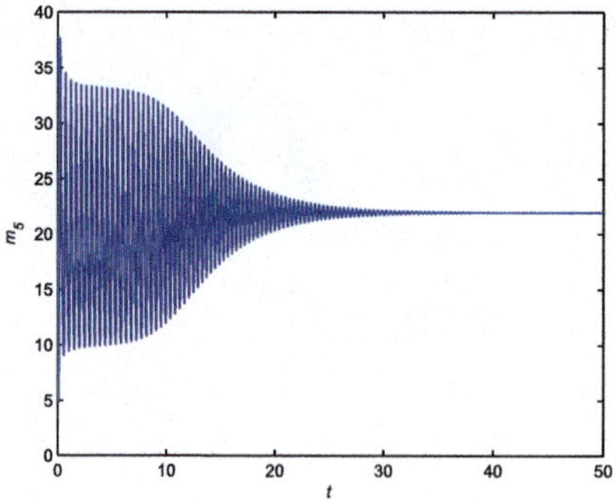

(a) $\alpha_1 = 0.807$

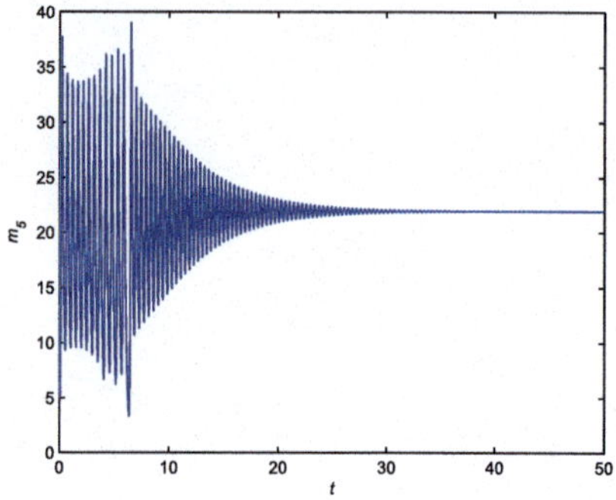

(b) $\alpha_1 = 0.810$

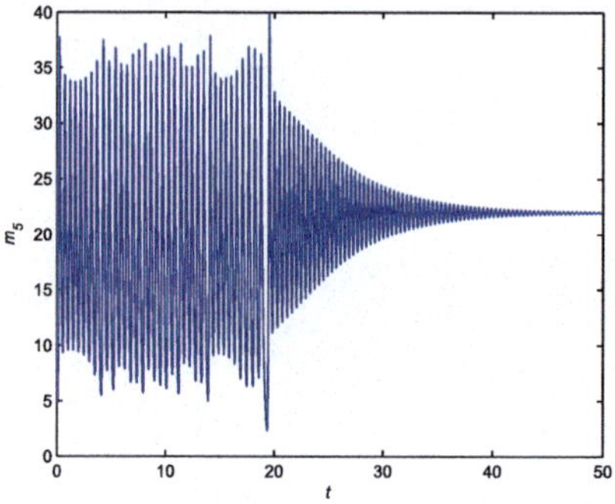

(c) $\alpha_1 = 0.811$

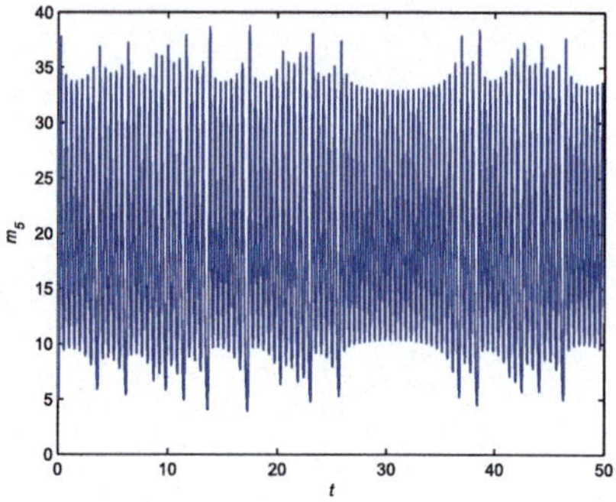

(d) $\alpha_1 = 0.812$

Fig. 13.6 Time histories of the new system at $\alpha_2 = \alpha_3 = 1$ with α_1 varying

we expand the region of $\alpha_2 \in [0.75, 0.85]$ with a step size of 0.0002, as illustrated in Fig. 13.7(b). The period-doubling bifurcations can be seen in Fig. 13.7(b). The phase portraits in Figs. 13.8(a)–(d) display period-1, period-2, period-4, and chaotic behaviors, respectively. Thus, Fig. 13.8 identifies a period doubling route to chaos. In this case, the lowest order for the new system to be chaotic is 4.61, where $\alpha_2 = 0.805$ and the LLE is 0.0461.

(4) $\alpha_1 = \alpha_2 = 1$, and let α_3 vary

The system is numerically analyzed over $\alpha_3 \in [0.8, 1]$ with a step size of 0.001. Phase portraits corresponding to $\alpha_3 = 0.830, 0.831, 0.90,$ and 0.950 are shown in Figs. 13.9(a)–(d), respectively. Results display that chaos exists in the range $\alpha_3 \in [0.831, 1]$. Figure 13.10 illustrates the time history of m_5. At $\alpha_1 = 0.820$, the state variables are stable at the fixed point as depicted in Fig. 13.10(a). When α increases, the system exhibits intermittent dynamical behavior in Figs. 13.10(b)–(c). As α is further increased, the motion becomes chaotic as shown for $\alpha = 0.831$, where the LLE is 0.0516. Numerical evidence displays the lowest order for this new system to be chaotic is 4.831, where $\alpha_3 = 0.831$.

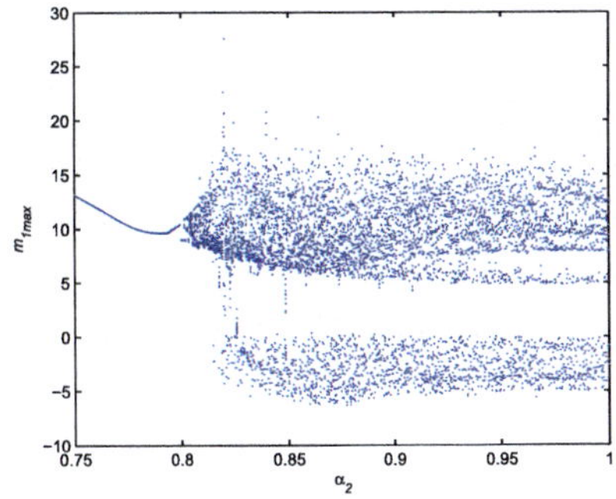

(a) $\alpha_2 \in (0.75, 1)$

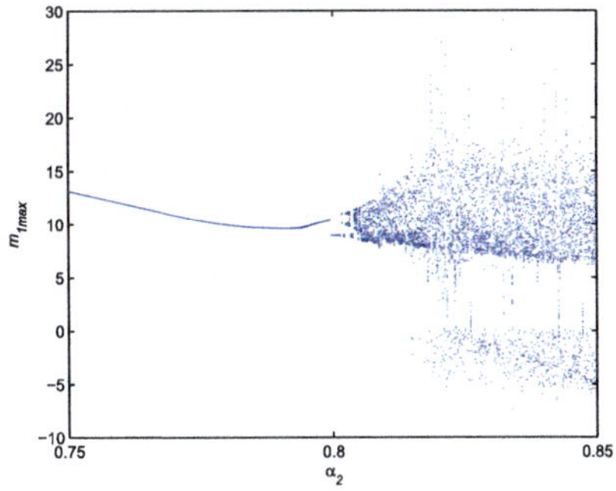

(b) $\alpha_2 \in (0.75, 0.85)$

Fig. 13.7 Bifurcation diagram of the new system at $\alpha_1 = \alpha_3 = 1$ with α_2 varying

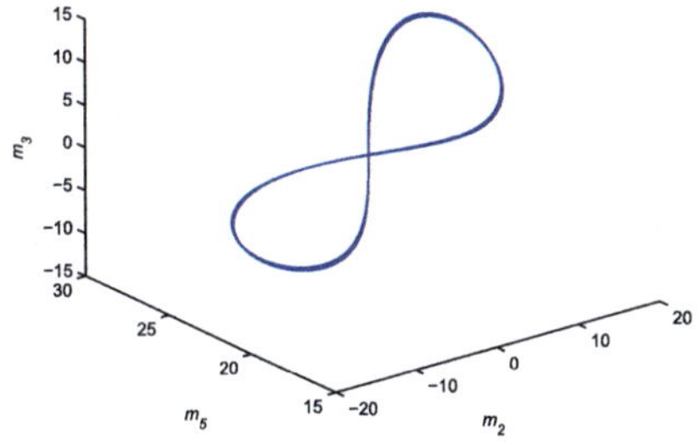

(a) $\alpha_2 = 0.801$

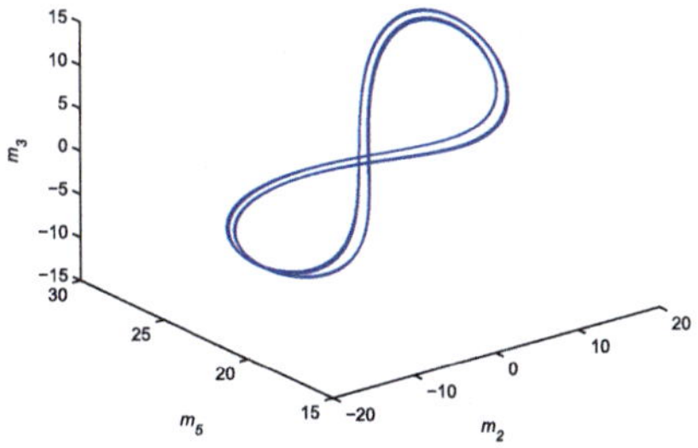

(b) $\alpha_2 = 0.802$

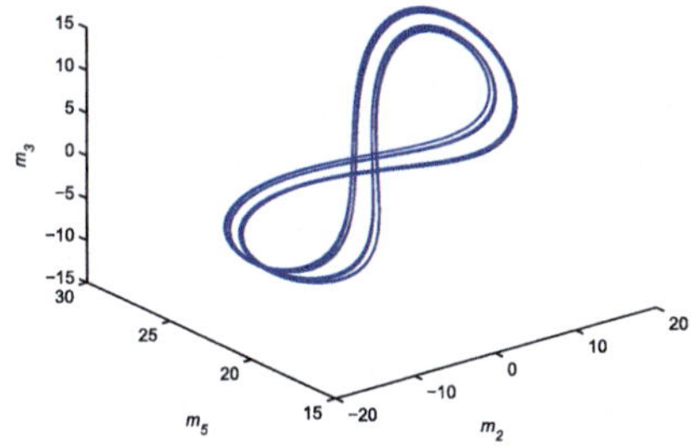

(c) $\alpha_2 = 0.804$

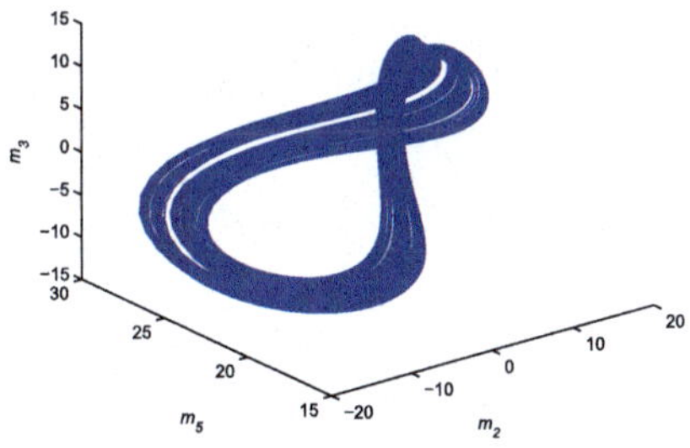

(d) $\alpha_2 = 0.805$

Fig. 13.8 Phase portraits of the new system at $\alpha_1 = \alpha_3 = 1$ with α_2 varying

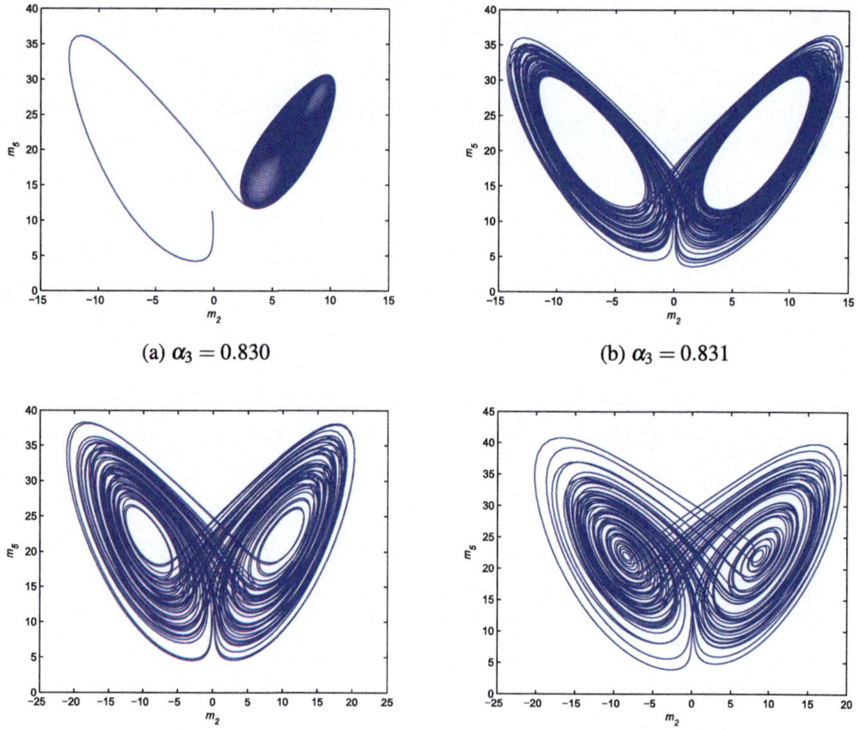

(a) $\alpha_3 = 0.830$ (b) $\alpha_3 = 0.831$

(c) $\alpha_3 = 0.900$ (d) $\alpha_3 = 0.950$

Fig. 13.9 Phase portraits of the new at $\alpha_1 = \alpha_2 = 1$ with α_3 varying

13.3 Anti-synchronization of Fractional-Order Complex Lü System

In this section, we investigate anti-synchronization between different fractional-order complex systems based on active control [215–219] and observer-based control [220–223], respectively. Firstly, we consider anti-synchronization of two fractional-order complex systems by means of active control method.

13.3.1 Active Control

The following fractional-order complex system is considered

$$D_*^{\alpha} x = Ax + f(x), \tag{13.3.1}$$

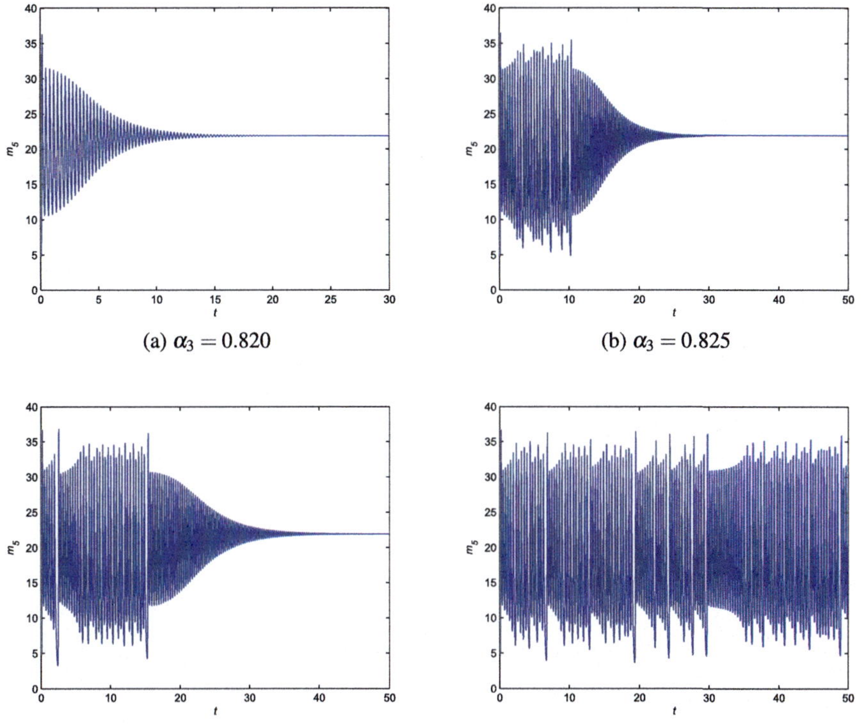

(a) $\alpha_3 = 0.820$ (b) $\alpha_3 = 0.825$

(c) $\alpha_3 = 0.830$ (d) $\alpha_3 = 0.831$

Fig. 13.10 Time histories showing the rout to chaos via intermittency for the new system at $\alpha_1 = \alpha_2 = 1$ with α_3 varying

where $x = (x_1, x_2, \ldots, x_n)^T = x^r + jx^i$ is the state complex vector, $f = (f_1, f_2, \ldots, f_n)^T$ is a vector of nonlinear complex functions and $A \in \mathbb{R}^{n\times n}$ is the matrix of system parameters. We consider system (13.3.1) as the master system and define the slave system as:

$$D_*^{\alpha} y = By + g(y) + u, \tag{13.3.2}$$

where $y = (y_1, y_2, \ldots, y_n)^T = y^r + jy^i$ is the state complex vector, $u = u^r + ju^i$ is designed controller, $u^r = (u_1, u_3, \ldots, u_{2n-1})^T$ and $u^i = (u_2, u_4, \ldots, u_{2n})^T$.

The following presents the main results of this section.

Theorem 13.1 *Anti-synchronization between systems* (13.3.1) *and* (13.3.2) *will be achieved, if the controller is designed as below:*

$$u(t) = u^r + ju^i = -f(x) - g(y) + (B - A)x - Ke. \tag{13.3.3}$$

The real and imaginary parts of (13.3.3) *are*

$$\begin{cases} u^r = -f^r(x) - g^r(y) + (B - A)x^r - Ke^r, \\ u^i = -f^i(x) - g^i(y) + (B - A)x^i - Ke^i, \end{cases} \tag{13.3.4}$$

here K is the control gain matrix, which satisfies $|arg(\lambda_i(B - K))| > \alpha\pi/2$ *for all the eigenvalues of* $B - K$.

Proof According to the definition of anti-synchronization, the error vector between systems (13.3.1) and (13.3.2) is defined as follows:

$$e(t) = y(t) + x(t),$$

The derivative of the aforementioned error vector is given by:

$$\begin{aligned} D_*^\alpha e(t) &= D_*^\alpha e^r(t) + jD_*^\alpha e^i(t) \\ &= D_*^\alpha y^r(t) + D_*^\alpha x^r(t) + j(D_*^\alpha y^i(t) + D_*^\alpha x^i(t)). \end{aligned}$$

Applying (13.3.1), (13.3.2) and (13.3.4) yields that

$$\begin{cases} D_*^\alpha e^r(t) = (B - K)e^r(t), \\ D_*^\alpha e^i(t) = (B - K)e^i(t). \end{cases} \tag{13.3.5}$$

The condition $|arg(\lambda_i(B - K))| > \alpha\pi/2$ satisfies the hypothesis of Theorem 1.3. Therefore, system (13.3.5) is asymptotically stable. Hence, anti-synchronization between different fractional-order complex systems is achieved by using the controller (13.3.3). This completes the proof. □

Remark 13.1 If $A = B$ and $f(.) = g(.)$, systems (13.3.1) and (13.3.2) become identical. Therefore, our scheme is also applicable to achieve anti-synchronization of two identical fractional-order chaotic complex systems.

In the following, we investigate the anti-synchronization behavior between the fractional-order complex Lü and Lorenz systems. It is assumed that the fractional-order complex Lü system drives the fractional-order complex Lorenz system. Consequently, the master system is defined as:

$$\begin{cases} D_*^\alpha x_1 = a_1(x_2 - x_1), \\ D_*^\alpha x_2 = -x_1x_3 + a_2x_2, \\ D_*^\alpha x_3 = \frac{1}{2}(\bar{x}_1x_2 + x_1\bar{x}_2) - a_3x_3. \end{cases} \tag{13.3.6}$$

The slave system is

$$\begin{cases} D_*^\alpha y_1 = b_1(y_2 - y_1) + U_1, \\ D_*^\alpha y_2 = b_2y_1 - y_2 - y_1y_3 + U_2, \\ D_*^\alpha y_3 = \frac{1}{2}(\bar{y}_1y_2 + y_1\bar{y}_2) - b_3y_3 + U_3. \end{cases} \tag{13.3.7}$$

The parameters of the master and slave systems are selected as $(a_1, a_2, a_3) = (42, 22, 5)$, $(b_1, b_2, b_3) = (10, 180, 1)$, and $\alpha_1 = \alpha_2 = \alpha_3 = 0.95$. To satisfy the condition $|\arg(\lambda_i(B - K))| > \alpha\pi/2$, the gain control matrix K is chosen as follows: $K = \begin{pmatrix} 0 & 0\,0 \\ 180 & 0\,0 \\ 0 & 0\,0 \end{pmatrix}$. Based on Theorem 13.2, the controller is designed as:

$$\begin{cases} U_1 = (b_1 - a_1)(x_2 - x_1), \\ U_2 = x_1x_3 + y_1y_3 + b_2x_1 - (a_2 + 1)x_2 - 180(y_1 + x_1), \\ U_3 = -\frac{1}{2}(\bar{x}_1x_2 + x_1\bar{x}_2) - \frac{1}{2}(\bar{y}_1y_2 + y_1\bar{y}_2) + (a_3 - b_3)x_3. \end{cases}$$

The numerical simulations are initialized with the master system at $x(0) = (-5 + 10j, 6 - j, 2)^T$ and the slave system at $y(0) = (2 - 3j, -8 + 4j, 3)^T$. Figures 13.11(a)–(e) describe state variables of drive system and response system, in which the state complex variables are demonstrated by real and imaginary part, respectively. The anti-synchronization errors converge asymptotically to zero in a very brief period, as shown in Fig. 13.11(f). The above results verify that anti-synchronization between fractional-order complex Lü system and Lorenz system has been achieved.

13.3.2 Observer-Based Control

In this section, a state observer is designed to achieve anti-synchronization of fractional-order complex chaotic systems. We consider the following fractional-order complex chaotic system:

$$D_*^{\alpha}x = Ax + Bf(x) + C, \tag{13.3.8}$$

where $x = (x_1, x_2, \ldots, x_n)^T \in \mathbb{C}^n$, $A \in \mathbb{R}^{n\times n}$ and $B \in \mathbb{R}^{n\times m}$ are real matrices, $C \in \mathbb{C}^{n\times 1}$, and $f = (f_1, f_2, \ldots, f_m)^T$ (f_i stands for complex nonlinear function) are column vectors.

To achieve anti-synchronization, system (13.3.8) is designated as the master system. Accordingly, we assume that system (13.3.8) has the following output:

$$s(x) = f(x) + Kx,$$

where $K \in \mathbb{R}^{n\times m}$ is a gain matrix. Thus, design the observer of fractional-order chaotic complex system (13.3.8) as the slave system

$$D_*^{\alpha}y = Ay + Bf(y) - C - B(s(x) + s(y)). \tag{13.3.9}$$

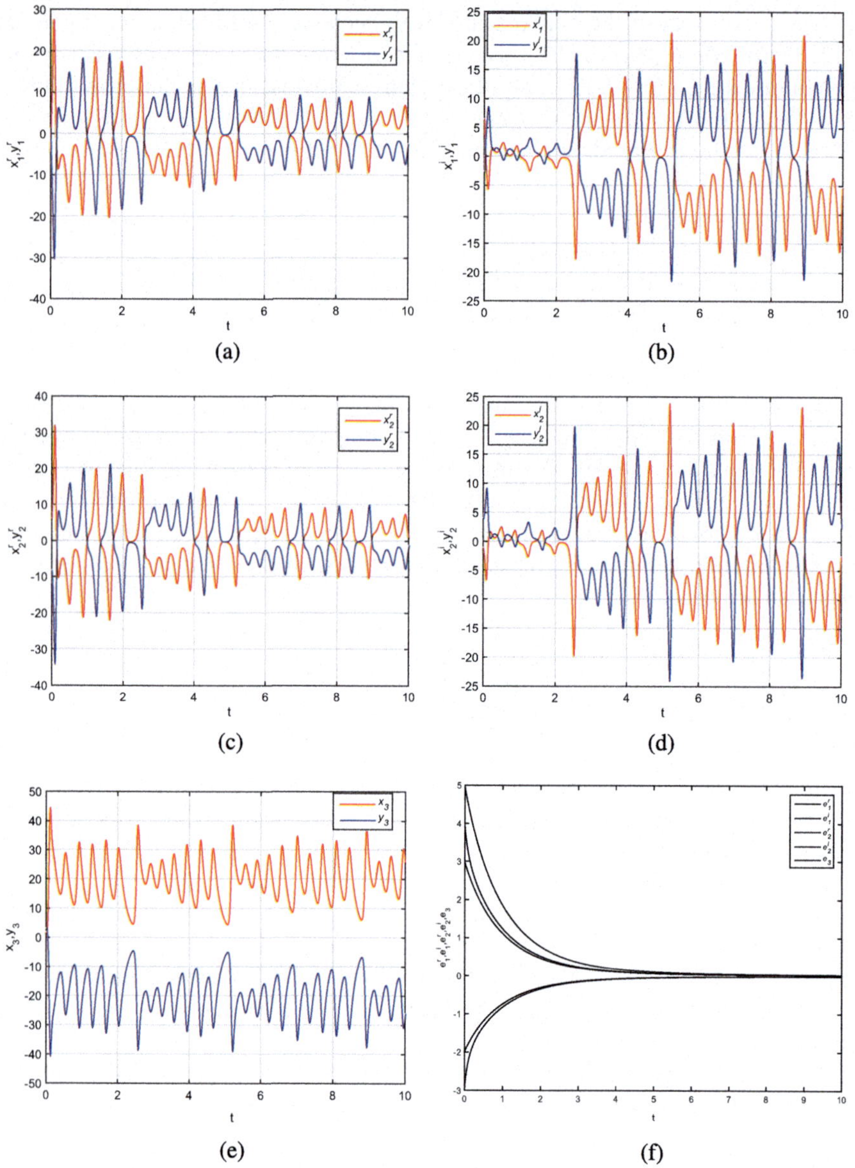

Fig. 13.11 State variables and anti-synchronization error of the master system (13.3.6) and the slave system (13.3.7) (time/s)

In the sequel, we define the synchronization error function as follows:

$$e(t) = y(t) + x(t).$$

Therefore, the error dynamical system is given by:

$$\begin{aligned} D_*^\alpha e &= D_*^\alpha y + D_*^\alpha x \\ &= Ae + B(f(x) + f(y)) - B(s(x) + s(y)) \\ &= (A - BK)e. \end{aligned}$$

where $A - BK$ is the linear time invariant matrix. For the sake of making systems (13.3.8) and (13.3.9) controllable, we choose the appropriate gain matrix K to satisfy $|arg(\lambda_l(A - BK))| > q\pi/2$, $(l = 1, 2, \ldots, \text{n})$. Thus, on the basis of Theorem 1.3, we can obtain $||e(t)|| \to 0$ as $t \to \infty$. Hence, system (13.3.8) and the observer (13.3.9) can realize anti-synchronization.

In the following, we investigate anti-synchronization behavior of the fractional-order complex Lü system. The fractional-order complex Lü system is considered as the master system, which is expressed as

$$\begin{pmatrix} D_*^\alpha x_1 \\ D_*^\alpha x_2 \\ D_*^\alpha x_3 \end{pmatrix} = \begin{pmatrix} -a & a & 0 \\ 0 & b & 0 \\ 0 & 0 & -c \end{pmatrix} \begin{pmatrix} x_1 \\ x_2 \\ x_3 \end{pmatrix} + \begin{pmatrix} 0 & 0 \\ 1 & 0 \\ 0 & 1 \end{pmatrix} \begin{pmatrix} -x_1 x_3 \\ \frac{1}{2}(\bar{x}_1 x_2 + x_1 \bar{x}_2) \end{pmatrix}, \tag{13.3.10}$$

where a, b, c are real constants. Take it that the output of system (13.3.10) can be expressed as $s(x) = f(x) + Kx$, where $K \in \mathbb{R}^{2\times 3}$.

Thus, the following state observer is designed for system (13.3.10):

$$\begin{pmatrix} D_*^\alpha y_1 \\ D_*^\alpha y_2 \\ D_*^\alpha y_3 \end{pmatrix} = \begin{pmatrix} -a & a & 0 \\ 0 & b & 0 \\ 0 & 0 & -c \end{pmatrix} \begin{pmatrix} y_1 \\ y_2 \\ y_3 \end{pmatrix} + \begin{pmatrix} 0 & 0 \\ 1 & 0 \\ 0 & 1 \end{pmatrix} \begin{pmatrix} -y_1 y_3 \\ \frac{1}{2}(\bar{y}_1 y_2 + y_1 \bar{y}_2) \end{pmatrix} \tag{13.3.11}$$

$$+ \begin{pmatrix} 0 & 0 \\ 1 & 0 \\ 0 & 1 \end{pmatrix} (s(x) + s(y)). \tag{13.3.12}$$

Then, the error dynamical system is derived from system (13.3.10) and observer (13.3.11) as follows

$$D_*^\alpha e = (A - BK)e = \left(\begin{pmatrix} -a & a & 0 \\ 0 & b & 0 \\ 0 & 0 & -c \end{pmatrix} - \begin{pmatrix} 0 & 0 \\ 1 & 0 \\ 0 & 1 \end{pmatrix} K \right) e.$$

Assume that the eigenvalues of matrix $A - BK$ are assigned as $(-42, -2, -10)$, satisfying $|\arg(\lambda_l(A - BK))| > q\pi/2$, $(l = 1, 2, 3)$. Based on the pole placement method, the feedback matrix K can be computed as

$$K = \begin{pmatrix} 0 & 24 & 6 \\ 0 & 0 & -2 \end{pmatrix}.$$

Thus, the Adams–Bashforth–Moulton predictor–corrector scheme is used to obtain simulation results illustrated with the initial condition $x(0) = (1 + 2j, 3 + 4j, 5)^T$, $y(0) = (-2 + 10j, 6 - j, 10)^T$. Choosing $\Phi = \text{diag}(1 + j, 1 - j, 1)$, the inverse matrix is computed as $\Phi^{-1} = \text{diag}((1 - j)/2, (1 + j)/2, 1)$. The state evolution of system (13.3.10) and observer (13.3.11) is depicted in Figs. 13.12(a)–(e). The state trajectories of the error system are demonstrated in Fig. 13.12(f), where it can be seen that the error system tends asymptotically to zero very quickly. Hence, the anti-synchronization of fractional-order complex Lü systems based on a nonlinear state observer can be realized.

13.4 Anti-synchronization of Fractional-Order Chaotic Complex Systems with Unknown Parameters via Adaptive Control

13.4.1 Mathematical Model and Problem Descriptions

Consider the following n-dimensional fractional-order chaotic complex system with the controller:

$$D_*^\alpha y = G(y)B + g(y) + U, \tag{13.4.1}$$

where $y = (y_1, y_2, \ldots, y_n)^T \in \mathbb{C}^n$ stands for a complex state vector, $G(y) \in \mathbb{C}^{n\times n}$ denotes a complex matrix, $B = (b_1, b_2, \ldots, b_n)^T$ represents the real (or complex) vector of unknown parameters, and $g : \mathbb{C}^n \to \mathbb{C}^n$ depicts the nonlinear term. $U = U^r + jU^i$ is a complex controller to be designed, where $U^r = (u_1, u_3, \ldots, u_{2n-1})^T$ and $U^i = (u_2, u_4, \ldots, u_{2n})^T$. System (13.4.1) is designated as the response system, while the drive system is defined by

$$D_*^\alpha x = h(x), \tag{13.4.2}$$

where $x = (x_1, x_2, \ldots, x_n)^T \in \mathbb{C}^n$ represents the complex state vector and $h : \mathbb{C}^n \to \mathbb{C}^n$ describes the nonlinear term.

This part investigates the adaptive anti-synchronization between two different fractional-order chaotic complex systems under a unified mathematical form. Anti-synchronization error is defined as $e(t) = y(t) + x(t)$, namely, $e^r(t) = y^r(t) + x^r(t)$ and $e^i(t) = y^i(t) + x^i(t)$. This part aims to design adaptive controller and parameter update laws such that anti-synchronization error vector $e(t)$ converges to zero as $t \to \infty$, i.e., $\lim_{t\to\infty} ||e(t)|| = 0$.

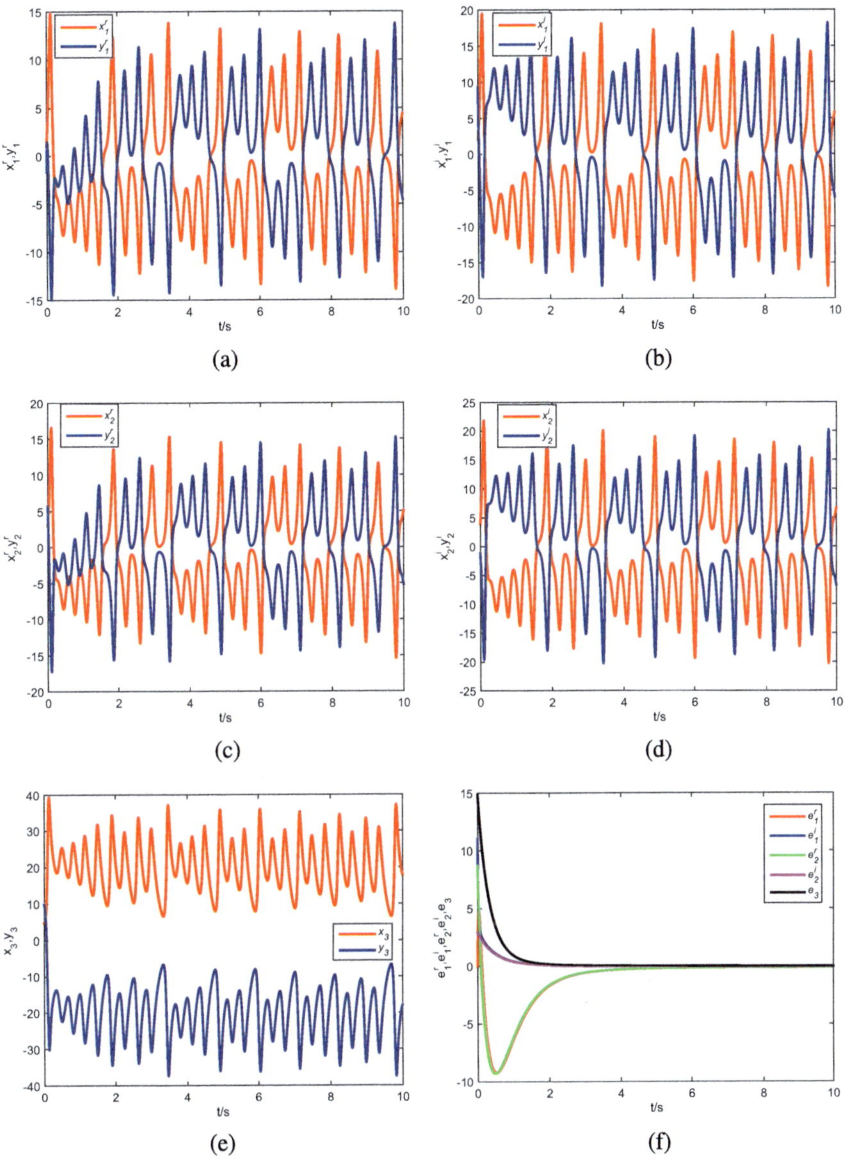

Fig. 13.12 State variables and anti-synchronization error of the master system (13.3.10) and the slave system (13.3.11) (time/s)

13.4.2 Adaptive Anti-synchronization Schemes

In the following, we study two cases with different situations of A in the drive system (13.4.2), i.e., A has known or unknown parameters.

I. The drive system without unknown parameters

We investigate two cases for parameter vector B in the response system (13.4.1), i.e., B is a real vector or a complex vector. We first analyze the case in which B is a real vector.

Theorem 13.2 *If the control law is designed as*

$$\begin{aligned} U &= U^r + jU^i = -G(y)\hat{B} - g(y) - h(x) + Ke \\ &= -G^r(y)\hat{B} - g^r(y) - h^r(x) + Ke^r - j\left(G^i(y)\hat{B} + g^i(y) + h^i(x) - Ke^i\right), \end{aligned} \tag{13.4.3}$$

and the adaptive laws are chosen as

$$\begin{cases} D_*^\alpha \hat{B} = (G^r(y))^T e^r + (G^i(y))^T e^i, \\ D_*^\alpha k_l = -\delta_l \left((e_l^r(t))^2 + (e_l^i(t))^2\right), \quad l = 1, 2, \ldots, n, \end{cases} \tag{13.4.4}$$

then anti-synchronization between systems (13.4.1) *and* (13.4.2) *can be achieved, where* $\hat{B}$ *is the estimated value of the unknown parameter vector* B, $K = \text{diag}(k_1, k_2, \ldots, k_n)$ *represents the real control strength matrix,* $\delta = \text{diag}(\delta_1, \delta_2, \ldots, \delta_n)$ *denotes a convergence factor matrix with* $\delta_l > 0$ $(l = 1, 2, \ldots, n)$.

Proof It follows from the fractional-order chaotic complex systems (13.4.1) and (13.4.2) that

$$D_*^\alpha e(t) = G^r(y)B + g^r(y) + U^r + h^r(x) + j\left(G^i(y)B + g^i(y) + U^i + h^i(x)\right). \tag{13.4.5}$$

Thus, the error system is obtained by inserting (13.4.3) into (13.4.5) and separating the real and imaginary parts, as follows:

$$\begin{cases} D_*^\alpha e^r(t) = G^r(y)\tilde{B} + Ke^r, \\ D_*^\alpha e^i(t) = G^i(y)\tilde{B} + Ke^i, \end{cases}$$

where $e^r(t) = (e_1^r(t), e_2^r(t), \ldots, e_n^r(t))^T$, $e^i(t) = (e_1^i(t), e_2^i(t), \ldots, e_n^i(t))^T$ and $\tilde{B} = B - \hat{B} = (\tilde{b}_1, \tilde{b}_2, \ldots, \tilde{b}_n)$ is the parameter estimation error. Furthermore, one has

$$\begin{cases} D_*^\alpha e_l^r(t) = G_l^r(y)\tilde{B} + k_l e_l^r, \\ D_*^\alpha e_l^i(t) = G_l^i(y)\tilde{B} + k_l e_l^i, \end{cases} \tag{13.4.6}$$

where $G_l^r(y)$ and $G_l^i(y)$ are the l-th row of $G^r(y)$ and $G^i(y)$, respectively. The following positive definite Lyapunov candidate function is defined as:

$$V(e,t) = \frac{1}{2}\sum_{l=1}^{n}\left((e_l^r(t))^2 + (e_l^i(t))^2\right) + \frac{1}{2}\sum_{l=1}^{n}\frac{1}{\delta_l}(k_l + L)^2 + \frac{1}{2}\tilde{B}^T\tilde{B},$$

where L is a positive constant. From Lemmas 1.2 and 1.3, it follows directly that

$$D_*^\alpha V \leq \sum_{l=1}^{n}\left(e_l^r D_*^\alpha e_l^r + e_l^i D_*^\alpha e_l^i\right) + \sum_{l=1}^{n}\frac{1}{\delta_l}(k_l + L) D_*^\alpha k_l + \tilde{B}^T D_*^\alpha \tilde{B}. \qquad (13.4.7)$$

In the case of (13.4.4) and (13.4.6), the relation (13.4.7) becomes

$$\begin{aligned} D_*^\alpha V &\leq \sum_{l=1}^{n}\left(e_l^r(G_l^r(y)\tilde{B} + k_l e_l^r) + e_l^i(G_l^i(y)\tilde{B} + k_l e_l^i)\right) + \tilde{B}^T D_*^\alpha \tilde{B} - \sum_{l=1}^{n}(k_l + L)\left((e_l^r)^2 + (e_l^i)^2\right) \\ &= \sum_{l=1}^{n}\left(e_l^r G_l^r(y)\tilde{B} + e_l^i G_l^i(y)\tilde{B}\right) - L\sum_{l=1}^{n}\left((e_l^r)^2 + (e_l^i)^2\right) + \tilde{B}^T D_*^\alpha \tilde{B}. \end{aligned}$$

On the other hand,

$$\begin{aligned} &\sum_{l=1}^{n}\left(e_l^r G_l^r(y)\tilde{B} + e_l^i G_l^i(y)\tilde{B}\right) \\ &= \left[G_1^r(y)\tilde{B}, G_2^r(y)\tilde{B}, \ldots, G_n^r(y)\tilde{B}\right]\begin{bmatrix} e_1^r \\ e_2^r \\ \vdots \\ e_n^r \end{bmatrix} + \left[G_1^i(y)\tilde{B}, G_2^i(y)\tilde{B}, \ldots, G_n^i(y)\tilde{B}\right]\begin{bmatrix} e_1^i \\ e_2^i \\ \vdots \\ e_n^i \end{bmatrix} \\ &= \tilde{B}^T\left[(G_1^r(y))^T, \ldots, (G_n^r(y))^T\right]\begin{bmatrix} e_1^r \\ \vdots \\ e_n^r \end{bmatrix} + \tilde{B}^T\left[(G_1^i(y))^T, \ldots, (G_n^i(y))^T\right]\begin{bmatrix} e_1^i \\ \vdots \\ e_n^i \end{bmatrix} \\ &= \tilde{B}^T\left[(G^r(y))^T, (G^i(y))^T\right]\begin{bmatrix} e^r \\ e^i \end{bmatrix} = \tilde{B}^T\left((G^r(y))^T e^r + (G^i(y))^T e^i\right). \end{aligned}$$

Thus, combining (13.4.4) and the latter equality implies that

$$D_*^\alpha V \leq -L\left((e^r)^T e^r + (e^i)^T e^i\right).$$

Based on Lemma 1.6, we deduce that

$$\lim_{t\to\infty} ||e^r(t)|| = 0 \quad \text{and} \quad \lim_{t\to\infty} ||e^i(t)|| = 0.$$

Hence, we have $\lim_{t\to\infty}||e(t)|| = 0$. Therefore, the designed controller (13.4.3) and the fractional adaption laws (13.4.4) enable the drive system (13.4.2) and the response system (13.4.1) to achieve anti-synchronization. The proof is complete. □

Next, we consider the case when B is a complex vector. Then B can be rewritten as $B = B^r + jB^i$ and the fractional-order system (13.4.1) can be represented as

$$D_*^{\alpha} y = G(y)B^r + jG(y)B^i + g(y) + U = G(y)B^r + Q(y)B^i + g(y) + U, \tag{13.4.8}$$

where $Q(y) = jG(y)$ describes a new $n \times n$ complex matrix. Similarly, applying the same procedure yields the following results.

Theorem 13.3 *If the control law is designed as*

$$\begin{aligned} U &= U^r + jU^i = -G(y)\hat{B}^r - Q(y)\hat{B}^i - g(y) - h(x) + Ke \\ &= -G^r(y)\hat{B}^r - Q^r(y)\hat{B}^i - g^r(y) - h^r(x) + Ke^r - j\left(G^i(y)\hat{B}^r + Q^i(y)\hat{B}^i + g^i(y) + h^i(x) - Ke^i\right), \end{aligned}$$

and the adaptive laws are chosen as

$$\begin{cases} D_*^{\alpha}\hat{B}^r = (G^r(y))^T e^r + (G^i(y))^T e^i, \\ D_*^{\alpha}\hat{B}^i = (Q^r(y))^T e^r + (Q^i(y))^T e^i, \\ D_*^{\alpha}k_l = -\delta_l\left((e_l^r(t))^2 + (e_l^i(t))^2\right), \quad l = 1, 2, \ldots, n, \end{cases}$$

then anti-synchronization between systems (13.4.2) *and* (13.4.8) *can be achieved, where $\hat{B}^r$ and $\hat{B}^i$ denote the estimated values of the unknown parameter vectors B^r and B^i, respectively, $K = \mathrm{diag}(k_1, k_2, \ldots, k_n)$ represents the real control strength matrix, $\delta = \mathrm{diag}(\delta_1, \delta_2, \ldots, \delta_n)$ is a convergence factor matrix with $\delta_l > 0$ $(l = 1, 2, \ldots, n)$.*

Proof The proof is analogous to that of Theorem 13.2, and hence is omitted. □

II. The drive system with unknown parameters

In what follows, the case where the drive system has unknown parameters is considered. If some unknown parameters exist in system (13.4.2), then system (13.4.2) can be expressed as:

$$D_*^{\alpha} x = F(x)A + f(x), \tag{13.4.9}$$

where $x = (x_1, x_2, \ldots, x_n)^T \in \mathbb{C}^n$ is the complex state vector, $F(x) \in \mathbb{C}^{n\times n}$ denotes a complex matrix, $A = (a_1, a_2, \ldots, a_n)^T$ represents the real (or complex) vector of unknown parameters, and $f : \mathbb{C}^n \to \mathbb{C}^n$ is a nonlinear complex vector function.

In the following, we investigate the case where A and B are real vectors.

Theorem 13.4 *If the control law is constructed as follows:*

$$\begin{aligned} U &= U^r + jU^i = -G(y)\hat{B} - g(y) - F(x)\hat{A} - f(x) + Ke \\ &= -G^r(y)\hat{B} - g^r(y) - F^r(x)\hat{A} - f^r(x) + Ke^r - j\left(G^i(y)\hat{B} + g^i(y) + F^i(x)\hat{A} + f^i(x) - Ke^i\right), \end{aligned} \tag{13.4.10}$$

and the adaptive laws are chosen as

$$\begin{cases} D_*^{\alpha}\hat{A} = (F^r(x))^T e^r + (F^i(x))^T e^i, \\ D_*^{\alpha}\hat{B} = (G^r(y))^T e^r + (G^i(y))^T e^i, \\ D_*^{\alpha}k_l = -\delta_l\left((e_l^r(t))^2 + (e_l^i(t))^2\right), \quad l = 1, 2, \ldots, n, \end{cases} \tag{13.4.11}$$

then the response system (13.4.1) *can anti-synchronize the drive system* (13.4.9), *where* $\hat{A}$ *and* $\hat{B}$ *denote the estimated values of the unknown parameter vectors A and B, respectively,* $K = \mathrm{diag}(k_1, k_2, \ldots, k_n)$ *is the real control strength matrix,* $\delta = \mathrm{diag}(\delta_1, \delta_2, \ldots, \delta_n)$ *is a convergence factor matrix with* $\delta_l > 0\,(l = 1, 2, \ldots, n)$.

Proof We derive the following error dynamical system from the fractional-order chaotic complex systems (13.4.1), (13.4.9), and the controller (13.4.10),

$$D_*^{\alpha}e(t) = G(y)B + g(y) + U + F(x)A + f(x) = F(x)\tilde{A} + G(y)\tilde{B} + Ke, \tag{13.4.12}$$

where $\tilde{A} = A - \hat{A} = (\tilde{a}_1, \tilde{a}_2, \ldots, \tilde{a}_n)$, $\tilde{B} = B - \hat{B} = (\tilde{b}_1, \tilde{b}_2, \ldots, \tilde{b}_n)$ are the errors between the true values and estimated values of unknown parameters. Thus, the error system is obtained by separating the real and imaginary parts of (13.4.12) as follows:

$$\begin{cases} D_*^{\alpha}e^r(t) = F^r(x)\tilde{A} + G^r(y)\tilde{B} + Ke^r, \\ D_*^{\alpha}e^i(t) = F^i(x)\tilde{A} + G^i(y)\tilde{B} + Ke^i, \end{cases}$$

where $e^r(t) = (e_1^r(t), e_2^r(t), \ldots, e_n^r(t))^T$ and $e^i(t) = (e_1^i(t), e_2^i(t), \ldots, e_n^i(t))^T$. Moreover, we have

$$\begin{cases} D_*^{\alpha}e_l^r(t) = F_l^r(x)\tilde{A} + G_l^r(y)\tilde{B} + k_l e_l^r, \\ D_*^{\alpha}e_l^i(t) = F_l^i(x)\tilde{A} + G_l^i(y)\tilde{B} + k_l e_l^i, \end{cases}$$

where $F_l^r(x)$, $F_l^i(x)$, $G_l^r(y)$, and $G_l^i(y)$ are the l-th row of $F^r(x)$, $F^i(x)$, $G^r(y)$, and $G^i(y)$, respectively. Consider the following positive definite Lyapunov candidate function:

$$V(e,t) = \frac{1}{2}\sum_{l=1}^{n}\left((e_l^r(t))^2 + (e_l^i(t))^2\right) + \frac{1}{2}\sum_{l=1}^{n}\frac{1}{\delta_l}(k_l + L)^2 + \frac{1}{2}\tilde{A}^T\tilde{A} + \frac{1}{2}\tilde{B}^T\tilde{B},$$

where L is a positive constant. Then,

$$
\begin{aligned}
D_*^{\alpha} V \leq & \sum_{l=1}^{n} \left(e_l^r D_*^{\alpha} e_l^r + e_l^i D_*^{\alpha} e_l^i \right) + \sum_{l=1}^{n} \frac{1}{\delta_l} (k_l + L) D_*^{\alpha} k_l + \tilde{A}^T D_*^{\alpha} \tilde{A} + \tilde{B}^T D_*^{\alpha} \tilde{B} \\
\leq & \sum_{l=1}^{n} \left(e_l^r (F_l^r(x)\tilde{A} + G_l^r(y)\tilde{B} + k_l e_l^r) \right) + \sum_{l=1}^{n} \left(e_l^i (F_l^i(x)\tilde{A} + G_l^i(y)\tilde{B} + k_l e_l^i) \right) \\
& - \sum_{l=1}^{n} (k_l + L) \left((e_l^r)^2 + (e_l^i)^2 \right) + \tilde{A}^T D_*^{\alpha} \tilde{A} + \tilde{B}^T D_*^{\alpha} \tilde{B} \\
= & \sum_{l=1}^{n} \left(e_l^r F_l^r(x)\tilde{A} + e_l^i F_l^i(x)\tilde{A} \right) + \sum_{l=1}^{n} \left(e_l^r G_l^r(y)\tilde{B} + e_l^i G_l^i(y)\tilde{B} \right) + \tilde{A}^T D_*^{\alpha} \tilde{A} \\
& + \tilde{B}^T D_*^{\alpha} \tilde{B} - L \sum_{l=1}^{n} \left((e_l^r)^2 + (e_l^i)^2 \right).
\end{aligned}
$$

Based on the adaptive laws (13.4.11) and

$$
\begin{cases}
\sum_{l=1}^{n} \left(e_l^r F_l^r(x)\tilde{A} + e_l^i F_l^i(x)\tilde{A} \right) = \tilde{A}^T \left((F^r(x))^T e^r + (F^i(x))^T e^i \right), \\
\sum_{l=1}^{n} \left(e_l^r G_l^r(y)\tilde{B} + e_l^i G_l^i(y)\tilde{B} \right) = \tilde{B}^T \left((G^r(y))^T e^r + (G^i(y))^T e^i \right),
\end{cases}
$$

we deduce that

$$
D_*^{\alpha} V \leq -L \left((e^r)^T e^r + (e^i)^T e^i \right).
$$

The rest of the proof is similar to that of Theorem 13.2 and so is omitted. This completes the proof. □

Thus, as a consequence of Theorem 13.4, the following corollary can be obtained.

Corollary 13.1 *If the structures of general fractional-order chaotic complex systems* (13.4.1) *and* (13.4.9) *are identical, i.e.,* $F(\cdot) = G(\cdot)$, $f(\cdot) = g(\cdot)$, *and* $A = B$, *the controller is constructed as*

$$
\begin{aligned}
U = U^r + jU^i &= -(F(y) + F(x))\hat{A} - f(y) - f(x) + Ke \\
&= -\left(F^r(y) + F^r(x) \right) \hat{A} - f^r(y) - f^r(x) + Ke^r \\
&\quad - j \left((F^i(y) + F^i(x))\hat{A} + f^i(y) + f^i(x) - Ke^i \right),
\end{aligned}
$$

and the adaptive laws are chosen as

$$
\begin{cases}
D_*^{\alpha} \hat{A} = (F^r(x))^T e^r + (F^i(x))^T e^i, \\
D_*^{\alpha} k_l = -\delta_l \left((e_l^r(t))^2 + (e_l^i(t))^2 \right), \quad l = 1, 2, \ldots, n,
\end{cases}
$$

then anti-synchronization of two identical general fractional-order chaotic complex systems is also achieved, where $\hat{A}$ *is the estimated value of the unknown parameter vector* A, $K = \mathrm{diag}(k_1, k_2, \ldots, k_n)$ *is the real control strength matrix,* $\delta = \mathrm{diag}(\delta_1, \delta_2, \ldots, \delta_n)$ *is a convergence factor matrix with* $\delta_l > 0\ (l = 1, 2, \ldots, n)$.

The following presents three cases with varying states of unknown parameters. When A is a real vector and B is a complex vector, we obtain the adaptive anti-synchronization scheme between systems (13.4.8) and (13.4.9) as follows.

Theorem 13.5 *If the control law is selected as*

$$\begin{aligned}U &= U^r + jU^i = -G(y)\hat{B}^r - Q(y)\hat{B}^i - g(y) - F(x)\hat{A} - f(x) + Ke\\ &= -G^r(y)\hat{B}^r - Q^r(y)\hat{B}^i - g^r(y) - F^r(x)\hat{A} - f^r(x) + Ke^r\\ &\quad - j\left(G^i(y)\hat{B}^r + Q^i(y)\hat{B}^i + g^i(y) + F^i(x)\hat{A} + f^i(x) - Ke^i\right),\end{aligned}$$

and the adaptive laws are chosen as

$$\begin{cases} D_*^\alpha \hat{A} = (F^r(x))^T e^r + (F^i(x))^T e^i,\\ D_*^\alpha \hat{B}^r = (G^r(y))^T e^r + (G^i(y))^T e^i, \quad D_*^\alpha \hat{B}^i = (Q^r(y))^T e^r + (Q^i(y))^T e^i,\\ D_*^\alpha k_l = -\delta_l\left((e_l^r(t))^2 + (e_l^i(t))^2\right), \quad l = 1, 2, \ldots, n,\end{cases}$$

then the response system (13.4.8) *can anti-synchronize the drive system* (13.4.9), *where $\hat{A}$, $\hat{B}^r$, and $\hat{B}^i$ denote the estimated values of the unknown parameter vectors A, B^r, and B^i, respectively, $K = \text{diag}(k_1, k_2, \ldots, k_n)$ represents the real control strength matrix, $\delta = \text{diag}(\delta_1, \delta_2, \ldots, \delta_n)$ is a convergence factor matrix with $\delta_l > 0$ $(l = 1, 2, \ldots, n)$.*

Proof The proof is similar to that of Theorem 13.4 and thus is omitted. □

When A is a complex vector and B a real vector, the response system is depicted in (13.4.1), and the drive system (13.4.9) can be expressed as:

$$D_*^\alpha x = F(x)A^r + jF(x)A^i + f(x) = F(x)A^r + P(x)A^i + f(x), \qquad (13.4.13)$$

where $P(x) = jF(x)$ is a new $n \times n$ complex matrix.

Theorem 13.6 *If the control law is designed as follows:*

$$\begin{aligned}U &= U^r + jU^i = -G(y)\hat{B} - g(y) - F(x)\hat{A}^r - P(x)\hat{A}^i - f(x) + Ke\\ &= -G^r(y)\hat{B} - g^r(y) - F^r(x)\hat{A}^r - P^r(x)\hat{A}^i - f^r(x) + Ke^r\\ &\quad - j\left(G^i(y)\hat{B} + g^i(y) + F^i(x)\hat{A}^r + P^i(x)\hat{A}^i + f^i(x) - Ke^i\right),\end{aligned}$$

and the adaptive laws are chosen as

$$\begin{cases} D_*^\alpha \hat{A}^r = (F^r(x))^T e^r + (F^i(x))^T e^i, \quad D_*^\alpha \hat{A}^i = (P^r(x))^T e^r + (P^i(x))^T e^i,\\ D_*^\alpha \hat{B} = (G^r(y))^T e^r + (G^i(y))^T e^i,\\ D_*^\alpha k_l = -\delta_l\left((e_l^r(t))^2 + (e_l^i(t))^2\right), \quad l = 1, 2, \ldots, n,\end{cases}$$

then the response system (13.4.1) *can anti-synchronize the drive system* (13.4.13), *where $\hat{A}^r$, $\hat{A}^i$, and $\hat{B}$ denote the estimated values of the unknown parameter vectors*

A^r, A^i, and B, respectively, $K = \mathrm{diag}(k_1, k_2, \ldots, k_n)$ represents the real control strength matrix, $\delta = \mathrm{diag}(\delta_1, \delta_2, \ldots, \delta_n)$ is a convergence factor matrix with $\delta_l > 0$ $(l = 1, 2, \ldots, n)$.

Proof The proof is similar to that of Theorem 13.5, and so is omitted. □

When A and B are complex vectors, the response system is equivalent to (13.4.8) and the drive system is (13.4.13). Then, the following anti-synchronization scheme between systems (13.4.8) and (13.4.13) can be obtained.

Theorem 13.7 *If the control law is chosen as follows:*

$$U = -G(y)\hat{B}^r - Q(y)\hat{B}^i - g(y) - F(x)\hat{A}^r - P(x)\hat{A}^i - f(x) + Ke,$$

and the adaptive laws are selected as

$$\begin{cases} D_*^\alpha \hat{A}^r = (F^r(x))^T e^r + (F^i(x))^T e^i, & D_*^\alpha \hat{A}^i = (P^r(x))^T e^r + (P^i(x))^T e^i, \\ D_*^\alpha \hat{B}^r = (G^r(y))^T e^r + (G^i(y))^T e^i, & D_*^\alpha \hat{B}^i = (Q^r(y))^T e^r + (Q^i(y))^T e^i, \\ D_*^\alpha k_l = -\delta_l \left((e_l^r(t))^2 + (e_l^i(t))^2\right), & l = 1, 2, \ldots, n, \end{cases}$$

then the response system (13.4.8) can anti-synchronize the drive system (13.4.13), where $\hat{A}^r$, $\hat{A}^i$, $\hat{B}^r$, and $\hat{B}^i$ denotes the estimated values of the unknown parameter vectors A^r, A^i, B^r, and B^i, respectively, $K = \mathrm{diag}(k_1, k_2, \ldots, k_n)$ represents the real control strength matrix, $\delta = \mathrm{diag}(\delta_1, \delta_2, \ldots, \delta_n)$ is a convergence factor matrix with $\delta_l > 0$ $(l = 1, 2, \ldots, n)$.

Proof The proof is similar to that of Theorem 13.4, and therefore is omitted. □

13.4.3 Numerical Simulations

To theoretically demonstrate anti-synchronization between two different fractional-order chaotic complex systems with unknown parameters, it is assumed that the fractional-order complex Lorenz system drives the fractional-order complex Lü system. Thus, the drive system is given by

$$\begin{cases} D_*^\alpha x_1 = a_1(x_2 - x_1), \\ D_*^\alpha x_2 = a_2 x_1 - x_2 - x_1 x_3, \\ D_*^\alpha x_3 = \frac{1}{2}(\bar{x}_1 x_2 + x_1 \bar{x}_2) - a_3 x_3, \end{cases} \tag{13.4.14}$$

where $x_1 = x_1^r + jx_1^i$ and $x_2 = x_2^r + jx_2^i$ denote complex state variables, x_3 is a real state variable, and

$$F(x) = \begin{pmatrix} x_2 - x_1 & 0 & 0 \\ 0 & x_1 & 0 \\ 0 & 0 & -x_3 \end{pmatrix}, \quad f(x) = \begin{pmatrix} 0 \\ -x_2 - x_1 x_3 \\ \frac{1}{2}(\bar{x}_1 x_2 + x_1 \bar{x}_2) \end{pmatrix}, \quad A = \begin{pmatrix} a_1 \\ a_2 \\ a_3 \end{pmatrix}.$$

The response system is defined as

$$\begin{cases} D_*^\alpha y_1 = b_1(y_2 - y_1) + U_1, \\ D_*^\alpha y_2 = -y_1 y_3 + b_2 y_2 + U_2, \\ D_*^\alpha y_3 = \frac{1}{2}(\bar{y}_1 y_2 + y_1 \bar{y}_2) - b_3 y_3 + U_3, \end{cases} \tag{13.4.15}$$

where y_1 and y_2 represent complex state variables, y_3 is a real state variable, $U_1 = u_1 + ju_2$, $U_2 = u_3 + ju_4$, $U_3 = u_5$ are control functions to be determined in the later discussion, and

$$G(y) = \begin{pmatrix} y_2 - y_1 & 0 & 0 \\ 0 & y_2 & 0 \\ 0 & 0 & -y_3 \end{pmatrix}, g(y) = \begin{pmatrix} 0 \\ -y_1 y_3 \\ \frac{1}{2}(\bar{y}_1 y_2 + y_1 \bar{y}_2) \end{pmatrix}, B = \begin{pmatrix} b_1 \\ b_2 \\ b_3 \end{pmatrix}.$$

Based on Theorem 13.4, the controller is constructed as follows

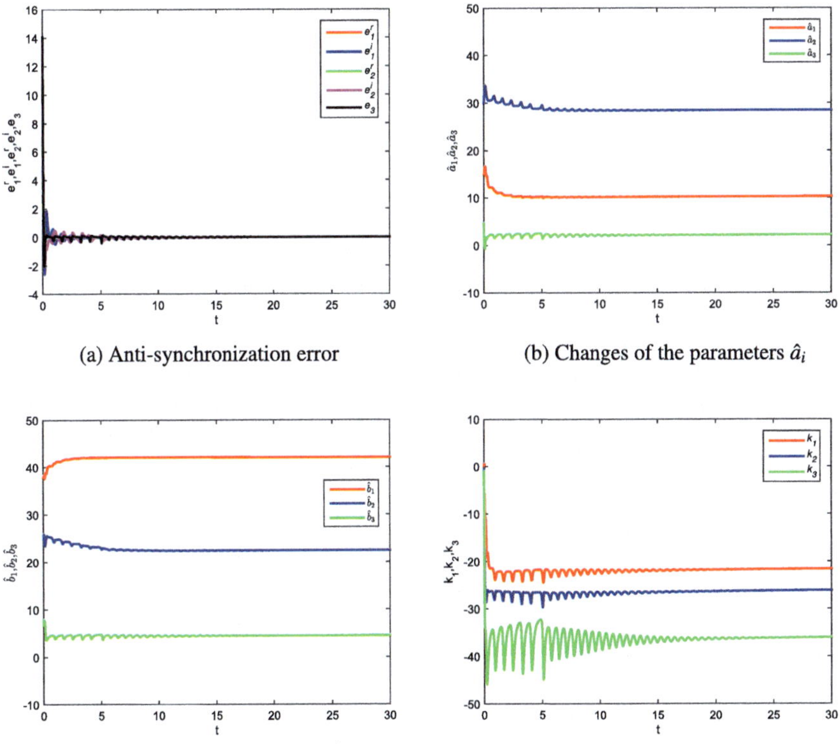

(a) Anti-synchronization error

(b) Changes of the parameters $\hat{a}_i$

(c) Changes of the parameters $\hat{b}_i$

(d) Changes of the parameters k_i

Fig. 13.13 Simulation results of anti-synchronization between systems (13.4.14) and (13.4.15) with $\delta_1 = 8$, $\delta_2 = 5$, and $\delta_3 = 6$

$$
\begin{cases}
u_1 = -\hat{b}_1(y_2^r - y_1^r) - \hat{a}_1(x_2^r - x_1^r) + k_1 e_1^r,\\
u_2 = -\hat{b}_1(y_2^i - y_1^i) - \hat{a}_1(x_2^i - x_1^i) + k_1 e_1^i,\\
u_3 = y_1^r y_3 - \hat{b}_2 y_2^r + x_2^r + x_1^r x_3 - \hat{a}_2 x_1^r + k_2 e_2^r,\\
u_4 = y_1^i y_3 - \hat{b}_2 y_2^i + x_2^i + x_1^i x_3 - \hat{a}_2 x_1^i + k_2 e_2^i,\\
u_5 = -y_1^r y_2^r - y_1^i y_2^i + \hat{b}_3 y_3 - x_1^r x_2^r - x_1^i x_2^i + \hat{a}_3 x_3 + k_3 e_3,
\end{cases}
$$

and the adaptive laws of parameters are chosen as

$$
\begin{cases}
D_*^\alpha \hat{a}_1 = (x_2^r - x_1^r)e_1^r + (x_2^i - x_1^i)e_1^i,\\
D_*^\alpha \hat{a}_2 = x_1^r e_2^r + x_1^i e_2^i,\\
D_*^\alpha \hat{a}_3 = -x_3 e_3,
\end{cases}
\quad
\begin{cases}
D_*^\alpha \hat{b}_1 = (y_2^r - y_1^r)e_1^r + (y_2^i - y_1^i)e_1^i,\\
D_*^\alpha \hat{b}_2 = y_2^r e_2^r + y_2^i e_2^i,\\
D_*^\alpha \hat{b}_3 = -y_3 e_3,
\end{cases}
$$

and

$$
\begin{cases}
D_*^\alpha k_1 = -\delta_1[(e_1^r)^2 + (e_1^i)^2],\\
D_*^\alpha k_2 = -\delta_2[(e_2^r)^2 + (e_2^i)^2],\\
D_*^\alpha k_3 = -\delta_3 e_3^2.
\end{cases}
$$

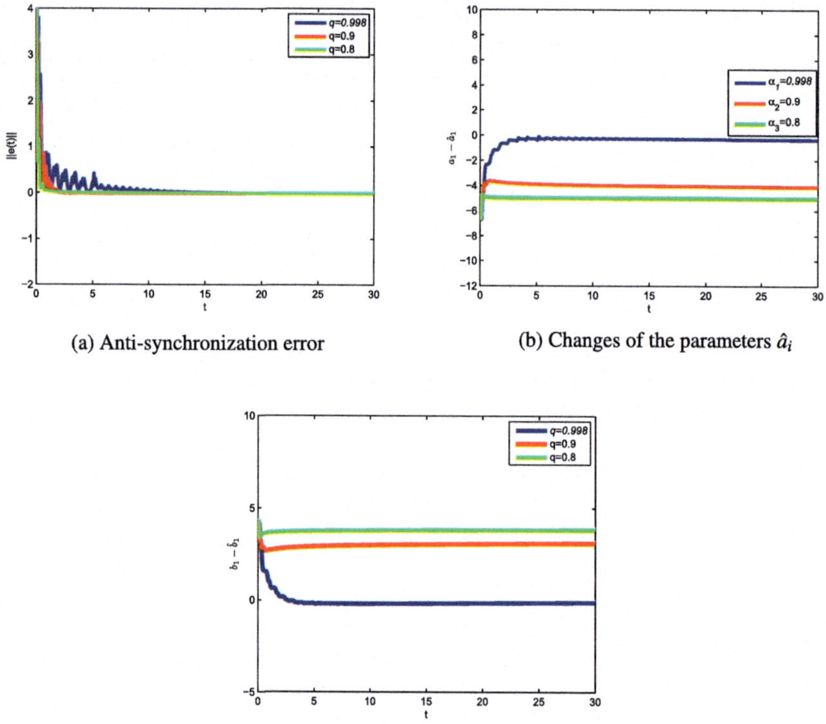

(a) Anti-synchronization error

(b) Changes of the parameters $\hat{a}_i$

(c) Changes of the parameters $\hat{b}_i$

Fig. 13.14 The related changes caused by different fractional order

Let $\alpha = 0.98$, $A = (10, 28, 8/3)^T$ and $B = (42, 22, 5)^T$. The initial values of the drive and response systems are chosen as $x(0) = (2 + 3j, 5 + 6j, 9)^T$ and $y(0) = (6 + 9j, 5 + 7j, 12)^T$. Furthermore, the initial values of the estimated parameters and control strength are set to $\hat{A}(0) = (15, 30, 5)^T$, $\hat{B}(0) = (38, 25, 8)^T$, and $K = \text{diag}(1, 1, 1)$. Figure 13.13 demonstrates the adaptive anti-synchronization results between systems (13.4.14) and (13.4.15). As shown in Fig. 13.13(a), the anti-synchronization errors converge asymptotically to zero. Figures 13.13(b)–(c) display the estimates $\hat{a}_1$, $\hat{a}_2$, $\hat{a}_3$ and $\hat{b}_1$, $\hat{b}_2$, $\hat{b}_3$ of the unknown parameters. Figure 13.13(d) clearly shows that the control coefficients k_l $(l = 1, 2, 3)$ converge to some constants as $t \to \infty$, respectively. As expected, the above results demonstrate the anti-synchronization has been achieved between two systems (13.4.14) and (13.4.15) with unknown parameters.

Remark 13.2 The fractional order α has a direct effect not only on the chaotic behavior of fractional-order nonlinear dynamical systems but also on the synchronization behavior of fractional-order systems. In this paper, we provide a more visible method to examine α's impact on the fractional-order chaotic complex systems with unknown parameters. To this end, we consider three cases including $\alpha_1 = 0.998$, $\alpha_2 = 0.9$, and $\alpha_3 = 0.8$. All of the other settings remain unchanged as stated in the above numerical example. From Fig. 13.14, we can conclude that the small value of the fractional order α would be harmful for identification of the unknown parameters although it has the beneficial effect on the synchronization state.

Chapter 14
Complex Modified Projective Synchronization of Fractional-Order Chaotic Complex Systems

Complex modified projective synchronization (CMPS) is a new type of complex synchronization based on complex chaotic systems that was proposed almost simultaneously in 2013 [163, 164]. CMPS means that state variables of the master system converge to state variables of the slave system with a complex constant scaling matrix [224, 225].

In this chapter, we first investigates CMPS of the commensurate fractional-order chaotic complex systems by combining stability results of the fractional-order systems and the pole placement method. Subsequently, CMPS of fractional-order chaotic complex systems with incommensurate orders is investigated based on the stability theory of incommensurate fractional-order systems and active control method.

14.1 CMPS of the Commensurate Fractional-Order Chaotic Complex Systems

In this section, CMPS of the commensurate fractional-order chaotic complex systems is investigated by designing a state observer. By means of the stability results of the fractional-order systems and the pole placement method, this paper proves the stability of fractional-order error systems and realizes CMPS. Finally, two numerical examples are studied to show the correctness of this new synchronization strategy.

14.1.1 Problem Description and Synchronization Scheme

Generally speaking, a fractional-order chaotic complex systems can be divided into two major parts: one is linear and the other is nonlinear. Accordingly, we investigate the following fractional-order chaotic complex systems:

S. Liu et al., *Chaos in Complex Dynamical System: Theory and Application*,
https://doi.org/10.1007/978-981-95-7762-0_14

$$D_*^\alpha x = Ax + Bf(x) + C, \tag{14.1.1}$$

where the state vector $x = (x_1, x_2, \ldots, x_n)^T = x^r + jx^i \in \mathbb{C}^n$, $A \in \mathbb{R}^{n\times n}$ and $B \in \mathbb{R}^{n\times m}$ are real matrices, $C \in \mathbb{R}^{n\times 1}$(or $C \in \mathbb{C}^{n\times 1}$), and $f = (f_1, f_2, \ldots, f_m)^T$ (f_i stands for complex nonlinear function) are column vectors.

To achieve CMPS, we regard system (14.1.1) as the master system, which is assumed to have the following output

$$y = f(x) + Kx, \tag{14.1.2}$$

where $K \in \mathbb{R}^{n\times m}$ is a gain matrix. Given an invertible constant matrix $\Phi = \Phi^r + j\Phi^i \in \mathbb{C}^{n\times n}$, its inverse matrix is denoted by Φ^{-1}. We design the slave system as an observer of the fractional-order chaotic complex system (14.1.1)

$$D_*^\alpha \hat{x} = \Phi^{-1}A\Phi\hat{x} + \Phi^{-1}Bf(\hat{x}) + \Phi^{-1}C + \Phi^{-1}B(y - \hat{y}), \tag{14.1.3}$$

with its output given by:

$$\hat{y} = f(\hat{x}) + K\Phi\hat{x}. \tag{14.1.4}$$

The synchronization error function is defined as

$$e(t) = x(t) - \Phi\hat{x}(t), \tag{14.1.5}$$

where Φ is called a complex scaling matrix.

Therefore, the definition of CMPS for systems (14.1.1) and (14.1.3) can now be stated as follows.

Definition 14.1 CMPS between systems (14.1.1) and (14.1.3) can be realized if there is a complex matrix $\Phi = \Phi^r + j\Phi^i = \text{diag}\{\varphi_1, \varphi_2, \ldots, \varphi_n\} \in \mathbb{C}^{n\times n}$ satisfying

$$\lim_{t\to\infty} ||e(t)||^2 = \lim_{t\to\infty} ||x(t) - \Phi\hat{x}(t)||^2 = 0,$$

where $\varphi_l \in \mathbb{C}$ ($l = 1, 2, \ldots, n$) is a constant.

We now analyze the process of CMPS based on a nonlinear state observer. Based on the error Eq. (14.1.5), we have

$$D_*^\alpha e = D_*^\alpha x - \Phi D_*^\alpha \hat{x}.$$

Incorporating the system (14.1.1) and the observer (14.1.3) gives

$$D_*^\alpha e = Ae + B(f(x) - f(\hat{x})) - B(y(t) - \hat{y}(t)).$$

Substituting (14.1.4) into the above equation, one can conclude that

$$D_*^\alpha e = (A - BK)e.$$

Separating real and imaginary parts, we have two real systems as follows:

$$D_*^\alpha e^r = (A - BK)e^r, \tag{14.1.6}$$

and

$$D_*^\alpha e^i = (A - BK)e^i, \tag{14.1.7}$$

where $A - BK$ is the linear time invariant matrix. For the sake of making systems (14.1.6) and (14.1.7) controllable, we choose the appropriate gain matrix K to satisfy $|arg(\lambda_l(A - BK))| > \alpha\pi/2$, $(l = 1, 2, \ldots, n)$. It follows from Theorem 1.3 that $e^r \to 0$ and $e^i \to 0$ as $t \to \infty$, which implies $e = e^r + je^i \to 0$ as $t \to \infty$. Thus, system (14.1.1) and the observer (14.1.3) can realize CMPS.

Remark 14.1 The eigenvalues of matrix $A - BK$ are independent of the complex scaling matrix Φ, so the complex scaling matrix Φ does not affect the controllability of the error systems (14.1.6) and (14.1.7). Therefore, the proposed method can arbitrarily adjust the complex scaling matrix in the synchronization process without worrying about the robustness of other synchronization methods.

Remark 14.2 In this synchronization scheme, we apply the pole placement method to determine the feedback gain matrix K satisfying $|\arg(\lambda_l(A - BK))| > \alpha\pi/2\,(l = 1, 2, \ldots, n)$.

14.1.2 Numerical Simulations

Next, the proposed theory is validated by investigating the CMPS of the following two cases.

I. CMPS of the fractional-order complex Lü systems

The master system is given by the following fractional-order complex Lü system

$$\begin{cases} D_*^\alpha x_1 = a(x_2 - x_1), \\ D_*^\alpha x_2 = -x_1x_3 + bx_2, \\ D_*^\alpha x_3 = \frac{1}{2}(\bar{x}_1x_2 + x_1\bar{x}_2) - cx_3, \end{cases} \tag{14.1.8}$$

where a, b, c are real constants.

Comparing system (14.1.8) with system (14.1.1), we easily have

$$A = \begin{pmatrix} -a & a & 0 \\ 0 & b & 0 \\ 0 & 0 & -c \end{pmatrix}, B = \begin{pmatrix} 0 & 0 \\ 1 & 0 \\ 0 & 1 \end{pmatrix}, C = \begin{pmatrix} 0 \\ 0 \\ 0 \end{pmatrix}, f(x) = \begin{pmatrix} -x_1x_3 \\ \frac{1}{2}(\bar{x}_1x_2 + x_1\bar{x}_2) \end{pmatrix}.$$

Let $\Phi = \text{diag}(\varphi_1, \varphi_2, \varphi_3)$ be an invertible complex scaling matrix. Then the inverse matrix is computed as $\Phi^{-1} = \text{diag}(\varphi_1^{-1}, \varphi_2^{-1}, \varphi_3^{-1})$. Assume the output of system (14.1.8) is $y = f(x) + Kx$, where $K \in \mathbb{R}^{2\times 3}$ is a gain matrix. Therefore, we design the following state observer of system (14.1.8):

$$\begin{aligned} \begin{pmatrix} D_*^\alpha \hat{x}_1 \\ D_*^\alpha \hat{x}_2 \\ D_*^\alpha \hat{x}_3 \end{pmatrix} &= \begin{pmatrix} -a & a\Phi_1^{-1}\Phi_2 & 0 \\ 0 & b & 0 \\ 0 & 0 & -c \end{pmatrix} \begin{pmatrix} \hat{x}_1 \\ \hat{x}_2 \\ \hat{x}_3 \end{pmatrix} \\ &+ \begin{pmatrix} \Phi_1^{-1} & 0 & 0 \\ 0 & \Phi_2^{-1} & 0 \\ 0 & 0 & \Phi_3^{-1} \end{pmatrix} \begin{pmatrix} 0 & 0 \\ 1 & 0 \\ 0 & 1 \end{pmatrix} \begin{pmatrix} -\hat{x}_1\hat{x}_3 \\ \frac{1}{2}(\bar{\hat{x}}_1\hat{x}_2 + \hat{x}_1\bar{\hat{x}}_2) \end{pmatrix} \\ &+ \begin{pmatrix} \Phi_1^{-1} & 0 & 0 \\ 0 & \Phi_2^{-1} & 0 \\ 0 & 0 & \Phi_3^{-1} \end{pmatrix} \begin{pmatrix} 0 & 0 \\ 1 & 0 \\ 0 & 1 \end{pmatrix} (y(t) - \hat{y}(t)). \end{aligned} \tag{14.1.9}$$

The CMPS error is expressed as $e(t) = x(t) - \Phi\hat{x}(t)$. By means of system (14.1.8) and the observer (14.1.9), we have the following error dynamical system:

$$D_*^\alpha e = (A - BK)e = \left(\begin{pmatrix} -a & a & 0 \\ 0 & b & 0 \\ 0 & 0 & -c \end{pmatrix} - \begin{pmatrix} 0 & 0 \\ 1 & 0 \\ 0 & 1 \end{pmatrix} K \right) e.$$

Assume that the eigenvalues of matrix $A - BK$ are assigned as $(-42, -2, -10)$, satisfying $|\arg(\lambda_l(A - BK))| > \alpha\pi/2, (l = 1, 2, 3)$. Based on the pole placement method, the feedback matrix K can be computed as

$$K = \begin{pmatrix} 0 & 24 & 0 \\ 0 & 0 & 5 \end{pmatrix}.$$

The simulation results is performed using the Adams–Bashforth–Moulton predictor–corrector scheme under the initial condition $x(0) = (10 + 6j, 7 + 8j, 5)^T$, $\hat{x}(0) = (2 + 9j, 5 + 10j, 2)^T$. With the parameters $\alpha = 0.96$, $\Phi = \text{diag}(1 + j, 1 - j, 1)$, the corresponding inverse matrix is $\Phi^{-1} = \text{diag}((1 - j)/2, (1 + j)/2, 1)$. Figures 14.1(a)–(e) describe the state evolution of system (14.1.8) and the observer (14.1.9). Figure 14.1(f) illustrates the state trajectories of the error system, which can be observed to converge rapidly to zero. This confirms the realization of CMPS for fractional-order complex Lü systems via a nonlinear state observer.

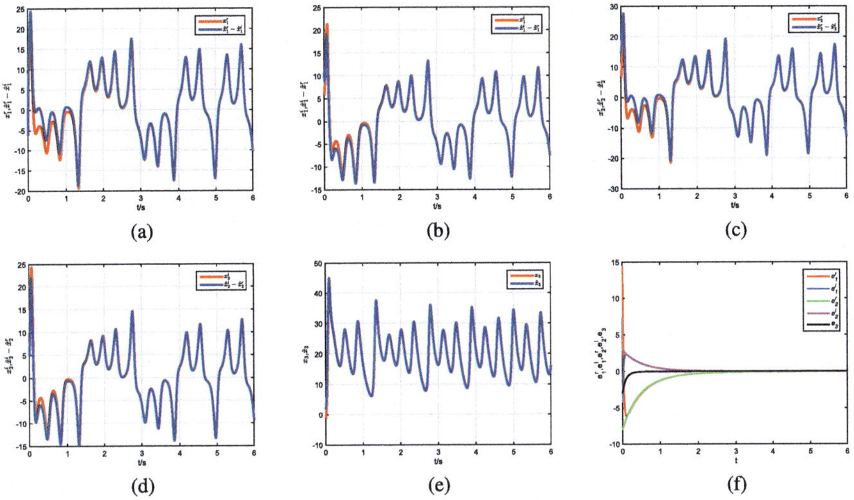

Fig. 14.1 State variables and synchronization error of drive system (14.1.7) and response (14.1.8)

II. CMPS of the fractional-order hyper-chaotic complex Lü system

The fractional-order hyper-chaotic complex Lü system was first introduced by Yang and Jiang [226], and is described as

$$\begin{cases} D_*^\alpha x_1 = \beta(x_2 - x_1) + x_4, \\ D_*^\alpha x_2 = \gamma x_2 - x_1 x_3 + x_4, \\ D_*^\alpha x_3 = \frac{1}{2}(\bar{x}_1 x_2 + x_1 \bar{x}_2) - \delta x_3, \\ D_*^\alpha x_4 = \frac{1}{2}(\bar{x}_1 x_2 + x_1 \bar{x}_2) - \sigma x_4, \end{cases} \tag{14.1.10}$$

where $\beta, \gamma, \delta, \sigma$ are real constants, and x_1, x_2, x_3, and x_4 are state variables. When $\alpha = 0.95$, $\beta = 42$, $\gamma = 25$, $\delta = 6$, and $\sigma = 5$, system (14.1.10) generates chaotic behavior (see Fig. 14.2).

Comparing system (14.1.10) with system (14.1.1), we easily obtain

$$A = \begin{pmatrix} -\beta & \beta & 0 & 1 \\ 0 & \gamma & 0 & 1 \\ 0 & 0 & -\delta & 0 \\ 0 & 0 & 0 & -\sigma \end{pmatrix}, \quad B = \begin{pmatrix} 0 & 0 & 0 \\ 1 & 0 & 0 \\ 0 & 1 & 0 \\ 0 & 0 & 1 \end{pmatrix},$$

$$C = \begin{pmatrix} 0 \\ 0 \\ 0 \\ 0 \end{pmatrix}, \quad f(x) = \begin{pmatrix} -x_1 x_3 \\ \frac{1}{2}(\bar{x}_1 x_2 + x_1 \bar{x}_2) \\ \frac{1}{2}(\bar{x}_1 x_2 + x_1 \bar{x}_2) \end{pmatrix}.$$

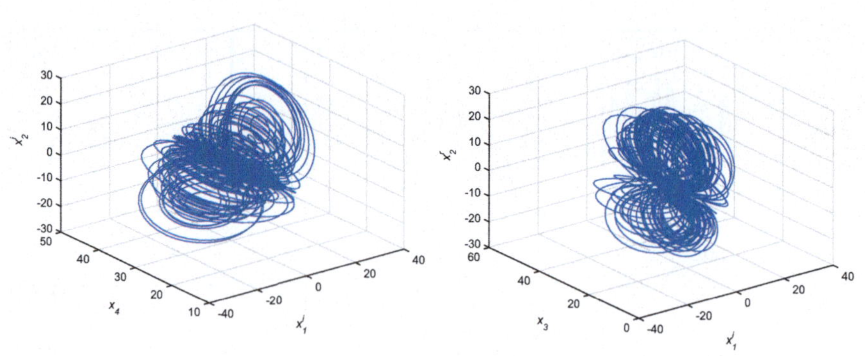

Fig. 14.2 The projection of chaotic attractor for system (14.1.10) with $\alpha = 0.95$, $\beta = 42$, $\gamma = 25$, $\delta = 6$, and $\sigma = 5$

Suppose that an invertible complex scaling matrix $\Phi = \mathrm{diag}(\varphi_1, \varphi_2, \varphi_3, \varphi_4)$. Then the inverse matrix is computed as $\Phi^{-1} = \mathrm{diag}(\varphi_1^{-1}, \varphi_2^{-1}, \varphi_3^{-1}, \varphi_4^{-1})$. Furthermore, the output of system (14.1.10) is assumed to be $y = f(x) + Kx$, where $K \in \mathbb{R}^{3\times 4}$ is a gain matrix. Thus, design the state observer of system (14.1.10) in the following form

$$\begin{pmatrix} D_*^\alpha \hat{x}_1 \\ D_*^\alpha \hat{x}_2 \\ D_*^\alpha \hat{x}_3 \\ D_*^\alpha \hat{x}_4 \end{pmatrix} = \begin{pmatrix} -\beta & \beta\varphi_1^{-1}\varphi_2 & 0 & \varphi_1^{-1}\varphi_4 \\ 0 & \gamma & 0 & \varphi_2^{-1}\varphi_4 \\ 0 & 0 & -\delta & 0 \\ 0 & 0 & 0 & -\sigma \end{pmatrix} \begin{pmatrix} \hat{x}_1 \\ \hat{x}_2 \\ \hat{x}_3 \\ \hat{x}_4 \end{pmatrix} \tag{14.1.11}$$

$$+ \begin{pmatrix} \varphi_1^{-1} & 0 & 0 & 0 \\ 0 & \varphi_2^{-1} & 0 & 0 \\ 0 & 0 & \varphi_3^{-1} & 0 \\ 0 & 0 & 0 & \varphi_4^{-1} \end{pmatrix} \begin{pmatrix} 0 & 0 & 0 \\ 1 & 0 & 0 \\ 0 & 1 & 0 \\ 0 & 0 & 1 \end{pmatrix} \begin{pmatrix} -\hat{x}_1\hat{x}_3 \\ \frac{1}{2}(\bar{\hat{x}}_1\hat{x}_2 + \hat{x}_1\bar{\hat{x}}_2) \\ \frac{1}{2}(\bar{\hat{x}}_1\hat{x}_2 + \hat{x}_1\bar{\hat{x}}_2) \end{pmatrix}$$

$$+ \begin{pmatrix} \varphi_1^{-1} & 0 & 0 & 0 \\ 0 & \varphi_2^{-1} & 0 & 0 \\ 0 & 0 & \varphi_3^{-1} & 0 \\ 0 & 0 & 0 & \varphi_4^{-1} \end{pmatrix} \begin{pmatrix} 0 & 0 & 0 \\ 1 & 0 & 0 \\ 0 & 1 & 0 \\ 0 & 0 & 1 \end{pmatrix} (y(t) - \hat{y}(t)), \tag{14.1.12}$$

where $\hat{x}_1$ and $\hat{x}_2$ are complex variables, $\hat{x}_3$ and $\hat{x}_4$ are real variables.

The error is defined as $e(t) = x(t) - \Phi\hat{x}(t)$. By means of system (14.1.10) and the observer (14.1.11), we obtain the following error system

$$D_*^\alpha e = (A - BK)e = \left(\begin{pmatrix} -\beta & \beta & 0 & 1 \\ 0 & \gamma & 0 & 1 \\ 0 & 0 & -\delta & 0 \\ 0 & 0 & 0 & -\sigma \end{pmatrix} - \begin{pmatrix} 0 & 0 & 0 \\ 1 & 0 & 0 \\ 0 & 1 & 0 \\ 0 & 0 & 1 \end{pmatrix} K \right) e.$$

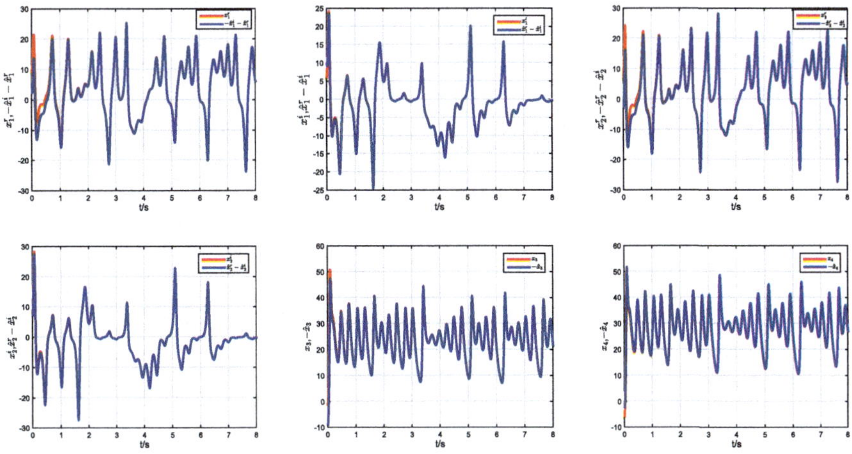

Fig. 14.3 The state evolution of system (14.1.10) and the observer (14.1.11)

The eigenvalues of matrix $A - BK$ are set to $(-42, -3, -8, -6)$, which satisfy $|\arg(\lambda_l(A - BK))| > \alpha\pi/2$, $(l = 1, 2, 3)$. Consequently, based on the pole placement method, we can calculate the gain matrix K as follows:

$$K = \begin{pmatrix} 0 & 28 & 0 & 0 \\ 0 & 0 & 2 & 0 \\ 0 & 0 & 5 & 1 \end{pmatrix}.$$

Simulation results are performed with the fractional derivative order $\alpha = 0.95$, and the initial condition $x(0) = (10 + 6j, 7 + 8j, -1, -6)^T$, $\hat{x}(0) = (2 - 8j, 5 - 2j, 9, 2)^T$. Choosing $\Phi = \text{diag}(-1 + j, -1 + j, -1, -1)$, the inverse matrix is computed as $\Phi^{-1} = \text{diag}((-1 - j)/2, (-1 - j)/2, -1, -1)$. Figure 14.3 presents the state evolution of system (14.1.10) and the observer (14.1.11). As shown in Fig. 14.4, the error system exhibits rapid asymptotic convergence to zero. It is verified that CMPS of fractional-order hyper-chaos complex Lü systems can be achieved through a nonlinear state observer.

14.2 CMPS of the Incommensurate Fractional-Order Chaotic Systems

This section aims to investigate CMPS between fractional-order chaotic nonlinear systems with incommensurate orders. By integrating the stability theory of incommensurate fractional-order systems and active control method, we derive specific control laws to achieve CMPS in three representative cases including fractional-order complex Lorenz system driving complex Chen system, fractional-order real

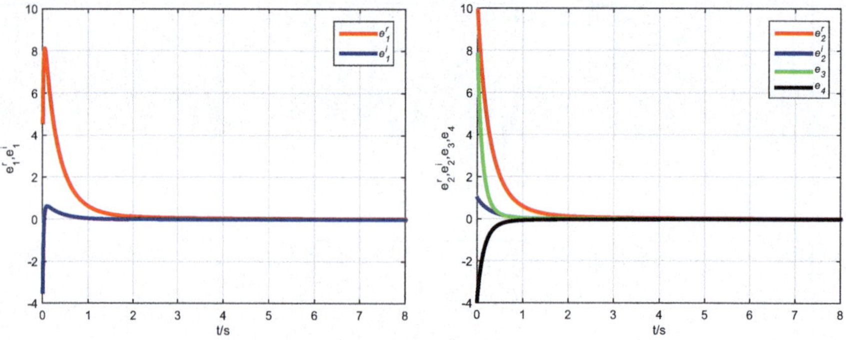

Fig. 14.4 The state trajectories of the error system

Rössler system [227] driving complex Chen system, and fractional-order complex Lorenz system driving real Lü system.

14.2.1 CMPS of Incommensurate Fractional-Order Chaotic Complex Systems

I. Mathematical model and problem descriptions

Let a fractional-order chaotic complex drive and response systems be respectively defined as follows. The drive system is

$$D_*^{\alpha} z = Cz + p(z), \tag{14.2.1}$$

and the response system is given by

$$D_*^{\alpha} y = By + g(y) + U, \tag{14.2.2}$$

where $\alpha = [\alpha_1, \alpha_2, \ldots, \alpha_n]^T$ denotes the fractional orders, $z = z^r + jz^i$, $y = y^r + jy^i \in \mathbf{C}^n$ are the complex state vectors of systems (14.2.1) and (14.2.2), respectively; $p = (p_1, p_2, \ldots, p_n)^T$ and $g = (g_1, g_2, \ldots, g_n)^T$ are vectors of nonlinear complex functions; $C, B \in \mathbb{R}^{n\times n}$ are two real matrices of system parameters; U is a nonlinear controller, $U = (U_1, U_2, \ldots, U_n)^T = U^r + jU^i$, $U^r = (u_1, u_3, \ldots, u_{2n-1})^T$, $U^i = (u_2, u_4, \ldots, u_{2n})^T$.

Based on two fractional-order chaotic complex systems with incommensurate orders, we introduce the definition of CMPS as follows:

Definition 14.2 For the drive system (14.2.1) and the response system (14.2.2), it is said to be CMPS if there exists a complex constant matrix $\Phi = \Phi^r + j\Phi^i \in \mathbf{C}^{n\times n}$ such that

$$\begin{aligned}\lim_{t\to\infty}||e||^2 &= \lim_{t\to\infty}||y-\Phi z||^2 \\ &= \lim_{t\to\infty}(||y^r-\Phi^r z^r+\Phi^i z^i||^2+||y^i-\Phi^r z^i-\Phi^i z^r||^2)=0,\end{aligned}$$

where $\Phi^r = \text{diag}\{\varphi_1, \varphi_3, \ldots, \varphi_{2n-1}\}, \Phi^i = \text{diag}\{\varphi_2, \varphi_4, \ldots, \varphi_{2n}\}, \varphi_1 + j\varphi_2, \varphi_3 + j\varphi_4, \ldots, \varphi_{2n-1} + j\varphi_{2n}$ are called complex scaling factors. $e = e^r + je^i$ is called the error vector between systems (14.2.1) and (14.2.2) for CMPS, $e^r = (e_1, e_3, \ldots, e_{2n-1})^T = y^r - \Phi^r z^r + \Phi^i z^i$, $e^i = (e_2, e_4, \ldots, e_{2n})^T = y^i - \Phi^r z^i - \Phi^i z^r$.

Control laws are now derived to achieve CMPS between two incommensurate fractional-order chaotic nonlinear systems. Subsequently, we consider that the fractional-order complex Lorenz system drives the fractional-order complex Chen system.

II. CMPS between the fractional-order complex Lorenz system and complex Chen system

The fractional-order complex Lorenz system is described by:

$$\begin{cases} D_*^{\alpha_1} z_1 = c_1(z_2 - z_1), \\ D_*^{\alpha_2} z_2 = c_2 z_1 - z_2 - z_1 z_3, \\ D_*^{\alpha_3} z_3 = \frac{1}{2}(\bar{z}_1 z_2 + z_1 \bar{z}_2) - c_3 z_3, \end{cases} \tag{14.2.3}$$

where z_1 and z_2 are complex variables, z_3 is a real variable; and c_i are real parameters $(i = 1, 2, 3)$.

The fractional-order complex Chen system reads as follows:

$$\begin{cases} D_*^{\alpha_1} y_1 = b_1(y_2 - y_1) + U_1, \\ D_*^{\alpha_2} y_2 = (b_2 - b_1) y_1 + b_2 y_2 - y_1 y_3 + U_2, \\ D_*^{\alpha_3} y_3 = \frac{1}{2}(\bar{y}_1 y_2 + y_1 \bar{y}_2) - b_3 y_3 + U_3, \end{cases} \tag{14.2.4}$$

here y_1 and y_2 are complex variables, y_3 is a real variable; b_i are real parameters $(i = 1, 2, 3)$; $U_1 = u_1 + ju_2, U_2 = u_3 + ju_4$ are complex functions and $U_3 = u_5$ is a real control function.

For the subsequent analysis, we assume $\Phi = diag(\varphi_1 + j\varphi_2, \varphi_3 + j\varphi_4, \varphi_5)$ and define the synchronization error as:

$$\begin{cases} e_1 = y_1^r - \varphi_1 z_1^r + \varphi_2 z_1^i, \\ e_2 = y_1^i - \varphi_1 z_1^i - \varphi_2 z_1^r, \\ e_3 = y_2^r - \varphi_3 z_2^r + \varphi_4 z_2^i, \\ e_4 = y_2^i - \varphi_4 z_2^r - \varphi_3 z_2^i, \\ e_5 = y_3 - \varphi_5 z_3. \end{cases}$$

Combining with systems (14.2.3) and (14.2.4), we obtain the error system as follows:

$$\begin{cases} D_*^{\alpha_1} e_1 = b_1 e_3 - b_1 e_1 + (b_1\varphi_3 - c_1\varphi_1) z_2^r - (b_1 - c_1)(\varphi_1 z_1^r - \varphi_2 z_1^i) \\ \quad +c_1\varphi_2 z_2^i - b_1\varphi_4 z_2^i + u_1, \\ D_*^{\alpha_1} e_2 = b_1 e_4 - b_1 e_2 + (b_1\varphi_4 - c_1\varphi_2) z_2^r - (b_1 - c_1)(\varphi_1 z_1^i + \varphi_2 z_1^r) \\ \quad -c_1\varphi_1 z_2^i + b_1\varphi_3 z_2^i + u_2, \\ D_*^{\alpha_2} e_3 = (b_2 - b_1) e_1 + [(b_2 - b_1)\varphi_1 - c_2\varphi_3] z_1^r - [(b_2 - b_1)\varphi_2 - c_2\varphi_4] z_1^i \\ \quad +b_2 e_3 - y_1^r y_3 + \varphi_3 z_1^r z_3 - \varphi_4 z_1^i z_3 + (b_2 + 1)(\varphi_3 z_2^r - \varphi_4 z_2^i) + u_3, \\ D_*^{\alpha_2} e_4 = (b_2 - b_1) e_2 + [(b_2 - b_1)\varphi_1 - c_2\varphi_3] z_1^i + [(b_2 - b_1)\varphi_2 - c_2\varphi_4] z_1^r \\ \quad +b_2 e_4 - y_1^i y_3 + \varphi_4 z_1^r z_3 + \varphi_3 z_1^i z_3 + (b_2 + 1)(\varphi_4 z_2^r + \varphi_3 z_2^i) + u_4, \\ D_*^{\alpha_3} e_5 = -b_3 e_5 + y_1^r y_2^r + y_1^i y_2^i - (b_3 - c_3)\varphi_5 z_3 - \varphi_5 z_1^r z_2^r - \varphi_5 z_1^i z_2^i + u_5. \end{cases}$$

Accordingly, the active control inputs $u_i(t)$ $(i = 1, 2, \ldots, 5)$ are defined as

$$\begin{cases} u_1(t) = -(b_1\varphi_3 - c_1\varphi_1) z_2^r - (c_1\varphi_2 - b_1\varphi_4) z_2^i + (b_1 - c_1)(\varphi_1 z_1^r - \varphi_2 z_1^i) + v_1(t), \\ u_2(t) = -(b_1\varphi_4 - c_1\varphi_2) z_2^r - (b_1\varphi_3 - c_1\varphi_1) z_2^i + (b_1 - c_1)(\varphi_1 z_1^i + \varphi_2 z_1^r) + v_2(t), \\ u_3(t) = -[(b_2 - b_1)\varphi_1 - c_2\varphi_3] z_1^r - \varphi_3 z_1^r z_3 - (b_2 + 1)(\varphi_3 z_2^r - \varphi_4 z_2^i) + \varphi_4 z_1^i z_3 \\ \quad +[(b_2 - b_1)\varphi_2 - c_2\varphi_4] z_1^i + y_1^r y_3 + v_3(t), \\ u_4(t) = -[(b_2 - b_1)\varphi_1 - c_2\varphi_3] z_1^i - \varphi_4 z_1^r z_3 - (b_2 + 1)(\varphi_4 z_2^r + \varphi_3 z_2^i) - \varphi_3 z_1^i z_3 \\ \quad -[(b_2 - b_1)\varphi_2 - c_2\varphi_4] z_1^r + y_1^i y_3 + v_4(t), \\ u_5(t) = -y_1^r y_2^r - y_1^i y_2^i + (b_3 - c_3)\varphi_5 z_3 + \varphi_5 z_1^r z_2^r + \varphi_5 z_1^i z_2^i + v_5(t), \end{cases}$$

which leads to

$$\begin{cases} D_*^{\alpha_1} e_1 = b_1 e_3 - b_1 e_1 + v_1(t), \\ D_*^{\alpha_1} e_2 = b_1 e_4 - b_1 e_2 + v_2(t), \\ D_*^{\alpha_2} e_3 = (b_2 - b_1) e_1 + b_2 e_3 + v_3(t), \\ D_*^{\alpha_2} e_4 = (b_2 - b_1) e_2 + b_2 e_4 + v_4(t), \\ D_*^{\alpha_3} e_5 = -b_3 e_5 + v_5(t). \end{cases} \tag{14.2.5}$$

To stabilize system (14.2.5), an appropriate feedback control is designed as

$$(v_1(t), v_2(t), v_3(t), v_4(t), v_5(t))^T = L(e_1(t), e_2(t), e_3(t), e_4(t), e_5(t))^T,$$

where $L = (l_{ij})_{5\times 5}$ is a 5×5 real matrix. Then the error system (14.2.5) is written as

$$D_*^{\alpha} e = \Xi e,$$

where

$$\Xi = \begin{pmatrix} l_{11} - b_1 & l_{12} & l_{13} + b_1 & l_{14} & l_{15} \\ l_{21} & l_{22} - b_1 & l_{23} & l_{24} + b_1 & l_{25} \\ l_{31} + b_2 - b_1 & l_{32} & l_{33} + b_2 & l_{34} & l_{35} \\ l_{41} & l_{42} + b_2 - b_1 & l_{43} & l_{44} + b_2 & l_{45} \\ l_{51} & l_{52} & l_{53} & l_{54} & l_{55} - b_3 \end{pmatrix}.$$

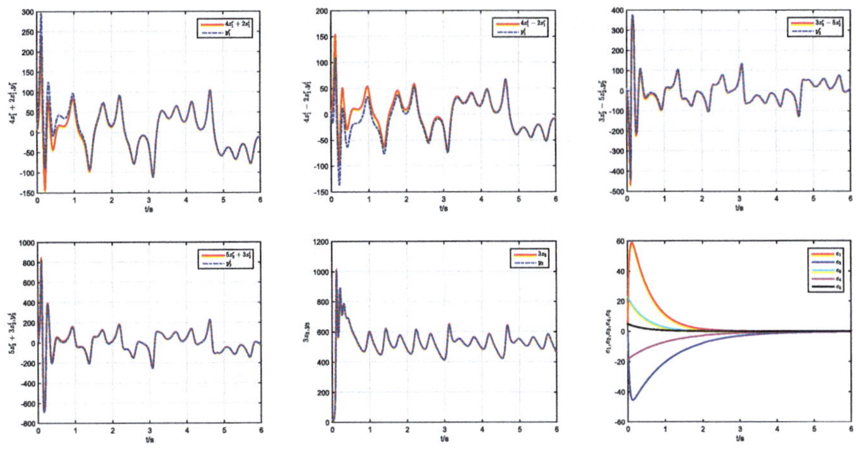

Fig. 14.5 State variables and errors of CMPS between systems (14.2.3) and (14.2.4)

To stabilize the error system, the matrix L should be chosen such that the resulting feedback system satisfies the conditions of Theorem 1.4. There is not a unique choice for such matrix L, a good choice can be as follows: $l_{11} = 15 < b_1, l_{22} = 10 < b_1, l_{33} = -30 < -b_2, l_{44} = -29 < -b_2, l_{55} = 1 < b_3, l_{31} = l_{42} = 7, l_{21} = l_{32} = l_{41} = l_{43} = l_{51} = l_{52} = l_{53} = l_{54} = 0$. Since $(b_1, b_2, b_3) = (35, 28, 3)$, $(\alpha_1, \alpha_2, \alpha_3) = (0.98, 0.99, 0.97)$, $m = 100$, $\gamma = 1/100$, then the characteristic equation can be obtained as

$$(\lambda^{98} + 20)(\lambda^{98} + 25)(\lambda^{99} + 2)(\lambda^{99} + 1)(\lambda^{97} + 2) = 0. \qquad (14.2.6)$$

It is clear that all roots of the (14.2.6) satisfy $|arg(\lambda)| > \gamma\pi/2$. According to Theorem 1.4, the error vector $e(t)$ asymptotically converges to zero as $t \to \infty$. Therefore, CMPS between the incommensurate fractional-order chaotic complex systems (14.2.3) and (14.2.4) is achieved.

In the numerical simulations, the initial values of systems (14.2.3) and (14.2.4) are set as $(z_1, z_2, z_3)^T = (1 + 2j, 3 + 4j, 5)^T, (y_1, y_2, y_3)^T = (10 + 8j, 12 + 9j, 20)^T$, parameters are selected as $(c_1, c_2, c_3) = (10, 180, 1)$, and a scaling matrix is taken as $\Phi = diag(4 - 2j, 3 + 5j, 3)$. Taking $l_{13} = -38, l_{15} = -1, l_{24} = -33, l_{12} = l_{14} = l_{23} = l_{25} = l_{34} = l_{35} = l_{45} = 0$, we obtain the CMPS error functions evolution by means of the Adams-Bashforth-Moulton predictor-correctors. Figure 14.5 displays all error states converge asymptotically to zero, i.e., CMPS between the incommensurate fractional-order complex Lorenz system and complex Chen system is achieved.

14.2.2 CMPS Between Fractional-Order Chaotic Real Drive Systems and Complex Response Systems

I. Mathematical model and problem descriptions

The fractional-order chaotic complex system (14.2.2) is regarded as the response system, and the fractional-order chaotic real drive system is given by:

$$D_*^{\alpha} x = Ax + f(x), \tag{14.2.7}$$

where $x = (x_1, x_2, \ldots, x_n)^T$ is a real state vector, $f = (f_1, f_2, \ldots, f_n)^T$ is the vector of nonlinear functions, and $A \in \mathbb{R}^{n\times n}$ is a real matrix of system parameters.

The error of CMPS with a complex scaling matrix $\Phi = \Phi^r + j\Phi^i \in \mathbf{C}^{n\times n}$ between systems (14.2.2) and (14.2.7) is described as:

$$e(t) = e^r(t) + je^i(t) = y - \Phi x,$$

i.e.

$$\begin{cases} e^r = y^r - \Phi^r x, \\ e^i = y^i - \Phi^i x. \end{cases}$$

Then, in the following, we design a controller to achieve CMPS between the fractional-order real Rössler system and complex Chen system with incommensurate orders.

II. CMPS between fractional-order real Rössler system and complex Chen system

The fractional-order complex Chen system is Eq. (14.2.4), while the fractional-order real Rössler system is depicted as follows:

$$\begin{cases} D_*^{\alpha_1} x_1 = -(x_2 + x_3), \\ D_*^{\alpha_2} x_2 = x_1 + a_1 x_2, \\ D_*^{\alpha_3} x_3 = a_2 + x_3(x_1 - a_3), \end{cases} \tag{14.2.8}$$

where x_i are real state variables and a_i are real parameters ($i = 1, 2, 3$). When $(a_1, a_2, a_3) = (0.4, 0.2, 10)$, $(\alpha_1, \alpha_2, \alpha_3) = (0.9, 0.91, 0.92)$, system (14.2.8) behaves chaotically.

With the scaling matrix $\Phi = diag(\varphi_1 + j\varphi_2, \varphi_3 + j\varphi_4, \varphi_5)$, the synchronization error is given by:

$$\begin{cases} e_1 = y_1^r - \varphi_1 x_1, \\ e_2 = y_1^i - \varphi_2 x_1, \\ e_3 = y_2^r - \varphi_3 x_2, \\ e_4 = y_2^i - \varphi_4 x_2, \\ e_5 = y_3 - \varphi_5 x_3. \end{cases}$$

By virtue of systems (14.2.4) and (14.2.8), we obtain the error system:

$$\begin{cases} D_*^{\alpha_1} e_1 = b_1 e_3 - b_1 e_1 + (b_1\varphi_3 + \varphi_1)x_2 - b_1\varphi_1 x_1 + \varphi_1 x_3 + u_1, \\ D_*^{\alpha_1} e_2 = b_1 e_4 - b_1 e_2 + (b_1\varphi_4 + \varphi_2)x_2 - b_1\varphi_2 x_1 + \varphi_2 x_3 + u_2, \\ D_*^{\alpha_2} e_3 = (b_2 - b_1)e_1 + b_2 e_3 + [(b_2 - b_1)\varphi_1 - \varphi_3]x_1 + (b_2 - a_1)\varphi_3 x_2 - y_1^r y_3 + u_3, \\ D_*^{\alpha_2} e_4 = (b_2 - b_1)e_2 + b_2 e_4 + [(b_2 - b_1)\varphi_2 - \varphi_4]x_1 + (b_2 - a_1)\varphi_4 x_2 - y_1^i y_3 + u_4, \\ D_*^{\alpha_3} e_5 = -b_3 e_5 + y_1^r y_2^r + y_1^i y_2^i + (a_3 - b_3)\varphi_5 x_3 - a_2\varphi_5 - \varphi_5 x_1 x_3 + u_5. \end{cases}$$

Active control functions $u_i(t)$ $(i = 1, 2, \ldots, 5)$ can be designed as

$$\begin{cases} u_1(t) = -(b_1\varphi_3 + \varphi_1)x_2 - \varphi_1 x_3 + b_1\varphi_1 x_1 + v_1(t), \\ u_2(t) = -(b_1\varphi_4 + \varphi_2)x_2 - \varphi_2 x_3 + b_1\varphi_2 x_1 + v_2(t), \\ u_3(t) = -[(b_2 - b_1)\varphi_1 - \varphi_3]x_1 - (b_2 - a_1)\varphi_3 x_2 + y_1^r y_3 + v_3(t), \\ u_4(t) = -[(b_2 - b_1)\varphi_2 - \varphi_4]x_1 - (b_2 - a_1)\varphi_4 x_2 + y_1^i y_3 + v_4(t), \\ u_5(t) = -y_1^r y_2^r - y_1^i y_2^i - (a_3 - b_3)\varphi_5 x_3 + \varphi_5 x_1 x_3 + a_2\varphi_5 + v_5(t), \end{cases}$$

which leads to

$$\begin{cases} D_*^{\alpha_1} e_1 = b_1 e_3 - b_1 e_1 + v_1(t), \\ D_*^{\alpha_1} e_2 = b_1 e_4 - b_1 e_2 + v_2(t), \\ D_*^{\alpha_2} e_3 = (b_2 - b_1)e_1 + b_2 e_3 + v_3(t), \\ D_*^{\alpha_2} e_4 = (b_2 - b_1)e_2 + b_2 e_4 + v_4(t), \\ D_*^{\alpha_3} e_5 = -b_3 e_5 + v_5(t). \end{cases}$$

Due to the term $v_i(t)$ are linear functions of the error terms $e_i(t)$ $(i = 1, 2, \ldots, 5)$, we define

$$(v_1(t), v_2(t), v_3(t), v_4(t), v_5(t))^T = L(e_1(t), e_2(t), e_3(t), e_4(t), e_5(t))^T,$$

where $L = (l_{ij})_{5\times5}$ is a 5×5 real matrix. Thus, we make a decision to select $l_{11} = 0 < b_1$, $l_{22} = 1 < b_1$, $l_{33} = -29 < -b_2$, $l_{44} = -30 < -b_2$, $l_{55} = 0 < b_3$, $l_{31} = l_{42} = 7$, $l_{21} = l_{32} = l_{41} = l_{43} = l_{51} = l_{52} = l_{53} = l_{54} = 0$. Since $(b_1, b_2, b_3) = (35, 28, 3)$, $(\alpha_1, \alpha_2, \alpha_3) = (0.9, 0.91, 0.92)$, $m = 100$, $\gamma = 1/100$, we have the following characteristic equation

$$(\lambda^{90} + 35)(\lambda^{90} + 34)(\lambda^{91} + 1)(\lambda^{91} + 2)(\lambda^{92} + 3) = 0. \tag{14.2.9}$$

All roots of Eq. (14.2.9) lie in the region $|arg(\lambda)| > \gamma\pi/2$. From Theorem 1.4, the error vector $e(t)$ asymptotically converges to zero as $t \to \infty$.

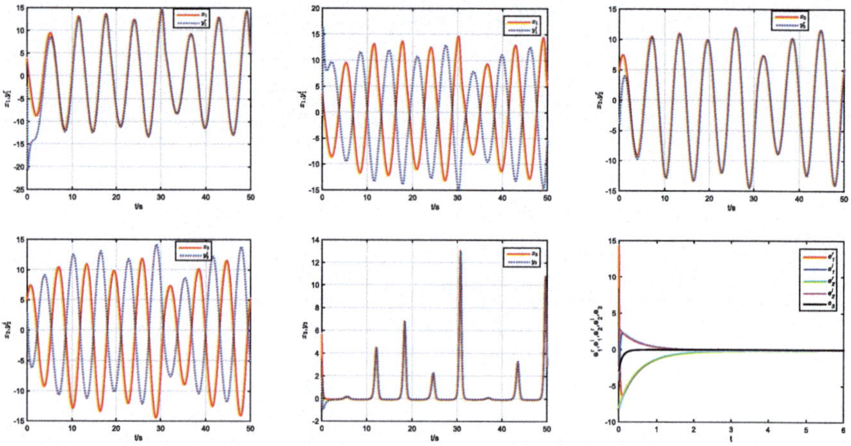

Fig. 14.6 State variables and errors of CMPS between systems (14.2.4) and (14.2.8)

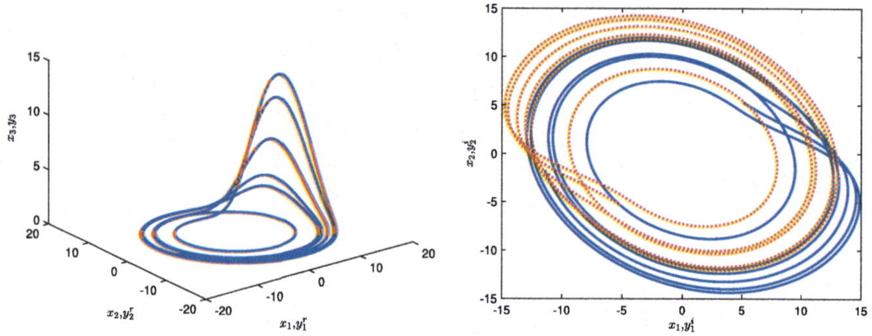

Fig. 14.7 State variables and errors of CMPS between systems (14.2.3) and (14.2.4)

The initial values of systems (14.2.8) and (14.2.4) are set to $(x_1, x_2, x_3)^T = (4, 5, 6)^T$ and $(y_1, y_2, y_3)^T = (5 - 8j, -4 + 3j, 2)^T$, respectively. With the scaling matrix set as $\Phi = diag(1 - j, 1 - j, 1)$, the initial errors become $(5 - 8j, -10 + 8j, -2)^T$. Selecting $l_{13} = 70, l_{24} = 70, l_{12} = l_{14} = l_{15} = l_{23} = l_{25} = l_{34} = l_{35} = l_{45} = 0$, we have simulation results as displayed in Figs. 14.6 and 14.7. State variables and errors of CMPS between systems (14.2.3) and (14.2.4) can be described in Fig. 14.6, where the solid line presents the states of drive system and the dotted line shows the states of response system. Figure 14.7 displays that the real parts (y_1^r, y_2^r, y_3) of the response system (14.2.4) completely synchronize (x_1, x_2, x_3) of the drive system (14.2.8), while the imaginary parts (y_1^i, y_2^i) of the response system (14.2.4) anti-synchronize (x_1, x_2) of the drive system (14.2.8). Thus, the incommensurate fractional-order chaotic real Rössler system and complex Chen system achieve CMPS.

14.2.3 CMPS Between Fractional-Order Chaotic Complex Drive Systems and Real Response Systems

I. Mathematical model and problem descriptions

The fractional-order chaotic complex system (14.2.1) is considered as the drive system, while the fractional-order chaotic real response system is of the following form:

$$D_*^{\alpha} w = Dw + q(w) + U, \tag{14.2.10}$$

where $w = (w_1, w_2, \ldots, w_n)^T$ is a real state vector, q is the vector of nonlinear functions, $D \in \mathbf{R}^{n\times n}$ is a coefficient matrix.

Since $w(t)$ is real, a real U is designed to ensure CMPS of real parts and suppress the imaginary parts of response system. With the scaling matrix $\Phi = \Phi^r + j\Phi^i \in \mathbf{C}^{n\times n}$, the corresponding error vector is given by

$$e = w - \Phi^r z^r + \Phi^i z^i.$$

Consequently, the complex drive system (14.2.2) and the real response system (14.2.10) are CMPS of real parts.

II. CMPS between fractional-order complex Lorenz system and real Lü system

We consider the fractional-order complex Lorenz system in Eq. (14.2.3), with the fractional-order real Lü system defined as follows:

$$\begin{cases} D_*^{\alpha_1} w_1 = d_1(w_2 - w_1) + U_1, \\ D_*^{\alpha_2} w_2 = -w_1 w_3 + d_2 w_2 + U_2, \\ D_*^{\alpha_3} w_3 = w_1 w_2 - d_3 w_3 + U_3, \end{cases} \tag{14.2.11}$$

where w_i are real state variables; d_i are real parameters and U_i are real control functions ($i = 1, 2, 3$). When $(\alpha_1, \alpha_2, \alpha_3) = (0.95, 0.96, 0.97)$ and $(d_1, d_2, d_3) = (35, 28, 3)$, chaotic behaviors can be displayed for system (14.2.11).

We adopt $\Phi = diag(\varphi_1 + j\varphi_2, \varphi_3 + j\varphi_4, \varphi_5)$ and define the synchronization error as:

$$\begin{cases} e_1 = w_1 - \varphi_1 z_1^r + \varphi_2 z_1^i, \\ e_2 = w_2 - \varphi_3 z_2^r + \varphi_4 z_2^i, \\ e_3 = w_3 - \varphi_5 z_3. \end{cases}$$

Then, we obtain the error system:

$$\begin{cases} D_*^{\alpha_1} e_1 = d_1 e_2 - d_1 e_1 + (d_1 \varphi_3 - c_1 \varphi_1) z_2^r - d_1 \varphi_4 z_2^i + c_1 \varphi_2 z_2^i \\ \qquad -(d_1 - c_1)(\varphi_1 z_1^r - \varphi_2 z_1^i) + U_1, \\ D_*^{\alpha_2} e_2 = d_2 e_2 + (d_2 + 1)(\varphi_3 z_2^r - \varphi_4 z_2^i) - w_1 w_3 + (\varphi_3 z_1^r - \varphi_4 z_1^i)(z_3 - c_2) + U_2, \\ D_*^{\alpha_3} e_3 = -d_3 e_3 + w_1 w_2 - (d_3 - c_3)\varphi_5 z_3 - \varphi_5 (z_1^r z_2^r + z_1^i z_2^i) + U_3. \end{cases}$$

Thus, the active control inputs $U_i(t)$ $(i = 1, 2, 3)$ are designed as

$$\begin{cases} U_1(t) = -(d_1 \varphi_3 - c_1 \varphi_1) z_2^r + (d_1 \varphi_4 - c_1 \varphi_2) z_2^i + (d_1 - c_1)(\varphi_1 z_1^r - \varphi_2 z_1^i) + v_1(t), \\ U_2(t) = -(d_2 + 1)(\varphi_3 z_2^r - \varphi_4 z_2^i) + w_1 w_3 - (\varphi_3 z_1^r - \varphi_4 z_1^i)(z_3 - c_2) + v_2(t), \\ U_3(t) = (d_3 - c_3)\varphi_5 z_3 + \varphi_5 (z_1^r z_2^r + z_1^i z_2^i) - w_1 w_2 + v_3(t). \end{cases}$$

And the error system becomes

$$\begin{cases} D_*^{\alpha_1} e_1 = d_1 e_2 - d_1 e_1 + v_1(t), \\ D_*^{\alpha_2} e_2 = d_2 e_2 + v_2(t), \\ D_*^{\alpha_3} e_3 = -d_3 e_3 + v_3(t). \end{cases}$$

Choosing $(v_1(t), v_2(t), v_3(t))^T = L(e_1(t), e_2(t), e_3(t))^T$ to satisfy the conditions of Theorem 1.4, where $L = (l_{ij})_{3\times 3}$ is a 3×3 real matrix. Thus we make a good choice as follows: $l_{11} = 30 < d_1$, $l_{22} = -36 < -d_2$, $l_{33} = 2 < d_3$, $l_{12} = -35$, $l_{13} = l_{23} = 0$. Since $(d_1, d_2, d_3) = (35, 28, 3)$, $(\alpha_1, \alpha_2, \alpha_3) = (0.95, 0.96, 0.97)$, $m = 100$, $\gamma = 1/100$, then the equation can be written as

$$(\lambda^{95} + 5)(\lambda^{96} + 8)(\lambda^{97} + 1) = 0. \qquad (14.2.12)$$

It can be demonstrated that all roots of the Eq. (14.2.12) lie in the region $|arg(\lambda)| > \gamma\pi/2$. Hence, the error vector $e(t)$ asymptotically converges to zero as $t \to \infty$.

Numerical simulations are conducted with the initial values $(z_1, z_2, z_3)^T = (2 + 3j, 5 + 6j, 9)^T$, $(w_1, w_2, w_3)^T = (1, 2, 3)^T$ for systems (14.2.3) and (14.2.11), and the complex constant scaling matrix $\Phi = diag(3 + j, 1 - 2j, -2)$, leading to initial errors are $(-2, -15, 21)^T$. With parameters $l_{21} = 2$, $l_{31} = 1$, $l_{32} = 4$, the results in Figs. 14.8 and 14.9 display that the errors of CMPS converge asymptotically to zero, i.e., fractional-order complex Lorenz system and real Lü system are CMPS of real parts.

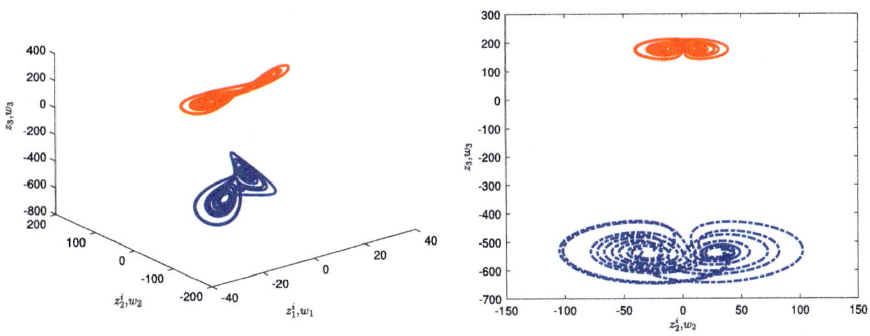

Fig. 14.8 Phase diagram of CMPS between systems (14.2.3) and (14.2.11)

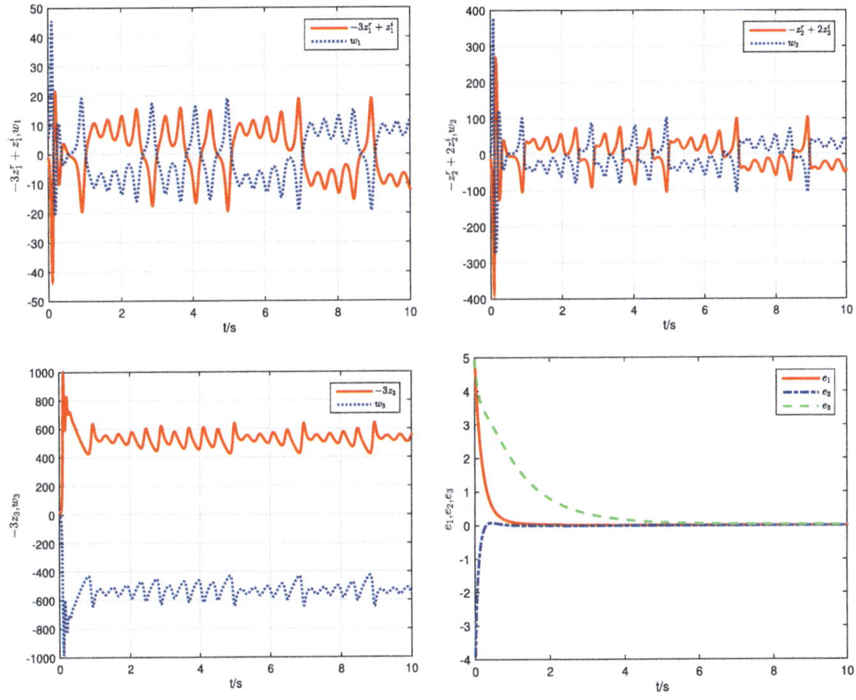

Fig. 14.9 State variables and errors of CMPS between systems (14.2.3) and (14.2.11)

Chapter 15
Modified Generalized Projective Synchronization of Fractional-Order Chaotic Systems

In the practical applications, a more form synchronization, called modified generalized projective synchronization (MGPS) [228], where the drive and response systems can be synchronized up to a transformation matrix that is not diagonal, will increase the complexity of the synchronization and further increase the diversity and the security of communications. This section introduces a fractional-order modified chaotic Lü system and aims to realize this new MGPS of fractional-order chaotic systems with different structure, where the drive and response systems could be asymptotically synchronized up to a desired transformation matrix, not a diagonal matrix.

15.1 A Novel Fractional-Order Chaotic System

15.1.1 A Novel Fractional-Order System Model

In 2006, Wang et al. introduced a new modified hyperchaotic Lü system [229], further investigated its dynamical behaviors, and physically implemented the system, which is described by

$$\begin{cases} \dot{x} = a(y - x + yz), \\ \dot{y} = -xz + by + w, \\ \dot{z} = xy - cz, \\ \dot{w} = -dx, \end{cases} \tag{15.1.1}$$

where $(x, y, z, w) \in \mathbb{R}^4$, and a, b, c, d are real constant parameters. The hyperchaotic attractors of system (15.1.1) for $a = 70$, $b = 15$, $c = 12$, $d = 5$ are plotted in Fig. 15.1.

In 2011, Mahmoud et al. studied a modified hyperchaotic complex Lü systems [230] given by

S. Liu et al., *Chaos in Complex Dynamical System: Theory and Application*,
https://doi.org/10.1007/978-981-95-7762-0_15

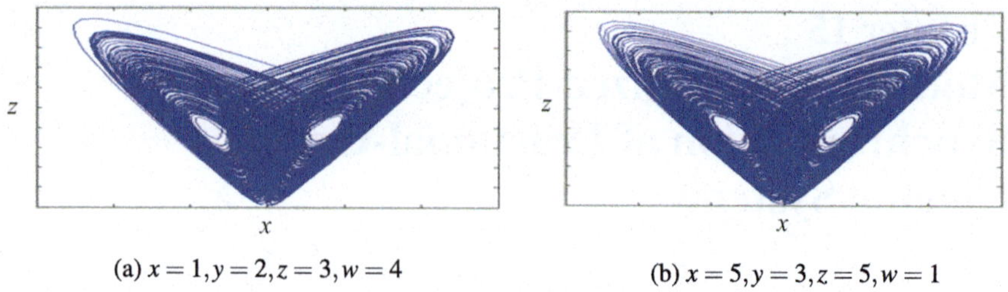

Fig. 15.1 Phase portraits in the x-z plane of the integer-order hyperchaotic real system (15.1.1) for $a = 70, b = 15, c = 12, d = 5$

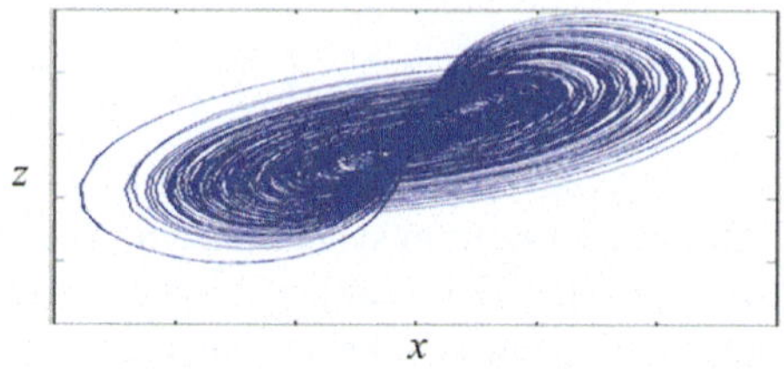

Fig. 15.2 Phase portraits in the $x - z$ plane of the hyperchaotic complex system (15.1.2) for $a = 70, b = 15, c = 12, d = 5$ and $x = 1, y = 2, z = 3, w = 4$

$$\begin{cases} \dot{x} &= a(y - x + yz), \\ \dot{y} &= -xz + by + w, \\ \dot{z} &= \frac{1}{2}(\overline{x}y + x\overline{y}) - cz, \\ \dot{w} &= -\frac{d}{2}(\overline{x} + x), \end{cases} \tag{15.1.2}$$

where $(x, y) \in \mathbb{C}^2$ are complex variables, $(z, w) \in \mathbb{R}^2$ are real variables, and a, b, c and d are real (or complex) positive constant parameters. The hyperchaotic attractors of (15.1.2) using the same choice of the parameters and initial conditions as in the real system (15.1.1) are plotted in Fig. 15.2. More detailed complex dynamics of the modified hyperchaotic complex Lü system can be seen in Ref. [230].

In the present paper, based on the above descriptions, we modify the derivative operator in Eq. (15.1.1) to be with respect to the fractional-order $\alpha (0 < \alpha \leq 1)$. Thus the fractional version of the hyperchaotic Lü system is given by

$$\begin{cases} D_*^\alpha x &= a(y - x + yz), \\ D_*^\alpha y &= -xz + by + w, \\ D_*^\alpha z &= xy - cz, \\ D_*^\alpha w &= -dx, \end{cases} \tag{15.1.3}$$

where a, b, c and d are real positive constant parameters. When $\alpha = 1$, system (15.1.3) reduces to the classical integer-order modified hyperchaotic Lü system (15.1.1).

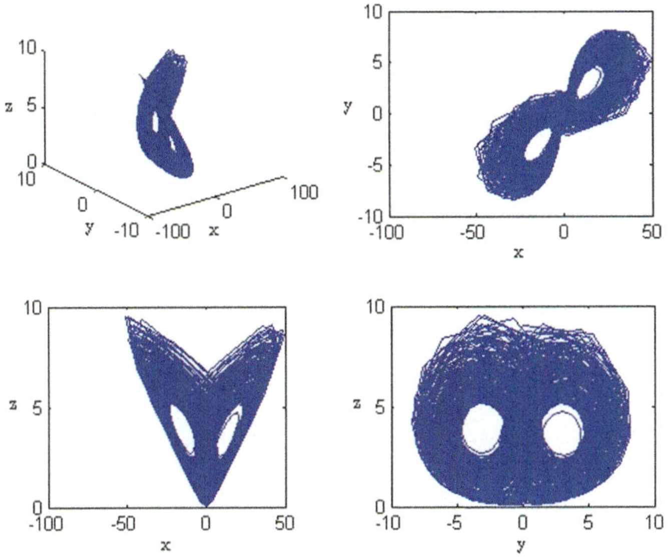

Fig. 15.3 Phase portraits of the fractional-order chaotic system (15.1.3) for $\alpha = 0.95$

15.1.2 Dynamic Analysis of the Fractional-Order Lü System

In the following simulations, the system parameters are always chosen as $a = 70$, $b = 15$, $c = 12$, $d = 5$. The simulation results demonstrate that chaos indeed exists in the fractional-order system (15.1.3) with order less than 4. For example, when $\alpha = 0.95$, a chaotic attractor is found, and the phase portraits of the system are displayed in Fig. 15.3. The Eq. (15.1.3) has no solution when $\alpha < 0.8320$. A chaotic attractor is found for $\alpha = 0.8321$, and its phase portrait is displayed in Fig. 15.4. Thus the lower limit of the fractional-order for this system to be chaotic is between $\alpha < 0.8320$ and $\alpha < 0.8321$, and the lowest order we found for this system to yield chaotic is 3.3284.

The largest Lyapunov exponent of the simulation time series [231] is positive only when $\alpha \geq 0.8321$, for example, $\alpha = 0.8321$, $\lambda_1 = 19.78383$. Obviously, the fractional-order system (15.1.3) is chaotic.

15.2 MGPS Between the Different Fractional-Order Systems

15.2.1 A General Method for MGPS of Fractional-Order Systems

In this section we apply stability analysis to fractional-order systems. Fractional-order differential equations are at least as stable as their integer order counter-

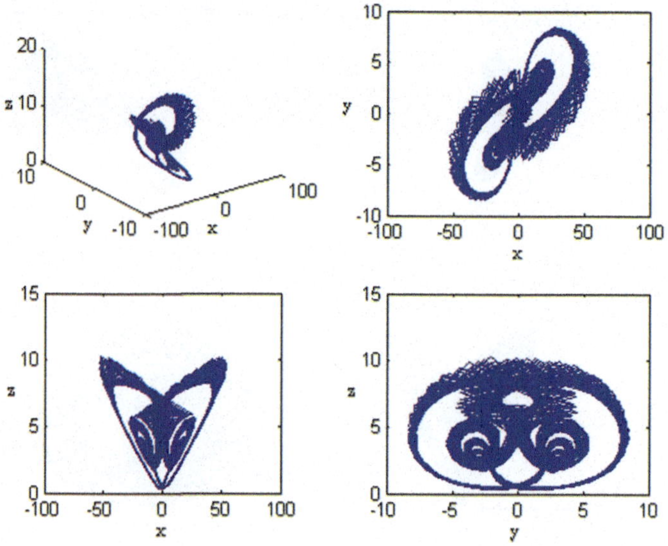

Fig. 15.4 Phase portraits of the fractional-order chaotic system (15.1.3) for $\alpha = 0.8321$

part because systems with memory are typically more stable than those without memory [232].

Consider the fractional-order chaotic drive and response systems as

$$D_*^\alpha X = f(X) \tag{15.2.1}$$

and

$$D_*^\alpha Y = g(Y) + U(X, Y), \tag{15.2.2}$$

where the state vectors $X, Y \in \mathbb{R}^n$, $f, g : \mathbb{R}^n \to \mathbb{R}^n$ are continuous vector functions and $U(X, Y) : \mathbb{R}^{2n} \to \mathbb{R}^n$ is a controller to be determined later.

Decompose the fractional-order systems (15.2.1) and (15.2.2) as

$$D_*^\alpha X = AX + F(X) \tag{15.2.3}$$

and

$$D_*^\alpha Y = BY + G(Y) + U(X, Y), \tag{15.2.4}$$

where $A, B : \mathbb{R}^{n\times n} \to \mathbb{R}^{n\times n}$ is the Jacobian matrix of the system at the origin and $F, G : \mathbb{R}^n \to \mathbb{R}^n$ are the nonlinear parts.

Remark 15.1 Fractional-order chaotic systems belong to the class individualized by (15.2.3). Examples include the fractional-order hyperchaotic modified Rössler system [233, 234], the Chen system [235], and the Lorenz system [236].

Definition 15.1 The drive system (15.2.3) and response system (15.2.4) are said to achieve MGPS if there exists a controller U such that $\lim_{t\to\infty} \|Y - CX\| = 0$, where $C : \mathbb{R}^{n\times n} \to \mathbb{R}^{n\times n}$ is a transformation matrix of the drive system (15.2.3).

Remark 15.2 In particular, if the transformation matrix is diagonal and all diagonal elements are the same, the MGPS is simplified to the GPS.

Hence, the error system is defined as

$$e = Y - CX, \tag{15.2.5}$$

which means that system (15.2.4) synchronizes with the projection of system (15.2.3). If $\lim_{t\to\infty} \|e\| = 0$, systems (15.2.3) and (15.2.4) substituted into (15.2.5) gives an error system that can be expressed as

$$D_*^\alpha = D_*^\alpha Y - CD_*^\alpha X = Be + BCX + G(Y) - CAX - CF(X) + U(X(Y)) \tag{15.2.6}$$

Then the discussions of MGPS between two coupled systems (15.2.3) and (15.2.4) can be translated into the analysis of the asymptotic stability of the zero solution of the error system (15.2.6). Next, a suitable controller is provided to ensure the asymptotic stability of the zero solution of the error system (15.2.6) based on the stability theorem of linear fractional-order systems.

Theorem 15.1 *Given a fractional-order drive system* (15.2.3) *and a response system* (15.2.4)*, there exists a suitable controller.*

$$U = CAX + CF(X) - BCX - G(Y) + Ke. \tag{15.2.7}$$

where $K \in R^{n\times n}$ is a gain matrix. MGPS between systems (15.2.3) *and* (15.2.4) *can be achieved if and only if all the eigenvalues of $B + K$ satisfy* $|\arg(\lambda_i(M))| > \alpha\pi/2(i = 1, 2, \ldots, n)$

Proof Substituting controller (15.2.7) into system (15.2.6), the error system (15.2.6) can be rewritten as

$$D_*^\alpha e = (B + K)e, \tag{15.2.8}$$

Due to Theorem 1.3, we arrive at the conclusion that system (15.2.9) is asymptotically stable if and only if all the eigenvalues $\lambda_i(B + K)$ satisfy $|\arg(\lambda_i(B + K))| > \alpha\pi/2(i = 1, 2, \ldots, n)$. That is, $\lim_{t\to\infty} \|e\| = 0$, or systems (15.2.3) and (15.2.4) realize MGPS. This completes the proof. □

15.2.2 MGPS Between the Fractional-Order Hyperchaotic Lorenz System and Lü system

In this section the MGPS behavior between two different fractional-order systems, the hyperchaotic Lorenz system and modified Lü system, is made. It is assumed that the fractional-order Lü system drives the hyperchaotic Lorenz system. Therefore, we define Lü system as a master system and the hyperchaotic Lorenz system as a slave system as follows. The master system described through Eq.(15.1.3) is

$$\begin{cases} D_*^\alpha x_1 = a(x_2 - x_1 + x_2 x_3), \\ D_*^\alpha x_2 = b x_2 + x_4 - x_1 x_3, \\ D_*^\alpha x_3 = -c x_3 + x_1 x_2, \\ D_*^\alpha x_4 = -d x_1, \end{cases} \tag{15.2.9}$$

The slave system described in [237] is

$$\begin{cases} D_*^\alpha y_1 = k(y_2 - y_1) + u_1, \\ D_*^\alpha y_2 = m y_1 + y_2 - y_1 y_3 - y_4 + u_2, \\ D_*^\alpha y_3 = y_1 y_2 - n y_3 + u_3, \\ D_*^\alpha y_4 = p y_2 y_3 + u_4, \end{cases} \tag{15.2.10}$$

where k, m, n, p are parameters and $y_i (i = 1, 2, 3, 4)$ are variables, $u_i(t)$ $(i = 1, 2, 3, 4)$ are the controller to be determined later.

Comparing systems (15.2.9) and (15.2.10) with systems (15.2.3) and (15.2.4), one has

$$A = \begin{pmatrix} -a & a & 0 & 0 \\ 0 & b & 0 & 1 \\ 0 & 0 & -c & 0 \\ -d & 0 & 0 & 0 \end{pmatrix}, \quad F(X) = \begin{pmatrix} a x_2 x_3 \\ -x_1 x_3 \\ x_1 x_2 \\ 0 \end{pmatrix}, \tag{15.2.11}$$

$$B = \begin{pmatrix} -k & k & 0 & 0 \\ m & 1 & 0 & -1 \\ 0 & 0 & n & 0 \\ 0 & 0 & 0 & 0 \end{pmatrix}, \quad G(Y) = \begin{pmatrix} 0 \\ -y_1 y_3 \\ y_1 y_2 \\ p y_2 y_3 \end{pmatrix}, \quad U(X, Y) = \begin{pmatrix} u_1(X, Y) \\ u_2(X, Y) \\ u_3(X, Y) \\ u_4(X, Y) \end{pmatrix}. \tag{15.2.12}$$

Here, choose a transformation matrix

$$C = \begin{pmatrix} 1 & 1 & 0 & 0 \\ 0 & 1 & 1 & 0 \\ 0 & 0 & 1 & 1 \\ 1 & 0 & 0 & 1 \end{pmatrix}. \tag{15.2.13}$$

The error system $e = Y - CX$ between systems (15.2.9) and (15.2.10) is

$$\begin{cases} e_1 = y_1 - (x_1 + x_2), \\ e_2 = y_2 - (x_2 + x_3), \\ e_3 = y_3 - (x_3 + x_4), \\ e_4 = y_4 - (x_1 + x_4). \end{cases} \quad (15.2.14)$$

According to Theorem 15.1 and Eq. (15.2.10), with a suitable controller $U = CAx + CF(x) - BCx - G(y) + Ke$, $K \in \mathbb{R}^{n\times n}$, the error dynamical system can be obtained in the form

$$D_*^\alpha e = (B + K)e. \quad (15.2.15)$$

Then Theorem 15.1 assures that there exists a feedback gain matrix K so that systems (15.2.9) and (15.2.10) can realize the synchronization. There is not a unique choice for such a matrix K. For example, when $\alpha = 0.95$, $(k, m, n, p) = (10, 28, 8/3, 0.1)$ and $(a, b, c, d) = (70, 15, 12, 5)$, we can set $B + K = diag(-1, -2, -1, -3)$ with the matrix

$$K = \begin{pmatrix} 9 & -10 & 0 & 0 \\ -28 & -3 & 0 & 1 \\ 0 & 0 & -11/3 & 0 \\ 0 & 0 & 0 & -3 \end{pmatrix}. \quad (15.2.16)$$

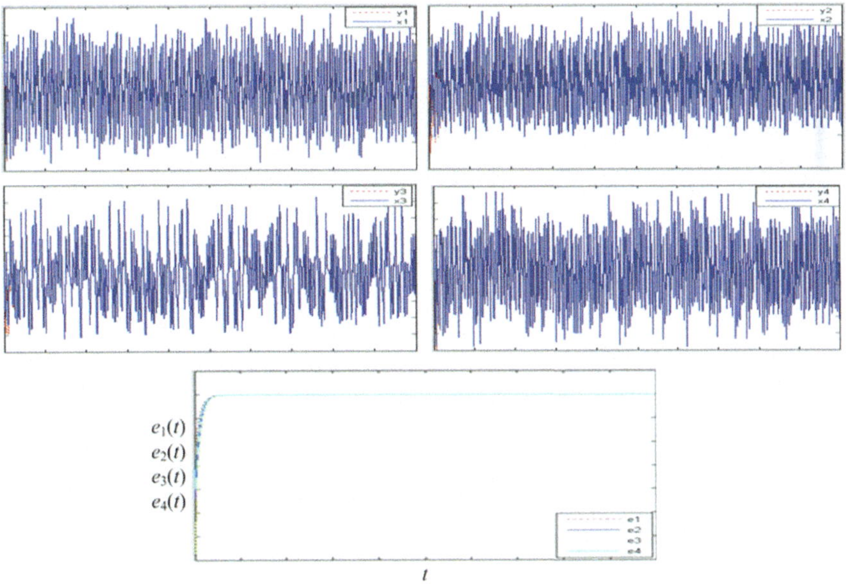

Fig. 15.5 MGPS of the novel fractional-order hyperchaotic Lorenz system (15.2.10) and Lü system (15.2.9) for $\alpha = 0$

so that all the eigenvalues $\lambda_i = (-1, -2, -1, -3)$ of $B + K$ satisfy $|arg(\lambda_i(B + K)| > 0.95\pi/2$, $(i = 1, 2, \ldots, n)$. In the numerical simulations, the initial values of the drive and response systems are arbitrarily chosen as $(x_1(0), x_2(0), x_3(0), x_4(0)) = (1, 2, 3, 4)$, and $y_1(0), y_2(0), y_3(0), y_4(0) = (0.3, 0.6, 0.3, 1)$. The time evolutions of the states are plotted in Figs. 15.5(a)–(d). Figure 15.5(e) displays the evolution of the MGPS error $e = (e_1, e_2, e_3, e_4)^T$ which tends to zero as $t \to \infty$, which implies that the error system (15.2.14) between the drive and the response systems (15.2.9)–(15.2.10) is globally and asymptotically stable. What should be mentioned is that the fractional order and control gains, i.e. the elements of K, play an important role in the synchronization rate.

Chapter 16
Combination Complex Synchronization of Fractional-Order Chaotic Complex Systems

Many researchers have made an endeavor to analyze synchronization of multi chaotic systems and present some new types of synchronization including combination synchronization [238–241], combination complex synchronization [242], combination-combination synchronization [243–246], compound synchronization [247–249], and so forth. Amongst the above-mentioned synchronization, combination complex synchronization, characterized by two scaling matrices that three chaotic systems synchronize proportionally, was proposed and studied on the basis of three integer-order chaotic complex systems with the same dimensions.

The aim of this chapter is to study combination complex synchronization of fractional-order chaotic nonlinear systems. This chapter discusses the general method to realize combination complex synchronization among three commensurate fractional-order complex systems in the sense of Lyapunov in detail. Combination complex synchronization of fractional-order chaotic complex systems with incommensurate orders is investigated based on the stability theory of incommensurate fractional-order systems and active control method.

16.1 Combination Complex Synchronization of the Commensurate Fractional-Order Chaotic Complex Systems

This section is devoted to investigating the problem of combination complex synchronization based on two fractional-order chaotic complex drive systems and one fractional-order chaotic complex response system. By choosing appropriate fractional-order Lyapunov functions and employing active control, we obtain some sufficient conditions for combination complex synchronization of three identical or

S. Liu et al., *Chaos in Complex Dynamical System: Theory and Application*,
https://doi.org/10.1007/978-981-95-7762-0_16

different fractional-order chaotic systems with complex variables. Corresponding numerical simulations confirm the validity and feasibility of the proposed synchronization schemes.

16.1.1 Problem Description and Synchronization Scheme

Two n-dimensional fractional-order chaotic complex drive systems read in the following form:

$$D_*^{\alpha} x(t) = f(x) \tag{16.1.1}$$

and

$$D_*^{\alpha} y(t) = g(y), \tag{16.1.2}$$

while an n-dimensional fractional-order chaotic complex response system is described as

$$D_*^{\alpha} z(t) = h(z) + U(x, y, z), \tag{16.1.3}$$

where $x = (x_1, x_2, \ldots, x_n)^T$, $y = (y_1, y_2, \ldots, y_n)^T$, and $z = (z_1, z_2, \ldots, z_n)^T$ are the complex state vectors of systems (16.1.1)–(16.1.3), respectively; f, g, and h are vectors of nonlinear complex functions; U is a nonlinear controller which will be designed later.

Next, we introduce the definition of combination complex synchronization among three fractional-order complex systems.

Definition 16.1 Two drive systems (16.1.1) and (16.1.2) combination complex synchronize with the response system (16.1.3) if there exist two complex constant matrices $\Phi = \Phi^r + j\Phi^i \in \mathbb{C}^{n\times n}$ and $\Psi = \Psi^r + j\Psi^i \in \mathbb{C}^{n\times n}$, such that

$$\lim_{t\to\infty} ||e(t)|| = \lim_{t\to\infty} ||z(t) - \Phi x(t) - \Psi y(t)|| = 0,$$

where $e(t)$ is the synchronization error vector.

Remark 16.1 The proposed combination complex synchronization scheme contains several kinds of synchronization found in previous papers. For example, complex projective synchronization will appear when the scaling matrix $\Phi = O$ or $\Psi = O$; combination synchronization can be carried out when $\Phi^i = O$ and $\Psi^i = O$; synchronization problem will degenerate into the chaos control problem when $\Phi = O$ and $\Psi = O$.

In what follows, we design suitable controllers to realize combination complex synchronization among three fractional-order chaotic complex systems in the sense of Lyapunov.

Theorem 16.1 *Two drive systems (16.1.1) and (16.1.2) combination complex synchronize with the response system (16.1.3) by the following control law:*

$$\begin{cases} U^r = -h^r(z) + \Phi^r f^r(x) - \Phi^i f^i(x) + \Psi^r g^r(y) - \Psi^i g^i(y) - Ke^r, \\ U^i = -h^i(z) + \Phi^i f^r(x) + \Phi^r f^r(x) + \Psi^i g^r(y) + \Psi^r g^i(y) - Ke^i, \end{cases} \tag{16.1.4}$$

where $K = \text{diag}(k_1, k_2, \ldots, k_n)$ *is the control strength matrix and all its elements are positive real numbers.*

Proof It follows from (16.1.1)–(16.1.4) that

$$D_*^\alpha e(t) = D_*^\alpha z(t) - \Phi D_*^\alpha x(t) - \Psi D_*^\alpha y(t) = -Ke(t).$$

Then, we have

$$\begin{cases} D_*^\alpha e^r = -Ke^r, \\ D_*^\alpha e^i = -Ke^i. \end{cases} \tag{16.1.5}$$

Define a positive definite Lyapunov candidate function as

$$V(t) = \frac{1}{2}\left(e^r(t)\right)^T e^r(t) + \frac{1}{2}\left(e^i(t)\right)^T e^i(t).$$

Thus, we have $V(t) = \frac{1}{2}||e(t)||^2$. Calculating the Caputo derivative of $V(t)$ and combining Lemma 1.6 with system (16.1.5), we conclude that

$$\begin{aligned} D_*^\alpha V &\le \frac{1}{2}\left((D_*^\alpha e^r(t))^T e^r(t) + (e^r(t))^T D_*^\alpha e^r(t)\right) + \frac{1}{2}\left((D_*^\alpha e^i(t))^T e^i(t) + (e^i(t))^T D_*^\alpha e^i(t)\right) \\ &= -(e^r(t))^T Ke^r(t) - (e^i(t))^T Ke^i(t) \\ &\le -k_{min}\left((e^r(t))^T e^r(t) + (e^i(t))^T e^i(t)\right) \\ &= -2k_{min} V(t), \end{aligned}$$

where $k_{min} = \min\{k_1, k_2, \ldots, k_n\}$. Taking into account Lemma 1.5 and $k_{min} > 0$, we have

$$V(t) \le V(0) E_\alpha(-2k_{min} t^\alpha).$$

Furthermore, we obtain

$$||e(t)||^2 = 2V(t) \le 2V(0) E_\alpha(-2k_{min} t^\alpha).$$

Therefore, we conclude $||e(t)|| \to 0$ as $t \to \infty$, i.e., the origin of the error dynamical system (16.1.5) is asymptotically stable. So, combination complex synchronization between systems (16.1.1)–(16.1.3) is achieved. This completes the proof. □

The following corollary can be easily obtained from Theorem 16.1.

Theorem 16.2 *If the structures of general complex chaotic systems are identical, i.e.,* $f(\cdot) = g(\cdot) = h(\cdot)$, *complex controller* $U_k = U_k^r + jU_k^i$ *can be constructed as*

$$\begin{cases} U^r = -f^r(z) + \Phi^r f^r(x) - \Phi^i f^i(x) + \Psi^r f^r(y) - \Psi^i f^i(y) - Ke^r, \\ U^i = -f^i(z) + \Phi^i f^r(x) + \Phi^r f^r(x) + \Psi^i f^r(y) + \Psi^r f^i(y) - Ke^i, \end{cases}$$

where $K = \text{diag}(k_1, k_2, \ldots, k_n)$ *is the control strength real matrix and* $k_i > 0$, *then we can achieve combination complex synchronization of three identical fractional-order complex chaotic systems.*

16.1.2 Numerical Simulations

To illustrate the effectiveness of the proposed synchronization scheme, two following examples are inserted in the text and their numerical simulations are performed.

I. Combination complex synchronization of three identical fractional-order complex systems

The first drive system takes the following form:

$$\begin{cases} D_*^\alpha x_1 = a(x_2 - x_1), \\ D_*^\alpha x_2 = -x_1 x_3 + b x_2, \\ D_*^\alpha x_3 = \frac{1}{2}(\bar{x}_1 x_2 + x_1 \bar{x}_2) - c x_3, \end{cases} \tag{16.1.6}$$

the second drive system is given as:

$$\begin{cases} D_*^\alpha y_1 = a(y_2 - y_1), \\ D_*^\alpha y_2 = -y_1 y_3 + b y_2, \\ D_*^\alpha y_3 = \frac{1}{2}(\bar{y}_1 y_2 + y_1 \bar{y}_2) - c y_3, \end{cases} \tag{16.1.7}$$

and the response system with the controllers is written in the form as:

$$\begin{cases} D_*^\alpha z_1 = a(z_2 - z_1) + u_1 + j u_2, \\ D_*^\alpha z_2 = -z_1 z_3 + b z_2 + u_3 + j u_4, \\ D_*^\alpha z_3 = \frac{1}{2}(\bar{z}_1 z_2 + z_1 \bar{z}_2) - c z_3 + u_5, \end{cases} \tag{16.1.8}$$

where a, b, and c are system parameters, x_i, y_i, and z_i $(i = 1, 2)$ are complex variables; x_3, y_3, and z_3 are real variables. u_i $(i = 1, 2, \ldots, 5)$ are real controllers to be determined.

For the convenience of the following discussions, we assume $\Phi = \mathrm{diag}(\varphi_1 + j\varphi_2, \varphi_3 + j\varphi_4, \varphi_5)$ and $\Psi = \mathrm{diag}(\phi_1 + j\phi_2, \phi_3 + j\phi_4, \phi_5)$ in our synchronization scheme. Thus, according to Theorem 16.2, controllers can be designed as follows:

$$\begin{cases} u_1 = (a-k_1)(z_1^r - \varphi_1 x_1^r + \varphi_2 x_1^i - \phi_1 y_1^r + \phi_2 y_1^i) + a(\varphi_1 x_2^r - \varphi_2 x_2^i + \phi_1 y_2^r - \phi_2 y_2^i) - a z_2^r, \\ u_2 = (a-k_1)(z_1^i - \varphi_2 x_1^r - \varphi_1 x_1^i - \phi_2 y_1^r - \phi_1 y_1^i) + a(\varphi_2 x_2^r + \varphi_1 x_2^i + \phi_2 y_2^r + \phi_1 y_2^i) - a z_2^i, \\ u_3 = z_1^r z_3 - (b+k_2)(z_2^r - \varphi_3 x_2^r + \varphi_4 x_2^i - \phi_3 y_2^r + \phi_4 y_2^i) - \varphi_3 x_1^r x_3 + \varphi_4 x_1^i x_3 - \phi_3 y_1^r y_3 + \phi_4 y_1^i y_3, \\ u_4 = z_1^i z_3 - (b+k_2)(z_2^i - \varphi_4 x_2^r - \varphi_3 x_2^i - \phi_4 y_2^r - \phi_3 y_2^i) - \varphi_4 x_1^r x_3 - \varphi_3 x_1^i x_3 - \phi_4 y_1^r y_3 - \phi_3 y_1^i y_3, \\ u_5 = -z_1^r z_2^r - z_1^i z_2^i + (c-k_3)(z_3 - \varphi_5 x_3 - \phi_5 y_3) + \varphi_5(x_1^r x_2^r + x_1^i x_2^i) + \phi_5(y_1^r y_2^r + y_1^i y_2^i). \end{cases}$$

In the simulation process, $\alpha = 0.98$, the initial values are chosen as $x(0) = (1 + 2j, 3 + 4j, 5)^T$, $y(0) = (-2 + 10j, 6 - j, 10)^T$, $z(0) = (-1 + 3j, -5 + 8j, 1)^T$, the scaling matrices are taken as $\Phi = \mathrm{diag}(1 + j, 1 + j, 1)$, $\Psi = \mathrm{diag}(1 + j, 1 + j, 1)$, and the control matrix $K = \mathrm{diag}(2, 2, 2)$. The synchronization process of systems (16.1.6)–(16.1.8) is described in Figs. 16.1(a)–(e), where the red line presents the states of drive systems and the blue line shows the states of response system. From Fig. 16.1(f), it can be seen that the errors of synchronization go to zero. As a result, two drive systems (16.1.6) and (16.1.7) approach combination complex synchronization with response system (16.1.8).

II. Combination complex synchronization of three different fractional-order complex systems

Assume that the fractional-order complex Lü system and the fractional-order complex Chen system drive the fractional-order complex Lorenz system. Two drive systems are depicted as follows:

$$\begin{cases} D_*^\alpha x_1 = a_1(x_2 - x_1), \\ D_*^\alpha x_2 = -x_1 x_3 + a_2 x_2, \\ D_*^\alpha x_3 = \frac{1}{2}(\bar{x}_1 x_2 + x_1 \bar{x}_2) - a_3 x_3, \end{cases} \tag{16.1.9}$$

$$\begin{cases} D_*^\alpha y_1 = b_1(y_2 - y_1), \\ D_*^\alpha y_2 = (b_2 - b_1) y_1 - y_1 y_3 + b_2 y_2, \\ D_*^\alpha y_3 = \frac{1}{2}(\bar{y}_1 y_2 + y_1 \bar{y}_2) - b_3 y_3, \end{cases} \tag{16.1.10}$$

and the response system reads in the following form:

$$\begin{cases} D_*^\alpha z_1 = c_1(z_2 - z_1) + u_1 + j u_2, \\ D_*^\alpha z_2 = c_2 z_1 - z_2 - z_1 z_3 + u_3 + j u_4, \\ D_*^\alpha z_3 = \frac{1}{2}(\bar{z}_1 z_2 + z_1 \bar{z}_2) - c_3 z_3 + u_5, \end{cases} \tag{16.1.11}$$

where a_i, b_i, and c_i $(i = 1, 2, 3)$ are system parameters, x_i, y_i, and z_i $(i = 1, 2)$ are complex variables; x_3, y_3, and z_3 are real variables. u_i $(i = 1, 2, \ldots, 5)$ are real controllers to be designed.

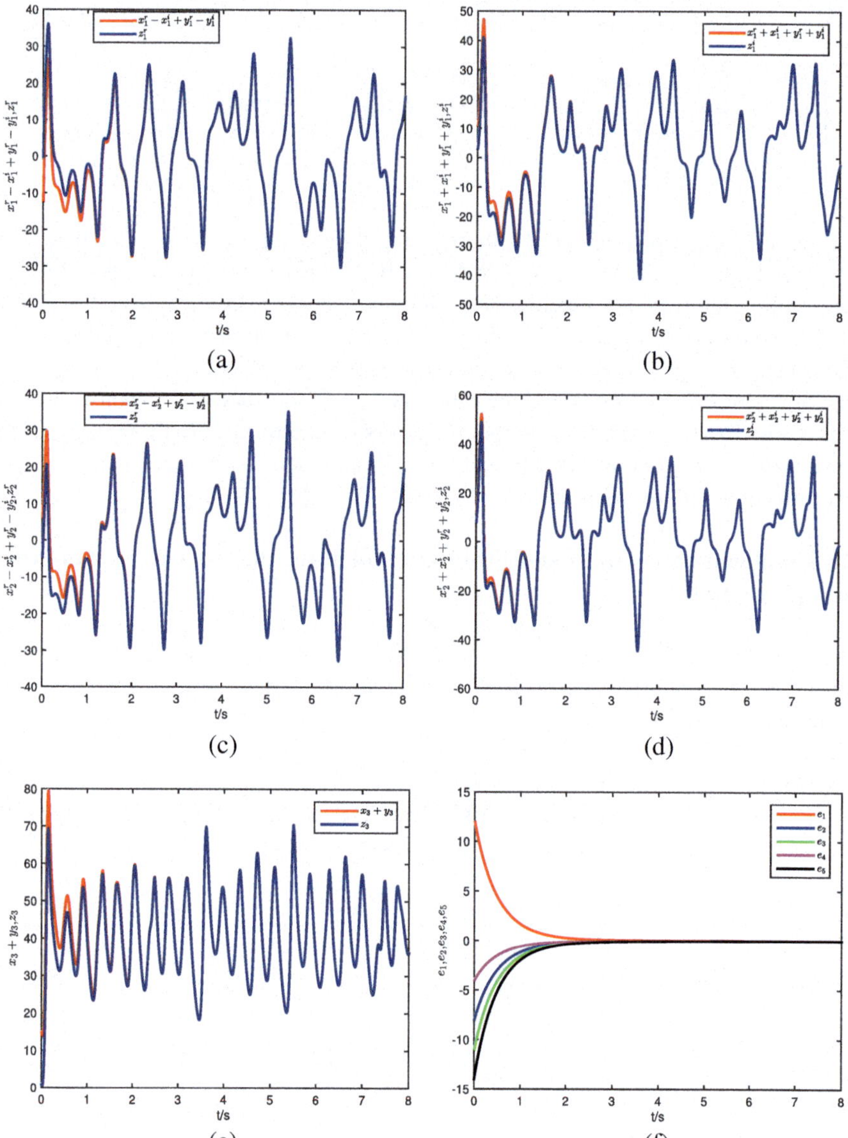

Fig. 16.1 State variables and synchronization error dynamics of drive systems (16.1.6), (16.1.7), and response system (16.1.8)

Assume that $\Phi = \mathrm{diag}(\varphi_1 + j\varphi_2, \varphi_3 + j\varphi_4, \varphi_5)$ and $\Psi = \mathrm{diag}(\phi_1 + j\phi_2, \phi_3 + j\phi_4, \phi_5)$. From Theorem 16.1, we obtain the designed controller as follows:

$$\begin{cases} u_1 = (k_1 - a_1)(\varphi_1 x_1^r - \varphi_2 x_1^i) + (k_1 - b_1)(\phi_1 y_1^r - \phi_2 y_1^i) + a_1(\varphi_1 x_2^r - \varphi_2 x_2^i) \\ \quad +b_1(\phi_1 y_2^r - \phi_2 y_2^i) - c_1 z_2^r + (c_1 - k_1)z_1^r, \\ u_2 = (k_1 - a_1)(\varphi_2 x_1^r + \varphi_1 x_1^i) + (k_1 - b_1)(\phi_2 y_1^r + \phi_1 y_1^i) + a_1(\varphi_2 x_2^r + \varphi_1 x_2^i) \\ \quad +b_1(\phi_2 y_2^r + \phi_1 y_2^i) - c_1 z_2^i + (c_1 - k_1)z_1^i, \\ u_3 = -\varphi_3 x_1^r x_3 + \varphi_4 x_1^i x_3 + (k_2 + a_2)(\varphi_3 x_2^r - \varphi_4 x_2^i) + (b_2 - b_1)(\phi_3 y_1^r - \phi_4 y_1^i) \\ \quad +(k_2 + b_2)(\phi_3 y_2^r - \phi_4 y_2^i) - \phi_3 y_1^r y_3 + \phi_4 y_1^i y_3 - c_2 z_1^r + z_1^r z_3 + (1 - k_2)z_2^r, \\ u_4 = -\varphi_4 x_1^r x_3 - \varphi_3 x_1^i x_3 + (k_2 + a_2)(\varphi_4 x_2^r + \varphi_3 x_2^i) + (b_2 - b_1)(\phi_4 y_1^r + \phi_3 y_1^i) \\ \quad +(k_2 + b_2)(\phi_4 y_2^r + \phi_3 y_2^i) - \phi_4 y_1^r y_3 - \phi_3 y_1^i y_3 - c_2 z_1^i + z_1^i z_3 + (1 - k_2)z_2^i, \\ u_5 = (c_3 - k_3)z_3 + \varphi_5(x_1^r x_2^r + x_1^i x_2^i) + (k_3 - a_3)\varphi_5 x_3 + \phi_5(y_1^r y_2^r + y_1^i y_2^i) \\ \quad +(k_3 - b_3)\phi_5 y_3 - z_1^r z_2^r - z_1^i z_2^i. \end{cases}$$

In the numerical simulations, $\alpha = 0.95$, the initial values are taken as $x(0) = (1 + 2j, 3 + 4j, 5)^T$, $y(0) = (6 + 9j, 5 + 7j, 12)^T$, and $z(0) = (2 + 3j, 5 + 6j, 9)^T$, respectively. The system parameters are chosen as $(a_1, a_2, a_3) = (42, 22, 5)$, $(b_1, b_2, b_3) = (35, 28, 3)$, and $(c_1, c_2, c_3) = (10, 180, 1)$. In the simulation process, we assume $\Phi = \mathrm{diag}(1 + j, 1 + j, 1)$, $\Psi = \mathrm{diag}(1 + j, 1 + j, 1)$, and $k_1 = k_2 = k_3 = 1$. The corresponding numerical results are shown in Fig. 16.2. The synchronization process of systems (16.1.9)–(16.1.11) is described in Figs. 16.2(a)–(e), where the red line presents the states of drive systems and the blue line shows the states of response system. Figure 16.2(f) reveals all error states converge asymptotically to zero, which demonstrates that drive systems (16.1.9) and (16.1.10) achieve combination complex synchronization with response system (16.1.11).

16.2 Combination Complex Synchronization of Incommensurate Fractional-Order Chaotic Systems

In this section, we investigate combination complex synchronization among three incommensurate fractional-order chaotic systems. Based on the stability theory of incommensurate fractional-order systems and the feedback control technique, we obtain some robust criteria on combination complex synchronization. Notably, the proposed combination complex synchronization can establish a link between the incommensurate fractional-order complex chaos and real chaos.

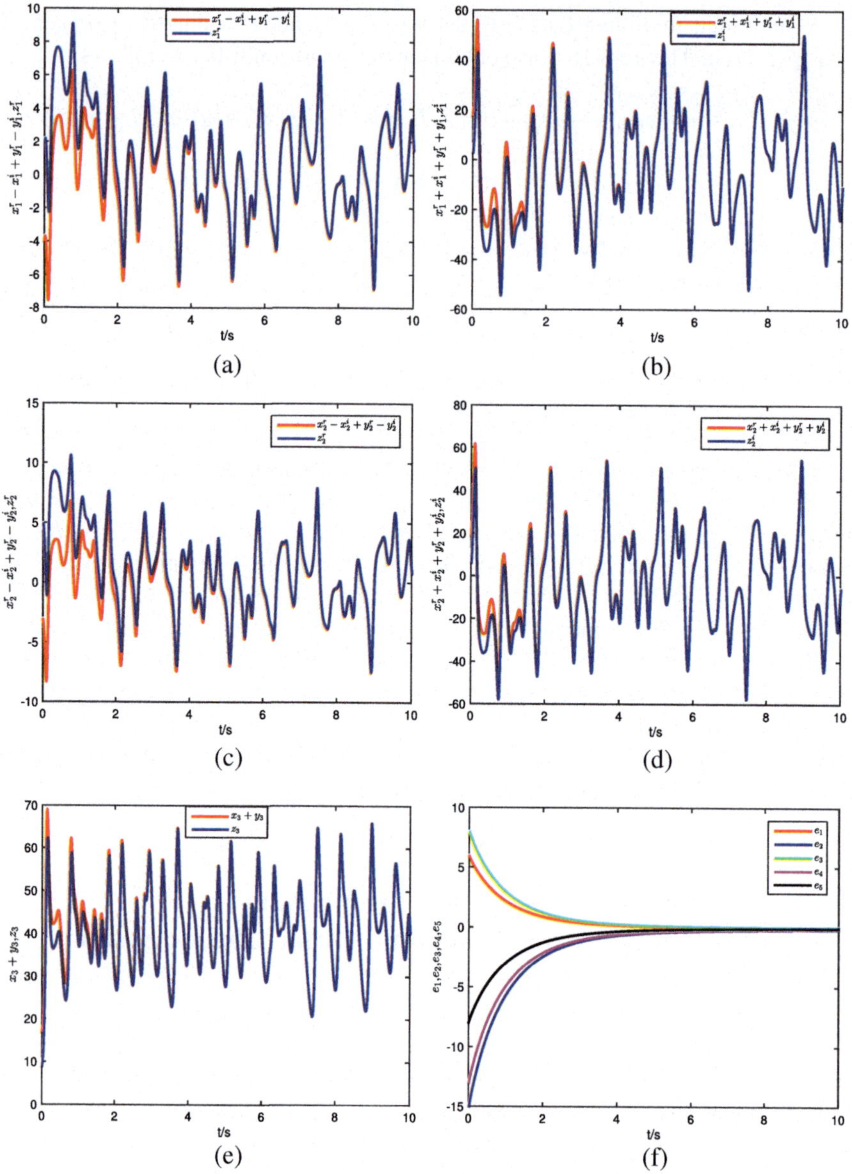

Fig. 16.2 State variables and synchronization error dynamics of drive systems (16.1.9), (16.1.10), and response system (16.1.11)

16.2.1 Combination Complex Synchronization Between Two Fractional-Order Real Drive Systems and One Fractional-Order Complex Response System

A fractional-order real chaotic system is taken as the first drive system

$$D_*^\alpha \omega = p(\omega), \tag{16.2.1}$$

the second real chaotic drive system is described as:

$$D_*^\alpha \upsilon = q(\upsilon), \tag{16.2.2}$$

while we define a fractional-order complex chaotic system as the response system:

$$D_*^\alpha z = h(z) + U, \tag{16.2.3}$$

where $\omega = (\omega_1, \omega_2, \ldots, \omega_n)^T$ and $\upsilon = (\upsilon_1, \upsilon_2, \ldots, \upsilon_n)^T$ denote real state vectors of systems (16.2.1) and (16.2.2), respectively, $z = z^r + jz^i \in \mathbb{C}^{n\times 1}$ is a complex state vector of system (16.2.3), $p = (p_1, p_2, \ldots, p_n)^T$ and $q = (q_1, q_2, \ldots, q_n)^T$ are vectors of nonlinear real functions, $h = (h_1, h_2, \ldots, h_n)^T$ is a nonlinear complex function, and U is a controller to be designed, $U = (U_1, U_2, \ldots, U_n)^T = U^r + jU^i$, $U^r = (u_1, u_3, \ldots, u_{2n-1})^T$, $U^i = (u_2, u_4, \ldots, u_{2n})^T$.

Definition 16.2 Two drive systems (16.2.1) and (16.2.2) are said to achieve combination complex synchronization with the response system (16.2.3) if there exist two complex matrices $\Phi = \Phi^r + j\Phi^i \in \mathbb{C}^{n\times n}$ and $\Psi = \Psi^r + j\Psi^i \in \mathbb{C}^{n\times n}$, such that

$$\lim_{t\to\infty} ||e(t)|| = \lim_{t\to\infty} ||z(t) - \Phi\omega(t) - \Psi\upsilon(t)|| = 0,$$

where $e = e^r + je^i$ is called the error vector, $e^r = (e_1, e_3, \ldots, e_{2n-1})^T = z^r - \Phi^r\omega - \Psi^r\upsilon$, $e^i = (e_2, e_4, \ldots, e_{2n})^T = z^i - \Phi^i\omega - \Psi^i\upsilon$.

In our synchronization scheme, we assume for convenience that Φ and Ψ are diagonal matrices of the form $\Phi = \mathrm{diag}(\varphi_1 + j\varphi_2, \varphi_3 + j\varphi_4, \ldots, \varphi_{2n-1} + j\varphi_{2n})$ and $\Psi = \mathrm{diag}(\phi_1 + j\phi_2, \phi_3 + j\phi_4, \ldots, \phi_{2n-1} + j\phi_{2n})$. If there exists $z_l \in \mathbb{R}$, then $\varphi_{2l} = \phi_{2l} = 0$ are chosen to avoid increasing a new imaginary part in the response system $(l = 1, 2, \ldots, n)$.

The primary objective of this paper is to achieve combination complex synchronization among three incommensurate fractional-order chaotic systems by designing appropriate controllers. Next, we study the incommensurate fractional-order real Chen system and real Rössler system drive fractional-order complex Lü system.

Consider the incommensurate fractional-order real Chen system as the first drive system:

$$\begin{cases} D_*^{\alpha_1}\omega_1 = d_1(\omega_2 - \omega_1), \\ D_*^{\alpha_2}\omega_2 = (d_2 - d_1)\omega_1 - \omega_1\omega_3 + d_2\omega_2, \\ D_*^{\alpha_3}\omega_3 = \omega_1\omega_2 - d_3\omega_3, \end{cases} \tag{16.2.4}$$

and the following incommensurate fractional-order real Rössler system is set to the second drive system:

$$\begin{cases} D_*^{\alpha_1}\upsilon_1 = -(\upsilon_2 + \upsilon_3), \\ D_*^{\alpha_2}\upsilon_2 = \upsilon_1 + \delta_1\upsilon_2, \\ D_*^{\alpha_3}\upsilon_3 = \delta_2 + \upsilon_3(\upsilon_1 - \delta_3). \end{cases} \tag{16.2.5}$$

The incommensurate fractional-order complex Lü system is defined as the response system:

$$\begin{cases} D_*^{\alpha_1}z_1 = c_1(z_2 - z_1) + U_1, \\ D_*^{\alpha_2}z_2 = c_2z_2 - z_1z_3 + U_2, \\ D_*^{\alpha_3}z_3 = \frac{1}{2}(\bar{z}_1z_2 + z_1\bar{z}_2) - c_3z_3 + U_3, \end{cases} \tag{16.2.6}$$

where z_1 and z_2 are complex variables, ω_i, υ_i and z_3 are real state variables, d_i, δ_i and c_i are real parameters ($i = 1, 2, 3$). $U_1 = u_1 + ju_2$, $U_2 = u_3 + ju_4$ are complex functions and $U_3 = u_5$ is a real control function.

Let $\Phi = \mathrm{diag}(\varphi_1 + j\varphi_2, \varphi_3 + j\varphi_4, \varphi_5)$ and $\Psi = \mathrm{diag}(\phi_1 + j\phi_2, \phi_3 + j\phi_4, \phi_5)$ be two given scaling matrices. The synchronization error is introduced as follows:

$$\begin{cases} e_1 = z_1^r - \varphi_1\omega_1 - \phi_1\upsilon_1, \\ e_2 = z_1^i - \varphi_2\omega_1 - \phi_2\upsilon_1, \\ e_3 = z_2^r - \varphi_3\omega_2 - \phi_3\upsilon_2, \\ e_4 = z_2^i - \varphi_4\omega_2 - \phi_4\upsilon_2, \\ e_5 = z_3 - \varphi_5\omega_3 - \phi_5\upsilon_3. \end{cases}$$

By combining systems (16.2.4) and (16.2.5) with (16.2.6), we obtain:

$$\begin{cases} D_*^{\alpha_1}e_1 = c_1e_3 - c_1e_1 + (c_1\varphi_3 - d_1\varphi_1)\omega_2 + (d_1 - c_1)\varphi_1\omega_1 + (\phi_1 + \phi_3c_1)\upsilon_2 \\ \qquad +\phi_1(\upsilon_3 - c_1\upsilon_1) + u_1, \\ D_*^{\alpha_1}e_2 = c_1e_4 - c_1e_2 + (c_1\varphi_4 - d_1\varphi_2)\omega_2 + (d_1 - c_1)\varphi_2\omega_1 + (\phi_2 + \phi_4c_1)\upsilon_2 \\ \qquad +\phi_2(\upsilon_3 - c_1\upsilon_1) + u_2, \\ D_*^{\alpha_2}e_3 = c_2e_3 + (c_2 - d_2)\varphi_3\omega_2 - (d_2 - d_1)\varphi_3\omega_1 - \phi_3\upsilon_1 + (c_2 - \delta_1)\phi_3\upsilon_2 \\ \qquad -z_1^rz_3 + \varphi_3\omega_1\omega_3 + u_3, \\ D_*^{\alpha_2}e_4 = c_2e_4 + (c_2 - d_2)\varphi_4\omega_2 - (d_2 - d_1)\varphi_4\omega_1 - \phi_4\upsilon_1 + (c_2 - \delta_1)\phi_4\upsilon_2 \\ \qquad -z_1^iz_3 + \varphi_4\omega_1\omega_3 + u_4, \\ D_*^{\alpha_3}e_5 = -c_3e_5 + (d_3 - c_3)\varphi_5\omega_3 + (\delta_3 - c_3 - \upsilon_1)\phi_5\upsilon_3 - \phi_5\delta_2 - \varphi_5\omega_1\omega_2 \\ \qquad +z_1^rz_2^r + z_1^iz_2^i + u_5. \end{cases}$$

Then, the active control functions $u_i(t)$ $(i = 1, 2, \ldots, 5)$ are defined as:

$$\begin{cases} u_1(t) = (d_1\varphi_1 - c_1\varphi_3)\omega_2 - (d_1 - c_1)\varphi_1\omega_1 - (\phi_1 + \phi_3 c_1)\upsilon_2 - \phi_1(\upsilon_3 - c_1\upsilon_1) + \beta_1(t), \\ u_2(t) = (d_1\varphi_2 - c_1\varphi_4)\omega_2 - (d_1 - c_1)\varphi_2\omega_1 - (\phi_2 + \phi_4 c_1)\upsilon_2 - \phi_2(\upsilon_3 - c_1\upsilon_1) + \beta_2(t), \\ u_3(t) = (d_2 - c_2)\varphi_3\omega_2 + (d_2 - d_1)\varphi_3\omega_1 + \phi_3\upsilon_1 - (c_2 - \delta_1)\phi_3\upsilon_2 + z_1^r z_3 - \varphi_3\omega_1\omega_3 + \beta_3(t), \\ u_4(t) = (d_2 - c_2)\varphi_4\omega_2 + (d_2 - d_1)\varphi_4\omega_1 + \phi_4\upsilon_1 - (c_2 - \delta_1)\phi_4\upsilon_2 + z_1^i z_3 - \varphi_4\omega_1\omega_3 + \beta_4(t), \\ u_5(t) = (c_3 - d_3)\varphi_5\omega_3 - (\delta_3 - c_3 - \upsilon_1)\phi_5\upsilon_3 + \phi_5\delta_2 + \varphi_5\omega_1\omega_2 - z_1^r z_2^r - z_1^i z_2^i + \beta_5(t). \end{cases}$$

which provides

$$\begin{cases} D_*^{\alpha_1} e_1 = c_1 e_3 - c_1 e_1 + \beta_1(t), \\ D_*^{\alpha_1} e_2 = c_1 e_4 - c_1 e_2 + \beta_2(t), \\ D_*^{\alpha_2} e_3 = c_2 e_3 + \beta_3(t), \\ D_*^{\alpha_2} e_4 = c_2 e_4 + \beta_4(t), \\ D_*^{\alpha_3} e_5 = -c_3 e_5 + \beta_5(t). \end{cases} \tag{16.2.7}$$

The synchronization error system described above is a linear system governed by active control inputs $\beta_l(t)$ $(l = 1, 2, \ldots, 5)$. Subsequently, we design an appropriate feedback control scheme to achieve stabilization of the error system. There are many possible choices for the control inputs $\beta_l(t)$ $(l = 1, 2, \ldots, 5)$. We set

$$(\beta_1(t), \beta_2(t), \beta_3(t), \beta_4(t), \beta_5(t))^T = L(e_1(t), e_2(t), e_3(t), e_4(t), e_5(t))^T,$$

where $L = (l_{kl})_{5\times5}$ is a 5×5 real matrix. So system (16.2.7) can be rewritten as

$$D_*^{\alpha} e = \Xi e,$$

where

$$\Xi = \begin{pmatrix} l_{11} - c_1 & l_{12} & l_{13} + c_1 & l_{14} & l_{15} \\ l_{21} & l_{22} - c_1 & l_{23} & l_{24} + c_1 & l_{25} \\ l_{31} & l_{32} & l_{33} + c_2 & l_{34} & l_{35} \\ l_{41} & l_{42} & l_{43} & l_{44} + c_2 & l_{45} \\ l_{51} & l_{52} & l_{53} & l_{54} & l_{55} - c_3 \end{pmatrix}.$$

The matrix L is not uniquely determined and one effective choice is as follows: $l_{13} = l_{24} = -42$, $l_{22} = 2$, $l_{33} = -24$, $l_{44} = -26$, $l_{55} = 2$, $l_{11} = l_{12} = l_{14} = l_{15} = l_{23} = l_{25} = l_{34} = l_{35} = l_{45} = 0$, and other elements of L can be chosen arbitrarily. Using $c_1 = 42$, $c_2 = 22$, $c_3 = 5$, $(\alpha_1, \alpha_2, \alpha_3) = (0.91, 0.94, 0.96)$, $m = 100$, and $\gamma = 1/100$, we have

$$(\lambda^{91} + 42)(\lambda^{91} + 40)(\lambda^{94} + 2)(\lambda^{94} + 4)(\lambda^{96} + 3) = 0. \tag{16.2.8}$$

Simple calculation shows that all roots of (16.2.8) satisfy $|\arg(\lambda)| > \gamma\pi/2$. According to Theorem 1.4, the error vector $e(t)$ asymptotically converges to zero as $t \to \infty$.

The initial values of systems (16.2.4), (16.2.5), and (16.2.6) are chosen as $(\omega_1, \omega_2, \omega_3)^T = (-6, -3, 2)^T$, $(\upsilon_1, \upsilon_2, \upsilon_3)^T = (4, 5, 6)^T$, and $(z_1, z_2, z_3)^T = (-1 + 3j, -5 + 8j, 1)^T$, respectively. We select the scaling matrices $\Phi = \mathrm{diag}(1 -$

$2j, 1+4j, 1)$ and $\Psi = \text{diag}(2-2j, 3+2j, 2)$, and set $l_{31} = l_{42} = l_{53} = 1$ and $l_{21} = l_{32} = l_{41} = l_{43} = l_{51} = l_{52} = l_{54} = 0$. The corresponding numerical results are presented in Fig. 16.3. The state variables of the two drive systems and the response system are displayed in Figs. 16.3(a)–(e). As shown in Fig. 16.3(f), the synchronization errors converge asymptotically to zero. Hence, the fractional-order real Chen system, real Rössler system, and complex Lü system with incommensurate order achieve combination complex synchronization.

16.2.2 Combination Complex Synchronization Between Two Fractional-Order Complex Drive Systems and One Fractional-Order Response System

Consider two fractional-order complex chaotic systems of the form:

$$D_*^{\alpha} x = f(x) \tag{16.2.9}$$

and

$$D_*^{\alpha} y = g(y), \tag{16.2.10}$$

with the fractional-order real chaotic response system is described by

$$D_*^{\alpha} \upsilon = q(\upsilon) + U, \tag{16.2.11}$$

where $x = x^r + jx^i \in \mathbb{C}^{n\times 1}$ and $y = y^r + jy^i \in \mathbb{C}^{n\times 1}$ denote complex state vectors $\upsilon = (\upsilon_1, \upsilon_2, \ldots, \upsilon_n)^T$ is a real state vector $f = (f_1, f_2, \ldots, f_n)^T$ and $g = (g_1, g_2, \ldots, g_n)^T$ represent vectors of nonlinear complex functions, $q = (q_1, q_2, \ldots, q_n)^T$ is a nonlinear real function.

Definition 16.3 Two complex drive systems (16.2.9), (16.2.10), and one real response system (16.2.11) achieve combination complex synchronization of real parts if there exist two complex matrices $\Phi = \Phi^r + j\Phi^i \in \mathbb{C}^{n\times n}$ and $\Psi = \Psi^r + j\Psi^i \in \mathbb{C}^{n\times n}$, such that

$$\lim_{t\to\infty} ||e(t)|| = \lim_{t\to\infty} ||\upsilon - \Phi^r x^r + \Phi^i x^i - \Psi^r y^r + \Psi^i y^i|| = 0,$$

where $e = \upsilon - \Phi^r x^r + \Phi^i x^i - \Psi^r y^r + \Psi^i y^i$ is called the error real vector.

Remark 16.2 As $\upsilon(t)$ is real, a real U is chosen to ensure combination complex synchronization of real parts and avoid increasing the imaginary parts of response system.

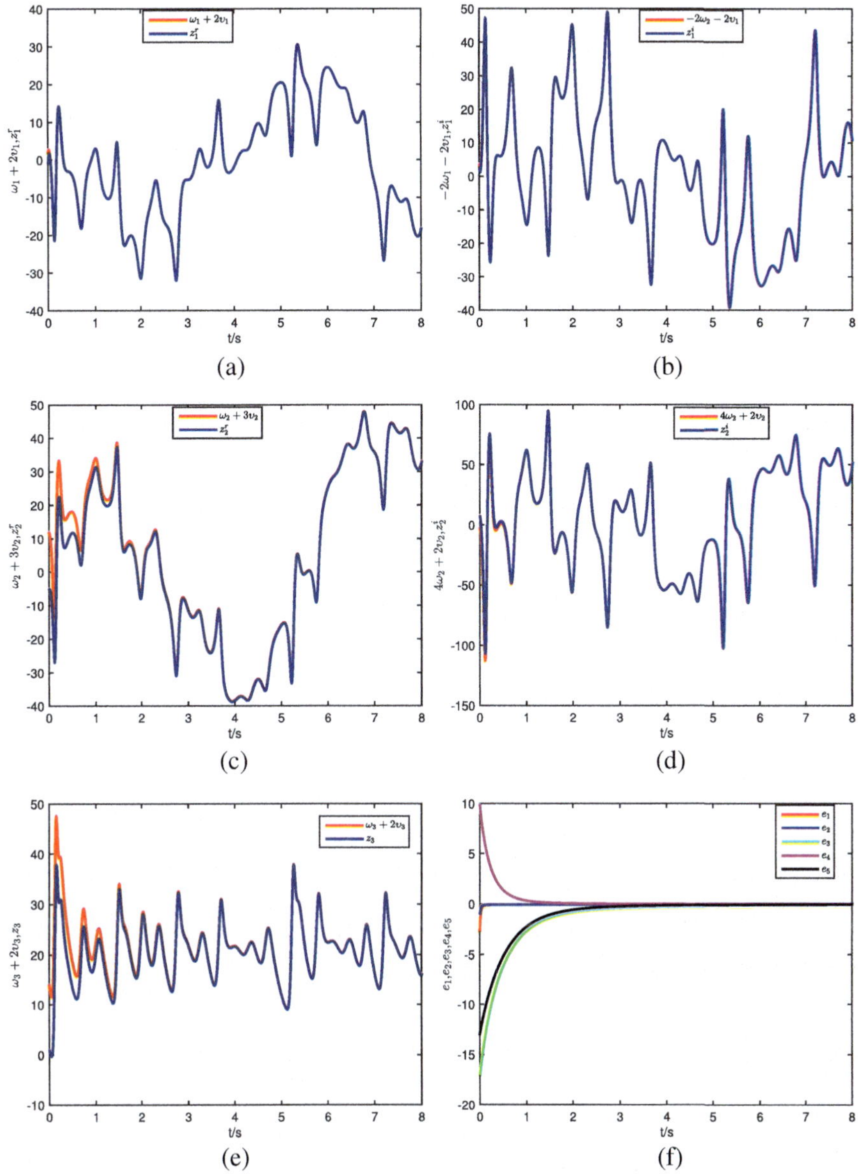

Fig. 16.3 State variables and synchronization error dynamics of drive systems (16.2.4), (16.2.5), and response system (16.2.6)

In the following, we investigate combination complex synchronization among the fractional-order complex Lorenz, complex Chen, and real Rössler systems of incommensurate orders.

The first drive system is chosen as the fractional-order complex Lorenz system:

$$\begin{cases} D_*^{\alpha_1} x_1 = a_1(x_2 - x_1), \\ D_*^{\alpha_2} x_2 = a_2 x_1 - x_2 - x_1 x_3, \\ D_*^{\alpha_3} x_3 = \frac{1}{2}(\bar{x}_1 x_2 + x_1 \bar{x}_2) - a_3 x_3, \end{cases} \tag{16.2.12}$$

and the fractional-order complex Chen system as the second drive system is depicted as:

$$\begin{cases} D_*^{\alpha_1} y_1 = b_1(y_2 - y_1), \\ D_*^{\alpha_2} y_2 = (b_2 - b_1) y_1 + b_2 y_2 - y_1 y_3, \\ D_*^{\alpha_3} y_3 = \frac{1}{2}(\bar{y}_1 y_2 + y_1 \bar{y}_2) - b_3 y_3. \end{cases} \tag{16.2.13}$$

We define the response system as the fractional-order real Rössler system:

$$\begin{cases} D_*^{\alpha_1} \upsilon_1 = -(\upsilon_2 + \upsilon_3) + U_1, \\ D_*^{\alpha_2} \upsilon_2 = \upsilon_1 + \delta_1 \upsilon_2 + U_2, \\ D_*^{\alpha_3} \upsilon_3 = \delta_2 + \upsilon_3(\upsilon_1 - \delta_3) + U_3, \end{cases} \tag{16.2.14}$$

where x_i and y_i are complex variables $(i = 1, 2)$, x_3, y_3 and υ_i $(i = 1, 2, 3)$ are real variables, a_i, b_i and δ_i are real parameters $(i = 1, 2, 3)$, and U_i are real control functions $(i = 1, 2, 3)$.

For two given scaling matrices $\Phi = \mathrm{diag}(\varphi_1 + j\varphi_2, \varphi_3 + j\varphi_4, \varphi_5)$ and $\Psi = \mathrm{diag}(\phi_1 + j\phi_2, \phi_3 + j\phi_4, \phi_5)$, the synchronization error can be expressed as

$$\begin{cases} e_1 = \upsilon_1 - \varphi_1 x_1^r + \varphi_2 x_1^i - \phi_1 y_1^r + \phi_2 y_1^i, \\ e_2 = \upsilon_2 - \varphi_3 x_2^r + \varphi_4 x_2^i - \phi_3 y_2^r + \phi_4 y_2^i, \\ e_3 = \upsilon_3 - \varphi_5 x_3 - \phi_5 y_3. \end{cases}$$

From systems (16.2.12), (16.2.13), and (16.2.14), the error dynamical system can be derived as:

$$\begin{cases} D_*^{\alpha_1} e_1 = -e_2 - e_3 - (\varphi_3 + a_1\varphi_1)x_2^r + (\varphi_4 + a_1\varphi_2)x_2^i + a_1(\varphi_1 x_1^r - \varphi_2 x_1^i) \\ \qquad -(\phi_3 + b_1\phi_1)y_2^r + (\phi_4 + b_1\phi_2)y_2^i + b_1(\phi_1 y_1^r - \phi_2 y_1^i) - \varphi_5 x_3 - \phi_5 y_3 + U_1, \\ D_*^{\alpha_1} e_2 = e_1 + (\varphi_1 - a_2\varphi_3)x_1^r - (\varphi_2 - a_2\varphi_4)x_1^i + (\delta_1 + 1)(\varphi_3 x_2^r - \varphi_4 x_2^i) \\ \qquad +\varphi_3 x_1^r x_3 - \varphi_4 x_1^i x_3 + [\phi_1 - (b_2 - b_1)\phi_3]y_1^r - [\phi_2 - (b_2 - b_1)\phi_4]y_1^i \\ \qquad +\delta_1 e_2 + (\delta_1 - b_2)(\phi_3 y_2^r - \phi_4 y_2^i) + \phi_3 y_1^r y_3 - \phi_4 y_1^i y_3 + U_2, \\ D_*^{\alpha_2} e_3 = (a_3 - \delta_3)\varphi_5 x_3 + (b_3 - \delta_3)\phi_5 y_3 + \delta_2 + \upsilon_1\upsilon_3 - \varphi_5(x_1^r x_2^r + x_1^i x_2^i) \\ \qquad -\delta_3 e_3 - \phi_5(y_1^r y_2^r + y_1^i y_2^i) + U_3. \end{cases}$$

Thus, we design the control inputs $U_i(t)$ $(i = 1, 2, 3)$ as:

$$\begin{cases} U_1(t) = (\varphi_3 + a_1\varphi_1)x_2^r - (\varphi_4 + a_1\varphi_2)x_2^i - a_1(\varphi_1 x_1^r - \varphi_2 x_1^i) + (\phi_3 + b_1\phi_1)y_2^r \\ \qquad -(\phi_4 + b_1\phi_2)y_2^i - b_1(\phi_1 y_1^r - \phi_2 y_1^i) + \varphi_5 x_3 + \phi_5 y_3 + \beta_1(t), \\ U_2(t) = (a_2\varphi_3 - \varphi_1)x_1^r + (\varphi_2 - a_2\varphi_4)x_1^i - (\delta_1 + 1)(\varphi_3 x_2^r - \varphi_4 x_2^i) \\ \qquad -\varphi_3 x_1^r x_3 + \varphi_4 x_1^i x_3 - [\phi_1 - (b_2 - b_1)\phi_3]y_1^r + [\phi_2 - (b_2 - b_1)\phi_4]y_1^i \\ \qquad -(\delta_1 - b_2)(\phi_3 y_2^r - \phi_4 y_2^i) - \phi_3 y_1^r y_3 + \phi_4 y_1^i y_3 + \beta_2(t), \\ U_3(t) = (\delta_3 - a_3)\varphi_5 x_3 - (b_3 - \delta_3)\phi_5 y_3 - \delta_2 - \upsilon_1\upsilon_3 + \varphi_5(x_1^r x_2^r + x_1^i x_2^i) \\ \qquad +\phi_5(y_1^r y_2^r + y_1^i y_2^i) + \beta_3(t), \end{cases}$$

which yields

$$\begin{cases} D_*^{\alpha_1} e_1 = -e_2 - e_3 + \beta_1(t), \\ D_*^{\alpha_2} e_2 = e_1 + \delta_1 e_2 + \beta_2(t), \\ D_*^{\alpha_3} e_3 = -\delta_3 e_3 + \beta_3(t). \end{cases}$$

On the basis of the fact that the term $\beta_i(t)$ are linear functions of the error terms $e_i(t)$ $(i = 1, 2, 3)$, we choose $(\beta_1(t), \beta_2(t), \beta_3(t))^T = L(e_1(t), e_2(t), e_3(t))^T$, where $L = (l_{kl})_{3\times 3}$ is a real matrix. Next, we choose that $l_{11} = l_{21} = -1$, $l_{22} = -2.4$, $l_{33} = 5$, $l_{31} = l_{32} = 0$, and l_{13}, l_{23} can be chosen arbitrarily. The characteristic equation is obtained with the values $(\delta_1, \delta_2, \delta_3) = (0.4, 0.2, 10)$, $(\alpha_1, \alpha_2, \alpha_3) = (0.91, 0.96, 0.99)$, $m = 100$, and $\gamma = 1/100$, as shown below:

$$(\lambda^{91} + 1)(\lambda^{96} + 2)(\lambda^{99} + 5) = 0. \tag{16.2.15}$$

It can be verified that all roots of (16.2.15) satisfy $|\arg(\lambda)| > \gamma\pi/2$. According to Theorem 1.4, the error vector $e(t)$ asymptotically converges to zero as $t \to \infty$.

The initial conditions for systems (16.2.12), (16.2.13), and (16.2.14) are set to $(x_1, x_2, x_3)^T = (3 + 4j, -5 + 12j, -7)^T$, $(y_1, y_2, y_3)^T = (6 - 9j, -8 + 6j, 1)^T$, and $(\upsilon_1, \upsilon_2, \upsilon_3)^T = (4, 5, 6)^T$, respectively. Taking two scaling matrices as $\Phi = \text{diag}(j, -3 + 2j, -1)$, $\Psi = \text{diag}(1, 3 + j, 2)$, and selecting $l_{13} = 1$, $l_{12} = l_{23} = 0$, the simulation results are shown in Fig. 16.4. Figures 16.4(a)–(c) depict the synchronization process of systems (16.2.12), (16.2.13), and (16.2.14), with the states of the two drive systems and the response system represented by red and blue lines, respectively. Figure 16.4(d) depicts the synchronization errors converge asymptotically to zero, i.e., the incommensurate fractional-order complex Lorenz system, complex Chen system, and real Rössler system can achieve combination complex synchronization of real parts.

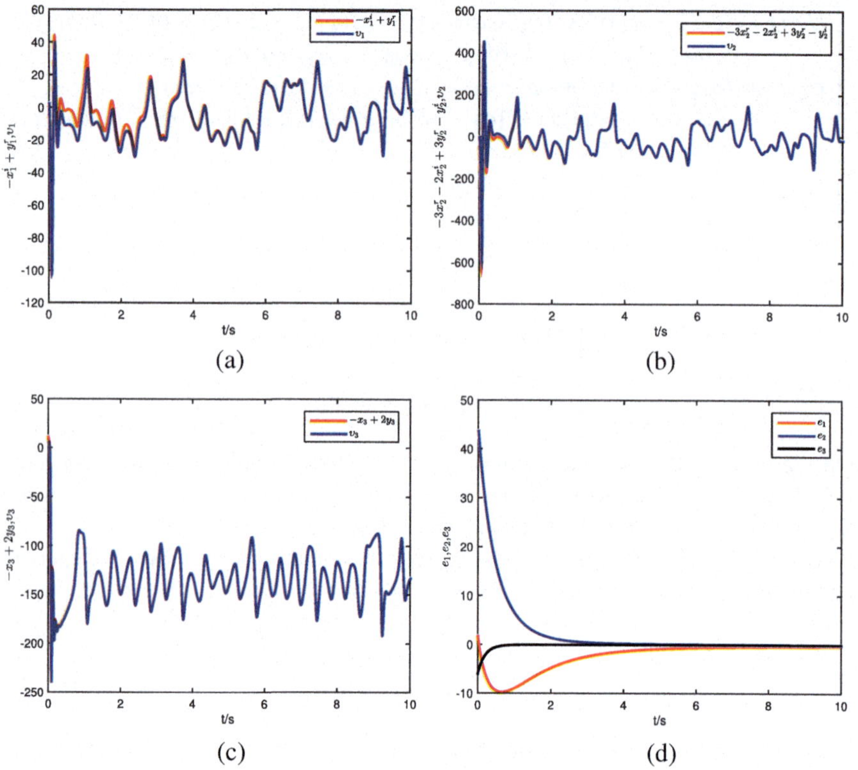

Fig. 16.4 State variables and synchronization error of drive systems (16.2.12), (16.2.13), and response system (16.2.14)

16.2.3 Combination Complex Synchronization of Incommensurate Fractional-Order Complex Chaotic Systems

Now, we study combination complex synchronization among fractional-order complex chaotic drive systems (16.2.9), (16.2.10) and response system (16.2.3).

Definition 16.4 Two drive systems (16.2.9) and (16.2.10) realize combination complex synchronization with the response system (16.2.3) if there exist two complex matrices $\Phi = \Phi^r + j\Phi^i \in \mathbb{C}^{n\times n}$ and $\Psi = \Psi^r + j\Psi^i \in \mathbb{C}^{n\times n}$, such that

$$\lim_{t\to\infty} ||e(t)|| = \lim_{t\to\infty} ||z(t) - \Phi x(t) - \Psi y(t)|| = 0,$$

where $e = e^r + je^i$ is called the error vector.

The following designs a controller to realize combination complex synchronization among the fractional-order complex Lorenz system, complex Lü system, and complex Chen system with incommensurate orders. The fractional-order complex Lorenz system (16.2.12) is defined as the first drive system, and the fractional-order complex Chen system (16.2.13) is taken as the second drive system, while the response system is the fractional-order complex Lü system described by (16.2.6).

Assume $\Phi = \mathrm{diag}(\varphi_1 + j\varphi_2, \varphi_3 + j\varphi_4, \varphi_5)$ and $\Psi = \mathrm{diag}(\phi_1 + j\phi_2, \phi_3 + j\phi_4, \phi_5)$, we produce the synchronization error as follows:

$$
\begin{cases}
e_1 = z_1^r - \varphi_1 x_1^r + \varphi_2 x_1^i - \phi_1 y_1^r + \phi_2 y_1^i, \\
e_2 = z_1^i - \varphi_1 x_1^i - \varphi_2 x_1^r - \phi_1 y_1^i - \phi_2 y_1^r, \\
e_3 = z_2^r - \varphi_3 x_2^r + \varphi_4 x_2^i - \phi_3 y_2^r + \phi_4 y_2^i, \\
e_4 = z_2^i - \varphi_3 x_2^i - \varphi_4 x_2^r - \phi_3 y_2^i - \phi_4 y_2^r, \\
e_5 = z_3 - \varphi_5 x_3 - \phi_5 y_3.
\end{cases}
$$

Based on systems (16.2.12), (16.2.13), and (16.2.6), the error dynamical system is obtained:

$$
\begin{cases}
D_*^{\alpha_1} e_1 = (c_1\varphi_3 - a_1\varphi_1)x_2^r + (a_1\varphi_2 - c_1\varphi_4)x_2^i + (a_1 - c_1)(\varphi_1 x_1^r - \varphi_2 x_1^i) + c_1 e_3 - c_1 e_1 \\
\qquad +(b_1 - c_1)(\phi_1 y_1^r - \phi_2 y_1^i) + (c_1\phi_3 - b_1\phi_1)y_2^r + (b_1\phi_2 - c_1\phi_4)y_2^i + u_1, \\
D_*^{\alpha_1} e_2 = (c_1\varphi_3 - a_1\varphi_1)x_2^i - (a_1\varphi_2 - c_1\varphi_4)x_2^r + (a_1 - c_1)(\varphi_2 x_1^r + \varphi_1 x_1^i) + c_1 e_4 - c_1 e_2 \\
\qquad +(b_1 - c_1)(\phi_2 y_1^r + \phi_1 y_1^i) + (c_1\phi_4 - b_1\phi_2)y_2^r - (b_1\phi_1 - c_1\phi_3)y_2^i + u_2, \\
D_*^{\alpha_2} e_3 = (c_2 + 1)(\varphi_3 x_2^r - \varphi_4 x_2^i) + a_2(\varphi_4 x_1^i - \varphi_3 x_1^r) + c_2 e_3 - z_1^r z_3 + (c_2 - b_2)(\phi_3 y_2^r - \phi_4 y_2^i) \\
\qquad -(b_2 - b_1)(\phi_3 y_1^r - \phi_4 y_1^i) + \varphi_3 x_1^r x_3 - \varphi_4 x_1^i x_3 + \phi_3 y_1^r y_3 - \phi_4 y_1^i y_3 + u_3, \\
D_*^{\alpha_2} e_4 = (c_2 + 1)(\varphi_3 x_2^i + \varphi_4 x_2^r) - a_2(\varphi_4 x_1^r + \varphi_3 x_1^i) + c_2 e_4 - z_1^i z_3 + (c_2 - b_2)(\phi_3 y_2^i + \phi_4 y_2^r) \\
\qquad -(b_2 - b_1)(\phi_3 y_1^i + \phi_4 y_1^r) + \varphi_3 x_1^i x_3 + \varphi_4 x_1^r x_3 + \phi_3 y_1^i y_3 + \phi_4 y_1^r y_3 + u_4, \\
D_*^{\alpha_3} e_5 = -c_3 e_5 + (a_3 - c_3)\varphi_5 x_3 + (b_3 - c_3)\phi_5 y_3 - \varphi_5(x_1^r x_2^r + x_1^i x_2^i) + z_1^r z_2^r \\
\qquad +z_1^i z_2^i - \phi_5(y_1^r y_2^r + y_1^i y_2^i) + u_5.
\end{cases}
$$

Therefore, the active control inputs $u_i(t)$ $(i = 1, 2, \ldots, 5)$ are defined as:

$$
\begin{cases}
u_1(t) = (a_1\varphi_1 - c_1\varphi_3)x_2^r - (a_1\varphi_2 - c_1\varphi_4)x_2^i - (a_1 - c_1)(\varphi_1 x_1^r - \varphi_2 x_1^i) - (b_1 - c_1)\phi_1 y_1^r \\
\qquad +(b_1 - c_1)\phi_2 y_1^i - (c_1\phi_3 - b_1\phi_1)y_2^r - (b_1\phi_2 - c_1\phi_4)y_2^i + \beta_1(t), \\
u_2(t) = (a_1\varphi_1 - c_1\varphi_3)x_2^i + (a_1\varphi_2 - c_1\varphi_4)x_2^r - (a_1 - c_1)(\varphi_2 x_1^r + \varphi_1 x_1^i) \\
\qquad -(b_1 - c_1)\phi_2 y_1^r - (b_1 - c_1)\phi_1 y_1^i - (c_1\phi_4 - b_1\phi_2)y_2^r \\
\qquad +(b_1\phi_1 - c_1\phi_3)y_2^i + \beta_2(t), \\
u_3(t) = (c_2 + 1)(\varphi_4 x_2^i - \varphi_3 x_2^r) - a_2(\varphi_4 x_1^i - \varphi_3 x_1^r) + \varphi_4 x_1^i x_3 + z_1^r z_3 \\
\qquad +(b_2 - b_1)(\phi_3 y_1^r - \phi_4 y_1^i) - (c_2 - b_2)(\phi_3 y_2^r - \phi_4 y_2^i) - \varphi_3 x_1^r x_3 \\
\qquad -\phi_3 y_1^r y_3 + \phi_4 y_1^i y_3 + \beta_3(t), \\
u_4(t) = -(c_2 + 1)(\varphi_3 x_2^i + \varphi_4 x_2^r) + a_2(\varphi_4 x_1^r + \varphi_3 x_1^i) - \varphi_3 x_1^i x_3 + z_1^i z_3 \\
\qquad -\varphi_4 x_1^r x_3 + (b_2 - b_1)(\phi_3 y_1^i + \phi_4 y_1^r) - (c_2 - b_2)(\phi_3 y_2^i + \phi_4 y_2^r) \\
\qquad -\phi_3 y_1^i y_3 - \phi_4 y_1^r y_3 + \beta_4(t), \\
u_5(t) = (c_3 - a_3)\varphi_5 x_3 - (b_3 - c_3)\varphi_5 y_3 - z_1^r z_2^r - z_1^i z_2^i + \varphi_5(x_1^r x_2^r + x_1^i x_2^i) \\
\qquad +\phi_5(y_1^r y_2^r + y_1^i y_2^i) + \beta_5(t),
\end{cases}
$$

which yields

$$\begin{cases} D_*^{\alpha_1} e_1 = c_1 e_3 - c_1 e_1 + \beta_1(t), \\ D_*^{\alpha_1} e_2 = c_1 e_4 - c_1 e_2 + \beta_2(t), \\ D_*^{\alpha_2} e_3 = c_2 e_3 + \beta_3(t), \\ D_*^{\alpha_2} e_4 = c_2 e_4 + \beta_4(t), \\ D_*^{\alpha_3} e_5 = -c_3 e_5 + \beta_5(t). \end{cases}$$

Consider the fact that the term $\beta_i(t)$ are linear functions of the error terms $e_i(t)$ $(i = 1, 2, \ldots, 5)$, we choose $(\beta_1(t), \beta_2(t), \beta_3(t), \beta_4(t), \beta_5(t))^T = L(e_1(t), e_2(t), e_3(t), e_4(t), e_5(t))^T$, where $L = (l_{ij})_{5\times 5}$ is a real matrix. L can be set as follows: $l_{11} = 40$, $l_{22} = 38$, $l_{33} = -23$, $l_{44} = -25$, $l_{55} = 3$, $l_{21} = l_{31} = l_{32} = l_{41} = l_{42} = l_{43} = l_{51} = l_{52} = l_{53} = l_{54} = 0$, and other elements can be chosen arbitrarily. Taking $(c_1, c_2, c_3) = (42, 22, 5)$, $(\alpha_1, \alpha_2, \alpha_3) = (0.92, 0.95, 0.98)$, $m = 100$, and $\gamma = 1/100$, the characteristic equation is given by:

$$(\lambda^{92} + 2)(\lambda^{92} + 4)(\lambda^{95} + 1)(\lambda^{95} + 3)(\lambda^{98} + 2) = 0. \tag{16.2.16}$$

It can be shown that all roots of (16.2.16) satisfy the condition $|\arg(\lambda)| > \gamma\pi/2$. According to Theorem 16.1.2, the error vector $e(t)$ asymptotically converges to zero as $t \to \infty$. Therefore, combination complex synchronization between the incommensurate fractional-order chaotic complex systems (16.2.12), (16.2.13), and (16.2.6) is achieved.

The numerical simulations are conducted with the initial conditions $(x_1, x_2, x_3)^T = (3 + 4j, -5 + 2j, -7)^T$, $(y_1, y_2, y_3)^T = (6 - 9j, -8 + 6j, 1)^T$, $(z_1, z_2, z_3)^T = (24 + 8j, -15 - 2j, 10)^T$, and with the scaling matrices $\Phi = \mathrm{diag}(3 + j, -1 + 2j, -2)$ and $\Psi = \mathrm{diag}(2 + j, 2 - j, 1)$. Further, taking $l_{13} = -40$, $l_{24} = -38$, $l_{35} = 1$, $l_{45} = -1$, and $l_{12} = l_{14} = l_{15} = l_{23} = l_{25} = l_{34} = 0$, we obtain simulation results as displayed in Fig. 16.5. The synchronization process of systems (16.2.12), (16.2.13), and (16.2.6) is described in Figs. 16.5(a)–(e), where the solid line presents the states of two drive systems and the dashed line shows the states of response system. Furthermore, Fig. 16.5(f) confirms that all error states asymptotically converge to zero. As expected, we approach combination complex synchronization among the incommensurate fractional-order complex Lorenz system, complex Chen system, and complex Lü system.

This chapter investigates combination complex synchronization of fractional-order chaotic complex systems. Firstly, combination complex synchronization of the commensurate fractional-order chaotic complex systems is investigated by designing a nonlinear observer in the sense of Lyapunov. Secondly, combination complex synchronization of fractional-order chaotic complex systems with incommensurate orders is investigated based on the stability theory of incommensurate fractional-order systems and active control method.

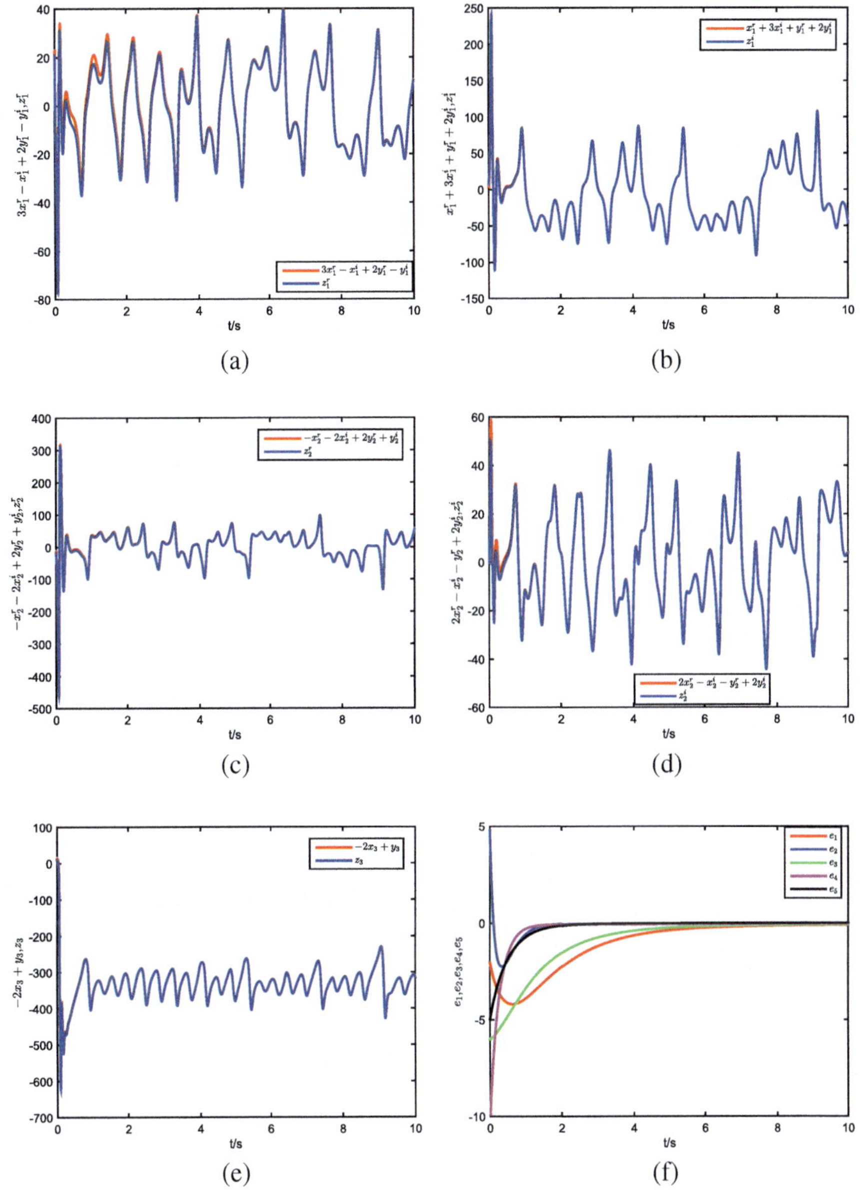

Fig. 16.5 State variables and synchronization error dynamics of drive systems (16.2.12), (16.2.13), and response system (16.2.6)

Chapter 17
Generalized Combination Complex Synchronization for the Fractional-Order Chaotic Complex Systems

In this chapter, we propose the generalized combination complex synchronization for fractional-order complex systems, which has two drive systems and one response system. Firstly, we discuss the general method to realize generalized combination complex synchronization among three commensurate fractional-order complex systems in detail. Secondly, generalized combination complex synchronization of fractional-order chaotic complex systems with incommensurate orders is investigated based on the stability theory of incommensurate fractional-order systems and active control method.

17.1 Generalized Combination Complex Synchronization of the Commensurate Fractional-Order Complex Systems

This section aims to present the generalized combination complex synchronization and design a scheme to realize the new form of chaos synchronization. By means of two scaling matrices, we establish a link between fractional-order chaotic real systems and chaotic complex systems with different dimensions. Two examples are performed to exhibit the feasibility and effectiveness of the proposed control technique.

S. Liu et al., *Chaos in Complex Dynamical System: Theory and Application*,
https://doi.org/10.1007/978-981-95-7762-0_17

17.1.1 Definition of Generalized Combination Complex Synchronization

The n_1-dimensional fractional-order chaotic complex system is taken as the first drive system

$$D_*^\alpha x = D_*^\alpha x^r + jD_*^\alpha x^i = Ax + f(x), \tag{17.1.1}$$

the n_2-dimensional fractional-order chaotic complex system is taken as the second drive system

$$D_*^\alpha y = D_*^\alpha y^r + jD_*^\alpha y^i = By + g(y), \tag{17.1.2}$$

while the response system is an n-dimensional complex system which is described as:

$$D_*^\alpha z = D_*^\alpha z^r + jD_*^\alpha z^i = Cz + h(z) + U(x, y, z), \tag{17.1.3}$$

where $x = x^r + jx^i \in \mathbb{C}^{n_1\times 1}$, $y = y^r + jy^i \in \mathbb{C}^{n_2\times 1}$, and $z = z^r + jz^i \in \mathbb{C}^{n\times 1}$ denote the state complex vectors, $A \in \mathbb{R}^{n_1\times n_1}$, $B \in \mathbb{R}^{n_2\times n_2}$, and $C \in \mathbb{R}^{n\times n}$ represent parameter matrices, while f, g, and h are nonlinear complex functions, U is a designed controller in the following.

We introduce the definition of generalized combination complex synchronization, which is based on two complex drive systems and one complex response system of different dimensions.

Definition 17.1 For two drive systems (17.1.1), (17.1.2), and one response system (17.1.3), it is said to be generalized combination complex synchronization if there exist two complex matrices $M_1 = M_1^r + jM_1^i \in \mathbb{C}^{n\times n_1}$ and $M_2 = M_2^r + jM_2^i \in \mathbb{C}^{n\times n_2}$, such that

$$\lim_{t\to\infty} ||e|| = \lim_{t\to\infty} ||z - M_1x - M_2y|| = 0,$$

where $||.||$ is the matrix norm, $e = e^r + je^i$ is called the error vector, $e^r = (e_1, e_3, \ldots, e_{2n-1})^T = z^r - M_1^r x^r + M_1^i x^i - M_2^r y^r + M_2^i y^i$, $e^i = (e_2, e_4, \ldots, e_{2n})^T = z^i - M_1^r x^i - M_1^i x^r - M_2^r y^i - M_2^i y^r$. the complex matrices M_1 and M_2 are called the scaling matrices.

Remark 17.1 If the dimensions of two drive systems (17.1.1) and (17.1.2) are equal to that of the response system (17.1.3), i.e., $n = n_1 = n_2$, then combination complex synchronization [247] will appear.

Our aim is to design suitable controllers such that the fractional-order chaotic nonlinear systems can achieve generalized combination complex synchronization.

17.1.2 A General Method for Generalized Combination Complex Synchronization

Theorem 17.1 *If the control law is chosen as follows:*

$$\begin{cases} U^r = -C(M_1^r x^r - M_1^i x^i + M_2^r y^r - M_2^i y^i) + M_1^r(Ax^r + f^r(x)) - M_1^i(Ax^i + f^i(x)) \\ \quad +M_2^r(By^r + g^r(y)) - M_2^i(By^i + g^i(y)) - h^r(z) - Ke^r, \\ U^i = -C(M_1^r x^i + M_1^i x^r + M_2^r y^i + M_2^i y^r) + M_1^r(Ax^i + f^i(x)) + M_1^i(Ax^r + f^r(x)) \\ \quad +M_2^r(By^i + g^i(y)) + M_2^i(By^r + g^r(y)) - h^i(z) - Ke^i, \end{cases} \tag{17.1.4}$$

where $K \in \mathbb{R}^{n\times n}$ is a control gain matrix, then we achieve generalized combination complex synchronization between two drive systems (17.1.1), (17.1.2), and one response system (17.1.3), if only if all eigenvalues of $C - K$ satisfy $|\arg(\lambda_l(C-K))| > \alpha\pi/2$ $(l = 1, 2, \ldots, n)$.

Proof The following error vector between systems (17.1.1), (17.1.2), and (17.1.3) is derived from Definition 17.1

$$\begin{aligned} e(t) &= e^r(t) + je^i(t) \\ &= z^r - M_1^r x^r + M_1^i x^i - M_2^r y^r + M_2^i y^i + j[z^i - M_1^r x^i - M_1^i x^r - M_2^r y^i - M_2^i y^r]. \end{aligned}$$

We easily obtain the error dynamical system as follows:

$$\begin{aligned} D_*^\alpha e(t) = {} & D_*^\alpha z^r - M_1^r D_*^\alpha x^r + M_1^i D_*^\alpha x^i - M_2^r D_*^\alpha y^r + M_2^i D_*^\alpha y^i \\ & +j[D_*^\alpha z^i - M_1^r D_*^\alpha x^i - M_1^i D_*^\alpha x^r - M_2^r D_*^\alpha y^i - M_2^i D_*^\alpha y^r]. \end{aligned}$$

Combining systems (17.1.1), (17.1.2), and (17.1.3), we have

$$\begin{cases} D_*^\alpha e^r(t) = Cz^r + h^r(z) + U^r - M_1^r(Ax^r + f^r(x)) + M_1^i(Ax^i + f^i(x)) - M_2^r(By^r + g^r(y)) \\ \quad +M_2^i(By^i + g^i(y)), \\ D_*^\alpha e^i(t) = Cz^i + h^i(z) + U^i - M_1^r(Ax^i + f^i(x)) - M_1^i(Ax^r + f^r(x)) - M_2^r(By^i + g^i(y)) \\ \quad -M_2^i(By^r + g^r(y)). \end{cases}$$

By substituting (17.1.4) into the above equality, the error dynamical system becomes:

$$\begin{cases} D_*^\alpha e^r(t) = (C-K)e^r, \\ D_*^\alpha e^i(t) = (C-K)e^i. \end{cases}$$

Since $|\arg(\lambda_i(C-K))| > \alpha\pi/2$, according to the stability theory of the fractional-order systems, the error vector $e(t)$ asymptotically converges to zero as $t \to \infty$. Therefore, generalized combination complex synchronization between systems (17.1.1), (17.1.2), and (17.1.3) is achieved. The proof is complete. □

We now turn to the problem of generalized combination complex synchronization between different dimensional fractional-order real and complex chaotic systems. The discussion focus on the following three cases.

Corollary 17.1 *Suppose that two drive systems (17.1.1) and (17.1.2) are chaotic real systems, i.e., $x \in \mathbb{R}^{n_1 \times 1}$, $y \in \mathbb{R}^{n_2 \times 1}$, and one response system is a chaotic complex system (17.1.3), i.e., $z \in \mathbb{C}^{n \times 1}$. For two given scaling matrices $M_1 \in \mathbb{C}^{n \times n_1}$ and $M_2 \in \mathbb{C}^{n \times n_2}$, the errors are defined as*

$$\begin{cases} e^r = z^r - M_1^r x - M_2^r y, \\ e^i = z^i - M_1^i x - M_2^i y. \end{cases}$$

Then generalized combination complex synchronization between two drive real systems (17.1.1), (17.1.2), and one response complex system (17.1.3) will occur with the designed controller

$$\begin{cases} U^r = -C(M_1^r x + M_2^r y) - h^r(z) + M_1^r(Ax + f(x)) + M_2^r(By + g(y)) - Ke^r, \\ U^i = -C(M_1^i x + M_2^i y) - h^i(z) + M_1^i(Ax + f(x)) + M_2^i(By + g(y)) - Ke^i, \end{cases}$$

where $K \in \mathbb{R}^{n \times n}$, all eigenvalues of $C - K$ satisfy $|\arg(\lambda_i(C - K))| > \alpha\pi/2$ $(i = 1, 2, \ldots, n)$.

Corollary 17.2 *Assume that system (17.1.1) is a chaotic real system and system (17.1.2) is a chaotic complex system, i.e., $x \in \mathbb{R}^{n_1 \times 1}$, $y \in \mathbb{C}^{n_2 \times 1}$, while the response system is a chaotic complex system (17.1.3), i.e., $z \in \mathbb{C}^{n \times 1}$. The synchronization errors with the given scaling matrices $M_1 \in \mathbb{C}^{n \times n_1}$ and $M_2 \in \mathbb{C}^{n \times n_2}$ are depicted as*

$$\begin{cases} e^r = z^r - M_1^r x - M_2^r y^r + M_2^i y^i, \\ e^i = z^i - M_1^i x - M_2^r y^i - M_2^i y^r. \end{cases}$$

Then two drive systems (17.1.1), (17.1.2) and one response complex system (17.1.3) can realize generalized combination complex synchronization with the following control law

$$\begin{cases} U^r = -C(M_1^r x + M_2^r y^r - M_2^i y^i) - h^r(z) + M_1^r(Ax + f(x)) - Ke^r \\ \qquad + M_2^r(By^r + g^r(y)) - M_2^i(By^i + g^i(y)), \\ U^i = -C(M_1^i x + M_2^r y^i + M_2^i y^r) - h^i(z) + M_1^i(Ax + f(x)) - Ke^i \\ \qquad + M_2^r(By^i + g^i(y)) + M_2^i(By^r + g^r(y)), \end{cases}$$

where $K \in \mathbb{R}^{n \times n}$, all eigenvalues of $C - K$ lie in the region $|\arg(\lambda_i(C - K))| > \alpha\pi/2$ $(i = 1, 2, \ldots, n)$.

Corollary 17.3 *Suppose that two drive systems (17.1.1) and (17.1.2) are chaotic complex systems, i.e., $x \in \mathbb{C}^{n_1 \times 1}$, $y \in \mathbb{C}^{n_2 \times 1}$, and the response system (17.1.3) is a chaotic real system, where $z \in \mathbb{R}^{n \times 1}$. Since $z(t)$ is real, we choose a real controller U to ensure synchronization of real parts and avoid increasing the imaginary parts* *of the response system. Therefore, the error vector is introduced as* $e = z - M_1^r x^r + M_1^i x^i - M_2^r y^r + M_2^i y^i$. *If a real controller is designed as*

$$\begin{aligned} U = & -h(z) - C(M_1^r x^r - M_1^i x^i + M_2^r y^r - M_2^i y^i) + M_1^r (Ax^r + f^r(x)) \\ & - M_1^i (Ax^i + f^i(x)) + M_2^r (By^r + g^r(y)) - M_2^i (By^i + g^i(y)) - Ke, \end{aligned}$$

where $K \in \mathbb{R}^{n\times n}$, *all eigenvalues of* $C - K$ *satisfy* $|\arg(\lambda_i(C - K))| > \alpha\pi/2$ $(i = 1, 2, \ldots, n)$, *then two drive complex systems (17.1.1), (17.1.2) and one response real system (17.1.3) realize generalized combination complex synchronization of real parts.*

17.1.3 Numerical Simulations

Two examples are provided to illustrate the validity and feasibility of the proposed schemes.

I. Synchronization Between Two Fractional-Order Hyperchaotic Real Drive Systems and a Fractional-Order Chaotic Complex Response System

Let us consider that the fractional-order hyperchaotic real Chen system and the fractional-order hyperchaotic real Rössler system drive the fractional-order chaotic complex Lorenz system. The fractional-order hyperchaotic real Chen system is considered as the first drive system

$$\begin{cases} D_*^\alpha x_1 = a_1(x_2 - x_1) + x_4, \\ D_*^\alpha x_2 = a_4 x_1 - x_1 x_3 + a_3 x_2, \\ D_*^\alpha x_3 = x_1 x_2 - a_2 x_3, \\ D_*^\alpha x_4 = x_2 x_3 + a_5 x_4, \end{cases} \tag{17.1.5}$$

and the second drive system is the fractional-order hyperchaotic real Rössler system

$$\begin{cases} D_*^\alpha y_1 = -y_2 - y_3, \\ D_*^\alpha y_2 = y_1 + b_1 y_2 + y_4, \\ D_*^\alpha y_3 = b_2 + y_1 y_3, \\ D_*^\alpha y_4 = -b_3 y_3 + b_4 y_4, \end{cases} \tag{17.1.6}$$

while the response system with the controller is written as follows:

$$\begin{cases} D_*^\alpha z_1 = c_1(z_2 - z_1) + U_1, \\ D_*^\alpha z_2 = c_2 z_1 - z_1 z_3 - z_2 + U_2, \\ D_*^\alpha z_3 = \frac{1}{2}(\bar{z}_1 z_2 + z_1 \bar{z}_2) - c_3 z_3 + U_3, \end{cases} \tag{17.1.7}$$

where $A = \begin{pmatrix} -a_1 & a_1 & 0 & 1 \\ a_4 & a_3 & 0 & 0 \\ 0 & 0 & -a_2 & 0 \\ 0 & 0 & 0 & a_5 \end{pmatrix}$, $B = \begin{pmatrix} 0 & -1 & -1 & 0 \\ 1 & b_1 & 0 & 1 \\ 0 & 0 & 0 & 0 \\ 0 & 0 & -b_3 & b_4 \end{pmatrix}$, $C = \begin{pmatrix} -c_1 & c_1 & 0 \\ c_2 & -1 & 0 \\ 0 & 0 & -c_3 \end{pmatrix}$, $f(x) = \begin{pmatrix} 0 \\ -x_1x_3 \\ x_1x_2 \\ x_2x_3 \end{pmatrix}$, $g(y) = \begin{pmatrix} 0 \\ 0 \\ b_2 + y_1y_3 \\ 0 \end{pmatrix}$, $h(z) = \begin{pmatrix} 0 \\ -z_1z_3 \\ \frac{1}{2}(\bar{z}_1 z_2 + z_1 \bar{z}_2) \end{pmatrix}$, z_1 and z_2 are complex variables, x_i, y_i, and z_3 are real variables ($i = 1, 2, 3, 4$), $U_1 = u_1 + ju_2$, $U_2 = u_3 + ju_4$, and $U_3 = u_5$ are complex and real control functions, respectively.

With the scaling matrices $M_1 = M_2 = \begin{pmatrix} 1+j & 0 & 0 & 0 \\ 0 & 1+j & 0 & 0 \\ 0 & 0 & 0 & -1 \end{pmatrix}$, the synchronization error is designed as follows:

$$\begin{cases} e_1^r = z_1^r - x_1 - y_1, \\ e_1^i = z_1^i - x_1 - y_1, \\ e_2^r = z_2^r - x_2 - y_2, \\ e_2^i = z_2^i - x_2 - y_2, \\ e_3 = z_3 + x_4 + y_4. \end{cases}$$

By choosing a control matrix as $K = \begin{pmatrix} 2 - c_1 & c_1 + 1 & 0 \\ c_2 - 2 & 3 & 1 \\ 0 & 0 & -c_3 + 5 \end{pmatrix}$, we have $C - K = \begin{pmatrix} -2 & -1 & 0 \\ 2 & -4 & -1 \\ 0 & 0 & -5 \end{pmatrix}$ with $(c_1, c_2, c_3) = (10, 180, 1)$. For $\alpha = 0.98$, all of the eigenvalues of $C - K$ are $\lambda_{1,2} = -4 \pm 2j$ and $\lambda_3 = -10$, which lie in the region $|\arg(\lambda_i(C - K))| > \alpha\pi/2$, $(i = 1, 2, 3)$.

According to Corollary 17.1, the control law is derived as follows:

$$\begin{cases} u_1 = (c_1 - a_1)x_1 + c_1y_1 + (a_1 - c_1)x_2 - (c_1 + 1)y_2 + x_4 - y_3 - (2 - c_1)e_1^r - (c_1 + 1)e_2^r, \\ u_2 = (c_1 - a_1)x_1 + c_1y_1 + (a_1 - c_1)x_2 - (c_1 + 1)y_2 + x_4 - y_3 - (2 - c_1)e_1^i - (c_1 + 1)e_2^i, \\ u_3 = (a_4 - c_2)x_1 + (1 - c_2)y_1 + (a_3 + 1)x_2 + (b_1 + 1)y_2 + y_4 - x_1x_3 + z_1^r z_3 \\ \quad -(c_2 - 2)e_1^r - 3e_2^r - e_3, \\ u_4 = (a_4 - c_2)x_1 + (1 - c_2)y_1 + (a_3 + 1)x_2 + (b_1 + 1)y_2 + y_4 - x_1x_3 + z_1^i z_3 \\ \quad -(c_2 - 2)e_1^i - 3e_2^i, \\ u_5 = -(c_3 + a_5)x_4 - (b_4 + c_3)y_4 - z_1^r z_2^r - z_1^i z_2^i - x_2x_3 + b_3y_3 - (5 - c_3)e_3. \end{cases}$$

In the following, $\alpha = 0.99$, the initial values of systems (17.1.5), (17.1.6), and (17.1.7) are taken as $x(0) = (3, -4, 2, 2)^T$, $y(0) = (-15, 9, -4, 18)^T$, $z(0) = (2+$

$3j, 5+6j, 9)^T$, and the parameters are chosen as $(a_1, a_2, a_3, a_4, a_5) = (35, 3, 12, 7, 0.5)$, $(b_1, b_2, b_3, b_4) = (0.25, 3, 0.5, 0.05)$, $(c_1, c_2, c_3) = (10, 180, 1)$. Figures 17.1 (a)–(e) demonstrate the synchronization process and error states of systems (17.1.5), (17.1.6), and (17.1.7). According to Fig. 17.1(f), it is obvious that all error states tend asymptotically to zero. Hence, the 4D fractional-order real Chen system, the 4D fractional-order real Rössler system, and the 3D fractional-order complex Lorenz system realize generalized combination complex synchronization.

II. Synchronization Between Two Fractional-Order Chaotic Complex Drive Systems and a Fractional-Order Hyperchaotic Real Response System

This section presents the design of the control law aimed at realizing generalized combination complex synchronization between the fractional-order complex Lü system, the fractional-order complex Chen system, and the fractional-order hyperchaotic real Rössler system. The first drive system is described as:

$$\begin{cases} D_*^\alpha x_1 = a_1(x_2 - x_1), \\ D_*^\alpha x_2 = a_2 x_2 - x_1 x_3, \\ D_*^\alpha x_3 = \frac{1}{2}(\bar{x}_1 x_2 + x_1 \bar{x}_2) - a_3 x_3, \end{cases} \tag{17.1.8}$$

the second drive system is given as:

$$\begin{cases} D_*^\alpha y_1 = b_1(y_2 - y_1), \\ D_*^\alpha y_2 = (b_2 - b_1) y_1 + b_2 y_2 - y_1 y_3, \\ D_*^\alpha y_3 = \frac{1}{2}(\bar{y}_1 y_2 + y_1 \bar{y}_2) - b_3 y_3, \end{cases} \tag{17.1.9}$$

while the response system with the controller is depicted as follows:

$$\begin{cases} D_*^\alpha z_1 = -z_2 - z_3 + U_1, \\ D_*^\alpha z_2 = z_1 + c_1 z_2 + z_4 + U_2, \\ D_*^\alpha z_3 = c_2 + z_1 z_3 + U_3, \\ D_*^\alpha z_4 = -c_3 z_3 + c_4 z_4 + U_4, \end{cases} \tag{17.1.10}$$

where $A = \begin{pmatrix} -a_1 & a_1 & 0 \\ 0 & a_2 & 0 \\ 0 & 0 & -a_3 \end{pmatrix}$, $B = \begin{pmatrix} -b_1 & b_1 & 0 \\ b_2 - b_1 & b_2 & 0 \\ 0 & 0 & -b_3 \end{pmatrix}$, $C = \begin{pmatrix} 0 & -1 & -1 & 0 \\ 1 & c_1 & 0 & 1 \\ 0 & 0 & 0 & 0 \\ 0 & 0 & -c_3 & c_4 \end{pmatrix}$,

$f(x) = \begin{pmatrix} 0 \\ -x_1 x_3 \\ \frac{1}{2}(\bar{x}_1 x_2 + x_1 \bar{x}_2) \end{pmatrix}$, $g(y) = \begin{pmatrix} -b_1 & b_1 & 0 \\ 0 & b_2 & 0 \\ 0 & 0 & -b_3 \end{pmatrix}$, $h(z) = \begin{pmatrix} 0 \\ 0 \\ c_2 + z_1 z_3 \\ 0 \end{pmatrix}$, x_i and y_i $(i = 1, 2)$ are complex variables, x_3, y_3, and z_i are real variables, U_i are real control functions, $(i = 1, 2, 3, 4)$.

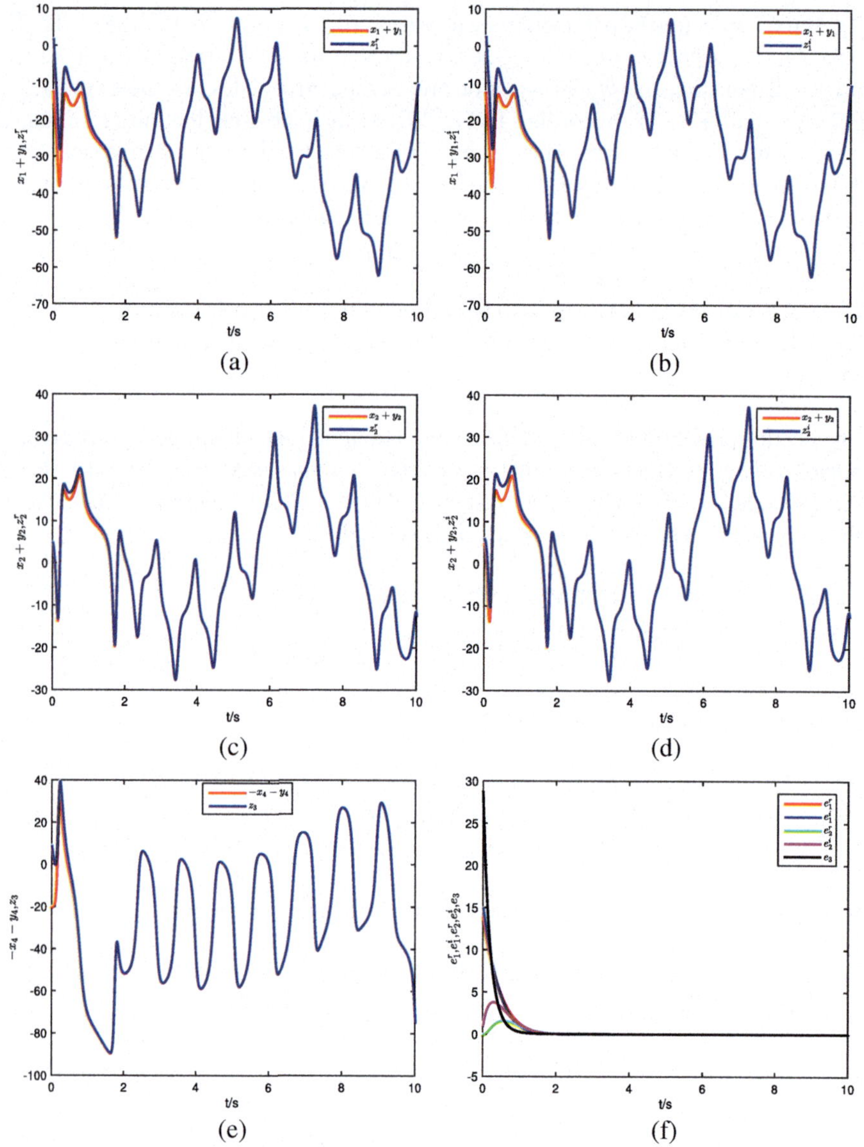

Fig. 17.1 State variables and synchronization error dynamics of drive systems (17.1.5), (17.1.6), and response system (17.1.7)

Suppose two scaling matrices $M_1 = M_2 = \begin{pmatrix} j & 0 & 0 \\ 0 & j & 0 \\ 0 & 0 & 1 \\ 0 & 0 & 1 \end{pmatrix} = \begin{pmatrix} 0 & 0 & 0 \\ 0 & 0 & 0 \\ 0 & 0 & 1 \\ 0 & 0 & 1 \end{pmatrix} + j \begin{pmatrix} 1 & 0 & 0 \\ 0 & 1 & 0 \\ 0 & 0 & 0 \\ 0 & 0 & 0 \end{pmatrix}$. The real synchronization error is given by:

$$\begin{cases} e_1 = z_1 + x_1^i + y_1^i, \\ e_2 = z_2 + x_2^i + y_2^i, \\ e_3 = z_3 - x_3 - y_3, \\ e_4 = z_4 - x_3 - y_3. \end{cases}$$

In order to satisfy $|\arg(\lambda_i(C - K))| > \alpha\pi/2$ $(i = 1, 2, 3, 4)$, we choose a control matrix as $K = \begin{pmatrix} -\lambda_1 & -1 & -1 & 0 \\ 1 & c_1 - \lambda_2 & 0 & 1 \\ 0 & 0 & -\lambda_3 & 0 \\ 0 & 0 & -c_3 & c_4 - \lambda_4 \end{pmatrix}$, where λ_i are real.

Applying Corollary 17.3 yields the following real controller:

$$\begin{cases} U_1 = a_1 x_1^i + b_1 y_1^i - (1 + a_1) x_2^i - (1 + b_1) y_2^i + x_3 + y_3 + \lambda_1 e_1 + e_2 + e_3, \\ U_2 = x_1^i + y_1^i - (b_2 - b_1) y_1^i + (c_1 - a_2) x_2^i + (c_1 - b_2) y_2^i - x_3 - y_3 + x_1^i x_3 + y_1^i y_3 \\ \qquad -e_1 - (c_1 - \lambda_2) e_2 - e_4, \\ U_3 = -c_2 - z_1 z_3 + x_1^r x_2^r + x_1^i x_2^i - a_3 x_3 - b_3 y_3 + y_1^r y_2^r + y_1^i y_2^i + \lambda_3 e_3, \\ U_4 = (c_3 - c_4 - a_3) x_3 + (c_3 - c_4 - b_3) y_3 + x_1^r x_2^r + x_1^i x_2^i + y_1^r y_2^r + y_1^i y_2^i \\ \qquad + c_3 e_3 - (c_4 - \lambda_4) e_4. \end{cases}$$

For the case of $\alpha = 0.98$, the initial values are $x(0) = (1 - 2j, 3 - 8j, 4)^T$, $y(0) = (6 - 9j, -8 + 6j, 1)^T$, $z(0) = (4, 5, 6, 2)^T$, and the parameters arehe real synchronization $(a_1, a_2, a_3) = (42, 22, 5)$, $(b_1, b_2, b_3) = (35, 28, 3)$, $(c_1, c_2, c_3, c_4) = (0.25, 3, 0.5, 0.05)$. we choose $\lambda_1 = -2$, $\lambda_2 = -1$, $\lambda_3 = -9$, and $\lambda_4 = -5$ such that all of the eigenvalues of $C - K$ lie in region $|\arg(\lambda_i(C - K))| > \alpha\pi/2$, $(i = 1, 2, 3, 4)$. Figures 17.2(a)–(d) describe the synchronization process of systems (17.1.8), (17.1.9), and (17.1.10). From Figs. 17.2(e)–(f), it can be observed that all error states converge asymptotically to zero. Thus, the fractional-order chaotic complex Lü system, the fractional-order chaotic complex Chen system, and the fractional-order hyperchaotic real Rössler achieve generalized combination complex synchronization.

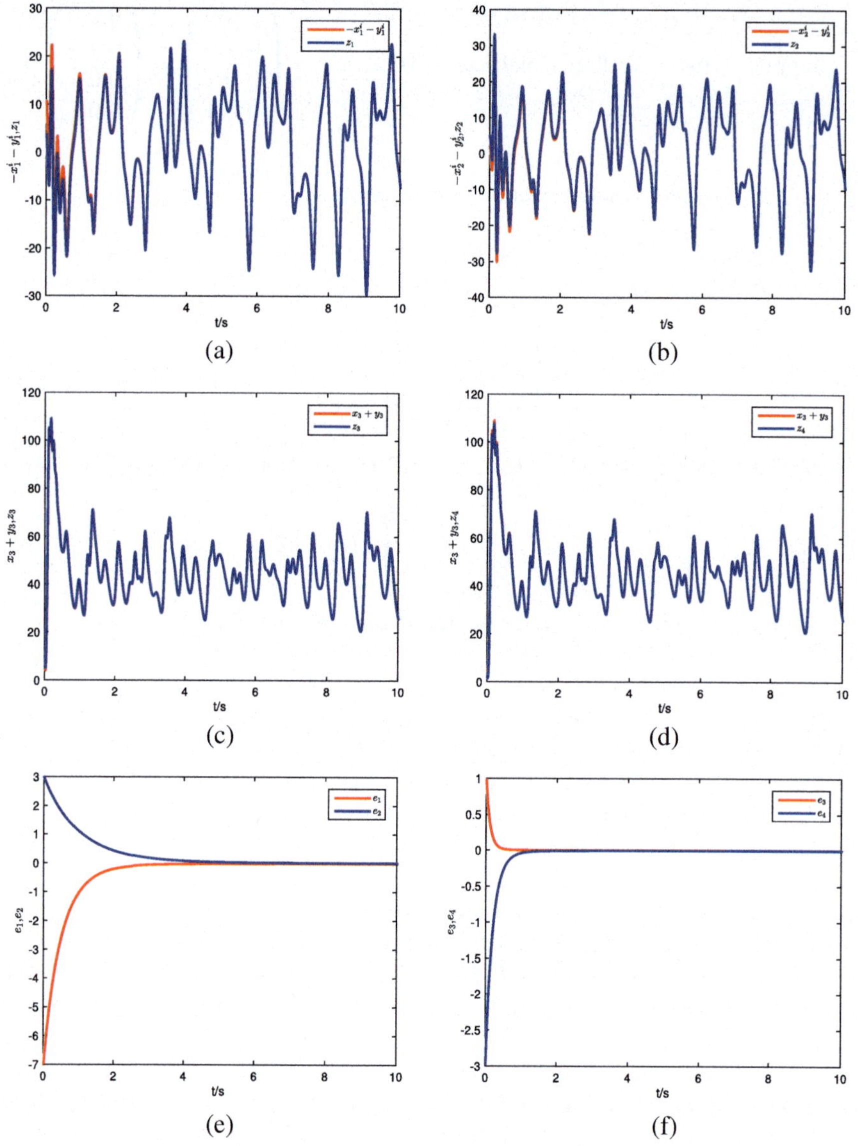

Fig. 17.2 State variables and synchronization error dynamics of drive systems (17.1.8), (17.1.9), and response system (17.1.10)

17.2 Generalized Combination Complex Synchronization Among Three Incommensurate Fractional-Order Chaotic Systems

This section investigates the problem of generalized combination complex synchronization among three incommensurate fractional-order chaotic systems.

17.2.1 *Mathematical Model and Problem Descriptions*

Consider a fractional-order complex chaotic system as the first drive system

$$D_*^{\alpha} x = Ax + f(x), \tag{17.2.1}$$

the second complex chaotic drive system is given as:

$$D_*^{\alpha} y = By + g(y), \tag{17.2.2}$$

while a fractional-order complex chaotic response system is configured as

$$D_*^{\alpha} z = Cz + h(z) + U(x, y, z), \tag{17.2.3}$$

where $\alpha = [\alpha_1, \alpha_2, \ldots, \alpha_n]^T$ denotes the fractional orders, $x = x^r + jx^i \in \mathbb{C}^{n_1 \times 1}$, $y = y^r + jy^i \in \mathbb{C}^{n_2 \times 1}$ and $z = z^r + jz^i \in \mathbb{C}^{n \times 1}$ are complex state vectors of systems (17.2.1), (17.2.2) and (17.2.3), respectively, $A \in \mathbb{R}^{n_1 \times n_1}$, $B \in \mathbb{R}^{n_2 \times n_2}$ and $C \in \mathbb{R}^{n \times n}$ are three real matrices of system parameters, $f = (f_1, f_2, \ldots, f_n)^T$, $g = (g_1, g_2, \ldots, g_n)^T$ and $h = (h_1, h_2, \ldots, h_n)^T$ are vectors of nonlinear complex functions, and U is a controller to be designed, $U = (U_1, U_2, \ldots, U_n)^T = U^r + jU^i$, $U^r = (u_1, u_3, \ldots, u_{2n-1})^T$, $U^i = (u_2, u_4, \ldots, u_{2n})^T$.

Definition 17.2 Two drive systems (17.2.1) and (17.2.2) are said to exhibit generalized combination complex synchronization with the response system (17.2.3) if there exist two complex matrices $M_1 = M_1^r + jM_1^i \in \mathbb{C}^{n \times n_1}$ and $M_2 = M_2^r + jM_2^i \in \mathbb{C}^{n \times n_2}$, such that

$$\lim_{t \to \infty} ||e(t)|| = \lim_{t \to \infty} ||z(t) - M_1 x(t) - M_2 y(t)|| = 0,$$

where $e = e^r + je^i$ is called the error vector, $e^r = (e_1, e_3, \ldots, e_{2n-1})^T = z^r - M_1^r x^r + M_1^i x^i - M_2^r y^r + M_2^i y^i$, $e^i = (e_2, e_4, \ldots, e_{2n})^T = z^i - M_1^r x^i - M_1^i x^r - M_2^r y^i - M_2^i y^r$.

17.2.2 Applications and Simulations

Let the incommensurate fractional-order hyperchaotic complex Lü system be the first drive system

$$\begin{cases} D_*^{\alpha_1} x_1 = a_1(x_2 - x_1) + x_4, \\ D_*^{\alpha_2} x_2 = a_2 x_1 - x_1 x_3 + x_4, \\ D_*^{\alpha_3} x_3 = \frac{1}{2}(\bar{x}_1 x_2 + x_1 \bar{x}_2) - a_3 x_3, \\ D_*^{\alpha_4} x_4 = \frac{1}{2}(\bar{x}_1 x_2 + x_1 \bar{x}_2) - a_4 x_4, \end{cases} \tag{17.2.4}$$

and the incommensurate fractional-order complex Lorenz system is defined as the second drive system:

$$\begin{cases} D_*^{\alpha_1} y_1 = b_1(y_2 - y_1), \\ D_*^{\alpha_2} y_2 = b_2 y_1 - y_2 - y_1 y_3, \\ D_*^{\alpha_3} y_3 = \frac{1}{2}(\bar{y}_1 y_2 + y_1 \bar{y}_2) - b_3 y_3. \end{cases} \tag{17.2.5}$$

The response system is the incommensurate fractional-order hyperchaotic complex Lü system:

$$\begin{cases} D_*^{\alpha_1} z_1 = c_1(z_2 - z_1) + z_4 + u_1, \\ D_*^{\alpha_2} z_2 = c_2 z_1 - z_1 z_3 + z_4 + u_2, \\ D_*^{\alpha_3} z_3 = \frac{1}{2}(\bar{z}_1 z_2 + z_1 \bar{z}_2) - c_3 z_3 + u_3, \\ D_*^{\alpha_4} z_4 = \frac{1}{2}(\bar{z}_1 z_2 + z_1 \bar{z}_2) - c_4 z_4 + u_4, \end{cases} \tag{17.2.6}$$

where x_l, y_l and z_l $(l = 1, 2)$ are complex variables, x_3, x_4, y_3, z_3 and z_4 are real state variables, a_l, b_l and c_l are real parameters $(l = 1, 2, 3)$. $u_1 = u_1^r + ju_1^i$ and $u_2 = u_2^r + ju_2^i$ are complex functions, u_3 and u_4 are real control functions.

For two given scaling matrices

$$M_1 = \begin{pmatrix} j & 0 & 0 & 0 \\ 0 & j & 0 & 0 \\ 0 & 0 & -1 & 0 \\ 0 & 0 & 0 & -1 \end{pmatrix}, \quad M_2 = \begin{pmatrix} 1-j & 0 & 0 \\ 0 & 1-j & 0 \\ 0 & 0 & 1 \\ 0 & 0 & 0 \end{pmatrix},$$

we define the synchronization error as follows:

$$\begin{cases} e_1^r = z_1^r + x_1^i - y_1^r - y_1^i, \\ e_1^i = z_1^i - x_1^r - y_1^i + y_1^r, \\ e_2^r = z_2^r + x_2^i - y_2^r - y_2^i, \\ e_2^i = z_2^i - x_2^r - y_2^i + y_2^r, \\ e_3 = z_3 + x_3 - y_3, \\ e_4 = z_4 + x_4. \end{cases}$$

Combining systems (17.2.4), (17.2.5), with (17.2.6), one can get

$$\begin{cases} D_*^{\alpha_1} e_1^r = c_1(z_2^r - z_1^r) + a_1(x_2^i - x_1^i) - b_1(y_2^r - y_1^r) - b_1(y_2^i - y_1^i) + z_4 + u_1^r, \\ D_*^{\alpha_1} e_1^i = c_1(z_2^i - z_1^i) - a_1(x_2^r - x_1^r) - b_1(y_2^i - y_1^i) + b_1(y_2^r - y_1^r) - x_4 + u_1^i, \\ D_*^{\alpha_2} e_2^r = c_2 z_2^r - z_1^r z_3 + z_4 + a_2 x_2^i - x_1^i x_3 - b_2(y_1^r + y_1^i) + (y_2^r + y_1^r) + y_1^r y_3 + y_1^i y_3 + u_2^r, \\ D_*^{\alpha_2} e_2^i = c_2 z_2^i - z_1^i z_3 - a_2 x_2^r + x_1^r x_3 - x_4 - b_2(y_1^i - y_1^r) + (y_2^i - y_2^r) + y_1^i y_3 + y_1^i y_3 + u_2^i, \\ D_*^{\alpha_3} e_3 = z_1^r z_2^r + z_1^i z_2^i - c_3 z_3 + x_1^r x_2^r + x_1^i x_2^i - a_3 x_3 - y_1^r y_2^r - y_1^i y_2^i + b_3 y_3 + u_3, \\ D_*^{\alpha_4} e_4 = z_1^r z_2^r + z_1^i z_2^i - c_4 z_4 + x_1^r x_2^r + x_1^i x_2^i - a_4 x_4 + u_4. \end{cases}$$

Then, we define active control functions $u_l(t)$ $(l = 1, 2, \ldots, 5)$ as

$$\begin{cases} u_1^r(t) = (a_1 - c_1)(x_1^i - x_2^i) + (c_1 - b_1)(y_1^r + y_1^i - y_2^r - y_2^i) + x_4 + \beta_1(t), \\ u_1^i(t) = (a_1 - c_1)(x_2^r - x_1^r) + (c_1 - b_1)(y_2^r - y_2^i + y_1^r - y_1^i) + x_4 + \beta_2(t), \\ u_2^r(t) = -(a_2 - c_2)x_2^i - (c_2 + 1)(y_2^r + y_2^i) + z_1^r z_3 + x_4 + x_1^i x_3 + b_2(y_1^r + y_1^i) - y_1^r y_3 \\ \qquad - y_1^i y_3 + \beta_3(t), \\ u_2^i(t) = (a_2 - c_2)x_2^r - (c_2 + 1)(y_2^i - y_2^r) + z_1^i z_3 + x_4 - x_1^r x_3 - b_2(y_1^r - y_1^i) - y_1^i y_3 \\ \qquad + y_1^r y_3 + \beta_4(t), \\ u_3(t) = (a_3 - c_3)x_3 - (b_3 - c_3)y_3 - z_1^r z_2^r - z_1^i z_2^i - x_1^r x_2^r - x_1^i x_2^i + y_1^r y_2^r + y_1^i y_2^i + \beta_5(t), \\ u_4(t) = (a_4 - c_4)x_4 - z_1^r z_2^r - z_1^i z_2^i - x_1^r x_2^r - x_1^i x_2^i + \beta_6(t), \end{cases}$$

which provides

$$\begin{cases} D_*^{\alpha_1} e_1^r = c_1 e_2^r - c_1 e_1^r + e_4 + \beta_1(t), \\ D_*^{\alpha_1} e_1^i = c_1 e_2^i - c_1 e_1^i + \beta_2(t), \\ D_*^{\alpha_2} e_2^r = c_2 e_2^r + e_4 + \beta_3(t), \\ D_*^{\alpha_2} e_2^i = c_2 e_2^i + \beta_4(t), \\ D_*^{\alpha_3} e_3 = -c_3 e_3 + \beta_5(t), \\ D_*^{\alpha_4} e_4 = -c_4 e_4 + \beta_6(t), \end{cases} \tag{17.2.7}$$

where $(\beta_1(t), \beta_2(t), \ldots, \beta_6(t))^T = L(e_1^r(t), e_1^i(t), \ldots, e_3(t))^T$, where $L = (l_{kl})_{6\times 6}$ is a real matrix. So system (17.2.7) can be rewritten as $D_*^{\alpha} e = \Xi e$, where

$$\Xi = \begin{pmatrix} l_{11} - c_1 & l_{12} & l_{13} + c_1 & l_{14} & l_{15} & l_{16} + 1 \\ l_{21} & l_{22} - c_1 & l_{23} & l_{24} + c_1 & l_{25} & l_{26} \\ l_{31} & l_{32} & l_{33} + c_2 & l_{34} & l_{35} & l_{36} + 1 \\ l_{41} & l_{42} & l_{43} & l_{44} + c_2 & l_{45} & l_{46} \\ l_{51} & l_{52} & l_{53} & l_{54} & l_{55} - c_3 & l_{56} \\ l_{61} & l_{62} & l_{63} & l_{64} & l_{65} & l_{66} - c_4 \end{pmatrix}.$$

To make the error system stable, the matrix L should be selected in such a way that the feedback system satisfies conditions of Lemma. There is not a unique choice for such matrix L, a good choice can be as follows: $l_{11} = 22$, $l_{22} = -50$, $l_{33} = -27$, $l_{44} = -26$, $l_{55} = 5$, $l_{66} = 2$, $l_{21} = l_{31} = l_{32} = l_{41} = l_{42} = l_{43} = l_{51} = l_{52} = l_{53} = l_{54} = l_{61} = l_{62} = l_{63} = l_{64} = l_{65} = 0$, and other elements of L can be chosen arbitrarily. Taking $(c_1, c_2, c_3, c_4) = (42, 25, 6, 5)$, $(\alpha_1, \alpha_2, \alpha_3, \alpha_4) = (0.98, 0.99, 0.91, 0.92)$, $m = 100$, and $\gamma = 1/100$, we obtain

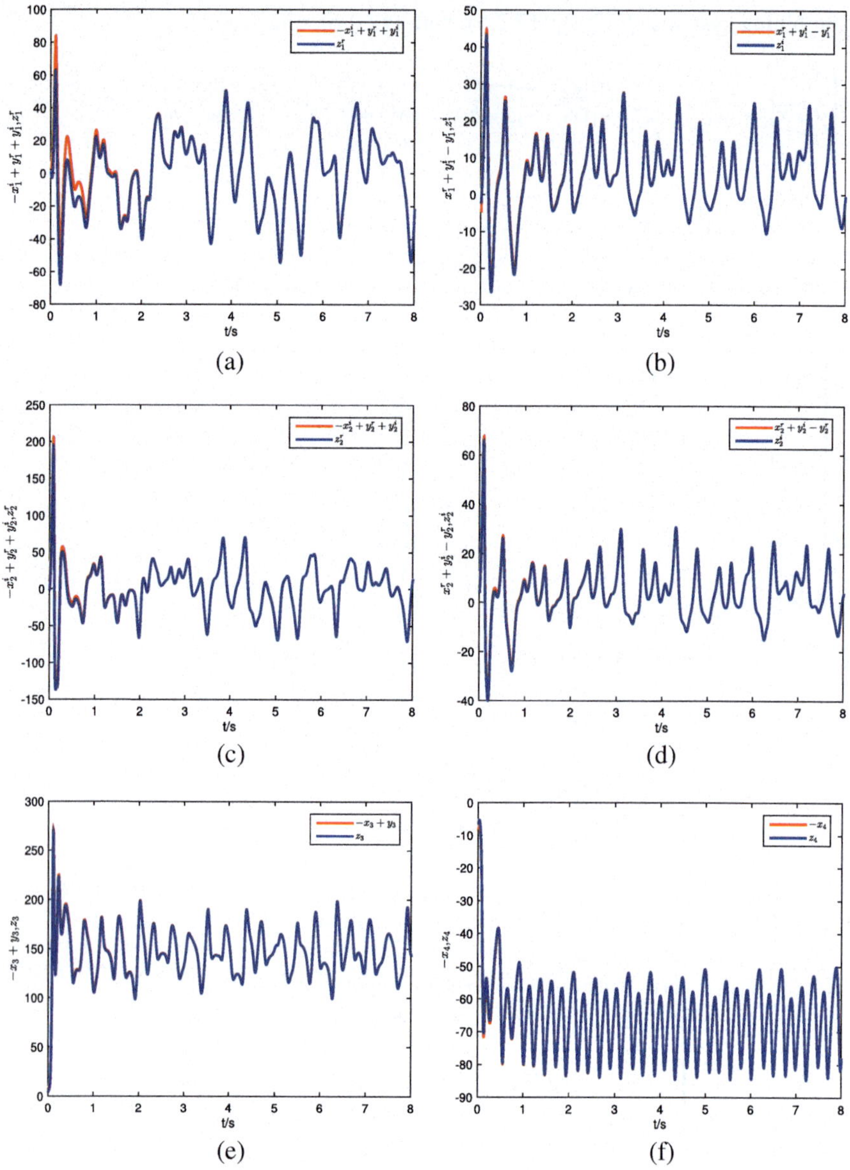

Fig. 17.3 State variables of drive systems (17.2.4), (17.2.5) and response system (17.2.6)

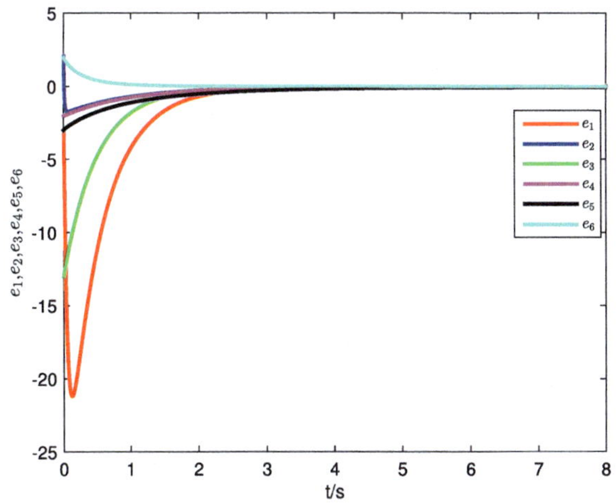

Fig. 17.4 Synchronization error dynamics of drive systems (17.2.4), (17.2.5) and response system (17.2.6)

$$(\lambda^{98} + 20)(\lambda^{98} + 25)(\lambda^{99} + 2)(\lambda^{99} + 1)(\lambda^{91} + 1)(\lambda^{92} + 2) = 0. \quad (17.2.8)$$

By a simple calculation, we conclude that all roots of (17.2.8) lie in the region $|\arg(\lambda)| > \gamma\pi/2$. The error vector $e(t)$ asymptotically converges to zero as $t \to \infty$.

In the numerical simulations, the initial values of systems (17.2.4), (17.2.5), and (17.2.6) are chosen as $x(0) = (-6 + 2j, 5 - j, 1, 8)^T$, $y(0) = (2 + 3j, 5 + 6j, 9)^T$ and $z(0) = (1 - 2j, -1 + 4j, 5, 6)^T$, respectively. Further, selecting $l_{12} = l_{13} = l_{14} = l_{15} = l_{16} = l_{23} = l_{24} = l_{25} = l_{26} = l_{34} = l_{35} = l_{36} = l_{45} = l_{46} = l_{56} = 0$, we have the corresponding numerical results as displayed in Figs. 17.3 and 17.4. Figures 17.3(a)–(f) depict the state variables of two drive systems and the response system. From Fig. 17.4, it can be observed that the errors of synchronization converge asymptotically to zero. Therefore, the fractional-order hyperchaotic complex Lü system, chaotic complex Lorenz system, and hyperchaotic complex Lü system with incommensurate order achieve generalized complex modified projective synchronization.

Remark 17.2 Similarly, the research method in this section can be used to realize generalized complex modified projective synchronization between the the incommensurate fractional-order complex system and the real system. Hence, we omit it.

Chapter 18
Synchronization of N-Coupled Fractional-Order Complex Chaotic Systems with Ring Connection

Coupled chaotic systems have been extensively studied in theoretical physics and engineering sciences [250, 251]. Multiple chaotic systems [252–254] can be coupled in a ring which makes them correlative. Synchronization of coupled chaotic systems with ring connection extends the traditional mode of synchronization from one-to-one into one-to-many [255–257]. This chapter aims to study synchronization schemes of coupled fractional-order chaotic systems with ring connection.

18.1 Synchronization and Antisynchronization of N-Coupled Fractional-Order Complex Chaotic Systems with Ring Connection via Unidirectional Coupling

18.1.1 Theoretical Analysis

Consider the following fractional-order complex chaotic system described as

$$\begin{cases} D^{\alpha} X_1 &= A_1 X_1 + G_1(X_1), \\ D^{\alpha} X_2 &= A_2 X_2 + G_2(X_2), \\ & \vdots \\ D^{\alpha} X_N &= A_N X_N + G_N(X_N), \end{cases} \tag{18.1.1}$$

where $X_1, X_2, \ldots, X_N$ $(N > 2)$ are complex state vectors and $X_k = X_k^r + jX_k^i$; $A_1, A_2, \ldots, A_N \in \mathbb{R}^{n\times n}$ are constant matrices; $G_k(X_k)$ are the continuous nonlinear complex functions ($k = 1, 2, \ldots, N$). We now couple multiple fractional-order complex chaotic systems in a unidirectional ring structure, as illustrated in Fig. 18.1. In this model, the state variables of the first complex chaotic system couple the Nth

S. Liu et al., *Chaos in Complex Dynamical System: Theory and Application*,
https://doi.org/10.1007/978-981-95-7762-0_18

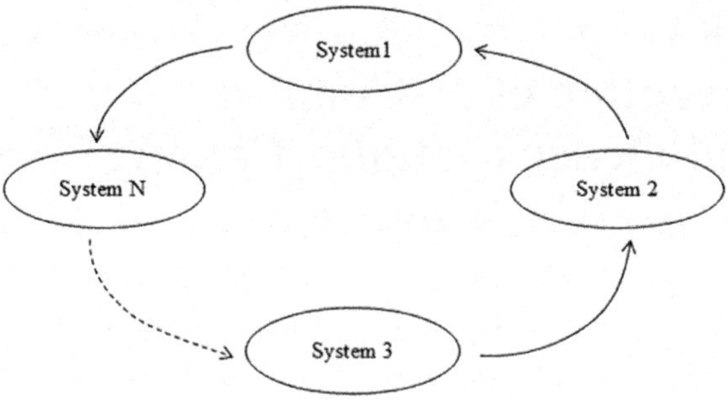

Fig. 18.1 Diagram of fractional-order complex chaotic systems with ring connection via unidirectional coupling

system, and the state variables of the second system couple the first,...analogically, until the state variables of the Nth system couple those of the $(N-1)$th. Therefore, we describe the N-coupled fractional-order complex chaotic systems with ring connection as:

$$\begin{cases} D^\alpha X_1 &= A_1X_1 + G_1(X_1) + Q_1(X_N - X_1), \\ D^\alpha X_2 &= A_2X_2 + G_2(X_2) + Q_2(X_1 - X_2), \\ &\vdots \\ D^\alpha X_N &= A_NX_N + G_N(X_N) + Q_N(X_{N-1} - X_N), \end{cases} \tag{18.1.2}$$

where $Q_k = \text{diag}(q_{k1}, q_{k2}, \ldots, q_{kn})$ are n-dimensional real diagonal matrices and $q_{kl} \geq 0$ are the ideal gains which represent the coupled parameters ($k = 1, 2, \ldots, N$, $l = 1, 2, \ldots, n$).

The first system is chosen as the master system, and controllers are added to the slave systems, respectively. Thus, system (18.1.2) is represented by:

$$\begin{cases} D^\alpha X_1 &= A_1X_1 + G_1(X_1) + Q_1(X_N - X_1), \\ D^\alpha X_2 &= A_2X_2 + G_2(X_2) + Q_2(X_1 - X_2) + U_1, \\ D^\alpha X_3 &= A_3X_3 + G_3(X_3) + Q_3(X_2 - X_3) + U_2, \\ &\vdots \\ D^\alpha X_N &= A_NX_N + G_N(X_N) + Q_N(X_{N-1} - X_N) + U_{N-1}. \end{cases} \tag{18.1.3}$$

The synchronization error vector is defined as $e_k = X_{k+1} - \delta X_k$ ($k = 1, 2, \ldots, N-1$, $\delta = 1, -1$). Obviously, when $\delta = 1$, the error system e_k becomes the complete synchronization error system. And when $\delta = -1$, the error system e_k is transformed into the anti-synchronization error system. The following error dynamical system is derived from system (18.1.3) and the definition of the synchronization error vector

$$\begin{aligned} D^\alpha e_1^r &= D^\alpha(X_2^r - \delta X_1^r) = D^\alpha X_2^r - \delta D^\alpha X_1^r \\ &= A_2X_2^r + G_2^r(X_2) + Q_2(X_1^r - X_2^r) + U_1^r - \delta(A_1X_1^r + G_1^r(X_1) + Q_1(X_N^r - X_1^r)) \end{aligned}$$

$$
\begin{aligned}
&= (A_2 - Q_2)e_1^r + (\delta(A_2 - A_1) + (1-\delta)Q_2)X_1^r + G_2^r(X_2) - \delta G_1^r(X_1) \\
&\quad -\delta Q_1(X_N^r - X_1^r) + U_1^r \\
&= (A_2 - Q_2)e_1^r + (\delta(A_2 - A_1) + (1-\delta)Q_2)X_1^r + G_2^r(X_2) - \delta G_1^r(X_1) + U_1^r \\
&\quad -\delta Q_1(X_N^r - \delta X_{N-1}^r + \delta X_{N-1}^r - \delta^2 X_{N-2}^r + \cdots + \delta^{N-2}X_2^r - \delta^{N-1}X_1^r + \delta^{N-1}X_1^r - X_1^r) \\
&= (A_2 - Q_2)e_1^r + (\delta(A_2 - A_1) + (1-\delta)Q_2)X_1^r + G_2^r(X_2) - \delta G_1^r(X_1) - \delta Q_1 e_{N-1}^r \\
&\quad -\delta^2 Q_1 e_{N-2}^r - \cdots - \delta^{N-1}Q_1 e_1^r - \delta(\delta^{N-1} - 1)Q_1 X_1^r + U_1^r \\
&= (A_2 - \delta^{N-1}Q_1 - Q_2)e_1^r - \delta^{N-2}Q_1 e_2^r - \cdots - \delta Q_1 e_{N-1}^r + (\delta(A_2 - A_1) + (1-\delta)Q_2 \\
&\quad -(\delta^N - \delta)Q_1)X_1^r + G_2^r(X_2) - \delta G_1^r(X_1) + U_1^r,
\end{aligned}
$$

$$
\begin{aligned}
D^\alpha e_2^r &= D^\alpha(X_3^r - \delta X_2^r) = D^\alpha X_3^r - \delta D^\alpha X_2^r \\
&= A_3 X_3^r + G_3^r(X_3) + Q_3(X_2^r - X_3^r) + U_2^r - \delta(A_2 X_2^r + G_2^r(X_2) + Q_2(X_1^r - X_2^r) + U_1^r) \\
&= (A_3 - Q_3)e_2^r + (\delta(A_3 - A_2) - (\delta-1)Q_3)X_2^r + G_3^r(X_3) - \delta G_2^r(X_2) + U_2^r - \delta U_1^r \\
&\quad +\delta Q_2(X_2^r - X_1^r) \\
&= (A_3 - Q_3)e_2^r + (\delta(A_3 - A_2) - (\delta-1)Q_3)X_2^r + G_3^r(X_3) - \delta G_2^r(X_2) + U_2^r - \delta U_1^r \\
&\quad +Q_2((\delta-1)X_2^r + X_2^r - \delta X_1^r) \\
&= Q_2 e_1^r + (A_3 - Q_3)e_2^r + (\delta(A_3 - A_2) + (\delta-1)(Q_2 - Q_3))X_2^r + G_3^r(X_3) - \delta G_2^r(X_2) \\
&\quad +U_2^r - \delta U_1^r, \\
&\quad \vdots
\end{aligned}
$$

$$
\begin{aligned}
D^\alpha e_{N-1}^r &= D^\alpha(X_N^r - \delta X_{N-1}^r) = D^\alpha X_N^r - \delta D^\alpha X_{N-1}^r \\
&= A_N X_N^r + G_N^r(X_N) + Q_N(X_{N-1}^r - X_N^r) + U_{N-1}^r - \delta(A_{N-1}X_{N-1}^r \\
&\quad +G_{N-1}^r(X_{N-1}) + Q_{N-1}(X_{N-2}^r - X_{N-1}^r) + U_{N-2}^r) \\
&= (A_N - Q_N)e_{N-1}^r + (\delta(A_N - A_{N-1}) - (\delta-1)Q_N)X_{N-1}^r + G_N^r(X_N) - \delta G_{N-1}^r(X_{N-1}) \\
&\quad +U_{N-1}^r - \delta U_{N-2}^r + \delta Q_{N-1}(X_{N-1}^r - X_{N-2}^r) \\
&= (A_N - Q_N)e_{N-1}^r + (\delta(A_N - A_{N-1}) - (\delta-1)Q_N)X_{N-1}^r + G_N^r(X_N) - \delta G_{N-1}^r(X_{N-1}) \\
&\quad +U_{N-1}^r - \delta U_{N-2}^r + Q_{N-1}((\delta-1)X_{N-1}^r + X_{N-1}^r - \delta X_{N-2}^r) \\
&= Q_{N-1}e_{N-2}^r + (A_N - Q_N)e_{N-1}^r + (\delta(A_N - A_{N-1}) + (\delta-1)(Q_{N-1} - Q_N))X_{N-1}^r \\
&\quad +G_N^r(X_N) - \delta G_{N-1}^r(X_{N-1}) + U_{N-1}^r - \delta U_{N-2}^r.
\end{aligned}
$$

Hence, the following form can be directly obtained:

$$
\begin{pmatrix} D^\alpha e_1^r \\ D^\alpha e_2^r \\ \vdots \\ D^\alpha e_{N-1}^r \end{pmatrix} = \Sigma \begin{pmatrix} e_1^r \\ e_2^r \\ \vdots \\ e_{N-1}^r \end{pmatrix}
+ \begin{pmatrix} (\delta(A_2 - A_1) + (1-\delta)Q_2 - (\delta^N - \delta)Q_1)X_1^r + G_2^r(X_2) - \delta G_1^r(X_1) + U_1^r \\ (\delta(A_3 - A_2) + (\delta-1)(Q_2 - Q_3))X_2^r + G_3^r(X_3) - \delta G_2^r(X_2) + U_2^r - \delta U_1^r \\ \vdots \\ (\delta(A_N - A_{N-1}) + (\delta-1)(Q_{N-1} - Q_N))X_{N-1}^r - \delta G_{N-1}^r(X_{N-1}) \\ +G_N^r(X_N) + U_{N-1}^r - \delta U_{N-2}^r \end{pmatrix},
$$

where

$$\Sigma = \begin{pmatrix} A_2 - \delta^{N-1}Q_1 - Q_2 & -\delta^{N-2}Q_1 & -\delta^{N-3}Q_1 & \cdots & -\delta^2 Q_1 & -\delta Q_1 \\ Q_2 & A_3 - Q_3 & 0 & \cdots & 0 & 0 \\ \vdots & \vdots & \vdots & \ddots & \vdots & \vdots \\ 0 & 0 & 0 & \cdots & Q_{N-1} & A_N - Q_N \end{pmatrix}.$$

Similarly, the imaginary part of the synchronization error vector is expressed as:

$$\begin{pmatrix} D^\alpha e_1^i \\ D^\alpha e_2^i \\ \vdots \\ D^\alpha e_{N-1}^i \end{pmatrix} = \Sigma \begin{pmatrix} e_1^i \\ e_2^i \\ \vdots \\ e_{N-1}^i \end{pmatrix} + \begin{pmatrix} (\delta(A_2 - A_1) + (1-\delta)Q_2 - (\delta^N - \delta)Q_1)X_1^i + G_2^i(X_2) - \delta G_1^i(X_1) + U_1^i \\ (\delta(A_3 - A_2) + (\delta - 1)(Q_2 - Q_3))X_2^i + G_3^i(X_3) - \delta G_2^i(X_2) + U_2^i - \delta U_1^i \\ \vdots \\ (\delta(A_N - A_{N-1}) + (\delta - 1)(Q_{N-1} - Q_N))X_{N-1}^i + G_N^i(X_N) \\ -\delta G_{N-1}^i(X_{N-1}) + U_{N-1}^i - \delta U_{N-2}^i \end{pmatrix},$$

where Σ is defined as in (18.1.4).

Thus, our aim is to design controllers $U_k = U_k^r + jU_k^i$ $(k = 1, 2, \ldots, N-1)$ such that the error dynamical system is not only transformed into a nonlinear system with anti-symmetric structure but also asymptotically stable at the origin.

18.1.2 Main Results

In previous literature [258–260], a direct design method was used to realize synchronization for integer-order chaotic systems with real variables. This method is flexible and practical for both design and implementation. Therefore, we will utilize this method to achieve synchronization and anti-synchronization of fractional-order complex chaotic systems.

We design the controllers U_k^r and U_k^i in the following form:

$$\begin{cases} U_1^r &= V_1^r - (\delta(A_2 - A_1) + (1-\delta)Q_2 - (\delta^N - \delta)Q_1)X_1^r - G_2^r(X_2) + \delta G_1^r(X_1), \\ U_2^r &= V_2^r - (\delta(A_3 - A_2) + (\delta - 1)(Q_2 - Q_3))X_2^r - G_3^r(X_3) + \delta G_2^r(X_2) + \delta U_1^r, \\ \vdots & \\ U_{N-1}^r &= V_{N-1}^r - (\delta(A_N - A_{N-1}) + (\delta - 1)(Q_{N-1} - Q_N))X_{N-1}^r - G_N^r(X_N) \\ & \quad + \delta G_{N-1}^r(X_{N-1}) + \delta U_{N-2}^r, \end{cases} \tag{18.1.4}$$

and

$$\begin{cases} U_1^i &= V_1^i - (\delta(A_2 - A_1) + (1-\delta)Q_2 - (\delta^N - \delta)Q_1)X_1^i - G_2^i(X_2) + \delta G_1^i(X_1), \\ U_2^i &= V_2^i - (\delta(A_3 - A_2) + (\delta - 1)(Q_2 - Q_3))X_2^i - G_3^i(X_3) + \delta G_2^i(X_2) + \delta U_1^i, \\ &\vdots \\ U_{N-1}^i &= V_{N-1}^i - (\delta(A_N - A_{N-1}) + (\delta-1)(Q_{N-1} - Q_N))X_{N-1}^i - G_N^i(X_N) \\ &\quad + \delta G_{N-1}^i(X_{N-1}) + \delta U_{N-2}^i, \end{cases} \tag{18.1.5}$$

where

$$\begin{pmatrix} V_1^r \\ V_2^r \\ \vdots \\ V_{N-1}^r \end{pmatrix} = \Theta \begin{pmatrix} e_1^r \\ e_2^r \\ \vdots \\ e_{N-1}^r \end{pmatrix}, \quad \begin{pmatrix} V_1^i \\ V_2^i \\ \vdots \\ V_{N-1}^i \end{pmatrix} = \Theta \begin{pmatrix} e_1^i \\ e_2^i \\ \vdots \\ e_{N-1}^i \end{pmatrix}, \tag{18.1.6}$$

and Θ is a real coefficient matrix. Then, the error dynamical systems can be expressed as:

$$D^\alpha e^r(t) = \Phi(t)e^r(t) \tag{18.1.7}$$

and

$$D^\alpha e^i(t) = \Phi(t)e^i(t), \tag{18.1.8}$$

where $\Phi(t) = \Sigma + \Theta$. Further, we have

$$D^\alpha e(t) = D^\alpha e^r(t) + jD^\alpha e^i(t) = \Phi(t)e(t). \tag{18.1.9}$$

Obviously, there is no a unique choice for such a matrix Θ to ensure the error dynamical system (18.1.9) is asymptotically stable at the origin. Without loss of generality, Θ is defined as a state-dependent real coefficient matrix. By transforming system (18.1.9) into a stable system with an antisymmetric structure, we derive the sufficient stability conditions, which lead to the following main result.

Theorem 18.1 *Synchronization and anti-synchronization of N-coupled fractional-order complex chaotic systems* (18.1.3) *will occur by using the control laws* (18.1.4) *and* (18.1.5), *if the state dependent real coefficient matrix $\Phi(t) = \Phi_1(t) + \Phi_2$ satisfies the following assumptions:*

$$\Phi_1^T(t) = -\Phi_1(t) \quad and \quad \Phi_2 = diag(-\varphi_1, -\varphi_2, \ldots, -\varphi_{N-1}), \tag{18.1.10}$$

where $\varphi_k = diag(\varphi_{k1}, \varphi_{k2}, \ldots, \varphi_{kn})$.

Proof A positive definite Lyapunov candidate function is constructed as

$$V = \frac{1}{2}\left(e^r(t)\right)^T e^r(t).$$

Differentiating along the system trajectory with respect to time and applying Lemma 1.4 yield:

$$D^\alpha V \leq \frac{1}{2}\left((D^\alpha e^r(t))^T e^r(t) + (e^r(t))^T D^\alpha e^r(t)\right) = \frac{1}{2}\left((e^r(t))^T(\Phi^T(t) + \Phi(t))e^r(t)\right).$$

Substituting (18.1.10) into the latter inequality, it follows that

$$\begin{aligned}
D^\alpha V &\leq \frac{1}{2}\left[(e^r(t))^T(\Phi_2^T + \Phi_2)e^r(t)\right] = (e^r(t))^T\Phi_2 e^r(t)\\
&= \left((e_1^r(t))^T, \ldots, (e_{N-1}^r(t))^T\right)\begin{pmatrix} -\varphi_1 & 0 & 0 \cdots & 0\\ 0 & -\varphi_2 & 0 \cdots & 0\\ \vdots & \vdots & \vdots \ddots & \vdots\\ 0 & 0 & 0 \cdots & -\varphi_{N-1}\end{pmatrix}\begin{pmatrix} e_1^r(t)\\ \vdots\\ e_{N-1}^r(t)\end{pmatrix}\\
&= -(e_1^r(t))^T\varphi_1 e_1^r(t) - (e_2^r(t))^T\varphi_2 e_2^r(t) - \cdots - (e_{N-1}^r(t))^T\varphi_{N-1}e_{N-1}^r(t)\\
&\leq -\varphi_{min}\left[(e_1^r(t))^T e_1^r(t) + (e_2^r(t))^T e_2^r(t) + \cdots + (e_{N-1}^r(t))^T e_{N-1}^r(t)\right]\\
&= -\varphi_{min}||e^r(t)||^2,
\end{aligned}$$

where $\varphi_{min} = \min_{1\leq k\leq N-1, 1\leq l\leq n}\{\varphi_{kl}\}$. According to Theorem 18.1, we deduce $\lim_{t\to\infty}||e^r(t)|| = 0$. Similarly, we consider a positive definite Lyapunov candidate function of the form:

$$W = \frac{1}{2}\left(e^i(t)\right)^T e^i(t),$$

then we have $D^\alpha W \leq -\varphi_{min}||e^i(t)||^2$ and $\lim_{t\to\infty}||e^i(t)|| = 0$. Hence, we conclude that $\lim_{t\to\infty}||e(t)|| = 0$, which implies the asymptotic stability of the equilibrium point in system (18.1.9). Consequently, we demonstrate that the N-coupled fractional-order complex chaotic systems (18.1.3) can achieve both synchronization and anti-synchronization. This completes the proof. □

Now, based on Theorem 18.1, we present the following corollary.

Corollary 18.1 *If the structures of general fractional-order complex chaotic systems are identical, i.e.,* $A_k = A_l = A$ *and* $G_k(\cdot) = G_l(\cdot) = G(\cdot)$, *complex controllers* $U_k = U_k^r + jU_k^i$ *can be provided as*

$$\begin{cases}
U_1 &= V_1 - ((1-\delta)Q_2 - (\delta^N - \delta)Q_1)X_1 - G(X_2) + \delta G(X_1),\\
U_2 &= V_2 - (\delta - 1)(Q_2 - Q_3)X_2 - G(X_3) + \delta G(X_2) + \delta U_1,\\
&\vdots\\
U_{N-1} &= V_{N-1} - (\delta - 1)(Q_{N-1} - Q_N)X_{N-1} - G(X_N) + \delta G(X_{N-1}) + \delta U_{N-2},
\end{cases}$$

where $V_k = V_k^r + jV_k^i$ $(k = 1, 2, \ldots, N-1)$ are defined as in (18.1.6) *and Θ is a real coefficient matrix. A similar analysis to the above-mentioned discussion leads to realize synchronization and anti-synchronization of N-coupled identical fractional-order complex chaotic systems with ring connection based on antisymmetric structure.*

As aforementioned analysis, the discussion of the error dynamical systems can be transformed into the analysis of the asymptotic stability of the zero solution of the error systems (18.1.7) and (18.1.8) by virtue of the complex control laws U_k^r and U_k^i, respectively. Then, a suitable real coefficient matrix Θ is selected to translate system (18.1.9) into a stable system with an antisymmetric structure. It should be noted that the choice of the coefficient matrix is crucial for achieving synchronization and anti-synchronization in the ring-connected, N-coupled fractional-order complex chaotic systems.

18.1.3 Numerical Simulations

To demonstrate the efficacy of the proposed synchronization scheme, two specific cases are introduced and numerically simulated. Chaos synchronization is simulated for three different fractional-order complex chaotic systems, which are the fractional-order complex Chen, the fractional-order complex Lorenz and the fractional-order complex Lü system.

$$\begin{cases} D^\alpha x_{11} = a_1(x_{12} - x_{11}) + q_{11}(x_{31} - x_{11}), \\ D^\alpha x_{12} = (a_2 - a_1)x_{11} + a_2 x_{12} - x_{11}x_{13} + q_{12}(x_{32} - x_{12}), \\ D^\alpha x_{13} = \frac{1}{2}(\bar{x}_{11}x_{12} + x_{11}\bar{x}_{12}) - a_3 x_{13} + q_{13}(x_{33} - x_{13}), \end{cases} \tag{18.1.11}$$

$$\begin{cases} D^\alpha x_{21} = b_1(x_{22} - x_{21}) + q_{21}(x_{11} - x_{21}) + u_{11}, \\ D^\alpha x_{22} = b_2 x_{21} - x_{22} - x_{21}x_{23} + q_{22}(x_{12} - x_{22}) + u_{12}, \\ D^\alpha x_{23} = \frac{1}{2}(\bar{x}_{21}x_{22} + x_{21}\bar{x}_{22}) - b_3 x_{23} + q_{23}(x_{13} - x_{23}) + u_{13}, \end{cases} \tag{18.1.12}$$

and

$$\begin{cases} D^\alpha x_{31} = c_1(x_{32} - x_{31}) + q_{31}(x_{21} - x_{31}) + u_{21}, \\ D^\alpha x_{32} = c_2 x_{32} - x_{31}x_{33} + q_{32}(x_{22} - x_{32}) + u_{22}, \\ D^\alpha x_{33} = \frac{1}{2}(\bar{x}_{31}x_{32} + x_{31}\bar{x}_{32}) - c_3 x_{33} + q_{33}(x_{23} - x_{33}) + u_{23}, \end{cases} \tag{18.1.13}$$

where

$$A_1 = \begin{pmatrix} -a_1 & a_1 & 0 \\ a_2 - a_1 & a_2 & 0 \\ 0 & 0 & -a_3 \end{pmatrix}, \quad A_2 = \begin{pmatrix} -b_1 & b_1 & 0 \\ b_2 & -1 & 0 \\ 0 & 0 & -b_3 \end{pmatrix}, \quad A_3 = \begin{pmatrix} -c_1 & c_1 & 0 \\ 0 & c_2 & 0 \\ 0 & 0 & -c_3 \end{pmatrix},$$

$G_k(X_k) = \begin{pmatrix} 0 \\ -x_{k1}x_{k3} \\ \frac{1}{2}(\bar{x}_{k1}x_{k2} + x_{k1}\bar{x}_{k2}) \end{pmatrix}$, $Q_k = \text{diag}(q_{k1}, q_{k2}, q_{k3})$ are the coupled matrices ($k = 1, 2, 3$), and $U_k = (u_{k1}, u_{k2}, u_{k3})^T$ are the control inputs ($k = 1, 2$). Let $\alpha = 0.96$, $(a_1, a_2, a_3) = (35, 28, 3)$, $(b_1, b_2, b_3) = (10, 180, 1)$, and $(c_1, c_2, c_3) = (42, 22, 4)$, in the absence of the controllers, fractional-order systems (18.1.8)–(18.1.13) without the coupling terms behave chaotically.

The synchronization error is defined as $e_k = X_{k+1} - \delta X_k$ ($k = 1, 2$). Then the error dynamical system is described as:

$$
\begin{aligned}
D^\alpha e(t) = &\begin{pmatrix} A_2 - \delta^2 Q_1 - Q_2 & -\delta Q_1 \\ Q_2 & A_3 - Q_3 \end{pmatrix} \begin{pmatrix} e_1(t) \\ e_2(t) \end{pmatrix} \\
&+ \begin{pmatrix} (\delta(A_2 - A_1) + (1 - \delta)Q_2 - (\delta^3 - \delta)Q_1)X_1 + G_2(X_2) - \delta G_1(X_1) + U_1 \\ (\delta(A_3 - A_2) + (\delta - 1)(Q_2 - Q_3))X_2 + G_3(X_3) - \delta G_2(X_2) + U_2 - \delta U_1 \end{pmatrix}.
\end{aligned}
$$

I. Synchronization of 3-Coupled Fractional-Order Complex Chaotic Systems

When $\delta = 1$, the error system reduces to the following complete synchronization error system.

$$
\begin{aligned}
D^\alpha e^r(t) = &\begin{pmatrix} A_2 - Q_1 - Q_2 & -Q_1 \\ Q_2 & A_3 - Q_3 \end{pmatrix} \begin{pmatrix} e_1^r(t) \\ e_2^r(t) \end{pmatrix} \\
&+ \begin{pmatrix} (A_2 - A_1)X_1^r + G_2^r(X_2) - G_1^r(X_1) + U_1^r \\ (A_3 - A_2)X_2^r + G_3^r(X_3) - G_2^r(X_2) + U_2^r - U_1^r \end{pmatrix}
\end{aligned}
$$

and

$$
\begin{aligned}
D^\alpha e^i(t) = &\begin{pmatrix} A_2 - Q_1 - Q_2 & -Q_1 \\ Q_2 & A_3 - Q_3 \end{pmatrix} \begin{pmatrix} e_1^i(t) \\ e_2^i(t) \end{pmatrix} \\
&+ \begin{pmatrix} (A_2 - A_1)X_1^i + G_2^i(X_2) - G_1^i(X_1) + U_1^i \\ (A_3 - A_2)X_2^i + G_3^i(X_3) - G_2^i(X_2) + U_2^i - U_1^i \end{pmatrix}.
\end{aligned}
$$

Choose the real coefficient matrix $\Theta = \begin{pmatrix} \Theta_1 \\ \Theta_2 \end{pmatrix}$ as follows:

$$
\Theta_1 = \begin{pmatrix} 0 & 0\,0\,0\,0\,0 \\ -b_1 - b_2 & 0\,0\,0\,0\,0 \\ 0 & 0\,0\,0\,0\,0 \end{pmatrix},
$$
$$
\Theta_2 = \begin{pmatrix} q_{11} - q_{21} & 0 & 0 & 0 & 0\,0 \\ 0 & q_{12} - q_{22} & 0 & -c_1 & 0\,0 \\ 0 & 0 & q_{13} - q_{23} & 0 & 0\,0 \end{pmatrix}.
$$

We design the complex controllers $U_k = (u_{k1}, u_{k2}, \ldots, u_{kn})^T$ ($k = 1, 2$) as:

$$\begin{cases} u_{11} = (b_1 - a_1)(x_{11} - x_{12}), \\ u_{12} = -(b_1 + b_2)e_{11} - (b_2 - a_2 + a_1)x_{11} + (a_2 + 1)x_{12} + x_{21}x_{23}, \\ u_{13} = (b_3 - a_3)x_{13} - x_{21}^r x_{22}^r - x_{21}^i x_{22}^i + x_{11}^r x_{12}^r + x_{11}^i x_{12}^i, \end{cases}$$

and

$$\begin{cases} u_{21} = (q_{11} - q_{21})e_{11} - (b_1 - c_1)(x_{21} - x_{22}) + (b_1 - a_1)(x_{11} - x_{12}), \\ u_{22} = (q_{12} - q_{22})e_{12} - (c_2 + 1)x_{22} - (b_1 + b_2)e_{11} - (b_2 - a_2 + a_1)x_{11} \\ \qquad -c_1 e_{21} + b_2 x_{21} + (a_2 + 1)x_{12} + x_{31}x_{33} - x_{11}x_{13}, \\ u_{23} = (q_{13} - q_{23})e_{13} - (b_3 - c_3)x_{23} - (a_3 - b_3)x_{13} - x_{31}^r x_{32}^r - x_{31}^i x_{32}^i \\ \qquad + x_{11}^r x_{12}^r + x_{11}^i x_{12}^i, \end{cases}$$

Then, we have $D^\alpha e^r(t) = (\Phi_1(t) + \Phi_2)e^r(t)$ and $D^\alpha e^i(t) = (\Phi_1(t) + \Phi_2)e^i(t)$, where $\Phi_2 = \text{diag}(-b_1 - q_{11} - q_{21}, -1 - q_{12} - q_{22}, -b_3 - q_{13} - q_{23}, -c_1 - q_{31}, c_2 - q_{32}, -c_3 - q_{33})$ and

$$\Phi_1(t) = \begin{pmatrix} 0 & b_1 & 0 & -q_{11} & 0 & 0 \\ -b_1 & 0 & 0 & 0 & -q_{12} & 0 \\ 0 & 0 & 0 & 0 & 0 & -q_{13} \\ q_{11} & 0 & 0 & 0 & c_1 & 0 \\ 0 & q_{12} & 0 & -c_1 & 0 & 0 \\ 0 & 0 & q_{13} & 0 & 0 & 0 \end{pmatrix}.$$

If the conditions

$$-b_1 - q_{11} - q_{21} < 0, \quad -1 - q_{12} - q_{22} < 0, \quad -b_3 - q_{13} - q_{23} < 0,$$
$$-c_1 - q_{31} < 0, \quad c_2 - q_{32} < 0, \quad -c_3 - q_{33} < 0$$

hold, then the error system is asymptotically stable under complex controllers U_k $(k = 1, 2)$ due to Theorem 18.1. Thus, complete synchronization of the 3-coupled fractional-order complex chaotic systems is achieved.

In the numerical simulations, the initial conditions of systems (18.1.11)–(18.1.13) are set to $X_1(0) = (1 + 4j, 3 - 5j, 6)^T$, $X_2(0) = (3 + 9j, 7 - 2j, 5)^T$, and $X_3(0) = (-10 + 8j, 5 - 3j, 8)^T$, respectively. Further, taking $q_{11} = q_{12} = 1$, $q_{13} = q_{21} = q_{22} = q_{23} = 3$, $q_{31} = 8$, $q_{32} = 30$, and $q_{33} = 50$, the Adams-Bashforth-Moulton predictor-corrector scheme is employed to obtain the simulation results shown in Figs. 18.2 and 18.3. Figure 18.2 displays the synchronization process of the 3-coupled fractional-order complex chaotic systems (18.1.11)–(18.1.13). As described in Fig. 18.3, the synchronization errors $e_k(t)$ $(k = 1, 2)$ converge to zero as $t \to \infty$, which incidates that complete synchronization is achieved in the 3-coupled fractional-order complex chaotic systems with a ring connection.

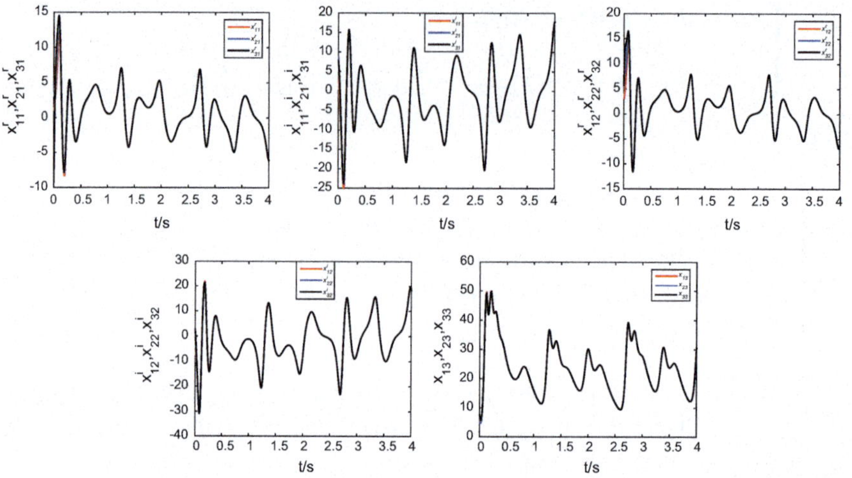

Fig. 18.2 The state trajectories x_{k1}, x_{k2}, and x_{k3} $(k = 1, 2, 3)$ for synchronization of 3-coupled fractional-order complex systems (18.1.11)–(18.1.13)

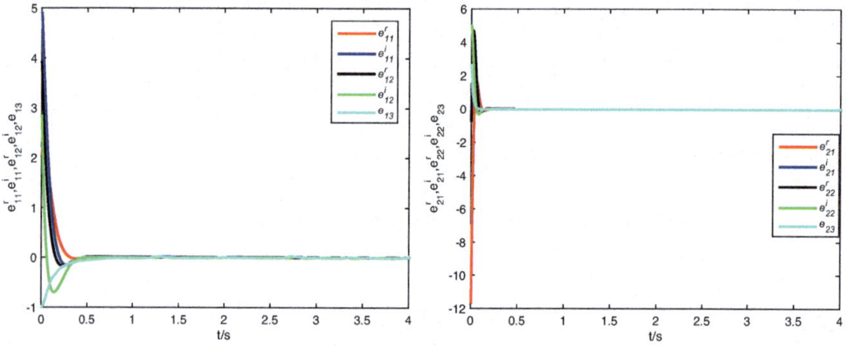

Fig. 18.3 The error dynamics of synchronization among 3-coupled fractional-order complex systems (18.1.11)–(18.1.13)

II. Anti-Synchronization of 3-Coupled Fractional-Order Complex Chaotic Systems

When $\delta = -1$, the error system becomes the following anti-synchronization error system.

$$
\begin{aligned}
D^{\alpha} e^r(t) = &\begin{pmatrix} A_2 - Q_1 - Q_2 & Q_1 \\ Q_2 & A_3 - Q_3 \end{pmatrix} \begin{pmatrix} e_1^r(t) \\ e_2^r(t) \end{pmatrix} \\
&+ \begin{pmatrix} (A_1 - A_2 + 2Q_2)X_1^r + G_2^r(X_2) + G_1^r(X_1) + U_1^r \\ (A_2 - A_3 - 2Q_2 + 2Q_3)X_2^r + G_3^r(X_3) + G_2^r(X_2) + U_2^r + U_1^r \end{pmatrix}
\end{aligned}
$$

and

$$D^{\alpha}e^{i}(t) = \begin{pmatrix} A_2 - Q_1 - Q_2 & Q_1 \\ Q_2 & A_3 - Q_3 \end{pmatrix} \begin{pmatrix} e_1^i(t) \\ e_2^i(t) \end{pmatrix}$$
$$+ \begin{pmatrix} (A_1 - A_2 + 2Q_2)X_1^i + G_2^i(X_2) + G_1^i(X_1) + U_1^i \\ (A_2 - A_3 - 2Q_2 + 2Q_3)X_2^i + G_3^i(X_3) + G_2^i(X_2) + U_2^i + U_1^i \end{pmatrix}.$$

Select the real coefficient matrix as follows $\Theta = \begin{pmatrix} \Theta_1 \\ \Theta_2 \end{pmatrix}$ as follows:

$$\Theta_1 = \begin{pmatrix} 0 & -b_1 & 0 & -q_{11} - q_{21} & 0 & 0 \\ -b_2 & 0 & 0 & 0 & -q_{12} - q_{22} & 0 \\ 0 & 0 & 0 & 0 & 0 & -q_{13} - q_{23} \end{pmatrix},$$
$$\Theta_2 = \begin{pmatrix} 0 & 0 & 0 & 0 & 0 & 0 \\ 0 & 0 & 0 & -c_1 & 0 & 0 \\ 0 & 0 & 0 & 0 & 0 & 0 \end{pmatrix}.$$

Therefore, we construct the complex controllers U_k $(k = 1, 2)$ as:

$$\begin{cases} u_{11} = -b_1 e_{12} - (q_{11} + q_{21})e_{21} - (b_1 - a_1 + 2q_{21})x_{11} - (a_1 - b_1)x_{12}, \\ u_{12} = -b_2 e_{11} - (q_{12} + q_{22})e_{22} - (a_2 - a_1 - b_2)x_{11} - (a_2 + 1 + 2q_{22})x_{12} + x_{21}x_{23} \\ \quad\quad + x_{11}x_{13}, \\ u_{13} = -(q_{13} + q_{23})e_{23} - (b_3 - a_3 + 2q_{23})x_{13} - x_{21}^r x_{22}^r - x_{21}^i x_{22}^i - x_{11}^r x_{12}^r - x_{11}^i x_{12}^i, \end{cases}$$

and

$$\begin{cases} u_{21} = b_1 e_{12} + (q_{11} + q_{21})e_{21} - (c_1 - b_1 - 2q_{21} + 2q_{31})x_{21} \\ \quad\quad -(b_1 - c_1)x_{22} + (b_1 - a_1 + 2q_{21})x_{11} + (a_1 - b_1)x_{12}, \\ u_{22} = -c_1 e_{21} + b_2 e_{11}(q_{12} + q_{22})e_{22} - b_2 x_{21} + (c_2 + 1 + 2q_{22} - 2q_{32})x_{22} \\ \quad\quad +(a_2 + 1 + 2q_{22})x_{12} + (a_2 - a_1 - b_2)x_{11} + x_{31}x_{33} - x_{11}x_{13}, \\ u_{23} = (q_{13} + q_{23})e_{23} + (b_3 - c_3 + 2q_{23} - 2q_{33})x_{23} + (b_3 - a_3 + 2q_{23})x_{13} \\ \quad\quad -x_{31}^r x_{32}^r - x_{31}^i x_{32}^i + x_{11}^r x_{12}^r + x_{11}^i x_{12}^i. \end{cases}$$

Then, we have $D^{\alpha}e^{r}(t) = (\Phi_1(t) + \Phi_2)e^{r}(t)$ and $D^{\alpha}e^{i}(t) = (\Phi_1(t) + \Phi_2)e^{i}(t)$, where $\Phi_2 = \text{diag}(-b_1 - q_{11} - q_{21}, -1 - q_{12} - q_{22}, -b_3 - q_{13} - q_{23}, -c_1 - q_{31}, c_2 - q_{32}, -c_3 - q_{33})$ and

$$\Phi_1(t) = \begin{pmatrix} 0 & 0 & 0 & -q_{21} & 0 & 0 \\ 0 & 0 & 0 & 0 & -q_{22} & 0 \\ 0 & 0 & 0 & 0 & 0 & -q_{23} \\ q_{21} & 0 & 0 & 0 & c_1 & 0 \\ 0 & q_{22} & 0 & -c_1 & 0 & 0 \\ 0 & 0 & q_{23} & 0 & 0 & 0 \end{pmatrix}.$$

If the conditions

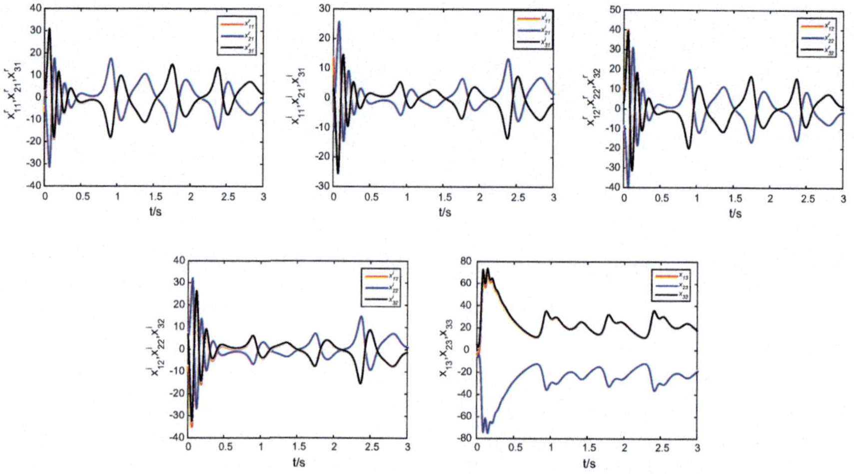

Fig. 18.4 The state trajectories x_{k1}, x_{k2}, and x_{k3} $(k = 1, 2, 3)$ for anti-synchronization of 3-coupled fractional-order complex systems (18.1.11)–(18.1.13)

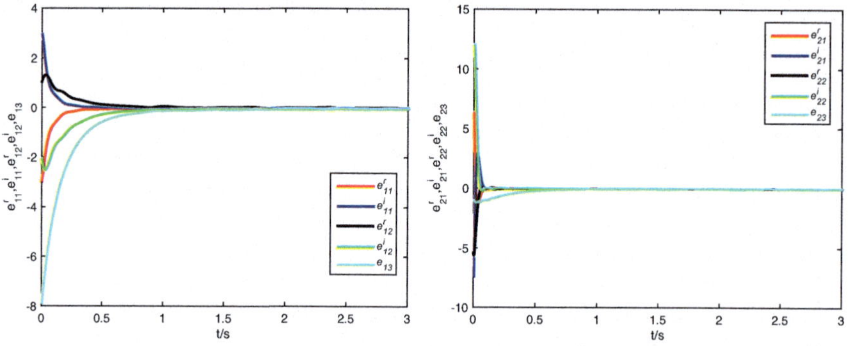

Fig. 18.5 The error dynamics of anti-synchronization among 3-coupled fractional-order complex systems (18.1.11)–(18.1.13)

$$-b_1 - q_{11} - q_{21} < 0, \quad -1 - q_{12} - q_{22} < 0, \quad -b_3 - q_{13} - q_{23} < 0,$$
$$-c_1 - q_{31} < 0, \quad c_2 - q_{32} < 0, \quad -c_3 - q_{33} < 0$$

are satisfied, Theorem 18.1 guarantees that the error system under the complex controllers U_k $(k = 1, 2)$ is asymptotically stable. Consequently, anti-synchronization is achieved in the 3-coupled fractional-order complex chaotic systems.

Let $\alpha = 0.95$. The initial values of systems (18.1.11)–(18.1.13) are chosen as $X_1(0) = (-7 + 15j, 9 - 7j, -2)^T$, $X_2(0) = (4 - 12j, -8 + 5j, -6)^T$, and $X_3(0) = (-2 + 10j, 6 - j, 10)^T$. Choose $q_{11} = q_{12} = q_{13} = 1$, $q_{21} = 1$, $q_{22} = 30$, $q_{23} = 10$, $q_{31} = 8$, $q_{32} = 50$, and $q_{33} = 10$. The simulation results are shown in Figs. 18.4 and 18.5. Figure 18.4 depicts the anti-synchronization process of the 3-coupled fractional-order complex chaotic systems (18.1.11)–(18.1.13). As shown in Fig. 18.5, the synchronization errors converge asymptotically to zero within a short transient period, confirming the successful achievement of anti-synchronization as theoretically expected.

18.2 Synchronization of Bidirectional N-Coupled Fractional-Order Chaotic Systems with Ring Connection

The following discusses the synchronization problem of N-coupled fractional-order chaotic systems with ring connection via bidirectional coupling. Based on the direct design method, we design the appropriate controllers to transform the fractional-order error dynamical system into a nonlinear system with antisymmetric structure. Finally, some numerical examples are presented to illustrate the validity of theoretical results.

18.2.1 Problem Descriptions

Consider a general form of N fractional-order chaotic systems defined in (18.1.1). We couple multiple fractional-order complex chaotic systems in a bidirectional ring structure, as illustrated in Fig. 18.6. Therefore, the bidirectional coupling scheme for N fractional-order chaotic systems (18.1.1) in a ring configuration is given by:

$$\begin{cases} D^{\alpha}X_1 &= A_1X_1 + G_1(X_1) + Q_1(X_2 + X_N - 2X_1), \\ D^{\alpha}X_2 &= A_2X_2 + G_2(X_2) + Q_2(X_1 + X_3 - 2X_2), \\ D^{\alpha}X_3 &= A_3X_3 + G_3(X_3) + Q_3(X_2 + X_4 - 2X_3), \\ &\vdots \\ D^{\alpha}X_N &= A_NX_N + G_N(X_N) + Q_N(X_{N-1} + X_1 - 2X_N), \end{cases} \tag{18.2.1}$$

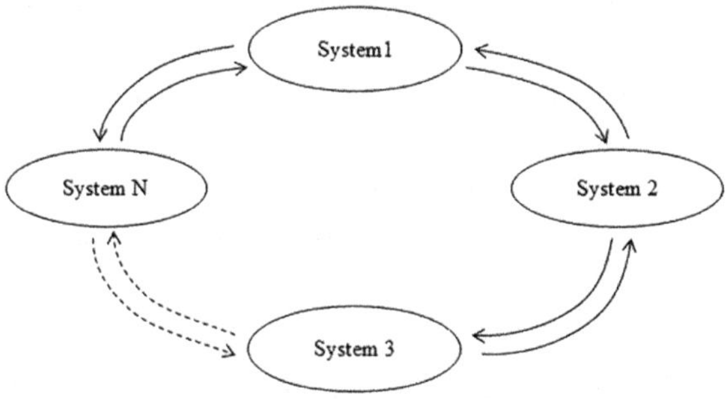

Fig. 18.6 The coupled mode of N systems via bidirectional coupling

where $Q_i = \text{diag}(q_{i1}, q_{i2}, \ldots, q_{in})$ are n-dimensional diagonal matrices and $q_{ij} \geq 0$ are the ideal gains which represent the coupled parameters.

Suppose that the first system is chosen as the aim system, and add the controller to the remaining systems. Thus, model (18.2.1) can be rewritten in the following form:

$$\begin{cases} D^\alpha X_1 = A_1X_1 + G_1(X_1) + Q_1(X_2 + X_N - 2X_1), \\ D^\alpha X_2 = A_2X_2 + G_2(X_2) + Q_2(X_1 + X_3 - 2X_2) + U_1, \\ D^\alpha X_3 = A_3X_3 + G_3(X_3) + Q_3(X_2 + X_4 - 2X_3) + U_2, \\ \quad\vdots \\ D^\alpha X_N = A_NX_N + G_N(X_N) + Q_N(X_{N-1} + X_1 - 2X_N) + U_{N-1}. \end{cases} \tag{18.2.2}$$

We define the synchronization error vector as $e_i = X_{i+1} - X_i \in \mathbb{R}^n$ $(i = 1, 2, \ldots, N-1)$. Then the error system is obtained

$$e = \begin{pmatrix} e_1 \\ e_2 \\ \vdots \\ e_{N-1} \end{pmatrix}.$$

By considering system (18.2.2), we derive the error dynamical system:

$$D^\alpha e = \begin{pmatrix} \Sigma_1 & \Sigma_2 \\ \Sigma_3 & \Sigma_4 \end{pmatrix} \begin{pmatrix} e_1 \\ e_2 \\ \vdots \\ e_{N-1} \end{pmatrix} + \begin{pmatrix} (A_2 - A_1)X_1 + G_2(X_2) - G_1(X_1) + U_1 \\ (A_3 - A_2)X_2 + G_3(X_3) - G_2(X_2) + U_2 - U_1 \\ \vdots \\ (A_N - A_{N-1})X_{N-1} + G_N(X_N) - G_{N-1}(X_{N-1}) + U_{N-1} - U_{N-2} \end{pmatrix},$$

where

$$\Sigma_1 = \begin{pmatrix} A_2 - 2Q_1 - Q_2 & Q_2 - Q_1 & -Q_1 & \cdots & -Q_1 & -Q_1 \\ Q_2 & A_3 - Q_3 - Q_2 & Q_3 & \cdots & 0 & 0 \\ \vdots & \vdots & \vdots & \ddots & \vdots & \vdots \\ 0 & 0 & 0 & \cdots & Q_{N-2} & A_{N-1} - Q_{N-1} - Q_{N-2} \end{pmatrix},$$

$$\Sigma_2 = [-Q_1 \ 0 \ 0 \ \cdots \ 0 \ Q_{N-1}]^T,$$
$$\Sigma_3 = [-Q_N \ -Q_N \ -Q_N \ \cdots \ -Q_N \ Q_{N-1} - Q_N],$$
$$\Sigma_4 = A_N - 2Q_N - Q_{N-1}.$$

Our work is to find the suitable and effective controllers U_i $(i = 1, 2, \ldots, N-1)$ such that the above error dynamical system is not only transformed into a nonlinear system with antisymmetric structure but also asymptotically stable at the origin.

18.2.2 Main Results

In the following, the direct design method is used to realize synchronization of bidirectional N-coupled fractional-order chaotic systems with ring connection based on antisymmetric structure.

First, the controllers U_i are designed as

$$\begin{cases} U_1 = V_1 - (A_2 - A_1)X_1 - G_2(X_2) + G_1(X_1), \\ U_2 = V_2 - (A_3 - A_2)X_2 - G_3(X_3) + G_2(X_2) + U_1, \\ \quad \vdots \\ U_{N-1} = V_{N-1} - (A_N - A_{N-1})X_{N-1} - G_N(X_N) + G_{N-1}(X_{N-1}) + U_{N-2}, \end{cases} \tag{18.2.3}$$

where

$$\begin{pmatrix} V_1 \\ V_2 \\ \vdots \\ V_{N-1} \end{pmatrix} = \Theta \begin{pmatrix} e_1 \\ e_2 \\ \vdots \\ e_{N-1} \end{pmatrix} \tag{18.2.4}$$

and Θ is a coefficient matrix. Then, the error dynamical system is described as:

$$D^{\alpha} e(t) = \Phi(t) e(t), \tag{18.2.5}$$

where

$$\Phi(t) = \begin{pmatrix} \Sigma_1 & \Sigma_2 \\ \Sigma_3 & \Sigma_4 \end{pmatrix} + \Theta.$$

Evidently, there are many possible choices for Θ to guarantee that the error dynamical system (18.2.5) is asymptotically stable at the origin. Without loss of generality, we define Θ as a state-dependent coefficient matrix. Therefore, the sufficient stability conditions of system (18.2.5) are produced by transforming it into a stable system with antisymmetric structure. The main result is depicted in the following.

Theorem 18.2 *Synchronization of bidirectional N-coupled fractional-order system* (18.2.2) *can be achieved by utilizing the control laws in Eq.* (18.2.3), *if the state dependent coefficient matrix $\Phi(t) = \Phi_1(t) + \Phi_2$ satisfies the hypotheses*

$$\Phi_1^T(t) = -\Phi_1(t) \quad and \quad \Phi_2 = diag(-\varphi_1, -\varphi_2, \ldots, -\varphi_{N-1}), \tag{18.2.6}$$

where $\varphi_i = diag(\varphi_{i1}, \varphi_{i2}, \ldots, \varphi_{in})$ $(i = 1, 2, \ldots, N-1)$ *and* $\varphi_{ij} > 0$ $(j = 1, 2, \ldots, n)$.

Proof A positive definite Lyapunov candidate function is defined as

$$V(e, t) = \frac{1}{2} e^T(t) e(t).$$

From Lemma 1.4 and system (18.2.5), we obtain

$$\begin{aligned} D^\alpha V(e, t) &\le \frac{1}{2} \left((D^\alpha e(t))^T e(t) + e^T(t) D^\alpha e(t) \right) \\ &= \frac{1}{2} e^T(t)(\Phi^T(t) + \Phi(t)) e(t). \end{aligned}$$

Substituting (18.2.6) into the latter inequality, one has

$$\begin{aligned} D^\alpha V(e, t) &\le \frac{1}{2} e^T(t)(\Phi_2^T + \Phi_2) e(t) = e^T(t) \Phi_2 e(t) \\ &= (e_1^T(t), e_2^T(t), \ldots, e_{N-1}^T(t)) \begin{pmatrix} -\varphi_1 & 0 & \cdots & 0 \\ 0 & -\varphi_2 & \cdots & 0 \\ \vdots & \vdots & \ddots & \vdots \\ 0 & 0 & \cdots & -\varphi_{N-1} \end{pmatrix} \begin{pmatrix} e_1(t) \\ e_2(t) \\ \vdots \\ e_{N-1}(t) \end{pmatrix} \\ &= -e_1^T(t)\varphi_1 e_1(t) - e_2^T(t)\varphi_2 e_2(t) - \cdots - e_{N-1}^T(t)\varphi_{N-1} e_{N-1}(t) \\ &\le -\varphi_{min}(e_1^T(t)e_1(t) + e_2^T(t)e_2(t) + \cdots + e_{N-1}^T(t)e_{N-1}(t)) \\ &= -\varphi_{min} ||e(t)||^2, \end{aligned}$$

where $\varphi_{min} = \min_{1 \le i \le N, 1 \le j \le n}\{\varphi_{ij}\}$. Based on Lemma 1.5, we obtain the closed-loop system (18.2.5) is globally asymptotically stable. Hence, synchronization is achieved for the bidirectional N-coupled fractional-order chaotic system. This completes the proof. □

The following corollary can be derived from Theorem 18.2.

Corollary 18.2 *If the structures of general fractional-order chaotic systems are identical, i.e.,* $A_i = A_j = A$ *and* $G_i(\cdot) = G_j(\cdot) = G(\cdot)$, *then the controllers* U_i *can be designed as*

$$\begin{cases} U_1 &= V_1 - G(X_2) + G(X_1), \\ U_2 &= V_2 - G(X_3) + G(X_2) + U_1, \\ &\vdots \\ U_{N-1} &= V_{N-1} - G(X_N) + G(X_{N-1}) + U_{N-2}, \end{cases}$$

where V_i $(i = 1, 2, \ldots, N-1)$ *are defined as in* (18.2.4) *and* Θ *is a coefficient matrix. Similarly to the aforementioned discussion, we can achieve synchronization of bidirectional* N*-coupled identical fractional-order chaotic systems with ring connection based on antisymmetric structure.*

18.2.3 Numerical Simulations

To demonstrate the efficacy of the proposed synchronization scheme, two sets of examples are presented and numerically simulated.

I. Synchronization of 3-Coupled Identical Fractional-Order Chaotic Systems

In this case, synchronization of three-coupled fractional-order Lü systems via bidirectional coupling is investigated. In 2006, Lu [261] studied the chaotic dynamics of fractional-order Lü system. We present the bidirectionally coupled fractional-order Lü systems with designed controllers as follows:

$$\begin{cases} D^{\alpha}x_{11} = a(x_{12} - x_{11}) + q_{11}(x_{21} + x_{31} - 2x_{11}), \\ D^{\alpha}x_{12} = bx_{12} - x_{11}x_{13} + q_{12}(x_{22} + x_{32} - 2x_{12}), \\ D^{\alpha}x_{13} = x_{11}x_{12} - cx_{13} + q_{13}(x_{23} + x_{33} - 2x_{13}), \end{cases}$$

$$\begin{cases} D^{\alpha}x_{21} = a(x_{22} - x_{21}) + q_{21}(x_{11} + x_{31} - 2x_{21}) + u_{11}, \\ D^{\alpha}x_{22} = bx_{22} - x_{21}x_{23} + q_{22}(x_{12} + x_{32} - 2x_{22}) + u_{12}, \\ D^{\alpha}x_{23} = x_{21}x_{22} - cx_{23} + q_{23}(x_{13} + x_{33} - 2x_{23}) + u_{13}, \end{cases}$$

and

$$\begin{cases} D^{\alpha}x_{31} = a(x_{32} - x_{31}) + q_{31}(x_{21} + x_{11} - 2x_{31}) + u_{21}, \\ D^{\alpha}x_{32} = bx_{32} - x_{31}x_{33} + q_{32}(x_{22} + x_{12} - 2x_{32}) + u_{22}, \\ D^{\alpha}x_{33} = x_{21}x_{22} - cx_{23} + q_{23}(x_{23} + x_{13} - 2x_{33}) + u_{23}, \end{cases}$$

where

$$A_1 = A_2 = A_3 = A = \begin{pmatrix} -a & a & 0 \\ 0 & b & 0 \\ 0 & 0 & -c \end{pmatrix}, \quad G_i(X_i) = \begin{pmatrix} 0 \\ -x_{i1}x_{i3} \\ x_{i1}x_{i2} \end{pmatrix} \quad (i = 1, 2, 3),$$

$Q_1 = \mathrm{diag}(q_{11}, q_{12}, q_{13})$, $Q_2 = \mathrm{diag}(q_{21}, q_{22}, q_{23})$, $Q_3 = \mathrm{diag}(q_{31}, q_{32}, q_{33})$ are the coupled matrices, $U_1 = (u_{11}, u_{12}, u_{13})^T$ and $U_2 = (u_{21}, u_{22}, u_{23})^T$ are the control inputs.

The synchronization error is defined as $e_i = X_{i+1} - X_i$ $(i = 1, 2)$, and the error dynamical system is derived as follows:

$$D^{\alpha}e = \begin{pmatrix} D^{\alpha}e_1 \\ D^{\alpha}e_2 \end{pmatrix} = \begin{pmatrix} A - 2Q_1 - Q_2 & Q_2 - Q_1 \\ Q_2 - Q_3 & A - 2Q_3 - Q_2 \end{pmatrix} \begin{pmatrix} e_1 \\ e_2 \end{pmatrix} + \begin{pmatrix} u_{11} \\ -x_{21}x_{23} + x_{11}x_{13} + u_{12} \\ x_{21}x_{22} - x_{11}x_{12} + u_{13} \\ u_{21} - u_{11} \\ -x_{31}x_{33} + x_{21}x_{23} + u_{22} - u_{12} \\ x_{31}x_{32} - x_{21}x_{22} + u_{23} - u_{13} \end{pmatrix},$$

where

$$A - 2Q_1 - Q_2 = \begin{pmatrix} -a - 2q_{11} - q_{21} & a & 0 \\ 0 & b - 2q_{12} - q_{22} & 0 \\ 0 & 0 & -c - 2q_{13} - q_{23} \end{pmatrix},$$

$$A - 2Q_3 - Q_2 = \begin{pmatrix} -a - 2q_{31} - q_{21} & a & 0 \\ 0 & b - 2q_{32} - q_{22} & 0 \\ 0 & 0 & -c - 2q_{33} - q_{23} \end{pmatrix},$$

$e_1 = (e_{11}, e_{12}, e_{13})^T$, and $e_2 = (e_{21}, e_{22}, e_{23})^T$. Choosing the coefficient matrix $\Theta = \begin{pmatrix} \Theta_1 \\ \Theta_2 \end{pmatrix}$ as follows:

$$\Theta_1 = \begin{pmatrix} 0 & 0 & 0 & 0 & 0 & 0 \\ -a & 0 & 0 & 0 & 0 & 0 \\ 0 & 0 & 0 & 0 & 0 & 0 \end{pmatrix}$$

and

$$\Theta_2 = \begin{pmatrix} q_{11} - 2q_{21} + q_{31} & 0 & 0 & 0 & 0 & 0 \\ 0 & q_{12} - 2q_{22} + q_{32} & 0 & -a & 0 & 0 \\ 0 & 0 & q_{13} - 2q_{23} + q_{33} & 0 & 0 & 0 \end{pmatrix},$$

the controllers U_1 and U_2 can be designed as follows:

$$U_1 = \begin{pmatrix} u_{11} \\ u_{12} \\ u_{13} \end{pmatrix} = \begin{pmatrix} 0 \\ -ae_{11} + x_{21}x_{23} - x_{11}x_{13} \\ -x_{21}x_{22} + x_{11}x_{12} \end{pmatrix},$$

$$U_2 = \begin{pmatrix} u_{21} \\ u_{22} \\ u_{23} \end{pmatrix} = \begin{pmatrix} (q_{11} - 2q_{21} + q_{31})e_{11} \\ (q_{12} - 2q_{22} + q_{32})e_{12} - a(e_{11} + e_{21}) + x_{31}x_{33} - x_{11}x_{13} \\ (q_{13} - 2q_{23} + q_{33})e_{13} - x_{31}x_{32} + x_{11}x_{12} \end{pmatrix}.$$

Then, we obtain the error system as $D^{\alpha}e(t) = (\Phi_1(t) + \Phi_2)e(t)$, where $\Phi_2 = \mathrm{diag}(-a - 2q_{11} - q_{21}, b - 2q_{12} - q_{22}, -c - 2q_{13} - q_{23}, -a - 2q_{31} - q_{21}, b - 2q_{32} - q_{22}, -c - 2q_{33} - q_{23})$ and

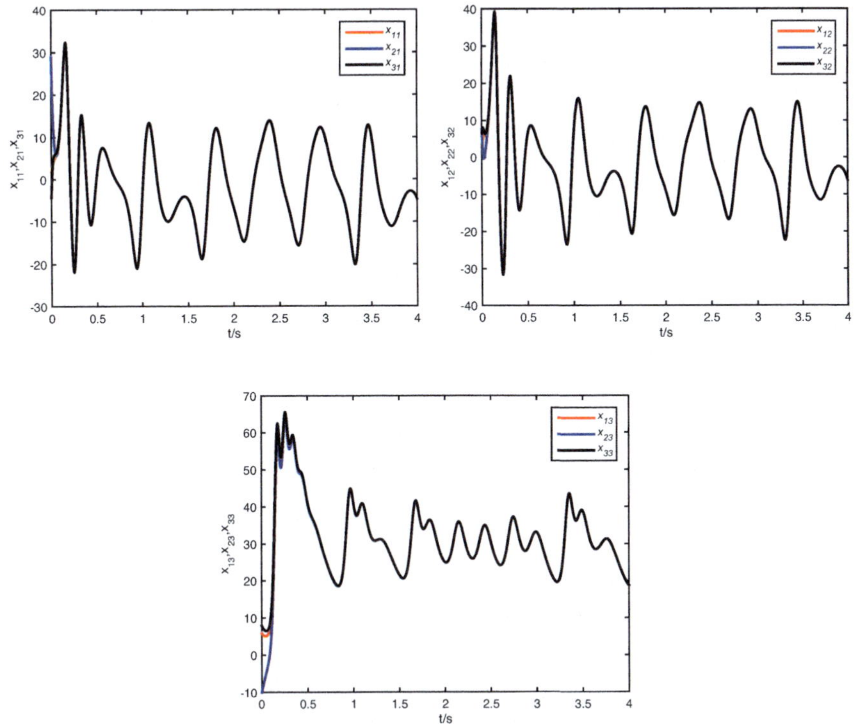

Fig. 18.7 The state trajectories x_{i1}, x_{i2}, and x_{i3} $(i = 1, 2, 3)$ of the fractional-order Lü system

$$\Phi_1(t) = \begin{pmatrix} 0 & a & 0 & q_{21} - q_{11} & 0 & 0 \\ -a & 0 & 0 & 0 & q_{22} - q_{12} & 0 \\ 0 & 0 & 0 & 0 & 0 & q_{23} - q_{13} \\ q_{11} - q_{21} & 0 & 0 & 0 & a & 0 \\ 0 & q_{12} - q_{22} & 0 & -a & 0 & 0 \\ 0 & 0 & q_{13} - q_{23} & 0 & 0 & 0 \end{pmatrix}.$$

Assume that conditions

$$\begin{aligned} -a - 2q_{11} - q_{21} < 0, \quad b - 2q_{12} - q_{22} < 0, \quad -c - 2q_{13} - q_{23} < 0, \\ -a - 2q_{31} - q_{21} < 0, \quad b - 2q_{32} - q_{22} < 0, \quad -c - 2q_{33} - q_{23} < 0 \end{aligned}$$

are satisfied. Therefore, based on Corollary 18.2, the error system is asymptotically stable with the controllers U_1 and U_2. That is, synchronization is achieved in the three-coupled fractional-order Lü systems via bidirectional coupling.

Numerical simulations are performed with the initial conditions set as $X_1(0) = (2, 3, 6)^T$, $X_2(0) = (30, 9, -10)^T$, $X_3(0) = (-5, 6, 8)^T$ and $\alpha = 0.98$. Choosing $(a, b, c) = (35, 28, 3)$, $q_{11} = q_{13} = q_{21} = q_{23} = q_{33} = 2$, $q_{12} = q_{22} = 23$, $q_{31} = 5$,

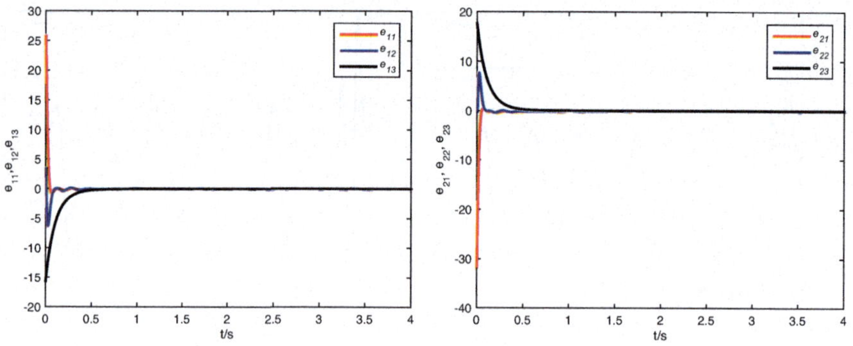

Fig. 18.8 The error dynamics of synchronization among three fractional-order Lü systems

and $q_{32} = 26$, we obtain simulation results as shown in Figs. 18.7 and 18.8. Figure 18.7 displays the state variables of the fractional-order Lü system with the designed controllers. As shown in Fig. 18.8, the synchronization errors converge asymptotically to zero in a quite short period, which verifies the realization of synchronization in the bidirectional 3-coupled fractional-order Lü systems.

II. Synchronization of 3-Coupled Nonidentical Fractional-Order Chaotic Systems

We now investigate synchronization of 3-coupled nonidentical fractional-order chaotic systems, which are the fractional-order Lorenz system [262], the fractional-order financial system [263], and the fractional-order PMSM system [264]. They are given by:

$$\begin{cases} D^{\alpha}x_{11} = a_1(x_{12} - x_{11}) + q_{11}(x_{21} + x_{31} - 2x_{11}), \\ D^{\alpha}x_{12} = a_2x_{11} - x_{12} - x_{11}x_{13} + q_{12}(x_{22} + x_{32} - 2x_{12}), \\ D^{\alpha}x_{13} = x_{11}x_{12} - a_3x_{13} + q_{13}(x_{23} + x_{33} - 2x_{13}), \end{cases} \tag{18.2.7}$$

$$\begin{cases} D^{\alpha}x_{21} = x_{23} + (x_{22} - b_1)x_{21} + q_{21}(x_{11} + x_{31} - 2x_{21}) + u_{11}, \\ D^{\alpha}x_{22} = 1 - b_2x_{22} - x_{21}^2 + q_{22}(x_{12} + x_{32} - 2x_{22}) + u_{12}, \\ D^{\alpha}x_{23} = -x_{21} - b_3x_{23} + q_{23}(x_{13} + x_{33} - 2x_{23}) + u_{13}, \end{cases} \tag{18.2.8}$$

and

$$\begin{cases} D^{\alpha}x_{31} = -x_{31} + x_{32}x_{33} + q_{31}(x_{21} + x_{11} - 2x_{31}) + u_{21}, \\ D^{\alpha}x_{32} = -x_{32} - x_{31}x_{33} + c_1x_{33} + q_{32}(x_{22} + x_{12} - 2x_{32}) + u_{22}, \\ D^{\alpha}x_{33} = c_2(x_{32} - x_{33}) + q_{33}(x_{23} + x_{13} - 2x_{33}) + u_{23}, \end{cases} \tag{18.2.9}$$

where

$$A_1 = \begin{pmatrix} -a_1 & a_1 & 0 \\ a_2 & -1 & 0 \\ 0 & 0 & -a_3 \end{pmatrix}, \quad A_2 = \begin{pmatrix} -b_1 & 0 & 1 \\ 0 & -b_2 & 0 \\ -1 & 0 & -b_3 \end{pmatrix}, \quad A_3 = \begin{pmatrix} -1 & 0 & 0 \\ 0 & -1 & c_1 \\ 0 & c_2 & -c_2 \end{pmatrix},$$
$$G_1(X_1) = \begin{pmatrix} 0 \\ -x_{11}x_{13} \\ x_{11}x_{12} \end{pmatrix}, \quad G_2(X_2) = \begin{pmatrix} x_{21}x_{22} \\ 1 - x_{21}^2 \\ 0 \end{pmatrix}, \quad G_3(X_3) = \begin{pmatrix} x_{32}x_{33} \\ -x_{31}x_{33} \\ 0 \end{pmatrix},$$

$Q_1 = \mathrm{diag}(q_{11}, q_{12}, q_{13})$, $Q_2 = \mathrm{diag}(q_{21}, q_{22}, q_{23})$, $Q_3 = \mathrm{diag}(q_{31}, q_{32}, q_{33})$ denote the coupled matrices, $U_1 = (u_{11}, u_{12}, u_{13})^T$ and $U_2 = (u_{21}, u_{22}, u_{23})^T$ represent the control inputs. The parameters are set as $\alpha = 0.99$, $(a_1, a_2, a_3) = (10, 28, 8/3)$, $(b_1, b_2, b_3) = (3, 0.1, 1)$, and $(c_1, c_2) = (50, 4)$. In the absence of the controllers, fractional-order systems (18.2.7)–(18.2.9) without the coupling terms behave chaotically, see [262–264] for more details.

The synchronization error is defined as $e_i = X_{i+1} - X_i$ $(i = 1, 2)$. Accordingly, the error dynamical system can thus be derived from systems (18.2.7)–(18.2.9) as

$$D^\alpha e = \begin{pmatrix} \Sigma_1 & \Sigma_2 \\ \Sigma_3 & \Sigma_4 \end{pmatrix} \begin{pmatrix} e_1 \\ e_2 \end{pmatrix} + \begin{pmatrix} (a_1 - b_1)x_{11} - a_1x_{12} + x_{13} + x_{21}x_{22} + u_{11} \\ -a_2x_{11} + (1 - b_2)x_{12} + 1 - x_{21}^2 + x_{11}x_{13} + u_{12} \\ -x_{11} + (a_3 - b_3)x_{13} - x_{11}x_{12} + u_{13} \\ (b_1 - 1)x_{21} - x_{23} + x_{32}x_{33} - x_{21}x_{22} + u_{21} - u_{11} \\ (b_2 - 1)x_{22} + c_1x_{23} - x_{31}x_{33} - 1 + x_{21}^2 + u_{22} - u_{12} \\ x_{21} + c_2x_{22} + (b_3 - c_2)x_{23} + u_{23} - u_{13} \end{pmatrix},$$

where

$$\Sigma_1 = A_2 - 2Q_1 - Q_2 = \begin{pmatrix} -b_1 - 2q_{11} - q_{21} & 0 & 1 \\ 0 & -b_2 - 2q_{12} - q_{22} & 0 \\ -1 & 0 & -b_3 - 2q_{13} - q_{23} \end{pmatrix},$$
$$\Sigma_2 = Q_2 - Q_1 = \begin{pmatrix} q_{21} - q_{11} & 0 & 0 \\ 0 & q_{22} - q_{12} & 0 \\ 0 & 0 & q_{23} - q_{13} \end{pmatrix},$$
$$\Sigma_3 = Q_2 - Q_3 = \begin{pmatrix} q_{21} - q_{31} & 0 & 0 \\ 0 & q_{22} - q_{32} & 0 \\ 0 & 0 & q_{23} - q_{33} \end{pmatrix},$$
$$\Sigma_4 = A_3 - 2Q_3 - Q_2 = \begin{pmatrix} -1 - 2q_{31} - q_{21} & 0 & 0 \\ 0 & -1 - 2q_{32} - q_{22} & c_1 \\ 0 & c_2 & -c_2 - 2q_{33} - q_{23} \end{pmatrix}.$$

Selecting the coefficient matrix $\Theta = \begin{pmatrix} \Theta_1 \\ \Theta_2 \end{pmatrix}$ as follows:

$$\Theta_1 = \begin{pmatrix} 0 & 0 & 0 & q_{31} - q_{21} & 0 & 0 \\ 0 & 0 & 0 & 0 & q_{32} - q_{22} & 0 \\ 0 & 0 & 0 & 0 & 0 & q_{33} - q_{23} \end{pmatrix}$$

and

$$\Theta_2 = \begin{pmatrix} q_{11}-q_{21} & 0 & 0 & 0\,0 & 0 \\ 0 & q_{12}-q_{22} & 0 & 0\,0 & -c_1-c_2 \\ 0 & 0 & q_{13}-q_{23} & 0\,0 & 0 \end{pmatrix},$$

we have the controllers in the form

$$\begin{cases} u_{11} = (q_{31}-q_{21})e_{21} - (a_1-b_1)x_{11} + a_1x_{12} - x_{13} - x_{21}x_{22}, \\ u_{12} = (q_{32}-q_{22})e_{22} + a_2x_{11} - (1-b_2)x_{12} - 1 + x_{21}^2 - x_{11}x_{13}, \\ u_{13} = (q_{33}-q_{23})e_{23} + x_{11} - (a_3-b_3)x_{13} + x_{11}x_{12}, \end{cases}$$

and

$$\begin{cases} u_{21} = (q_{11}-q_{21})e_{11} + (q_{31}-q_{21})e_{21} - (b_1-1)x_{21} + x_{23} - x_{32}x_{33} - (a_1-b_1)x_{11} \\ \qquad +a_1x_{12} - x_{13}, \\ u_{22} = (q_{12}-q_{22})e_{12} + (q_{32}-q_{22})e_{22} - (c_1+c_2)e_{23} - (b_2-1)x_{22} - c_1x_{23} \\ \qquad +a_2x_{11} - (1-b_2)x_{12} + x_{31}x_{33} - x_{11}x_{13}, \\ u_{23} = (q_{13}-q_{23})e_{13} + (q_{33}-q_{23})e_{23} - x_{21} - c_2x_{22} - (b_3-c_2)x_{23} + x_{11} \\ \qquad -(a_3-b_3)x_{13} + x_{11}x_{12}. \end{cases}$$

The error system is then obtained as $D^\alpha e(t) = (\Phi_1(t) + \Phi_2)e(t)$, where $\Phi_2 = \text{diag}(-b_1-2q_{11}-q_{21}, -b_2-2q_{12}-q_{22}, -b_3-2q_{13}-q_{23}, -1-2q_{31}-q_{21}, -1-2q_{32}-q_{22}, -c_2-2q_{33}-q_{23})$ and

$$\Phi_1(t) = \begin{pmatrix} 0 & 0 & 1 & q_{31}-q_{11} & 0 & 0 \\ 0 & 0 & 0 & 0 & q_{32}-q_{12} & 0 \\ -1 & 0 & 0 & 0 & 0 & q_{33}-q_{13} \\ q_{11}-q_{31} & 0 & 0 & 0 & 0 & 0 \\ 0 & q_{12}-q_{32} & 0 & 0 & 0 & -c_2 \\ 0 & 0 & q_{13}-q_{33} & 0 & c_2 & 0 \end{pmatrix}.$$

If conditions

$$\begin{aligned} &-b_1-2q_{11}-q_{21}<0, \quad -b_2-2q_{12}-q_{22}<0, \quad -b_3-2q_{13}-q_{23}<0, \\ &-1-2q_{31}-q_{21}<0, \quad -1-2q_{32}-q_{22}<0, \quad -c_2-2q_{33}-q_{23}<0 \end{aligned}$$

are satisfied, then the error system is asymptotically stable under the designed controllers by means of Theorem 18.2. Consequently, systems (18.2.7)–(18.2.9) achieve synchronization.

The simulation results is performed under the initial condition $X_1(0) = (-2, 3, 1)^T$, $X_2(0) = (5, -4, -1)^T$, $X_3(0) = (-4, 2, 3)^T$, and $\alpha = 0.96$. We set $q_{11} = q_{12} = q_{13} = 1$, $q_{21} = q_{22} = q_{23} = 6$, $q_{11} = q_{12} = q_{13} = q_{31} = q_{32} = q_{33} = 8$, and obtain the simulation results as shown in Figs. 18.9 and 18.10. It can be seen from Fig. 18.9 that the state variables of the 3-coupled fractional-order systems (18.2.7)–(18.2.9). Figure 18.10 displays the errors of synchronization converge

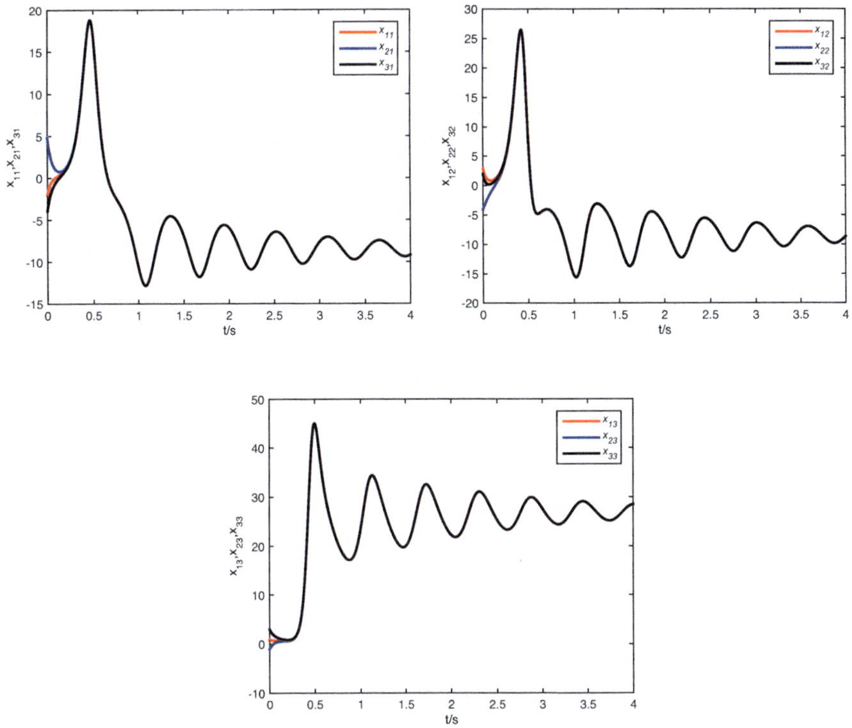

Fig. 18.9 The state trajectories x_{i1}, x_{i2}, and x_{i3} $(i = 1, 2, 3)$ of 3-coupled fractional-order chaotic systems (18.2.7)–(18.2.9)

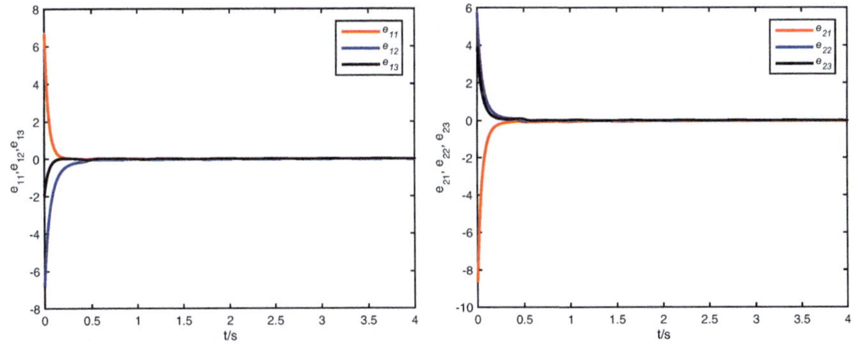

Fig. 18.10 The error dynamics of synchronization among 3-coupled fractional-order chaotic systems (18.2.7)–(18.2.9)

asymptotically to zero in a quite short period. As expected, synchronization of bidirectional 3-coupled nonidentical fractional-order chaotic systems can be realized.

This chapter discusses synchronization of N-coupled fractional-order chaotic systems with ring connection. Firstly, we analyze synchronization and anti-synchronization of N-coupled general fractional-order complex chaotic systems, which are depicted by a unified mathematical expression with ring connection via unidirectional coupling. Secondly, we investigate the synchronization of N-coupled fractional-order chaotic systems with ring connection via bidirectional coupling, utilizing the direct design method.

Chapter 19
A Novel Hybrid Color Image Encryption Algorithm Using Two Complex Chaotic Systems

Certain fundamental characteristics of chaotic systems, including their sensitivity to initial conditions and pseudorandom behavior, align well with the requirements for secure cryptosystems [265–268]. This has led to significant advancements in chaotic image encryption [269–272], establishing it as an emerging area within image encryption research [273]. In this chapter, we introduce a novel hybrid encryption algorithm for color images utilizing two complex chaotic systems. The proposed scheme, which integrates both complex Chen and complex Lorenz systems, offers an expanded key space to enhance resistance against potential cryptographic attacks.

19.1 Theoretical Basis

19.1.1 Complex Chen System

The complex Chen system is introduced by

$$\begin{cases} \dot{z_1} = a_1(z_2 - z_1), \\ \dot{z_2} = (c_1 - a_1)z_1 - z_1 z_3 + c_1 z_2, \\ \dot{z_3} = \frac{1}{2}(\bar{z_1} z_2 + z_1 \bar{z_2}) - b_1 z_3. \end{cases} \tag{19.1.1}$$

In system (19.1.1), the positive real parameters a_1, b_1, and c_1 determine the system's dynamic behavior. Here, $z_1 = z_1^r + jz_1^i$ and $z_2 = z_2^r + jz_2^i$ represent complex variables, while z_3 is a real variable. For parameter values $a_1 = 27$, $b_1 = 4.5$, and $c_1 = 23$, the complex Chen system exhibits highly complex and enhanced chaotic characteristics, with its corresponding attractors illustrated in Fig. 19.1. The Lyapunov exponents and bifurcation behavior of the system are further presented in Fig. 19.2.

S. Liu et al., *Chaos in Complex Dynamical System: Theory and Application*,
https://doi.org/10.1007/978-981-95-7762-0_19

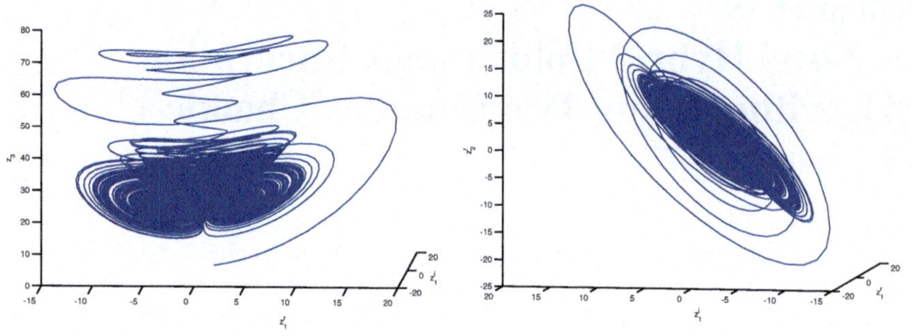

Fig. 19.1 The attractors of complex Chen system

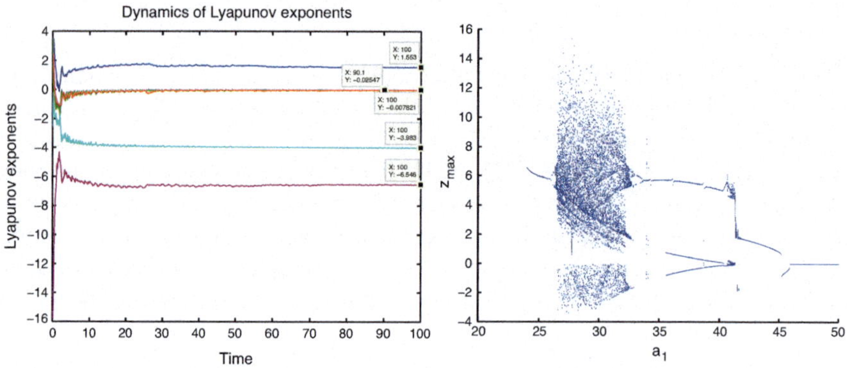

(a) The Lyapunov exponent

(b) The bifurcation diagram

Fig. 19.2 The Lyapunov exponent and bifurcation of complex Chen system

19.1.2 Complex Lorenz System

The complex Lorenz system is described as

$$\begin{cases} \dot{w_1} = a_2(w_2 - w_1), \\ \dot{w_2} = c_2 w_1 - w_2 - w_1 w_3, \\ \dot{w_3} = -b_2 w_3 + \frac{1}{2}(\bar{w_1} w_2 + w_1 \bar{w_2}). \end{cases} \tag{19.1.2}$$

In this system, a_1, b_2 and c_3 are positive real parameters, while $w_1 = w_1^r + jw_1^i$ and $w_2 = w_2^r + jw_2^i$ denote complex variables, and w_3 is a real variable. For parameter values $a_1 = 14$, $b_2 = 3.7$ and $c_2 = 35$, the chaotic attractors of the complex Lorenz system are depicted in Fig. 19.3. The corresponding Lyapunov exponents and bifurcation diagram are presented in Fig. 19.4.

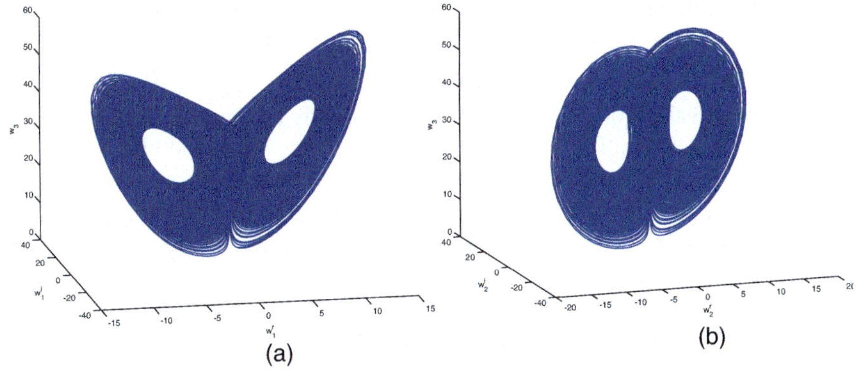

Fig. 19.3 The attractors of complex Lorenz system

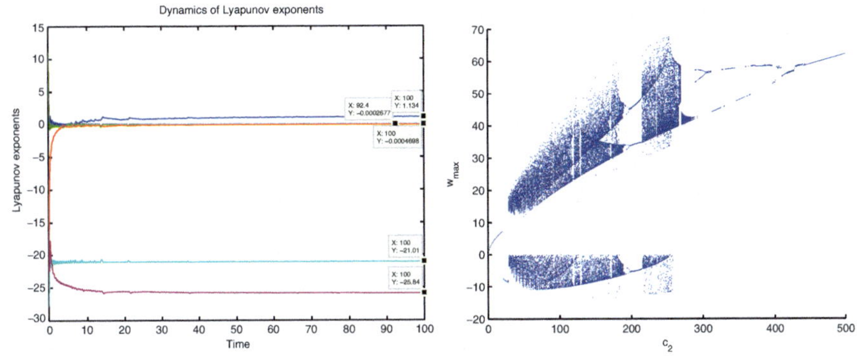

(a) The Lyapunov exponent

(b) The bifurcation diagram

Fig. 19.4 The Lyapunov exponent and bifurcation of complex Lorenz system

19.1.3 Hamming Distance

The Hamming distance measures the number of positions at which two sequences of equal length differ. For two sequences $\alpha = (\alpha_1, \alpha_2, \ldots, \alpha_n)$ and $\beta = (\beta_1, \beta_2, \ldots, \beta_n)$, the Hamming distance $H(\alpha, \beta)$ is defined as:

$$\begin{cases} H(\alpha, \beta) &= \sum_{i=1}^{n} h(\alpha_i, \beta_i); \\ h(\alpha_i, \beta_i) &= \begin{cases} 0, & \alpha_i = \beta_i \\ 1, & \alpha_i \neq \beta_i. \end{cases} \end{cases} \tag{19.1.3}$$

19.2 Algorithm Description

The main idea to encrypt a plain image is to permutate the positions of pixels and to conceal the values of pixels via different methods commonly. The encryption of a plain image typically involves permuting pixel positions and obscuring pixel values through various techniques. This section presents several key innovations in our approach. First, two complex chaotic systems are employed to generate random sequences for image encryption, significantly expanding the key space and strengthening security against potential attacks. Second, our scheme fully utilizes the two-dimensional structure of images, offering an advantage over traditional encryption methods. Third, an additional processing step is introduced to mix the RGB channels with chaotic sequences, achieving comprehensive multi-level encryption in combination with preceding steps. The 512 × 512 "Lena" image serves as a demonstration example throughout the encryption process. Figure 19.5 provides a schematic overview of the encryption procedure.

19.2.1 Generation of the Secret Key

Based on the above definition, we get H_{RG}, H_{RB}, H_{GB} for three channels $R,\ G,\ B$. System (19.2.1) shows the generation of the secret key.

$$\begin{cases} H &= [H_{RG},\ H_{RB},\ H_{GB},\ H_{RG},\ H_{RB},\ H_{GB}]; \\ Key1 &= \begin{cases} Key_{Chen}; \\ Key_{Lorenz}; \end{cases} \\ Key2 &= mod(H,\ Key1). \end{cases} \tag{19.2.1}$$

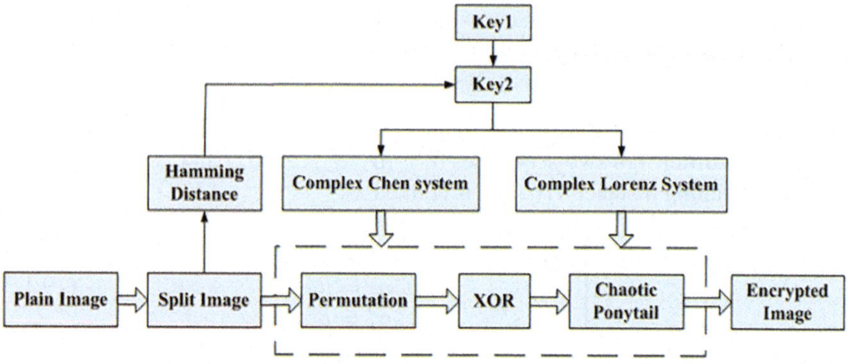

Fig. 19.5 The steps of the encryption approach

In this context, Key_{Chen} and Key_{Lorenz} represent two six-element sequences selected by the message sender, while $Key2$ serves as the initial value for the two complex chaotic systems.

During secret key generation, the Hamming distance values vary for each encryption process depending on the plain image. The message receiver can recalculate these values after decryption to verify information integrity. These features collectively contribute to achieving a higher level of security.

19.2.2 Pseudorandom Number Sequences

The solutions of systems (19.1.1) and (19.1.2) generate pseudorandom sequences; however, these sequences consist of floating-point numbers and are not directly applicable to image encryption schemes. To make them suitable, the pixel values in each channel must be converted to integers in the range [0, 255]. The integer pseudorandom sequences derived from system (19.1.1) or (19.1.2) are obtained as follows:

$$t_i = fix(mod(s_i \times 10^{12}, 256)). \tag{19.2.2}$$

Here, s_i represents a solution from system (19.1.1) or (19.1.2), and the mod operator returns the remainder after division.

19.2.3 Pixels Permutation

The objective of this stage in our encryption algorithm is to permute the positions of pixels within each color channel. For clarity of explanation, the R channel is used as an example to demonstrate the encryption procedure. The same methodology applies to the other two channels, albeit using distinct pseudorandom number sequences generated by the two complex chaotic systems.

Let $P_{M \times N}$, an array of size $M \times N$, represent the data of the R channel in the plain image, and let t_x and t_y be two pseudorandom number sequences of lengths M and N, respectively. Two sequences, pos_x and pos_y, are used to record the element positions of t_x, t_y and $P_{M \times N}$. The sequences t_x and t_y are then sorted according to the magnitude of their elements, yielding updated sequences pos_x and pos_y. Finally, the array $P_{M \times N}$ is rearranged by the new pos_x and pos_y. The new step is visually illustrated in Fig. 19.6. Subsequently, the array $P_{M \times N}$ is reshaped as a one-dimensional array P_n. Consider a pseudorandom number sequence t_z and its corresponding element position index pos_z, both of length ($nPixels = M \times N$). The same method described above can then be applied to process P_n.

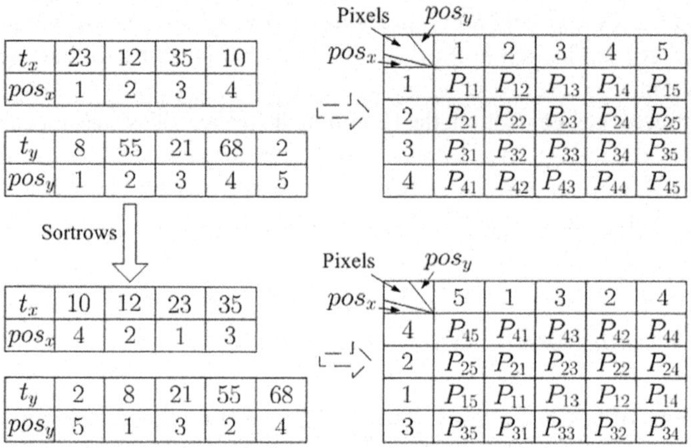

Fig. 19.6 The process of the permutation

19.2.4 XOR Operation

The XOR operation provides a simple yet effective method for both encrypting and decrypting information. In this step, it is utilized to conceal pixel information by performing bitwise XOR operations between pixel values and pseudorandom number sequences.

Consider three pseudorandom number sequences t_R, t_G, and t_B, generated from system (19.2.2). A bitwise XOR operation is then applied to each corresponding channel of the image using these sequences.

19.2.5 Chaotic Ponytail

The final encryption step involves mixing the RGB channels using chaotic sequences. This process is vividly termed "Chaotic Ponytail" due to its resemblance to the weaving of a ponytail. By applying this method, the distinct boundaries between the RGB channels are effectively disrupted. Therefore, it becomes virtually impossible to decrypt the image by starting from one channel and proceeding to another. The procedure for this encryption step is outlined as follows.

(1) To achieve variable segment lengths in the "ponytail" structure, a function is designed to generate a random element length sequence M_{mix}:

$$M_{mix} = FunMixRGB(u_i, zoom), \tag{19.2.3}$$

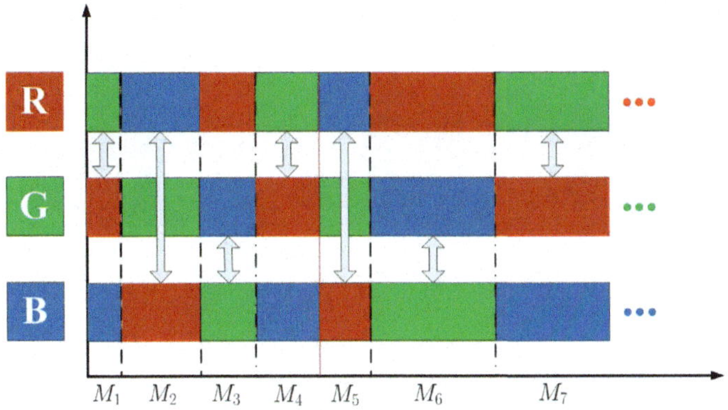

Fig. 19.7 The process of the chaotic ponytail process

where u_i represents a solution from the complex Chen system, and i is set to 5 in this context. The parameter $zoom$ constrains the maximum allowable step length. (2) The step of braiding the ponytail is described as follows.

Step1. Exchange the data in R and G channel with the length of first element in M_{mix};
Step2. Exchange the data in R and B channel with the length of next element in M_{mix};
Step3. Exchange the data in G and B channel with the length of next element in M_{mix};
Step4. Repeat the step from 1 to 3 until all the pixels were exchanged;

The fundamental encryption principle and procedural steps of the "Chaotic Ponytail" method are illustrated in Fig. 19.7.

19.3 Security Analysis

Security represents the primary consideration in any cryptosystem. This section evaluates the encryption performance and its security through multiple aspects.

19.3.1 Histogram Analysis

An image histogram represents the intensity distribution of pixels in a color channel. A uniformly flat histogram in an encrypted image indicates effective pixel permutation and demonstrates strong resistance against statistical attacks [274].

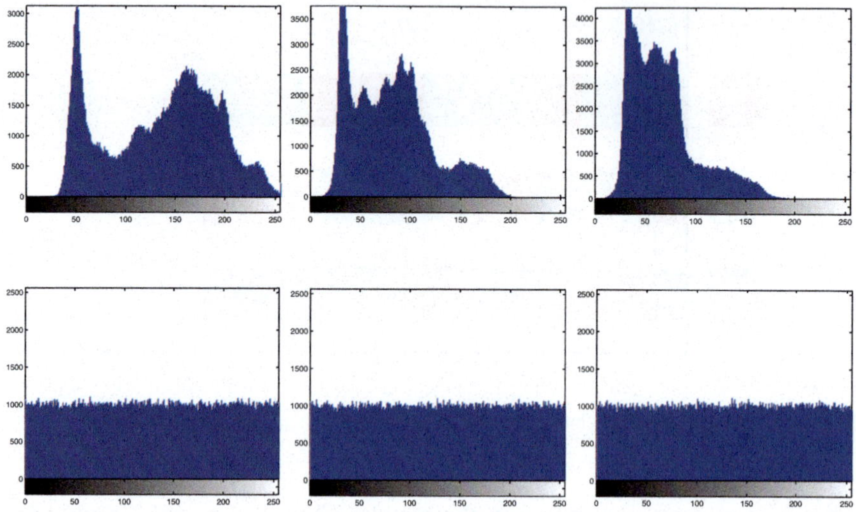

Fig. 19.8 Grayscale histograms of plain image and ciphered image

The histograms of both the plain image and the encrypted image for each channel are displayed in Fig. 19.8. Figures 19.8(a)–(c) display the histograms of the plain image in the R, G, and B channels, respectively, showing irregular distributions. The remaining figures below exhibit uniformly flat distributions resulting from the encryption process. Clearly, no useful information can be deduced from the encrypted image, thereby ensuring strong security against statistical attacks.

19.3.2 Key Space and Sensitivity Analysis

The proposed encryption algorithm, utilizing two complex chaotic systems, enhances algorithmic complexity and significantly expands the key space. Although several parameters in systems (19.2.2) and (19.2.3) may influence encryption results, only the initial values are considered here for simplicity. The two complex chaotic systems collectively contain ten parameters that can be freely assigned. Assuming a value range from 10^{-4} to 10^4 with a precision of 10^{-15}, the resulting key space is calculated as $(2 \times 10^4 \times 10^{15})^{10} \approx 10^{194}$. According to [275], the baseline for a secure key space is $2^{100} \approx 10^{30}$. Therefore, the key space of our encryption algorithm is sufficiently large to resist brute-force attacks.

To analyze sensitivity, the Baboon image is encrypted using two slightly different initial values differing by 10^{-15} [276]. The differences between the two resulting encrypted images are stored as a new image, shown in Fig. 19.9. The black pixels indicate regions where the two encrypted images are identical. The extensive non-black areas demonstrate the high sensitivity of our algorithm to minor changes in initial conditions.

19.3.3 Correlation Analysis

The correlation coefficient serves as an indicator to evaluate the randomness of an image. In this analysis, it is used to measure the correlation between adjacent pixels in both the plain and encrypted images. To compute the correlation coefficient, 1000 pairs of adjacent pixels are randomly selected from the images along the horizontal, vertical, and diagonal directions. The formula for calculating the correlation coefficient [277] is given as follows:

$$\begin{aligned} cov(x,y) &= \frac{1}{N}\sum_{i=1}^{N} E(x_i - E(x))(y_i - E(y)), \\ r_{xy} &= \frac{cov(x,y)}{\sqrt{D(x)D(y)}}, \\ E(x) &= \frac{1}{N}\sum_{i=1}^{N} x_i, \\ D(x) &= \frac{1}{N}\sum_{i=1}^{N}(x_i - E(x))^2, \end{aligned} \tag{19.3.1}$$

where x and y represent the values of two adjacent pixels randomly selected along the three directions. Here, N is set to 1000, as previously mentioned.

Figures 19.10 and 19.11 illustrate the correlation distributions of adjacent pixels in the horizontal direction for each channel of the plain and encrypted images, respectively. Additional quantitative results are provided in Table 19.1. Evidently, the plain

Fig. 19.9 The difference between encrypted results of Baboon with close initial values

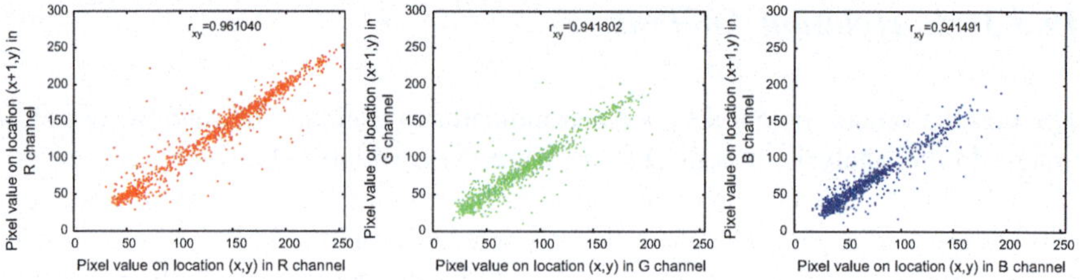

Fig. 19.10 Correlation of each channel in plain image

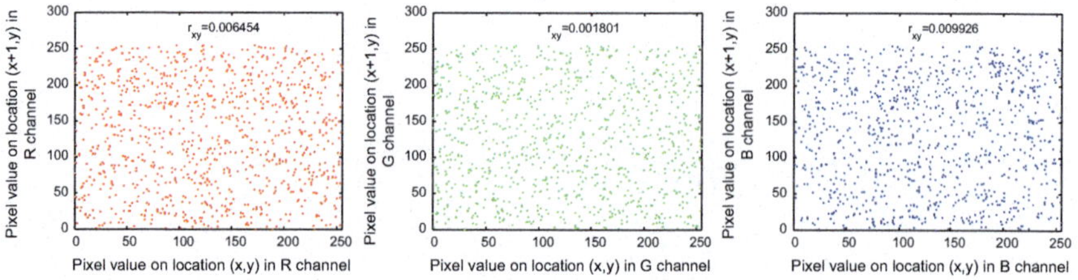

Fig. 19.11 Correlation of each channel in encrypted image

Table 19.1 Correlation coefficients of two adjacent pixels in three typical directions of encrypted image

Orientation	R	G	B
Horizontal	0.006454	0.001801	0.009926
Vertical	0.005386	0.004798	0.004394
Diagonal	0.006142	0.002753	0.003562

image exhibits strong correlations in all three directions, with correlation coefficients approaching one. In contrast, the encrypted image shows correlation coefficients near zero, indicating high randomness. These results demonstrate the effectiveness of the proposed algorithm in randomizing pixel relationships.

19.4 Execution Time Analysis

To assess computational efficiency, tests were conducted using Matlab 7.1 under a Windows XP Professional operating system, with hardware configuration comprising an Intel Celeron 2.7 GHz CPU, 2 GB RAM, and a 500 GB hard drive. The color images "Lena" and "Baboon," both of size 512×512, were encrypted ten times each using the proposed algorithm. The average execution times are summarized in Table 19.2.

The computational cost inevitably increases when employing both complex Chen and complex Lorenz systems, compared to schemes based on one-dimensional

Table 19.2 The average execution times

Files	Permutation time (ms)	Diffusion time (ms)	Mixing time (ms)
Lena	18.1	3.0	475.2
Baboon	18.0	3.4	468.8

chaotic systems. Nevertheless, the execution time remains acceptable for our current implementation, as evidenced by the results in Table 19.2. For improved efficiency, a future release version will be rewritten in C++, which offers significantly better performance than Matlab [278].

19.5 Noise Interference Analysis

19.5.1 Gaussian Noise Analysis

In this section, Gaussian noise with zero mean and varying variances is introduced to evaluate the algorithm's robustness against noise. The function used to add Gaussian noise is defined as follows:

$$ImgG = imnoise(Img, \text{“}gaussian\text{”}, 0, var), \tag{19.5.1}$$

where Img and $ImgG$ represent the encrypted image data and its noise-affected version, respectively. The parameter var denotes the noise variance, with values ranging from 0.01 to 1. Figure 19.12 displays the encrypted Lena and Baboon images corrupted by Gaussian noise at variance levels of 0.4, 0.7, and 0.9. Notably, the decrypted images remain recognizable even at a variance of 0.9, demonstrating the algorithm's robustness to noise interference.

19.5.2 Occlusion Noise Analysis

Data loss during transmission and storage is a common occurrence, which can be modeled as occlusion noise in the decryption process. As illustrated in Fig. 19.13(a), we simulate a scenario where 50% of the pixel data on the left side is lost and replaced by black pixels, corresponding to a 25% data loss in the encrypted image. Despite this significant occlusion, the decrypted image remains clearly recognizable, as shown in Fig. 19.13(b). These results confirm that the proposed algorithm exhibits strong stability and robustness against various types of noise interference.

Fig. 19.12 Decrypted image with Gaussian noise

Fig. 19.13 Encrypted and Decrypted images with occlusion noise

Chapter 20
Image Encryption Based on Hybrid Projective Synchronization of Chaotic Complex Systems

This chapter examines image encryption utilizing hybrid projective synchronization within chaotic complex systems [279–283]. The discussion is structured around three key aspects: first, the hybrid projective synchronization of the complex Duffing-Holmes oscillator is analyzed; second, the encryption-decryption algorithm is developed and examined; finally, the security of the proposed algorithm is thoroughly evaluated.

20.1 Hybrid Projective Synchronization of Complex Duffing-Holmes Oscillators

The hybrid projective synchronization of complex Duffing-Holmes oscillators is investigated from two perspectives: proposal and characteristic analysis of the system, along with the design of synchronization controllers and numerical simulations.

20.1.1 Proposal and Characteristic Analysis of the Complex Duffing-Holmes Oscillator

I. Proposal of the Complex Duffing-Holmes Oscillator

The chaotic behavior of the complex Duffing-Holmes oscillator is derived from the real Duffing-Holmes oscillator [284] and is expressed as follows:

$$\ddot{z} + a\dot{z} - bz + cz|z|^2 = d'cos(wt), \tag{20.1.1}$$

where $d' = \sqrt{2}d \exp(i\frac{\pi}{4})$, a, b, c and d being positive parameters, and $z(t) = x(t) + iy(t)$ representing a complex variable. Clearly, Eq. (20.1.1) defines a determin-

S. Liu et al., *Chaos in Complex Dynamical System: Theory and Application*,
https://doi.org/10.1007/978-981-95-7762-0_20

istic complex Duffing-Holmes system, which can be decomposed into two coupled real systems as follows:

$$\ddot{x} + a\dot{x} - bx + cx(x^2 + y^2) = dcos(wt) \tag{20.1.2}$$

and

$$\ddot{y} + a\dot{y} - by + cy(x^2 + y^2) = dcos(wt). \tag{20.1.3}$$

By introducing the new variables $u_1(t) = x(t)$, $u_2(t) = \dot{x}(t)$, $u_3(t) = y(t)$, $u_4(t) = \dot{y}(t)$ and $u_5(t) = t$, Eqs. (20.1.2) and (20.1.3) can be reformulated as a system of first-order autonomous ordinary differential equations as follows:

$$\begin{cases} \dot{u}_1 = u_2, \\ \dot{u}_2 = -au_2 + bu_1 - c(u_1^2 + u_3^2)u_1 + dcos(wu_5), \\ \dot{u}_3 = u_4, \\ \dot{u}_4 = -au_4 + bu_3 - c(u_1^2 + u_3^2)u_3 + dcos(wu_5), \\ \dot{u}_5 = 1. \end{cases} \tag{20.1.4}$$

II. Characteristic Analysis

To demonstrate the presence of chaotic attractors in the deterministic complex Duffing-Holmes oscillator (20.1.1), the maximal (local) Lyapunov exponent must be computed, as any system exhibiting at least one positive Lyapunov exponent is defined to be chaotic.

The Jacobian matrix of system (20.1.4) is denoted as

$$J_{pq} = \begin{pmatrix} 0 & 1 & 0 & 0 & 0 \\ b - c(u_1^2 + u_3^2) - 2cu_1^2 & -a & -2cu_1u_3 & 0 & -dwsin(wu_5) \\ 0 & 0 & 0 & 1 & 0 \\ -2cu_1u_3 & 0 & b - c(u_1^2 + u_3^2) - 2cu_3^2 & -a & -dwsin(wu_5) \\ 0 & 0 & 0 & 0 & 0 \end{pmatrix}.$$

Following the definition of Lyapunov exponents, the Lyapunov exponents for system (20.1.4) are obtained as follows:

$$\lambda_p = \lim_{t \to \infty} \frac{1}{t} \log \frac{\|\delta u'_{p1}(t)\|}{\|\delta u'_{p1}(0)\|}, \quad (p = 1,\ 2,\ 3,\ 4,\ 5). \tag{20.1.5}$$

Using the parameters $a = 0.25, b = 1, c = 1, d = 0.2$ and $w = 1$, along with the initial conditions $t_0 = 0, u_1(0) = x(0) = u_3(0) = y(0) = 0.2, u_2(0) = \dot{x}(0) = 0.3, u_4(0) = \dot{y}(0) = 0.2$, the computed local Lyapunov exponents are $\lambda_1 = 2.22$, $\lambda_2 = \lambda_3 = 0, \lambda_4 = -8.05$ and $\lambda_5 = -12.67$. The presence of a positive Lyapunov exponent confirms that system (20.1.1) exhibits chaotic behavior for this parameter set. The chaotic attractor of the system under these parameters and initial values

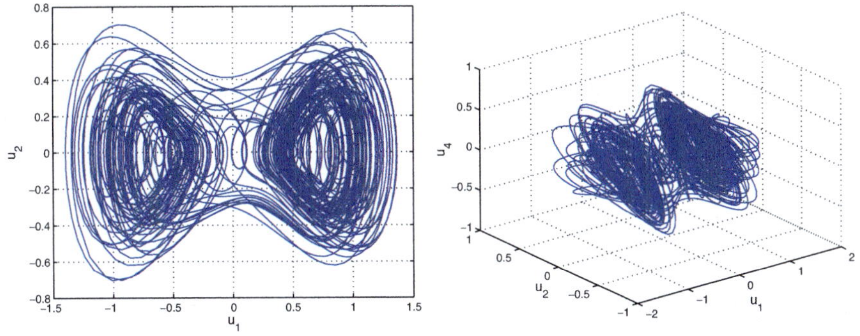

Fig. 20.1 Chaotic behavior and attractors of system (20.1.4) in different phase planes and projections

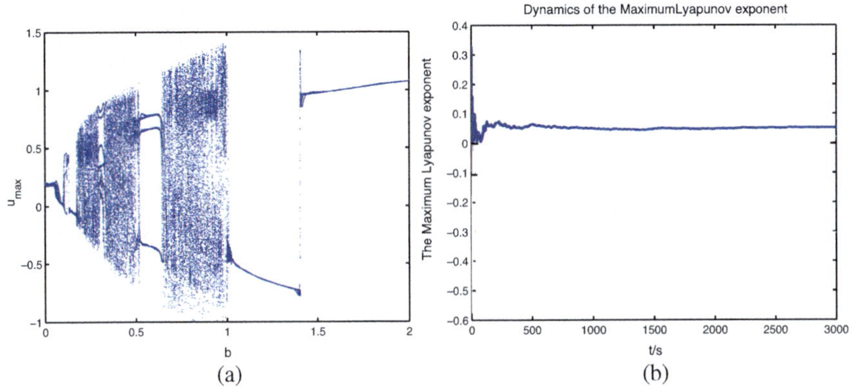

Fig. 20.2 The bifurcation and Maximum Lyapunov exponent about parliament b

is depicted in Fig. 20.1 across various phase planes and projections. Additionally, Fig. 20.2 presents a bifurcation diagram with respect to parameter b.

Using the same parameters and initial values as in Fig. 20.1, the calculated Lyapunov exponents are $\lambda_1 = 0.05$, $\lambda_2 = 0.025$, $\lambda_3 = 0$, $\lambda_4 = -0.285$, $\lambda_5 = -0.3$. This confirms that the motion on the attractor of system (20.1.1) with the selected parameters a, b, c and d is chaotic, as evidenced by the presence of two positive Lyapunov exponents. Using the same parameters and initial conditions as in Fig. 20.1, the numerically computed maximal Lyapunov exponent as a function of time t for the attractor depicted in Fig. 20.1 is presented as follows: The result is plotted in Fig. 20.2. Clearly, the maximum Lyapunov exponent of this attractor is positive, confirming that the motion on the attractor exhibits chaotic behavior.

20.1.2 Controller Design and Simulations of Hybrid Projective Synchronization for Complex Duffing-Holmes Oscillators

In a recent study, Hu et al. introduced a novel synchronization scheme termed hybrid projective synchronization (HPS) in Ref. [284], which describes the synchronization behavior between two dynamical systems as follows:

$$\begin{cases} \dot{\mathbf{z}}(t) &= H(\mathbf{z}, t) \leftarrow \textit{drive system}, \\ \dot{\mathbf{w}}(t) &= L(\mathbf{w}, t) \leftarrow \textit{response system}, \end{cases} \tag{20.1.6}$$

where $\mathbf{z} = [z_1, z_2, \ldots, z_n]^T$ and $\mathbf{w} = [w_1, w_2, \ldots, w_n]^T$ are the state complex vectors.

Definition 20.1 If there exists a real constant matrix $\mathbf{D} = diag\{d_1, d_2, \ldots, d_n\}$, such that $\lim_{t\to\infty} \|\mathbf{w}(t) - \mathbf{D}\mathbf{z}(t)\| = 0$, then $\mathbf{z}$ synchronize $\mathbf{w}$ up to a scaling matrix $\mathbf{D}$, such synchronization is called as hybrid projective synchronization (HPS).

Remark 20.1 All the dynamical states from two synchronized systems in the sense of HPS can be regulated separately by the scale factors, since there are many scaling factors in the scaling matrix $\mathbf{D}$. In the application of secure communications, this feature can be used for achieving fast communication and enhancing security in chaotic encryptation schemes.

The following subsections will separately address the HPS of complex Duffing-Holmes oscillators for cases with known parameters and unknown parameters.

I. HPS with Known Parameters

In this section, we investigate the HPS of the complex Duffing-Holmes system with known parameters using nonlinear control techniques. This approach proves effective and convenient for achieving HPS, as it eliminates the need to compute a Lyapunov function.

The complex system (20.1.1) can be reformulated as a set of four real first-order differential equations as follows:

$$\begin{cases} \dot{u}_1 = u_2, \\ \dot{u}_2 = -au_2 + bu_1 - c(u_1^2 + u_3^2)u_1 + dcos(wt), \\ \dot{u}_3 = u_4, \\ \dot{u}_4 = -au_4 + bu_3 - c(u_1^2 + u_3^2)u_3 + dcos(wt), \end{cases} \tag{20.1.7}$$

We consider two identical systems of the form (20.1.7) with known parameters a, b, c, d and w. The drive and response systems are defined as follows, beginning with the drive system:

$$\begin{cases} \dot{u}_{11} = u_{21}, \\ \dot{u}_{21} = -au_{21} + bu_{11} - c(u_{11}^2 + u_{31}^2)u_{11} + dcos(wt), \\ \dot{u}_{31} = u_{41}, \\ \dot{u}_{41} = -au_{41} + bu_{31} - c(u_{11}^2 + u_{31}^2)u_{31} + dcos(wt), \end{cases} \quad (20.1.8)$$

the response system with control is

$$\begin{cases} \dot{u}_{12} = u_{22} + v_1, \\ \dot{u}_{22} = -au_{22} + bu_{12} - c(u_{12}^2 + u_{32}^2)u_{12} + dcos(wt) + v_2, \\ \dot{u}_{32} = u_{42} + v_3, \\ \dot{u}_{42} = -au_{42} + bu_{32} - c(u_{12}^2 + u_{32}^2)u_{32} + dcos(wt) + v_4, \end{cases} \quad (20.1.9)$$

where v_1, v_2, v_3 and v_4 represent the control inputs to be designed, enabling the two systems to achieve hybrid projective synchronization via a nonlinear controller.

Following the definition of HPS, the state errors are defined as:

$$\begin{cases} e_1 = u_{12} - d_1 u_{11}, \\ e_2 = u_{22} - d_2 u_{21}, \\ e_3 = u_{32} - d_3 u_{31}, \\ e_4 = u_{42} - d_4 u_{41}, \end{cases} \quad (20.1.10)$$

and the derivative of state errors is

$$\begin{cases} \dot{e}_1 = \dot{u}_{12} - d_1 \dot{u}_{11}, \\ \dot{e}_2 = \dot{u}_{22} - d_2 \dot{u}_{21}, \\ \dot{e}_3 = \dot{u}_{32} - d_3 \dot{u}_{31}, \\ \dot{e}_4 = \dot{u}_{42} - d_4 \dot{u}_{41}. \end{cases} \quad (20.1.11)$$

We submit systems (20.1.8) and (20.1.9) into formula (20.1.11) and obtain

$$\begin{cases} \dot{e}_1 = e_2 + (d_2 - d_1)u_{21} + v_1, \\ \dot{e}_2 = -ae_2 + be_1 + b(d_1 - d_2)u_{11} - c[(u_{12}^2 + u_{32}^2)u_{12} - d_2(u_{11}^2 + u_{31}^2)u_{11}] \\ \qquad +dcos(wt)(1 - d_2) + v_2, \\ \dot{e}_3 = e_4 + (d_4 - d_3)u_{41} + v_3, \\ \dot{e}_4 = -ae_4 + be_3 + b(d_3 - d_4)u_{31} - c[(u_{12}^2 + u_{32}^2)u_{32} - d_4(u_{11}^2 + u_{31}^2)u_{31}] \\ \qquad +dcos(wt)(1 - d_4) + v_4. \end{cases} \quad (20.1.12)$$

Then, the controller of the system (20.1.9) is designed as

$$\begin{cases} v_1 = -e_2 - (d_2 - d_1)u_{21} - k_1 e_1, \\ v_2 = ae_2 - be_1 - b(d_1 - d_2)u_{11} + c[(u_{12}^2 + u_{32}^2)u_{12} - d_2(u_{11}^2 + u_{31}^2)u_{11}] \\ \qquad -dcos(wt)(1 - d_2) - k_2 e_2, \\ v_3 = -e_4 - (d_4 - d_3)u_{41} - k_3 e_3, \\ v_4 = ae_4 - be_3 - b(d_3 - d_4)u_{31} + c[(u_{12}^2 + u_{32}^2)u_{32} - d_4(u_{11}^2 + u_{31}^2)u_{31}] \\ \qquad -dcos(wt)(1 - d_4) - k_4 e_4, \end{cases} \quad (20.1.13)$$

where $k_n > 0$ $(n = 1, 2, 3, 4)$ are real constants that regulate the convergence rate for achieving HPS in the complex Duffing-Holmes oscillators. Substituting the controller (20.1.13) into the error system (20.1.12) yields:

$$\begin{cases} \dot{e}_1 = -k_1 e_1, \\ \dot{e}_2 = -k_2 e_2, \\ \dot{e}_3 = -k_3 e_3, \\ \dot{e}_4 = -k_4 e_4. \end{cases} \tag{20.1.14}$$

The error system (20.1.14) can be expressed as a linear system: $\dot{E} = AE$, where $E = [e_1, e_2, e_3, e_4]^T$ and A=diag$\{-k_1, -k_2, -k_3, -k_4\}$. Since all eigenvalues of the closed-loop system (20.1.14) are $-k_1$, $-k_2$, $-k_3$ and $-k_4$, which possess negative real parts, the error states e_1, e_2, e_3 and e_4 converge to zero as $t \to \infty$. Consequently, HPS between system (20.1.8) and system (20.1.9) is achieved.

To validate the designed controller (20.1.13), numerical simulations are performed using the fourth-order Runge-Kutta method. The scaling matrix is set as $\mathbf{D} = \text{diag}\{0.5, 1, \text{-}1, 2\}$, and the control coefficients are chosen as $k_1 = 0.2$, $k_2 = 0.5$, $k_3 = 1$ and $k_4 = 2$. Systems (20.1.8) and (20.1.9) share the parameters $a = 0.25$, $b = 1$, $c = 1$, $d = 0.2$, $w = 1$, with initial conditions $(u_{11}(0), u_{21}(0), u_{31}(0), u_{41}(0)) = (0.2, 0.3, 0.2, 0.2)$ for the drive system and $(u_{12}(0), u_{22}(0), u_{32}(0), u_{42}(0)) = (-0.3, 0.1, 0.2, 0.2)$ for the response system. Thus, the initial conditions of system (20.1.12) are $(e_1(0),\ e_2(0),\ e_3(0),\ e_4(0)) = (-0.4,\ -0.2,\ 0.4,\ 0.2)$. The simulation results are presented in Figs. 20.3 and 20.4. Figure 20.3 demonstrates that chaos HPS between systems (20.1.8) and (20.1.9) is achieved within a short time interval. As shown in Fig. 20.4, the HPS errors $e_1(t)$, $e_2(t)$, $e_3(t)$, and $e_4(t)$ asymptotically converge to zero as $t \to \infty$, consistent with the theoretical analysis. On the other hand, the control coefficients $k_n(n = 1,\ 2,\ 3,\ 4)$ can be preselected to adjust the convergence rate for achieving hybrid projective synchronization in the complex Duffing-Holmes oscillators. Larger values of these coefficients result in faster synchronization.

II. HPS with Unknown Parameters

This section addresses the hybrid projective synchronization of the complex Duffing-Holmes system (20.1.1) with unknown parameters using adaptive control techniques.

Similarly, systems (20.1.8) and (20.1.9) with uncertain parameters are selected as the drive and response systems, respectively. Since the parameters a, b, c and d are unknown a priori, systems (20.1.8) and (20.1.9) are considered uncertain. A nonlinear controller vector $V = [v_1, v_2, v_3, v_4]^T$ must be designed to achieve hybrid projective synchronization between these systems despite the parameter uncertainties. The corresponding error dynamical system is derived as follows:

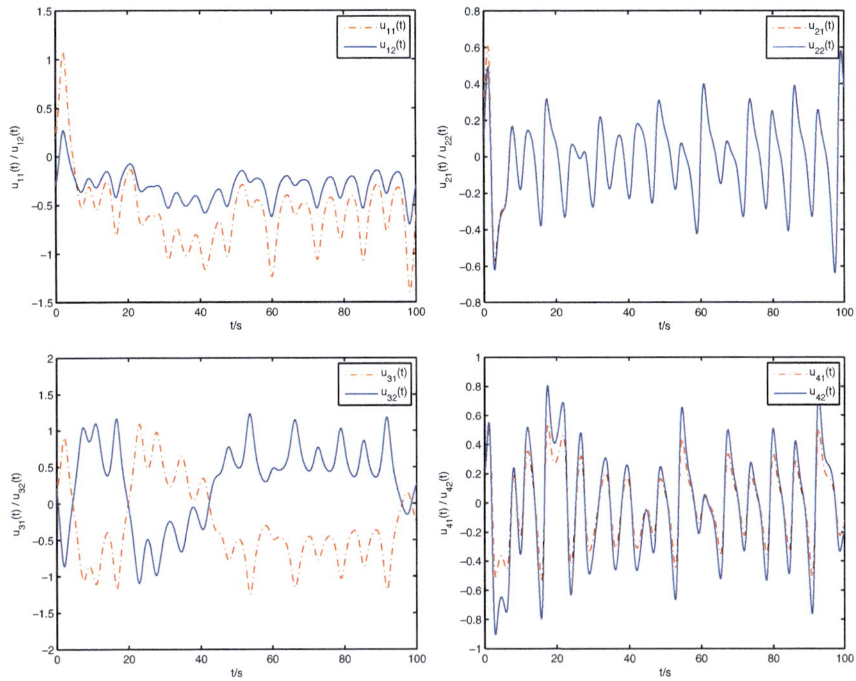

Fig. 20.3 The HPS of systems (20.1.8) and (20.1.9) with the nonlinear controller (20.1.13)

$$\begin{cases} \dot{e}_1 = e_2 + (d_2 - d_1)u_{21} + v_1, \\ \dot{e}_2 = -ae_2 + be_1 + b(d_1 - d_2)u_{11} - c[(u_{12}^2 + u_{32}^2)u_{12} - d_2(u_{11}^2 + u_{31}^2)u_{11}] \\ \qquad +dcos(wt)(1 - d_2) + v_2, \\ \dot{e}_3 = e_4 + (d_4 - d_3)u_{41} + v_3, \\ \dot{e}_4 = -ae_4 + be_3 + b(d_3 - d_4)u_{31} - c[(u_{12}^2 + u_{32}^2)u_{32} - d_4(u_{11}^2 + u_{31}^2)u_{31}] \\ \qquad +dcos(wt)(1 - d_4) + v_4, \end{cases} \tag{20.1.15}$$

where the state errors are defined as $e_1 = u_{12} - d_1u_{11},\ e_2 = u_{22} - d_2u_{21},\ e_3 = u_{32} - d_3u_{31}$ and $e_4 = u_{42} - d_4u_{41}$. The objective is to design a controller V and a parameter update law such that system (20.1.9) with uncertain parameters achieves hybrid projective synchronization with system (20.1.8), i.e.,

$$\lim_{t\to\infty} \|E\| = \lim_{t\to\infty} \sum_{l=1}^{4} \|u_{l1} - d_l u_{l2}\| = 0,$$

where $E = [e_1,\ e_2,\ e_3,\ e_4]^T$.

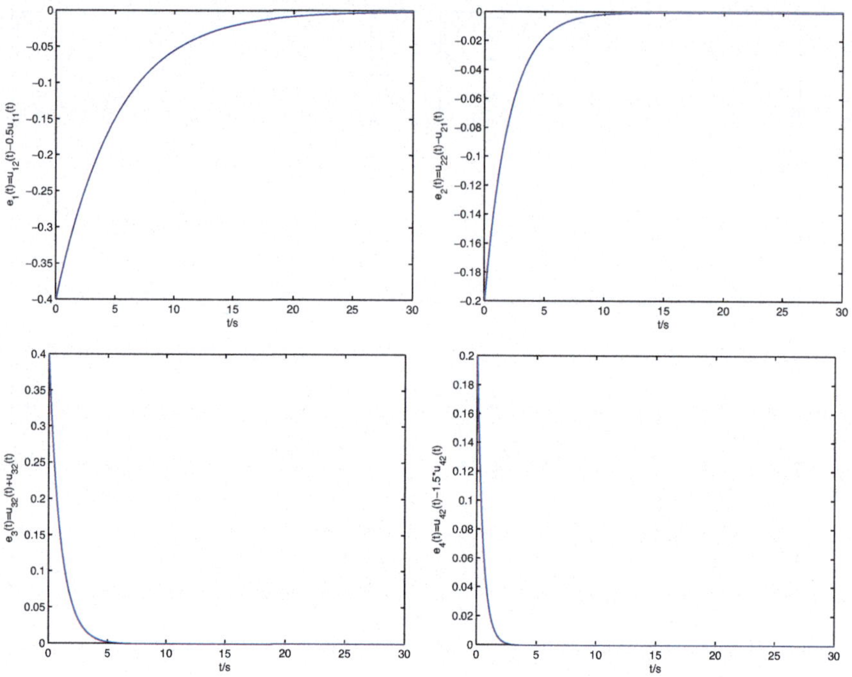

Fig. 20.4 Dynamics of the error system (20.1.12) under the nonlinear controller (20.1.13)

The parameter errors are defined as

$$\tilde{a} = a - \hat{a},\ \tilde{b} = b - \hat{b},\ \tilde{c} = c - \hat{c},\ \tilde{d} = d - \hat{d},$$

where $\hat{a},\ \hat{b},\ \hat{c},\ \hat{d}$ are estimate values of the unknown parameters $a,\ b,\ c,\ d$, separately.

For systems (20.1.8) and (20.1.9) without the controller V if the initial condition $(u_{11}(0),\ u_{21}(0), u_{31}(0),\ u_{41}(0)) \neq (u_{12}(0),\ u_{22}(0),\ u_{32}(0),\ u_{42}(0))$, then the trajectories of two systems will quickly separate each other and become irrelevant. However, when the controller is implemented, systems (20.1.8) and (20.1.9) will achieve hybrid projective synchronization under arbitrary initial conditions. To this end, we design the adaptive controller and parameter update law as follows:

$$\begin{cases} v_1 = -(d_2 - d_1)u_{21} - e_1 - e_2, \\ v_2 = -e_2 + \hat{a}e_2 - \hat{b}[e_1 + (d_1 - d_2)u_{11}] + \hat{c}[(u_{12}^2 + u_{32}^2)u_{12} - \ d_2(u_{11}^2 + u_{31}^2)u_{11}] \\ \qquad -\hat{d}cos(wt)(1 - d_2), \\ v_3 = -(d_4 - d_3)u_{41} - e_3 - e_4, \\ v_4 = -e_4 + \hat{a}e_4 - \hat{b}[e_3 + (d_3 - d_4)u_{31}] + \hat{c}[(u_{12}^2 + u_{32}^2)u_{32} - \ d_4(u_{11}^2 + u_{31}^2)u_{31}] \\ \qquad -\hat{d}cos(wt)(1 - d_4), \end{cases} \tag{20.1.16}$$

and

$$\begin{cases} \dot{\hat{a}} = -e_2^2 - e_4^2, \\ \dot{\hat{b}} = e_1e_2 + e_2(d_1 - d_2)u_{11} + e_3e_4 + e_4(d_3 - d_4)u_{31}, \\ \dot{\hat{c}} = -(u_{12}^2 + u_{32}^2)(u_{12}e_2 + u_{32}e_4) + (u_{11}^2 + u_{31}^2)(d_2u_{11}e_2 + d_4u_{31}e_4), \\ \dot{\hat{d}} = [(1 - d_2)e_2 + (1 - d_4)e_4]cos(wt). \end{cases} \tag{20.1.17}$$

Choose a Lyapunov function as

$$V(E, \tilde{a}, \tilde{b}, \tilde{c}, \tilde{d}) = (1/2)(e_1^2 + e_2^2 + e_3^2 + e_4^2 + \tilde{a}^2 + \tilde{b}^2 + \tilde{c}^2 + \tilde{d}^2),$$

then the time derivative of V along the trajectory of the error system (20.1.15), under the designed controller (20.1.16) and parameter update law (20.1.17), is given by:

$$\begin{aligned} \dot{V} &= e_1\dot{e}_1 + e_2\dot{e}_2 + e_3\dot{e}_3 + e_4\dot{e}_4 + \tilde{a}\dot{\tilde{a}} + \tilde{b}\dot{\tilde{b}} + \tilde{c}\dot{\tilde{c}} + \tilde{d}\dot{\tilde{d}} \\ &= e_1[e_2 + (d_2 - d_1)u_{21} + \mu_1] + e_2[-ae_2 + be_1 + b(d_1 - d_2)u_{11} \\ &\quad - c[(u_{12}^2 + u_{32}^2)u_{12} - d_2(u_{11}^2 + u_{31}^2)u_{11}] + dcos(wt)(1 - d_2) + \mu_2] \\ &\quad + e_3[e_4 + (d_4 - d_3)u_{41} + \mu_3] + e_4[-ae_4 + be_3 + b(d_3 - d_4)u_{31} \\ &\quad - c[(u_{12}^2 + u_{32}^2)u_{32} - d_4(u_{11}^2 + u_{31}^2)u_{31}] + dcos(wt)(1 - d_4) + \mu_4] \\ &\quad - \tilde{a}\dot{\hat{a}} - \tilde{b}\dot{\hat{b}} - \tilde{c}\dot{\hat{c}} - \tilde{d}\dot{\hat{d}}, \\ &= -(e_1^2 + e_2^2 + e_3^2 + e_4^2) \le 0. \end{aligned}$$

Clearly, while V is positive definite and $\dot{V}$ is negative semi-definite, this alone does not guarantee the asymptotic stability of the origin in error system (20.1.15) under controller (20.1.16). Since $\dot{V} \le 0$, we get $e_1, e_2, e_3, e_4 \in L_\infty$ and $\tilde{a}, \tilde{b}, \tilde{c}, \tilde{d} \in L_\infty$ From the fact that

$$\int_0^t \|E\| = -(1/k)(V(t) - V(0)) \le V(0)/k,$$

we get $e_1, e_2, e_3, e_4 \in L_2$. Based on the error system (20.1.15) with the designed controller (20.1.16), we have $\dot{e}_1, \dot{e}_2, \dot{e}_3, \dot{e}_4 \in L_\infty$. Applying Barbalat's lemma, it follows that $e_1, e_2, e_3, e_4 \to 0$, as $t \to \infty$, that is,

$$\lim_{t\to\infty} \|E\| = \lim_{t\to\infty} \sum_{l=1}^{n} \|e_l(t, e_l(0))\| = 0.$$

Therefore, with the controller (20.1.16) and parameter update law (20.1.17), the response complex system (20.1.9) asymptotically achieves hybrid projective synchronization with the drive complex system (20.1.8).

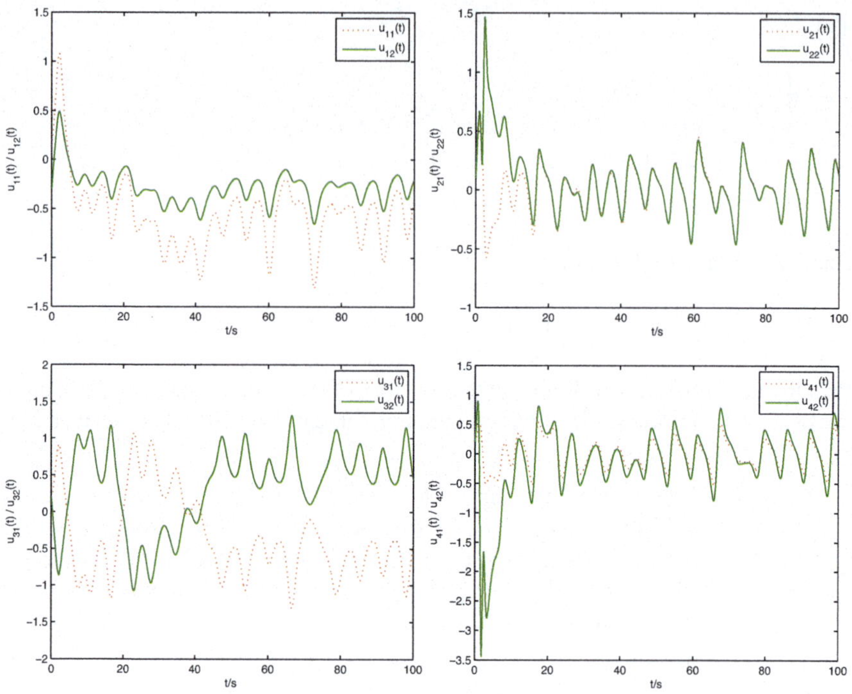

Fig. 20.5 The HPS of uncertain systems (20.1.8) and (20.1.9) with the adaptive controller (20.1.16) and parameter update law (20.1.17)

In the simulations, the fourth-order Runge-Kutta method is employed to numerically solve systems (20.1.8) and (20.1.9) with the designed controller (20.1.16). The scaling matrix **D** and the initial conditions for systems (20.1.8) and (20.1.9) remain identical to those used in the simulations of Sect. 4.1. Meanwhile, the true values of the unknown parameters for both the drive and response systems are chosen as $a = 0.25$, $b = c = 1$ and $d = 0.2$. The numerical simulation results are presented in Figs. 20.5, 20.6 and 20.7. As shown in Fig. 20.5, the response system (20.1.9) with uncertain parameters achieves HPS with the unknown drive system (20.1.8) within a short time interval. Figure 20.6 demonstrates that the HPS errors asymptotically converge to zero as time t increases, consistent with the theoretical analysis. Furthermore, Fig. 20.7 illustrates that the parameter estimates $\hat{a}$, $\hat{b}$, $\hat{c}$ and $\hat{d}$ converge to the true values $a = 0.25$, $b = 1$, $c = 1$ and $d = 0.2$, respectively, as $t \to \infty$.

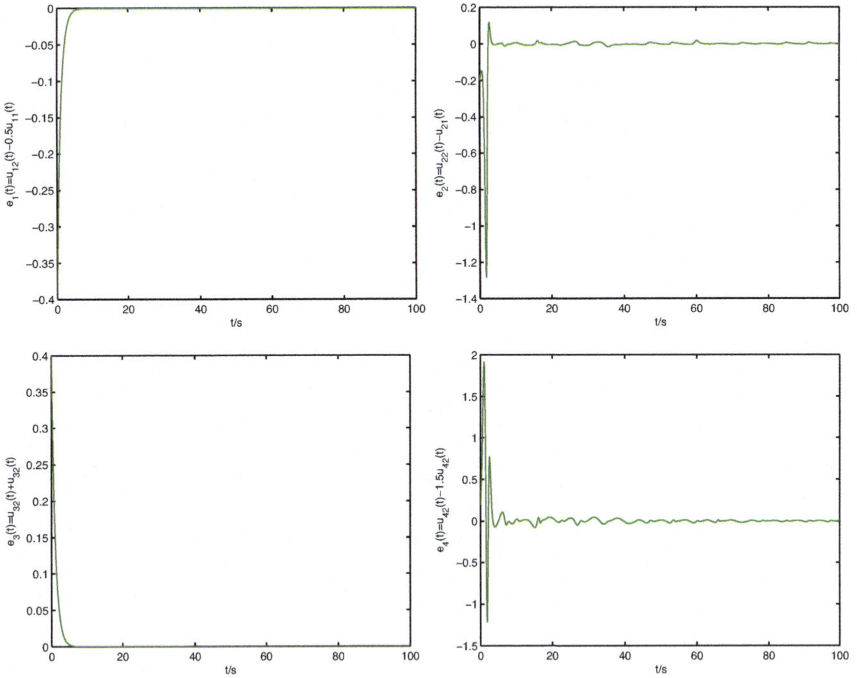

Fig. 20.6 Dynamics of the error system (20.1.15) under the controller (20.1.16) and parameter update law (20.1.17)

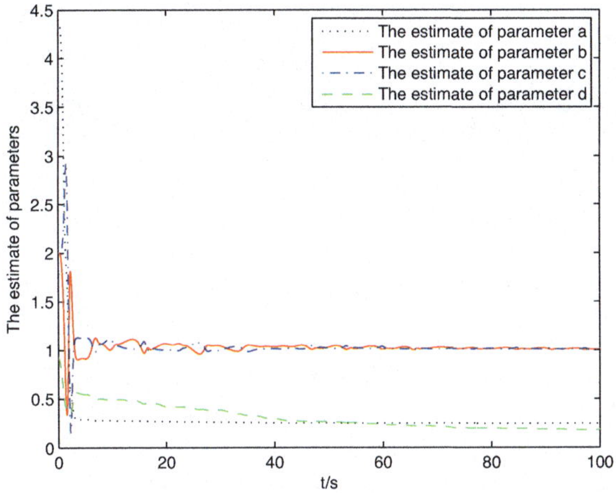

Fig. 20.7 Dynamics of the estimates $\hat{a}$, $\hat{b}$, $\hat{c}$, $\hat{d}$ of uncertain parameters a, b, c, d

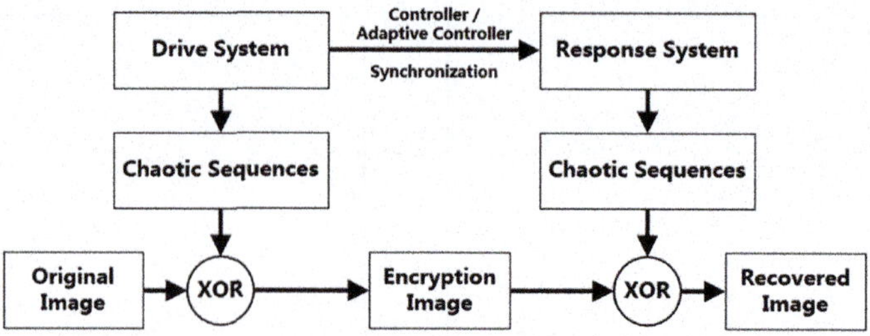

Fig. 20.8 The process of the encryption and decryption

Fig. 20.9 Simulation result

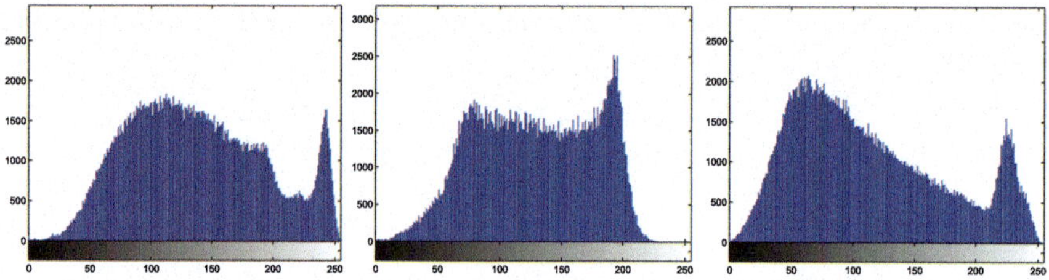

Fig. 20.10 Histogram analysis of the original image

20.2 Encryption-Decryption Algorithm Description

In this section, the proposed algorithms are applied to color image encryption. The encryption and decryption processes, based on XOR operations, are illustrated in Fig. 20.8. Three sequences generated by the drive system are used to perform XOR operations with the image pixel values for encryption, while the corresponding three sequences from the response system are utilized for decryption. Simulation results are presented in Fig. 20.9.

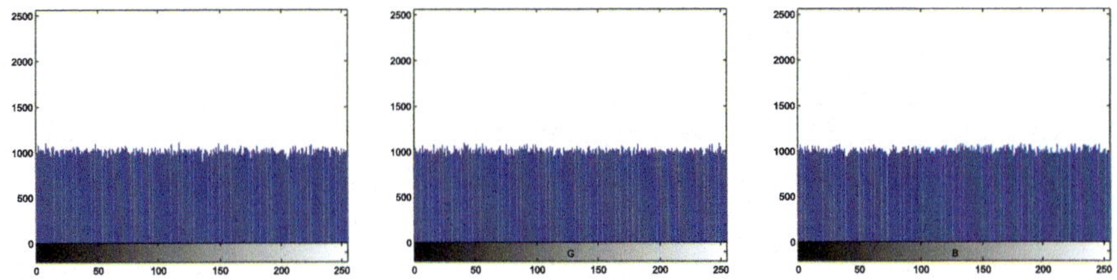

Fig. 20.11 Histogram analysis of the encrypted image

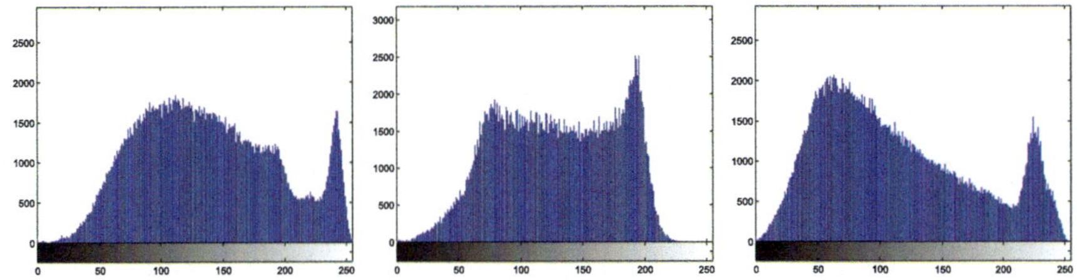

Fig. 20.12 Histogram analysis of the decrypted image

20.3 Security Analysis

The histograms of the original, encrypted, and decrypted images for the R, G, and B channels are displayed in Figs. 20.10, 20.11, and 20.12, respectively. These results confirm the strong capability of the proposed algorithm to resist statistical attacks.

References

1. E. N. Lorenz. Deterministic nonperiodic flows [J]. Journal of Atmospheric Science, 1963, 29, 130–141.
2. E. N. Lorenz, K. Haman. The essence of chaos [J]. Pure and Applied Geophysics, 1996, 147(3): 598–599.
3. J. Gleick. Chaos: Making a new science [M]. Penguin, 2008.
4. J. C. Sprott. Chaos and time-series analysis [M]. Oxford University Press, 2003.
5. R. M. May. Simple mathematical models with very complicated dynamics [J]. Nature, 1976, 261(5560): 459–467.
6. B. B. Mandelbrot. The fractal geometry of nature/Revised and enlarged edition [M]. New York, 1983.
7. G. R. Chen, X. N. Dong. From chaos to order: methodologies, perspectives and applications [M]. World Scientific, 1998.
8. J. Guckenheimer, P. Holmes. Nonlinear oscillations, dynamical systems, and bifurcations of vector fields [M]. Springer Science & Business Media, 2013.
9. J. M. T. Thompson, H. B. Stewart. Nonlinear dynamics and chaos [M]. John Wiley & Sons, 2002.
10. S. Boccaletti, C. Grebogi, Y. C. Lai, et al. The control of chaos: theory and applications [J]. Physics Reports, 2000, 329(3): 103–197.
11. K. M. Cuomo, A. V. Oppenheim. Circuit implementation of synchronized chaos with applications to communications [J]. Physical Review Letters, 1993, 71(1): 65.
12. L. Kocarev, U. Parlitz. General approach for chaotic synchronization with applications to communication [J]. Physical Review Letters, 1995, 74(25): 5028.
13. S. H. Strogatz. Sync: How order emerges from chaos in the universe, nature, and daily life [M]. Grand Central Publishing, 2012.
14. R. L. Devaney. An introduction to chaotic dynamical systems [M]. CRC Press, 2018.
15. M. L. West. Theogony and works and days [M]. OUP Oxford, 2008.
16. H. Poincaré. Sur le problème des trois corps et les équations de la dynamique [J]. Acta Mathematica, 1890, 13(1): A3–A270.
17. B. L. Hao. Chaos and Fractals [M]. Shanghai Scientific & Technical Publishers, 2004.
18. T. Y. Li, J. A. Yorke. Period three implies chaos [J]. The American Mathematical Monthly, 1975, 82(10): 985–992.
19. J. D. Farmer, E. Ott, J. A. Yorke. The dimension of chaotic attractors [J]. Physica D: Nonlinear Phenomena, 1983, 7(1–3): 153–180.
20. E. Ott. Chaos in dynamical systems [M]. Cambridge University Press, 2002.

S. Liu et al., *Chaos in Complex Dynamical System: Theory and Application*,
https://doi.org/10.1007/978-981-95-7762-0

21. R. C. Hilborn. Chaos and nonlinear dynamics: an introduction for scientists and engineers [M]. Oxford University Press, 2000.
22. K. T. Alligood, D. Tim, J. A. Sauer, et al. Chaos: an introduction to dynamical systems [J]. SIAM Review, 1998, 40(3): 732–732.
23. H. S. Steven. Nonlinear dynamics and chaos: with applications to physics, biology, chemistry, and engineering [M]. Chapman and Hall/CRC, 2024.
24. L. M. Pecora, T. Carroll. Synchronization in chaotic systems [J]. Physical Review Letters, 1990, 64(8): 821–824.
25. G. M. Mahmoud, E. E. Mahmoud. Complete synchronization of chaotic complex nonlinear systems with uncertain parameters [J]. Nonlinear Dynamics, 2010, 62: 875–882.
26. G. M. Mahmoud, M. A. Al-Kashif, S. A. Aly. Basic properties and chaotic synchronization of complex Lorenz system [J]. International Journal of Modern Physics, 2007, 18(02): 253–265.
27. G. M. Mahmoud, T. Bountis, M. A. Al-Kashif, et al. Dynamical properties and synchronization of complex nonlinear equations for detuned lasers [J]. Dynamic Systems, 2009, 24(1): 63–79.
28. L. Glass, M.C. Mackay. From clocks to chaos: the rhythms of life [M]. Princeton NJ: Prineeton University Press, 1988.
29. P. Liu, S. Liu. Anti-synchronization between different chaotic complex systems [J]. Physica Scripta, 2011, 83(6): 65006–65014.
30. C.M. Kim, S. Rim, W. H. Kye, et al. Anti-synchronization of chaotic oseillators [J]. Physics Letters A, 2003, 320(1): 39–46.
31. J. Hu, S. H. Chen, L. Chen. Adaptive control for anti-synchronization of Chua's chaotic system [J]. Physics Letters A, 2005, 339(6): 455–460.
32. R. Mainieri, J. Rehacek. Projective synchronization in three-dimensional chaotic systems [J]. Physical Review Letters, 1999, 82(15): 3042–3045.
33. L. Guo, Z. Xu. Projective synchronization in drive-response networks via impulsive Control [J]. Chinese Physics Letters, 2008, 25: 2816–2819.
34. L. Guo, Z. Xu, M. Hu. Adaptive projective synchronization with different sealing factors in networks [J]. Chinese Physics B, 2008, 17: 4067–4072.
35. G. M. Mahmoud, E. E. Mahmoud. Lag synchronization of hyperchaotic complex nonlinear systems[J]. Nonlinear Dynamics, 2012, 67: 1613–1622.
36. E. M. Shahverdiev, S. Sivaprakasam, K. A. Shore. Lag synchronization in time-delayed systems [J]. Physics Letters A, 2002, 292(6): 320–324.
37. S. Taherion, Y.C. Lai. Observability of lag synchronization of coupled chaotic oseillators [J]. Physical Review E, 1999, 59(6): 6247–6250.
38. C. D. Li, X. F. Liao. Lag synchronization of Rössler system and Chua circuit via a sealar signal [J]. Physics Letters A, 2004,329(4–5): 301–308.
39. M. G. Rosenblum, A. S. Pikovsky, J. Kurths. From phase to lag synchronization in coupled chaoticoscillators [J]. Physical Review Letters, 1997, 78(22): 4193–4196.
40. G. M. Mahmoud, E. E. Mahmoud. Phase and antiphase synchronization of two identical hyperchaotic complex nonlinear systems [J]. Nonlinear Dynamics, 2010, 61: 141–152.
41. M. C. Ho, Y. C. Hung, C. H. Chou. Phase and anti-phase synchronization of two chaotic systems by using active control [J]. Physics Letters A, 2002, 296(1): 43–48.
42. G. L. Wen, D. Xu. Nonlinear observer control for full-state projective syn-chronization in chaotic continuous-time systems [J]. Chaos, Solitons and Fractals, 2005, 26(1): 71–77.
43. F. F. Zhang, J. X. Liu, Z. X. Wang, et al. N-systems function projective combination synchronization-a review of real and complex continuous time chaos synchronization [J]. IEEE Access, 2019, 7: 179320–179338.
44. X.J. Wu, Y. Lu. Generalized projective synchronization of the fractional-order Chen hyperchaotic system [J]. Nonlinear Dynamics, 2009, 57(1): 25–35.
45. Y. G. Yu, H. X. Li. Adaptive generalized funetion projective synchronization of uncertain chaotic systems [J]. Nonlinear Analysis: Real World Applications, 2010, 11(4): 2456–2464.
46. X. j. Wu, H. Wang, H. T. Lu. Hyperchaotie secure communication via generalized function projective synchronization [J]. Nonlinear Analysis: Real World Applications, 2011, 12(2): 1288–1299.

47. B. Li, X. S. Zhao. Generalized funetion projective synchronization of two different hyperchaotic systems with unknown parameters [J]. Nonlinear Analysis: Real World Applications, 2011, 12(5): 2607–2615.
48. B. X. Wang, Z. M. Zhu, S. M. Zhu, et al. Ordinary Differential Equation [M]. Higher Education Press, 2020.
49. H. K. Khalil, J. W. Grizzle. Nonlinear systems [M]. Upper Saddle River, NJ: Prentice hall, 2002.
50. E. Ott, C. Grebogi, J. A. Yorke. Controlling chaos [J]. Physical Review Letters, 1990, 64(11): 1196–1199.
51. G. R. Chen, X. F. Wang. The Theory and Methods of Chaotic System Control [M]. Science Press, 2003.
52. A. J. Chorin, J. E. Marsden. A mathematical introduction to fluid mechanics [M]. New York: Springer-Verlag, 1990.
53. B. R. Munson, T. H. Okiishi, W. W. Huebsch, et al. Fluid mechanics [M]. Singapore: Wiley, 2013.
54. I. Podlubny. Fractional Differential Equations [M]. Academic Press, New York, 1999.
55. N. Aguila-Camacho, M. A. Duarte-Mermoud, J. A. Gallegos. Lyapunov functions for fractional order system [J]. Communications in Nonlinear Science and Numerical Simulation, 2014, 19(9): 2951–2957.
56. S. Liang, R. C. Wu, L. P. Chen. Adaptive pinning synchronization in fractional-order uncertain complex dynamical networks with delay [J]. Physica A, 2016, 444: 49–62.
57. S. Liu, W. Jiang, X. Y. Li, et al. Lyapunov staility analysis of fractional nonlinear systems [J]. Applied Mathematics Letters, 2016, 51: 13–19.
58. J. H. Luo, H. Liu, J. F. Yang. Robust synchronization of uncertain fractional order chaotic systems [J]. IEICE Transactions on Fundamentals, 2015, E98-A: 2109–2116.
59. D. Matignon. Stability results for fractional differential equations with applications to control processing [J]. Computational Engineering in Systems Applications, 1996, 2(1): 963–968.
60. W. Deng, C. Li, J. Lü. Stability analysis of linear fractional differential system with multiple time delays [J]. Nonlinear Dynamics, 2007, 48: 409–416.
61. Y. Li, Y. Q. Chen, I. Podlubny. Stability of fractional-order nonlinear dynamic systems: Lyapunov direct method and generalized Mittag-Leffler stability [J]. Computers & Mathematics with Applications, 2010, 59(5): 1810–1821.
62. Y. Li, Y. Q. Chen, I. Podlubny, Mittag-Leffler stability of fractional order nonlinear dynamic systems [J]. Automatica, 2009, 45(8): 1965–1969.
63. A. Charef, H. H. Sun, Y. Y. Tsao, et al. Fractal system as represented by singularity function [J]. IEEE Transactions on Automatic Control, 1992, 37(9): 1465–1470.
64. K. Diethelm, N. J. Ford, A. D. Freed. A predictor-corrector approch for the numerical solution of fractional differential equations [J]. Nonlinear Dynamics, 2002, 29(3): 3–22.
65. K. Diethelm, N. J. Ford, A. D. Freed. Detailed error analysis for a fractional Adams method [J]. Numer. Algorithms, 2004, 36: 31–52.
66. Q. Song, J. Cao. Synchronization and anti-synchronization for chaotic systems [J]. Chaos, Solitons & Fractals, 2007, 33(3): 929–939.
67. V. Sundarapandian. Analysis and anti-synchronization of a novel chaotic system via active and adaptive controllers [J]. Journal of Engineering Science and Technology Review, 2013, 6(4): 45–52.
68. M. Rehan. Synchronization and anti-synchronization of chaotic oscillators under input saturation [J]. Applied Mathematical Modelling, 2013, 37(10–11): 6829–6837.
69. C.K. Ahn. An $H\infty$ approach to anti-synchronization for chaotic systems [J]. Physics Letters A, 2009, 373(20): 1729–1733.
70. J. Hu, S. Chen, L. Chen. Adaptive control for anti-synchronization of Chua's chaotic system [J]. Physics Letters A, 2005, 339(6): 455–460.
71. A. M. Lyapunov. The general problem of the stability of motion [J]. International Journal of Control, 1992, 55(3): 531–534.

72. J. Z. Wang, Z. Q. Chen, Z. Z. Yuan. A new chaotic system and analysis of its properties [J]. Acta Physica Sinica, 2006, 55(8): 3956–3963.
73. J. Xu, G. Cai. Adaptive backstepping control of new chaotic system [J]. Journal of Jiangsu University of Science and Technology: Natural Science Edition, 2007, 21: 86–89.
74. J. Xu, G. Cai. Nonlinear feedback synchronization of new chaotic system [J]. Journal of Jiangsu University of Science and Technology: Natural Science Edition, 2008, 22: 86–90.
75. A. Wolf, J. Swift, H. Swinney, et al. Determining Lyapunov exponents from a time series [J]. Physica D, 1985, 16(3): 285–317.
76. A. C. Fowler, J. D. Gibbon, M. J. McGuinness. The complex Lorenz equations [J]. Physica D, 1982, 4(2): 139–163.
77. A. Rauh, L. Hannibal and N. Abraham. Global stability properties of the complex Lorenz model [J]. Physica D, 1996, 99(1): 45–58.
78. Y. Zhang, J. Sun. Chaotic synchronization and anti-synchronization based on suitable separation [J]. Physics Letters A, 2004, 330(6): 442–447.
79. G. M. Mahmoud, T. Bountis, E. E. Mahmoud. Active control and global synchronization for complex Chen and Lü systems [J]. International Journal of Bifurcation and Chaos, 2007, 17(12): 4295–4308.
80. G. M. Mahmoud, S. A. Aly. Periodic attractors of complex damped non-linear systems [J]. International Journal of Non-Linear Mechanics, 2000, 35(2): 309–323.
81. A. Al-Qahtani, H. B. Khenous, S. Aly. Synchronization of impulsive real and complex van der pol oscillators [J]. Applied Mathematics, 2015, 6(6): 922–932.
82. A. H. S. Salas. Analytic solution to the generalized complex Duffing equation and its application in soliton theory [J]. Applicable Analysis, 2021, 100(13): 2867–2872.
83. K. Konno, H. Tateno. Duffing's equation in complex times and chaos [J]. Progress of Theoretical Physics, 1984, 72(5): 1047–1049.
84. G. M. Mahmoud, T. Bountis, G. M. Abdei-Latif, et al. Chaos synchronization of two different chaotic complex Chen and Lü systems [J]. Nonlinear Dynamics, 2009, 55(1): 43–53.
85. G. M. Mahmoud, E. E. Mahmoud, M. E. Ahmed. A hyperchaotic complex Chen system and its dynamics [J]. International Journal of Applied Mathematics and Statistics, 2007, 12(D07): 90–100.
86. T. L. Liao, S. H. Tsai. Adaptive synchronization of chaotic systems and its application to secure communications [J]. Chaos Soliton & Fractals, 2000, 11(9): 1387–1396.
87. T. L. Liao, N. S. Huang. An observer based approach for chaotic synchronization and secure communication [J]. IEEE Transactions on Circuits and Systems, 1999, 46(9): 1144–1149.
88. M. Feki. An adaptive chaos synchronization scheme applied to secure communication [J]. Chaos, Solitons & Fractals, 2003, 18(1): 141–148.
89. Y. Yu, S. Zhang. Adaptive backsteping synchronization of uncertain chaotic system [J]. Chaos, Solitons & Fractals, 2004, 21(3): 643–649.
90. M. T. Yassen. Controlling chaos and synchronization for new chaotic system using linear feedback control [J]. Chaos, Solitons & Fractals, 2005, 26(3): 913–920.
91. H. Zhang, W. Huang, Z. Wang, et al. Adaptive synchronization between two different chaotic systems with unknown parameters [J]. Physics Letters A, 2006, 350(5–6): 363–366.
92. U. Parlitz, L. Kocarev, T. Stojanovski, et al. Encoding messages using chaotic synchronization [J]. Physical Review E, 1996, 53(5): 4351–4361.
93. H. N. Agiza, M. T. Yassen. Synchronization of Rössler and Chen chaotic dynamical systems using active control [J]. Physics Letters A, 2001, 278(4): 191–197.
94. M. Ho, Y. Hung. Synchronization of two different systems by using generalized active control [J]. Physics Letters A, 2002, 301(5–6): 424–428.
95. L. Huang, R. Feng and M. Wang. Synchronization of chaotic systems via nonlinear control [J]. Physics Letters A, 2004, 320(4): 271–275.
96. P. N. Paraskevopoulos. Modern control engineering [M]. New York: Marcel Dekker, 2002.
97. G. M. Mahmoud, S. A. Aly and A. A. Farghaly. On chaos synchronization of a complex two coupled dynamos system [J]. Chaos Solitons & Fractals, 2007, 33(1): 178–187.

98. M. Hu, Z. Xu, R. Zhang. Full state hybrid projective synchronization in continuous-time chaotic (hyperchaotic) systems [J]. Communications in Nonlinear Science and Numerical Simulation, 2008, 13(2): 456–464.
99. Y. Chu, Y. X. Chang, J. G. Zhang, et al. Full state hybrid projective synchronization in hyperchaotic systems [J]. Chaos, Solitons & Fractals, 2009, 42(3): 1502–1510.
100. F. Sun, H. Peng, Q. Luo, et al. Parameter identification and projective synchronization between different chaotic systems [J]. Chaos, 2009, 19(2): 023109–02314.
101. H. Peng, L. Li, Y. Yang, et al. Conditions of parameter identification from time series [J]. Physical Review E, 2011, 83(3): 036202–036209.
102. M. Hu, Z. Xu, R. Zhang, A. Hu, et al. Parameters identification and adaptive full state hybrid projective synchronization of chaotic (hyper-chaotic) systems [J]. Physics Letters A, 2007, 361(3): 231–237.
103. M. Hu, Z. Xu, R. Zhang, et al. Adaptive full state hybrid projective synchronization of chaotic systems with the same and different order [J]. Physics Letters A, 2007, 365(4): 315–327.
104. G. Tao. Adaptive control design and analysis [M]. John Wiley & Sons, 2003.
105. I. D. Landau, R. Lozano, M. M'Saad, et al. Adaptive control: algorithms, analysis and applications [M]. Springer Science & Business Media, 2011.
106. C. Hua, X. Guan. Adaptive control for chaotic systems [J]. Chaos, Solitons & Fractals, 2004, 22(1): 55–60.
107. Y. Chu, Y. X. Chang, J. G. Zhang, et al. Full state hybrid projec-tive synchronization in hyperchaotic systems [J]. Chaos, Solitons & Fractals, 2009, 42(3): 1502–1510.
108. G. M. Mahmoud, E. E. Mahmoud. Synchronization and control of hyper-chaotic complex Lorenz system [J]. Mathematics and Computers in Simulation, 2010, 80(12): 2286–2296.
109. H. Du, Q. Zeng and C. Wang. Modified function projective synchronization of chaotic system [J]. Chaos, Solitons & Fractals, 2009, 42(4): 2399–2404.
110. K. S. Sudheer, M. Sabir. Modified function projective synchronization of hyperchaotic systems through Open-Plus-Closed-Loop coupling [J]. Physics Letters A, 2010, 374(19–20): 2017–2023.
111. H. Nabil, H. Tayeb. A secure communication scheme based on generalized modified projective synchronization of a new 4-D fractional-order hyperchaotic system [J]. Physica Scripta, 2024, 99(9): 095203.
112. K. S. Sudheer, M. Sabir. Adaptive modified function projective synchronization between hyperchaotic Lorenz system and hyperchaotic Lü system with uncertain parameters [J]. Physics Letters A, 2009, 373(41): 3743–3748.
113. C. W. Wu, T. Yang, L. O. Chua. On adaptive synchronization and control of nonlinear dynamical systems [J]. International Journal of Bifurcation and Chaos, 1996, 6(03): 455–471.
114. J. Zhou, J. Lu, J. Lu. Adaptive synchronization of an uncertain complex dynamical network [J]. IEEE Transactions on Automatic Control, 2006, 51(4): 652–656.
115. X. F. Li, A. C. Leung, X. J. Liu, et al. Adaptive synchronization of identical chaotic and hyperchaotic systems with uncertain parameters [J]. Nonlinear Analysis, 2010, 11(4): 2215–2223.
116. R. Bellman. Stability theory of differential equations [M]. Courier Corporation, 2008.
117. G. Tao. A simple alternative to the Barbalat lemma [J]. IEEE Transactions on Automatic Control. 1997, 42(5):698.
118. P. Liu, S. T. Liu. Anti-synchronization between different chaotic complex systems [J]. Physica Scripta, 2011, 83(6): 065006.
119. G. M. Mahmoud, S. A. Aly, M.A. Al-Kashif. Dynamical properties and chaos synchronization of a new chaotic complex nonlinear system [J]. Nonlinear Dynamics, 2008, 51: 171–181.
120. G. Mahmoud, M. Ahmed, E. Mahmoudb. Analysis of hyperchaotic complex Lorenz systems [J]. International Journal of Modern Physics C, 2008, 19(10): 1477–1494.
121. G. Mahmoud, M. Al-Kashif, A. Farghaly. Chaotic and hyperchaotic attractors of a complex nonlinear system [J]. Journal of Physics A: Mathematical and Theoretical, 2008, 41(5): 055104.
122. S. Zheng, G. G. Dong, Q. S. Bi. Adaptive modified function projective synchronization of hyperchaotic systems with unknown parameters [J]. Communications in Nonlinear Science and Numerical Simulation, 2010, 15(11): 3547–3556.

123. G. Y. Fu. Robust adaptive modified function projective synchronization of different hyperchaotic systems subject to external disturbance [J]. Communications in Nonlinear Science and Numerical Simulation, 2012, 17(6): 2602–2608.
124. S. Zheng. Adaptive modified function projective synchronization of unknown chaotic systems with different order [J]. Applied Mathematics and Computation, 2012, 218(10): 5891–5899.
125. P. Liu, S. T. Liu. Adaptive modified function projective synchronization of general uncertain chaotic complex system [J]. Physics Scripts, 2012, 85: 035005.
126. L. F. Alexander, Y. P. Alexander. Speed gradient control of chaotic continuous-time systems [J]. IEEE Transactions on circuits and systems-I: Fundamental theory and applications, 1996, 43(11): 907–913.
127. P. Liu, S. T. Liu. Robust adaptive full state hybrid synchronization of chaotic complex systems with unknown parameters and external disturbances [J]. Nonlinear Dynamics, 2012, 70(1): 585–599.
128. H. Zhang, X. K. Ma, M. Li. Controlling and tracking hyperchaotic Rössler system via active backstepping design[J]. Chaos, Solitions & Fractals, 2005, 26: 353–361.
129. E. E. Mahmoud, Dynamics and synchronization of new hyperchaotic complex Lorenz system [J]. Mathematical and Computer Modelling, 2012, 55: 1951–1962.
130. H. Y. Du, Q. S. Zeng, C. H. Wang. Function projective synchronization of different chaotic systems with uncertain parameters [J]. Physics Letters A, 2008, 372(33): 5402–5410.
131. H. Zhu. Adaptive modified function projective synchronization of a new chaotic complex system with uncertain parameters [J]. ICCRD: The 3rd International Conference on Computer Research and Development, 2011, 4: 451–455.
132. P. Liu, S. Liu, X. Li. Adaptive modified function projective synchronization of general uncertain chaotic complex systems [J]. Physica Scripta, 2012, 85(3): 035005.
133. H. Y. Du, Q. S. Zeng, C. H. Wang. Modified function projective synchronization of chaotic system [J]. Chaos, Solitons & Fractals, 2009, 42(4): 2399–2404.
134. Y. Chen, X. Li. Function projective synchronization between two identical chaotic systems [J]. International Journal of Modern Physics C, 2007, 18(05): 883–888.
135. K. S. Sudheer, M. Sabir. Switched modified function projective synchronization of hyperchaotic Qi system with uncertain parameters [J]. Communications in Nonlinear Science and Numerical Simulation, 2010, 15(12): 4058–4064.
136. S. Zheng, G. G. Dong, Q. S. Bi. Adaptive modified function projective synchronization of hyperchaotic systems with unknown parameters [J]. Communications in Nonlinear Science and Numerical Simulation, 2011, 15(11): 3547–3556.
137. G. Y. Fu. Robust adaptive modified function projective synchronization of different hyperchaotic systems subject to external disturbance [J]. Communications in Nonlinear Science and Numerical Simulation, 2012, 17(6): 2602–2608.
138. Y. G. Yu, H. X. Li. Adaptive generalized function projective synchronization of uncertain chaotic systems [J]. Nonlinear Analysis, 2010, 11(4): 2456–2464.
139. X. J. Wu, H. T. Lu. Generalized function projective (lag, anticipated and complete) synchronization between two different complex networks with nonidentical nodes [J]. Communications in Nonlinear Science and Numerical Simulation, 2012, 17(7): 3005–3021.
140. H. Y. Du, Q. S. Zeng. Function projective synchronization in coupled chaotic systems [J]. Nonlinear Analysis, 2010, 11(2): 705–712.
141. P. Zhou, W. Zhu. Function projective synchronization for fractional-order chaotic systems [J]. Nonlinear Analysis, 2011, 12(2): 811–816.
142. I. M. Olga, A. K. Alexey, E. H. Alexander. Generalized synchronization of chaos for secure communication: Remarkable stability to noise [J]. Physics Letters A, 2010, 374(29): 2925–2931.
143. A. A. Zaher, A. Abu-Rezq. On the design of chaos-based secure communication systems [J]. Communications in Nonlinear Science and Numerical Simulation, 2011, 16(9): 3721–3737.
144. O. I. Moskalenko, A. A. Koronovskii and A. E. Hramov. Generalized synchronization of chaos for secure communication: Remarkable stability to noise [J]. Physics Letters A, 2010, 374(29): 2925–2931.

145. C. Y. Chee, D. L. Xu. Chaos-based M-nary digital communication technique using controlled projective synchronization [J]. IEE Proceedings-Circuits, Devices and Systems, 2006, 153(4): 357–360.
146. K. Ikeda. Multiple-valued stationary state and its instability of the transmitted light by a ring cavity system [J]. Optics Communications, 1979, 30(2), 257–261.
147. K. Pyragas. Continuous control of chaos by self-controlling feedback [J]. Physics Letters A, 1992, 170(6), 421–428.
148. T. L. Liao, N. S. Huang. An observer-based approach for chaotic synchronization with applications to secure communications [J]. IEEE Transactions on Circuits and Systems I: Fundamental Theory and Applications, 1999, 46(9): 1144–1150.
149. M. Chen, D. Zhou, Y. Shang. Y. A new observer-based synchronization scheme for private communication [J]. Chaos, Solitons & Fractals, 2005, 24(4): 1025–1030.
150. C. Li, X. Liao, K. W. Wong, et al. Lag synchronization of hyperchaos with application to secure communications [J]. Chaos, Solitons & Fractals, 2005, 23(1): 183–193.
151. M. C. Mackey, L. Glass. Oscillation and chaos in physiological control systems [J]. Science, 1977, 197(4300): 287–289.
152. J. D. Farmer. Chaotic attractors of an infinite-dimensional dynamical system [J]. Physica D: Nonlinear Phenomena, 1982, 4(3): 366–393.
153. J. J. Wei, D. J. Fan. Hopf bifurcation analysis in a Mackey-Glass system [J]. International Journal of Bifurcation and Chaos, 2007, 17(06): 2149–2157.
154. D. Biswas, T. Banerjee. Time-delayed chaotic dynamical systems [M]. Cham: Springer Nature, 2018.
155. D. V. Senthilkumar, M. Lakshmanan. Transition from anticipatory to lag synchronization via complete synchronization in time-delay systems [J]. Physical Review E, 2005, 71(1): 016211.
156. M. G. Rosenblum, A. S. Pikovsky, J. Kurths. Phase to lag synchronization in coupled chaotic oscillators [J]. Physical Review Letters, 1997, 78: 4193–4196.
157. S. Taherion, Y. C. Lai. Observability of lag synchronization of coupled chaotic oscillators [J]. Physical Review E, 59(6): 86247.
158. E. M. Shahverdiev, S. Sivaprakasam, K.A. Shore. Lag synchronization in time-delayed systems [J]. Physics Letters A, 2002, 292(6): 320–324.
159. C. D. Li, X. F. Liao. Lag synchronization of Rössler system and Chua circuit via a scalar signal [J]. Physics Letters A, 2004, 329(4–5): 301–308.
160. G. M. Mahmoud, E. E. Mahmoud. Lag synchronization of hyperchaotic complex nonlinear systems [J]. Nonlinear Dynamics, 2012, 67: 1613–1622.
161. N. W. Fang. Eigenvalues of a class of diagonally dominant matrics and application [J]. Journal of Anhui University. Natural Science Edition, 1995, 1: 18–22.
162. J. Hale. Theory of Functional Differential Equations [M]. Springer-Verlag, Berlin, 1977.
163. G. M. Mahmoud and E. E. Mahmoud. Complex modified projective synchronization of two chaotic complex nonlinear systems [J]. Nonlinear Dynamics, 2013, 73(4): 2231–2240.
164. F. F. Zhang, S. T. Liu, W. Y. Yu. Modified projective synchronization with complex scaling factors of uncertain real chaos and complex chaos [J]. Chinese Physics B, 2013, 22(12): 120505.
165. H. Delavari, M. Mohadeszadeh. Adaptive modified hybrid robust projective synchronization between identical and nonidentical fractional-order complex chaotic systems with fully unknown parameters [J]. Journal of Computational and Nonlinear Dynamics, 2016, 11(4): 041023.
166. J. Liu, S. T. Liu, J. C. Sprott. Adaptive complex modified hybrid function projective synchronization of different dimensional complex chaos with uncertain complex parameters [J]. Nonlinear Dynamics, 2016, 83: 1109–1121.
167. A. Loria. Master-slave synchronization of fourth-order Lü chaotic oscillators via linear output feedback [J]. IEEE Transactions on Circuits and Systems, 2010, 57(3): 213–217.
168. M. Moghtadaei, M. R. H. Golpayegani. Complex dynamic behaviors of the complex Lorenz system [J]. Scientia Iranica, 2012, 19(3): 733–738.

169. J. Sun, Y. Shen. Adaptive anti-synchronization of chaotic complex systems and chaotic real systems with unknown parameters [J]. Journal of Vibration and Control, 2016, 22(13): 2992–3003.
170. G. L. Cai, L. Yao, P. Hu, et al. Adaptive full state hybrid function projective synchronization of financial hyperchaotic systems with uncertain parameters [J]. Discrete and Continuous Dynamical Systems-B, 2013, 18(8): 2019–2028.
171. H. C. Wei, X. C. Zheng. The Matrix Theory in Engineering [M]. China University of Petroleum Press, Dongying, China, 1999.
172. J. W. Sun, G. Z. Cui, Y. F. Wang, et al. Combination complex synchronization of three chaotic complex systems [J]. Nonlinear Dynamics, 2015, 79: 953–965.
173. X. B. Zhou, M. R. Jiang, Y. Q. Huang, Combination synchronization of three identical or different nonlinear complex hyperchaotic systems [J]. Entopy, 2013, 15: 3746–3761.
174. W. N. Zhou, Y. H. Xu, H. Q. Lu, et al. On dynamic analysis of a new chaotic attractor [J]. Physics Letters A, 2008, 372(36): 5773–5777.
175. S. F. Alazzawi. Study of dynamical properties and effective of a state u for hyperchaotic Pan systems [J]. AL-Rafidain Journal of Computer Sciences and Mathematics, 2013, 10(3): 89–99.
176. P. Frederickson, J. L. Kaplan, J. A. Yorke. The Lyapunov dimension of strange attractors [J]. Journal of Differential Equations, 1983, 44(2): 185–207.
177. N. F. Rulkov, M. M. Sushchik, L. S. Tsimring, et al. Generalized synchronization of chaos in directionally coupled chaotic systems [J]. Physical Review E, 1995, 51(2): 980.
178. G. P. Jiang, W. K. S. Tang, G. Chen. A simple global synchronization criterion for coupled chaotic systems [J]. Chaos, Solitons & Fractals, 2003, 15(5): 925–935.
179. J. Lü, T. Zhou, S. Zhang. Chaos synchronization between linearly coupled chaotic systems [J]. Chaos, Solitons & Fractals, 2002, 14(4): 529–541.
180. I. M. Kyprianidis, I. N. Stouboulos. Chaotic synchronization of three coupled oscillators with ring connection [J]. Chaos, Solitons & Fractals, 2003, 17(2): 327–336.
181. Y. G. Yu, S. C. Zhang. Global synchronization of three coupled chaotic systems with ring connection [J]. Chaos, Solitons & Fractals, 2005, 24(5): 1233–1242.
182. X. Y. Chen, J. L. Qiu, Q. Song, et al. Synchronization of coupled chaotic systems with ring connection based on special antisymmetric structure [J]. Abstract and Applied Analysis, 2013, 2013(1): 680604.
183. X. Y. Chen, C. Y. Wang, J. L. Qiu. Synchronization and anti-synchronization of N different coupled chaotic systems with ring connection [J]. International Journal of Modern Physics C, 2014, 25(05): 1440011.
184. X. Y. Chen, J. L. Qiu, J. D. Cao, et al. Hybrid synchronization behavior in an array of coupled chaotic systems with ring connection [J]. Neurocomputing, 2016, 173(3): 1299–1309.
185. X. Y. Chen, J. H. Park, J. D. Cao, et al. Sliding mode synchronization of multiple chaotic systems with uncertainties and disturbances [J]. Applied Mathematics and Computation, 2017, 308: 161–173.
186. X. Y. Chen, J. H. Park, J. D. Cao, et al. Adaptive synchronization of multiple uncertain coupled chaotic systems via sliding mode control [J]. Neurocomputing, 2018, 273: 9–21.
187. N. Cai, Y. W. Jing and S. Y. Zhang. Generalized projective synchronization of different chaotic systems based on antisymmetric structure [J]. Chaos, Solitons & Fractals, 2009, 42(2): 1190–1196.
188. B. Liu, Y. M. Zhou, M. Jiang, et al. Synchronizing chaotic systems using control based on tridiagonal structure [J]. Chaos, Solitons & Fractals, 2009, 39(5): 2274–2281.
189. B. Liu, Z. K. Zhang. Stability of nonlinear systems with tridiagonal structure and its applications [J]. Acta Automatica Sinica, 2007, 33(4): 442–445.
190. X. Y. Wang, H. Zhang. Backstepping-based lag synchronization of a complex permanent magnet synchronous motor system [J]. Chinese Physics B, 2013, 22(4): 048902.
191. F. Zhang, C. Mu, X. Wang, et al. Solution bounds of a new complex PMSM system [J]. Nonlinear Dynamics, 2013, 74(4): 1041–1051.

192. X. Zhou, M. Jiang, X. Cai. Finite-Time chaos control of a complex permanent magnet synchronous motor system [J]. Abstract and Applied Analysis. 2014, 2014(1): 976042.
193. S. H. Chen, J. Liu, J. Xie, et al. Tracking control and synchronization of chaotic systems based upon sampled-data feedback [J]. Chinese Physics, 2002, 11(3): 233–237.
194. F. Y. Sun. Global Chaos synchronization between two new different chaotic systems via active control [J]. Chinese Physics Letters, 2006, 23(1): 32–34.
195. S. T. Liu, P. Liu. Adaptive anti-synchronization of chaotic complex nonlinear systems with unknown parameters [J]. Nonlinear Analysis, 2011, 12(6): 3046–3055.
196. W. G. Yu. Finite-time stabilization of three-dimensional chaotic systems based on CLF [J]. Physics Letters A, 2010, 374(30): 3021–3024.
197. R. W. Guo, U. E. Vincent. Finite time stabilization of chaotic systems via single input [J]. Physics Letters A, 2010, 375(2): 119–124.
198. J. Lu, X. Wu, J. Lü. Synchronization of a unified chaotic system and the application in secure communication [J]. Physics letters A, 2002, 305(6): 365–370.
199. J. M. V. Grzybowski, M. Rafikov, J. M. Balthazar. Synchronization of the unified chaotic system and application in secure communication [J]. Communications in Nonlinear Science and Numerical Simulation, 2009, 14(6): 2793–2806.
200. Z. Y. Wu, J. Q. Duan and X. C. Fu. Complex projective synchronization in coupled chaotic complex dynamical systems [J]. Nonlinear Dynamics, 2012, 69(3): 771–779.
201. P. Liu, S. T. Liu. Robust adaptive full state hybrid synchronization of chaotic complex systems with unknown parameters and external disturbances [J]. Nonlinear Dynamics, 2012, 70(1): 585–599.
202. J. H. Lu, G. R. Chen, D. Z. Cheng, et al. Bridge the gap between the Lorenz System and the Chen System [J]. World Scientific, 2002, 12(12): 2919–2926.
203. T. T. Hartley, C. F. Lorenzo, and H. K. Qammer. Chaos in a fractional order Chua's system [J]. IEEE Transactions on Circuits and Systems I: Fundamental Theory and Applications, 1995, 42(8): 485–490.
204. C. Li, G. Chen. Chaos and hyperchaos in the fractional order Rössler equations [J]. Physica A: Statistical Mechanics and its Applications, 2004:341: 55–61.
205. I. Grigorenko, E. Grigorenko. Chaotic dynamics of the fractional Lorenz system [J]. Physical Review Letters, 2003, 91(3): 034101.
206. L. Pan, W. Zhou, J. Fang, et al. Synchronization and antisynchronization of new uncertain fractional-order modified unified chaotic systems via novel active pinning control [J]. Communications in Nonlinear Science and Numerical Simulation, 2010, 15(12): 3754–3762.
207. G. Si, Z. Sun, Y. Zhang, et al. Projective synchronization of different fractional-order chaotic systems with nonidentical orders [J]. Nonlinear Analysis: Real World Applications, 2012, 13(4): 1761–1771.
208. S. Xu, X. Wang, X. Ye. A new fractional-order chaos system of Hopfield neural network and its application in image encryption [J]. Chaos, Solitons and Fractals, 2022, 157: 111889.
209. J. Wang, C. Dong. A new 3D fractional-order chaotic system with complex dynamics [J]. Physica Scripta, 2023, 99(1): 015217.
210. L. Ren, S. Muhsen, S. Shateyi, et al. Dynamical behaviour, control, and boundedness of a fractional-order chaotic system [J]. Fractal and Fractional, 2023, 7(7): 492.
211. K. Sun, S. He, H. Wang. Solution and characteristic analysis of fractional-Order chaotic systems [M]. Springer Nature, 2022.
212. K. B. Oldham, J. Spanier. The Fractional Calculus [M]. Academic Press, San Diego, 1974.
213. P. M. Pardalos, V. Yatsenko. Optimization approach to the estimation and control of Lyapunov exponents [J]. Journal of Optimization Theory and its Applications, 2006, 128(1): 29–48.
214. S. P. Nair, D. S. Shiau, P. M. Pardalos, et al. An investigation of EEG dynamics in an animal model of temporal lobe epilepsy using the maximum Lyapunov exponent [J]. Experimental Neurology, 2009, 216(1): 115–121.
215. C. R. Fuller, A. H. von Flotow. Active control of sound and vibration [J]. IEEE Control Systems Magazine, 2002, 15(6): 9–19.

216. J. Ghaboussi, A. Joghataie. Active control of structures using neural networks [J]. Journal of Engineering Mechanics, 1995, 121(4): 555–567.
217. S. K. Agrawal, M. Srivastava, S. Das. Synchronization of fractional order chaotic systems using active control method [J]. Chaos, Solitons & Fractals, 2012, 45(6): 737–752.
218. A. G. Radwan, K. Moaddy, K. N. Salama, et al. Control and switching synchronization of fractional order chaotic systems using active control technique [J]. Journal of advanced research, 2014, 5(1): 125–132.
219. M. T. Yassen. Chaos synchronization between two different chaotic systems using active control [J]. Chaos, Solitons & Fractals, 2005, 23(1): 131–140.
220. L. A. Jones, J. H. Lang. A state observer for the permanent-magnet synchronous motor [J]. IEEE transactions on Industrial Electronics, 1989, 36(3): 374–382.
221. M. Soroush. Nonlinear state-observer design with application to reactors [J]. Chemical Engineering Science, 1997, 52(3): 387–404.
222. M. M. Dalla, A. Germani, C. Manes. A state observer for nonlinear dynamical systems [J]. Nonlinear analysis, Theory, Methods & Applications, 1997, 30: 4485–4496.
223. X. Chen, H. Kano. A new state observer for perspective systems [J]. IEEE Transactions on Automatic Control, 2002, 47(4): 658–663.
224. E. E. Mahmoud. Complex complete synchronization of two nonidentical hyperchaotic complex nonlinear systems [J]. Mathematical Methods in the Applied Sciences, 2014, 37(3): 321–328.
225. E. E. Mahmoud, S. M. Abo-Dahab. Dynamical properties and complex anti synchronization with applications to secure communications for a novel chaotic complex nonlinear model [J]. Chaos, Solitons & Fractals, 2018, 106: 273–284.
226. L. X. Yang, J. Jiang. Complex dynamical behavior and modified projective synchronization in fractional-order hyper-chaotic complex Lü system [J]. Chaos, Solitons & Fractals, 2015, 78: 267–276.
227. C. G. Li, G. R. Chen. Chaos and hyperchaos in the fractional-order Rössler equations [J]. Physica A, 2004, 341: 55–61.
228. X. J. Wu, H. Wang, H. Lu. Modified generalized projective synchronization of a new fractional-order hyperchaotic system and its application to secure communication [J]. Nonlinear Analysis: Real World Applications, 2012, 13(3): 1441–1450.
229. G. Y. Wang, X. Zhang, Y. Zheng, Y. X. Li. A new modified hyperchaotic Lü system [J]. Physica A, 2006, 371(2): 1913–1926.
230. G. M. Mahmoud, M. E. Ahmed, N. Sabor. On autonomous and nonautonomous modified hyperchaotic complex Lü systems [J]. International Journal of Bifurcation and Chaos, 2011, 21(7): 1913–1926.
231. M. T. Rosenstein, J. J. Collins and C. J. De Luca. A practical method for calculating largest Lyapunov exponents from small data sets [J]. Physica D, 1993, 65(1–2): 117–134.
232. M. S. Tavazoei, M. Haeri. A necessary condition for double scroll attractor existence in fractional-order systems [J]. Phys. Lett. A, 2007, 367(1–2): 102–113.
233. C. G. Li, G. R. Chen. Chaos and hyperchaos in the fractional-order Rössler equations [J]. Physica A, 2004, 341: 55–61.
234. X. J. Wu, Y. Lu. Generalized projective synchronization of the fractional-order Chen hyperchaotic system [J]. Nonlinear Dynamics, 2009, 57(1–2): 25–35.
235. G. M. Mahmoud, E. E. Mahmoud and A. A. Arafa. On projective synchronization of hyperchaotic complex nonlinear systems based on passive theory for secure communications [J]. Physica Scripta, 2013, 87(5): 055002.
236. G. M. Mahmoud, T. Bountis. The dynamics of systems of complex nonlinear oscillators: a review [J]. International Journal of Bifurcation and Chaos, 2004, 14(11): 3821–3846.
237. G. M. Mahmoud, E. E. Mahmoud, A. A. Arafa. On projective synchronization of hyperchaotic complex nonlinear systems based on passive theory for secure communications [J]. Physica Scripta, 2013, 87(5): 055002.
238. R. Z. Luo, Y. L. Wang, S. C. Deng. Combination synchronization of three classic chaotic systems using active backstepping design [J]. Chaos, 2011, 21(4): e043114.

239. Z. Y. Wu, X. C. Fu. Combination synchronization of three different order nonlinear systems using active backstepping design [J]. Nonlinear Dynamics, 2013, 73: 1863–1872.
240. X. B. Zhou, M. R. Jiang, Y. Q. Huang. Combination synchronization of three identical or different nonlinear complex chaotic systems [J]. Entopy, 2013, 15(9): 3746–3761.
241. U. E. Vincent, A. O. Saseyi, P. V. E. McClintock. Multi-switching combination synchronization of chaotic systems [J]. Nonlinear Dynamics, 2015, 80(1): 845–854.
242. J. W. Sun, G. Z. Cui, Y. F. Wang, Y. Shen, et al. Combination complex synchronization of three chaotic complex systems [J]. Nonlinear Dynamics, 2015, 79: 953–965.
243. J. W. Sun, Y. Shen, G. D. Zhang, et al. Combination-combination synchronization among four identical or different chaotic systems[J]. Nonlinear Dynamics, 2013, 73: 1211–1222.
244. X. B. Zhou, M. R. Jiang, Y. Q. Huang. Combination-combination synchronization of four nonlinear complex hyperchaotic systems [J]. Abstract and Applied Analysis, 2014, 2014: 953265.
245. J. Sun, Y. Shen, X. Wang, et al. Finite-time combination-combination synchronization of four different chaotic systems with unknown parameters via sliding mode control [J]. Nonlinear Dynamics, 2014, 76(1): 383–397.
246. A. Khan. Chaotic analysis and combination-combination synchronization of a novel hyperchaotic system without any equilibria [J]. Chinese Journal of Physics, 2018, 56(1): 238–251.
247. J. W. Sun, Y. Shen, Q. Yin, et al. Compound synchronization of four memristor chaotic oscillator systems and secure communication [J]. Chaos: An Interdisciplinary Journal of Nonlinear Science, 2013, 23(1).
248. J. W. Sun, Q. Yin, Y. Shen. Compound synchronization for four chaotic systems of integer order and fractional order [J]. Europhysics Letters, 2014, 106(4): 40005.
249. J. W. Sun, Y. Shen, G. Cui. Compound synchronization of four chaotic complex systems [J]. Advances in Mathematical Physics, 2015, 2015(1): 921515.
250. G. P. Jiang, W. K. S. Tang, G. Chen. A simple global synchronization criterion for coupled chaotic systems [J]. Chaos, Solitons & Fractals, 2003, 15(5): 925–935.
251. M. Peña, E. Kalnay. Separating fast and slow modes in coupled chaotic systems [J]. Nonlinear Processes in Geophysics, 2004, 11(3): 319–327.
252. L. H. Zhang. Cryptanalysis of the public key encryption based on multiple chaotic systems [J]. Chaos, Solitons & Fractals, 2008, 37(3): 669–674.
253. M. Khan, T. Shah, H. Mahmood, et al. An efficient method for the construction of block cipher with multi-chaotic systems [J]. Nonlinear Dynamics, 2013, 71(3): 489–492.
254. X. Y. Wang, Q. Yu. A block encryption algorithm based on dynamic sequences of multiple chaotic systems [J]. Communications in Nonlinear Science and Numerical Simulation, 2009, 14(2): 574–581.
255. Y. G. Yu, S. C. Zhang. Global synchronization of three coupled chaotic systems with ring connection [J]. Chaos, Solitons & Fractals, 2005, 24(5): 1233–1242.
256. J. A. Lu, X. P. Han, Y. T. Li, et al. Adaptive coupled synchronization among multi-Lorenz systems family [J]. Chaos Solitons & Fractals, 2007, 31(4): 866–878.
257. X. Chen, J. Qiu, Q. Song, A. Zhang. Synchronization of N coupled chaotic systems with ring connection based on special antisymmetric structure [J]. Abstract and Applied Analysis, 2013, 680604.
258. N. Cai, Y. Jing, S. Zhang. Generalized projective synchronization of different chaotic systems based on antisymmetric structure [J]. Chaos Solitons & Fractals, 2009, 42(2): 1190–1196.
259. B. Liu, Y. Zhou, M. Jiang, et al. Synchronizing chaotic systems using control based on tridiagonal structure [J]. Chaos Solitons & Fractals, 2009, 39(5): 2274–2281.
260. B. Liu, Z. K. Zhang. Stability of nonlinear systems with tridiagonal structure and its applications [J]. Acta Automatica Sinica, 2007, 33(4): 442–445.
261. J. G. Lu. Chaotic dynamics of the fractional-order Lü system and its synchronization [J]. Physics Letters A, 2006, 354(4): 305–311.
262. I. Grigorenko, E. Grigorenko. Chaotic dynamics of the fractional Lorenz system [J]. Physical Review Letters, 2003, 91(3): 034101.

263. W. C. Chen. Nonlinear dynamics and chaos in a fractional-order financial system [J]. Chaos Solitons & Fractals, 2008, 36(5): 1305–1314.
264. C. L. Li, S. M. Yu, X. S. Luo. Fractional-order permanent magnet synchronous motor and its adaptive chaotic control [J]. Chinese Physics B, 2012, 21(10): 100506.
265. Z. H. Guan, F. Huang, W. Guan. Chaos-based image encryption algorithm [J]. Physics letters A, 2005, 346(1–3): 153–157.
266. S. Liu, J. Sun, Z. Xu. An improved image encryption algorithm based on chaotic system [J]. Journal of Computers, 2009, 4(11): 1091–1100.
267. Y. Zhou, L. Bao, C. L. P. Chen. A new 1D chaotic system for image encryption [J]. Signal Processing, 2014, 97: 172–182.
268. B. Norouzi, S. Mirzakuchaki. A fast color image encryption algorithm based on hyper-chaotic systems [J]. Nonlinear Dynamics, 2014, 78(2): 995–1015.
269. X. Wang, Y. Zhang, X. Bao. A novel chaotic image encryption scheme using DNA sequence operations [J]. Optics and Lasers in Engineering, 2015, 73: 53–61.
270. X. Wang, S. Gu, Y. Zhang. Novel image encryption algorithm based on cycle shift and chaotic system [J]. Optics and Lasers in Engineering, 2015, 68: 126–134.
271. X. Wang, L. Liu, Y. Zhang. A novel chaotic block image encryption algorithm based on dynamic random growth technique [J]. Optics and Lasers in Engineering, 2015, 66: 10–18.
272. Z. Liu, Y. Zhang, W. Liu, et al. Optical color image hiding scheme based on chaotic mapping and Hartley transform [J]. Optics and Lasers in Engineering, 2013, 51(8): 967–972.
273. M. Sharma, M. K. Kowar. Image encryption techniques using chaotic schemes: a review [J]. International Journal of Engineering Science and Technology, 2010, 2(6): 2359–2363.
274. Y. Zhou, W. Cao, C. Philip. Image encryption using binary bitplane [J]. Signal Processing, 2014, 100: 197–207.
275. S. J. Xu, X. B. Chen, R. Zhang, et al. An improved chaotic cryptosystem based on circular bit shift and XOR operations [J]. Physics Letters A, 2012, 376(10–11): 1003–1010.
276. H. J. Liu, X. Y. Wang, A. Kadir. Image encryption using DNA complementary rule and chaotic maps [J]. Applied Soft Computing, 2012, 12(5): 1457–1466.
277. Y. Liu, X. Tong, S. Hu. A family of new complex number chaotic maps based image encryption algorithm [J]. Signal Processing: Image Communication, 2013, 28(10): 1548–1559.
278. Y. Q. Zhang, X. Y. Wang. A symmetric image encryption algorithm based on mixed linear Cnonlinear coupled map lattice [J]. Information Sciences, 2014, 273: 329–351.
279. M. A. Ansari, D. Arora, S. P. Ansari. Chaos control and synchronization of fractional order delay-varying computer virus propagation model [J]. Mathematical Methods in the Applied Sciences, 2016, 39(5): 1197–1205.
280. A. Tamasevicius. Reproducible analogue circuit for chaotic synchronization [J]. Electron Letter, 1997, 33(13): 1105–1106.
281. B. Blasius, A. Huppert, L. Stone. Complex dynamics and phase synchronization in spatially extended ecological systems [J]. Nature, 1999, 399(6734): 354–359.
282. Y. Xu, C. Xie, Y. Wang, et al. Chaos projective synchronization of the chaotic finance system with parameter switching perturbation and input time-varying delay [J]. Mathematical Methods in the Applied Sciences, 2015, 38(17): 4279–4288.
283. M. F. Hu, Z. Y. Xu and R. Zhang. Full state hybrid projective synchronization in continuous-time chaotic (hyperchaotic) systems [J]. Communications in Nonlinear Science and Numerical Simulation, 2008, 13(2): 456–464.
284. H. Yau, C. Shieh. Chaos synchronization using fuzzy logic controller [J]. Nonlinear Analysis: Real World Applications, 2008, 9(4): 1800–1810.
285. J. Xu, G. Cai. Adaptive backstepping control of new determining Lyapunov exponents from a time series [J]. Physica D, 1985, 16(3): 285–317.

The manufacturer's authorised representative in the EU is Springer
Nature Customer Service Centre GmbH, Europaplatz 3, 69115 Heidelberg,
Germany. If you have any concerns regarding our products, please
contact ProductSafety@springernature.com

Printed and bound by CPI Group (UK) Ltd, Croydon, CR0 4YY
11/10/2026
02231395-0001